Signal Processing Techniques in Wireless Communications

Signal Processing Techniques in Wireless Communications

Edited by **Rapheal Dagget**

Willford Press
New York

Published by Willford Press,
118-35 Queens Blvd., Suite 400,
Forest Hills, NY 11375, USA
www.willfordpress.com

Signal Processing Techniques in Wireless Communications
Edited by Rapheal Dagget

International Standard Book Number: 978-1-68285-220-0 (Hardback)

Printed in China.

Contents

Preface

Signal processing is the technology which is mainly concerned with processing and transmission of information via electronic signals. This book discusses various signal processing techniques in wireless networks. Some of the topics included in this book such as coding for wireless systems, security and cryptography for wireless, communication tools and methods for signal processing, wireless traffic and routing, are of utmost significance and bound to provide incredible insights to readers. The main objective of this book is to present emerging challenges and applications of signal processing techniques in wireless communication networks. It aims to serve as a resource guide for students and experts alike and contribute to the growth of the discipline.

The information contained in this book is the result of intensive hard work done by researchers in this field. All due efforts have been made to make this book serve as a complete guiding source for students and researchers. The topics in this book have been comprehensively explained to help readers understand the growing trends in the field.

I would like to thank the entire group of writers who made sincere efforts in this book and my family who supported me in my efforts of working on this book. I take this opportunity to thank all those who have been a guiding force throughout my life.

Editor

1

Deployment analysis and optimization of heterogeneous networks under the spectrum underlay strategy

Anabel Martínez-Vargas* and Ángel G Andrade

Abstract

To meet the growing mobile traffic demand, deployment of heterogeneous networks (HetNets) is proposed as a solution to enhance network capacity. It will enable the coexistence of several systems consisting of Macro, Micro, and Femto or WiFi layers. However, the main issue is the coexistence due to mutual interference. To guarantee that HetNet performance will not be compromised due to the interference, some deployment parameters should be controlled, such as primary exclusion zones and density of users. In this context, the present work provides deployment analysis and optimization for a Macro-Femto scenario, considering interference constraints to maximize the data rate of the HetNet. To fulfill this, we characterize some spectrum underlay scenarios in which HetNets would be expected to operate, setting different signal-to-interference-plus-noise ratio (SINR) thresholds. They are evaluated using a solution procedure based on the socio-cognitive particle swarm optimization (SCPSO) algorithm. Considering the best scenarios found, we show deployment parameters to achieve maximum system throughput, such as density of femto-users and primary exclusion zones of macro-users.

Keywords: Deployment parameters; Heterogeneous network; Spectrum underlay

1 Introduction

Architecture of traditional cellular networks has been undergoing dramatic changes due to the intense consumer demand for mobile data. Recently, the heterogeneous networks (HetNets) have emerged as the network architecture to support the growing demand for data services. They have a hierarchical architecture, comprising an overlay of a macrocell network with additional tiers of small cells (picos, femtos, or WiFi access points). This enlarges the pool of available spectrum resources for mobile users, since they are reused across the multiple tiers in the HetNet, harnessing also an additional spectrum in unlicensed bands by integrating WiFi hotspots (femtocells) in the network. The use of WiFi in conjunction with the Macrocell allows relieving congestion on it. From the above and the cost point of view, WIFi has become an integral part of the operator's strategy [1] However, by reusing resources, potentially destructive interference in the HetNet is generated.

* Correspondence: martinez.anabel@uabc.edu.mx
College of Engineering, MyDCI, Autonomous University of Baja California (UABC), Blvd. Benito Juárez, Mexicali 21280, México

Unlike macro base stations (macro-BSs), which are installed according to detail network planning, operators have limited control over femto base stations (femto-BSs) since they are deployed by the end-users. This leads to two kinds of interference between macro and femto cells: cross-tier (the aggressor and the victim of interference belong to different tiers) and intra-tier (the aggressor and the victim of interference belong to the same tier) [2]. Cognitive radio (CR) technology can address those issues, providing to femto-BS capabilities to sense the environment, interpret the received signal from the macro-BS and the surrounding femto-BSs, and intelligently allocate spectrum resources while cross-tier and the intra-tier interference are controlled. The set of techniques to share and allocate spectrum resources through CR technology is known as dynamic spectrum access, where the users with low priority are known as secondary users (femtocells), whereas the members of the prioritized user group are named primary users (macrocell) [3]. To access the cellular channels (or primary channels), a secondary user performs one of the following dynamic spectrum access techniques:

transmit simultaneously with the primary user as long as the resulting interference is constrained (spectrum underlay), or exploit an unused channel of the primary user (spectrum overlay) [2]. Therefore, the underlay scheme has a more efficient reuse of spectral resources since one primary channel is exploited simultaneously by two or more users. Furthermore, the spectrum sensing operation of the overlay scheme requires complicated computation at the secondary transmitter.

Identifying deployment parameters for HetNets help to prevent the generation of harmful interference between heterogeneous wireless systems that interact with each other. Interference is considered harmful if it causes disruption in user's service. In general, examples of deployment parameters for HetNets are [4,5] primary exclusion zones (radius of protection of a primary receiver/transmitter), transmission power limits on secondary users, number of admitted secondary users, acceptable level of interference in primary receptor, and monitoring time. The deployment parameters depend on the dynamic spectrum access technique (overlay or underlay) that a secondary user performs to exploit a primary band. In the case of deploying a HetNet with spectrum underlay access, it is relevant to set a limit on the transmission power of secondary users, primary exclusion zones, permissible number of secondary users, and acceptable level of interference in the primary receptor. The detection threshold and monitoring time are factors to overlay access mode. The first refers to the maximum value to be set to detect the primary signal; therefore, the primary channel status is known (busy or idle). In contrast, the second refers to the time that a secondary user should expend to detect the presence of the primary user.

In this paper, we characterize some spectrum underlay scenarios in which HetNets would be expected to operate under different signal-to-interference-plus-noise ratio (SINR) thresholds. They are evaluated using our proposed access control and channel assignment algorithm presented in [6], which is a solution procedure with less complexity based on an improved version of binary particle swarm optimization (BPSO) algorithm called socio-cognitive particle swarm optimization (SCPSO) [7]. Considering the best scenarios, the deployment parameters to achieve maximum system throughput, such as the number of admitted secondary users coexisting with primary users and primary exclusion zones, are presented. Particularly, we focus on the Macro-Femto scenario in which the secondary network (femtocell) operates in an unlicensed band; while the primary network (macrocell) is composed by devices transmitting on Long Term Evolution Release 8 and beyond (LTE-Release 8).

Some works have addressed the specifications of network deployment using mainly statistical models such as in [8] in which from a log-normal distribution, the authors find radius of exclusion zones and density of secondary users. In this context, in [9], an outage-based distributed user removal algorithm to devise the number of secondary users in a spectrum underlay HetNet is presented. In [10], a tool for estimating the protection distance around the primary transmitter by means of gamma and normal distributions is proposed. Although the

Table 1 Symbols used in the access control and channel assignment algorithm

Symbols	Meaning
$SINR_u$	The *SINR* at each secondary link *u*
P_u	The transmit power of secondary transmitter *u*
P_k	The transmit power of secondary transmitter *k*
P_v	The transmit power of primary transmitter *v*
lds(u)	The link distance of secondary link *u*
dss(k,u)	The distance from secondary transmitter *k* to secondary receiver *u*
dps(v,u)	The distance from primary transmitter *v* to secondary receiver *u*
k	The index of active secondary transmitters
Φ	The set of active secondary transmitters
γ	The path loss exponent (a value between 2 and 4)
Sl	The number of secondary links
$SINR_v$	The *SINR* at each primary link *v*
ldp(v)	The link distance of primary link *v*
dps(k,v)	The distance from secondary transmitter *k* to primary receiver *v*
Pl	The number of primary links
c'_u	The data rate of the secondary link
c^*_v	The data rate of the primary link
B	The primary channel bandwidth
α	The SINR threshold for the secondary network
β	The SINR threshold for the primary network
S	The swarm
X_i	The position of the *i*th particle
V_i	The velocity of the *i*th particle
V_{max}	The maximum velocity
P_i	The best position ever visited by the *i*th particle
X'_i	The candidate channel allocation vector of the *i*th particle
P'_i	The best channel allocation for secondary links find so far by the *i*th particle
g	The index of the best particle in the swarm
$T_{\max}$	The number of iterations
c_1, c_2, c_3	The cognitive, social and socio-cognitive factors
w, w^1	The inertia weights
PC	The number of primary channels to share
Spectrum status	The channel allocation vector for primary links
P_g	The best particle in the swarm

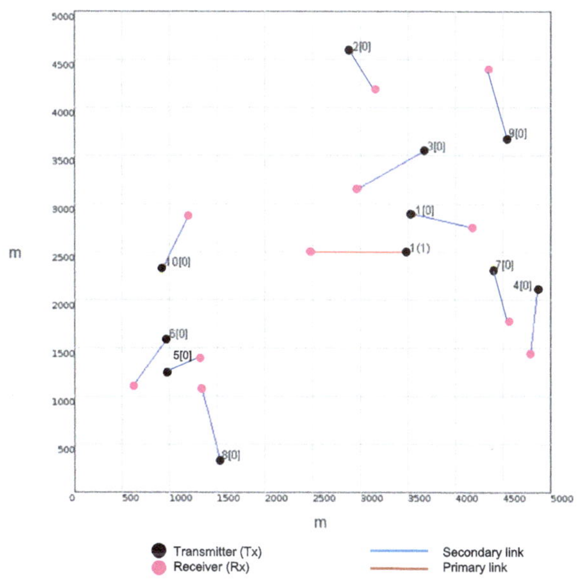

Figure 1 A snapshot sample before applying the access control and channel assignment algorithm based on SCPSO.

above works provide deployment parameters to guarantee primary network protection, our framework can suggest deployment parameters that ensure that the maximum throughput in the HetNet will be achieved, providing not only primary network protection but also secondary network protection. In others words, achieving the maximum throughput in the HetNet is possible as long as the deployment guidelines provided by our framework are applied. On the other hand, in [11], deployment issues on macro-pico heterogeneous networks are studied such as locations, density, and transmit power of pico eNodeBs (PeNBs) to offload traffic from the macro to pico cells to maximize overall throughput. To fulfill this, the cell range extension (CRE) is applied. Although it optimizes the aforementioned parameters, it first requires carefully setting the CRE bias value that implies also its optimization for a given scenario.

The novelty of our work is that it can suggest radius of exclusion zone and density of secondary users to achieve the maximum throughput in the HetNet with spectrum underlay, unlike the works cited above. The aim in works [8-11] is only to devise exclusion zones or permissible density of secondary users to offer protection to primary users. To the best of our knowledge, there is no effort to suggest exclusion zones and density of secondary users to achieve the maximum throughput in the context of an underlay spectrum sharing network. Instead, there are considerable efforts to let more wireless devices operate opportunistically within unused frequency bands referred as white spaces (overlay strategy).

The rest of this paper is organized as follows. The access control and channel algorithm is given in Section 2.

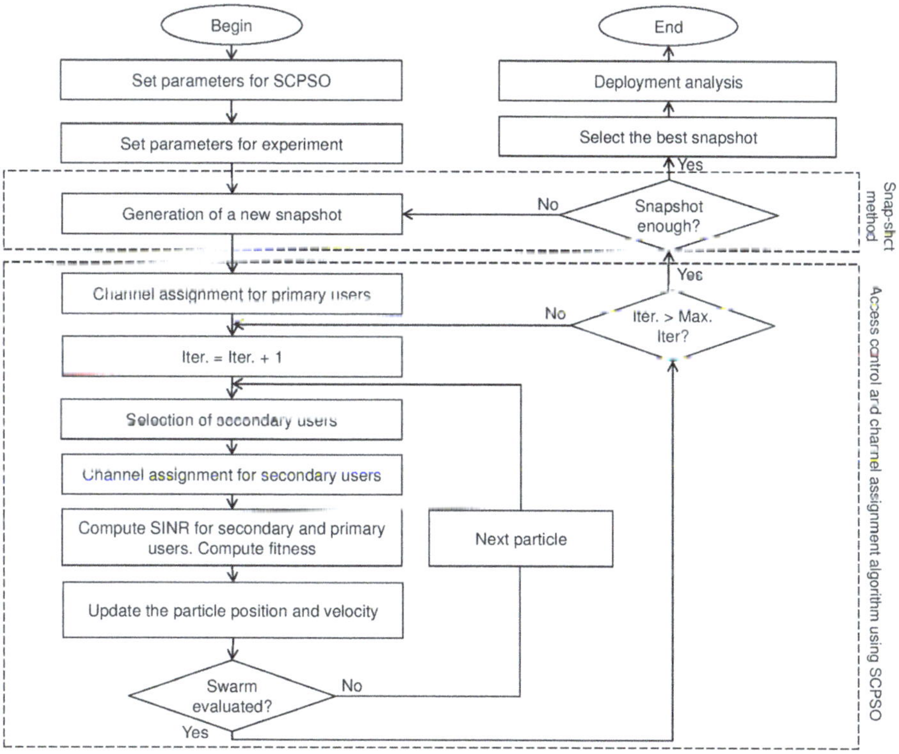

Figure 2 The process of simulation.

Section 3 describes the simulation framework. Then, in Section 4, the deployment parameters of HetNets are presented. Finally, Section 5 gives the conclusion and future work.

2 The access control and channel assignment algorithm

To describe the development of the proposed algorithm, we used the symbols as defined in Table 1.

In the context of HetNets with spectrum underlay, when a secondary transmitter requests for a primary channel, they must be able to check if mutual interference among secondary users and primary users does not arise to the level of harmful interference. In other words, the HetNet must determine the new interference situation in the system if a primary channel is allocated to the requested secondary transmitter, to guarantee the quality of service (QoS) not only at the primary network but also at the secondary network. To fulfill this, the SINR is the most commonly studied model. In the SINR model, the energy of a signal fades with the distance to the power of the path loss parameter. If the signal strength received by a device divided by the interfering strength of other simultaneous transmissions (plus the fixed background noise) is above some reception threshold, then the receiver successfully receives the message; otherwise, it does not [12].

The SINR at each secondary link u is given by:

$$\mathrm{SINR}_u = \frac{P_u/lds(u)^\gamma}{\sum_{k\in\Phi} P_k/dss(k,u)^\gamma + P_v/dps(v,u)^\gamma}, 1 \leq u \leq Sl \tag{1}$$

where P_u is the transmit power of secondary transmitter u, P_k is the transmit power of secondary transmitter k, P_v is the transmit power of primary transmitter v, $lds(u)$ is the link distance of secondary link u, $dss(k,u)$ is the distance from secondary transmitter k to secondary receiver u, $dps(v,u)$ is the distance from primary transmitter v to secondary receiver u, k is the index of active secondary transmitters, Φ is the set of active secondary transmitters, excluding the secondary transmitter u, γ is the path loss exponent (a value between 2 and 4), and Sl is the number of secondary links deployed in the area.

On the other hand, to calculate the SINR at each primary link v, the following equation is considered:

$$\mathrm{SINR}_v = \frac{P_v/ldp(v)^\gamma}{\sum_{k\in\Phi} P_k/dps(k,v)^\gamma}, 1 \leq v \leq Pl \tag{2}$$

where P_v is the transmit power of primary transmitter v, P_k is the transmit power of secondary transmitter k, $ldp(v)$ is the link distance of primary link v, $dps(k,v)$ is the distance from secondary transmitter k to primary receiver v, and Pl is the number of primary links presented in the area. For purpose of our analysis, we only have one primary link; therefore, $Pl = 1$ and the index $v = 1$.

Table 2 Parameters used for SCPSO

Parameters	Values
Swarm size S =	40
Number of iterations T_{max} =	150
Cognitive, social and socio-cognitive factors c_1, c_2, c_3 =	2, 2, 12
Inertia weight w =	0.721
Maximum velocity V_{max} =	[−6,6]

Data rate contributions of the secondary links and primary links are calculated according to Equations 3 and 4, respectively. The data rate depends on primary channel bandwidth B that secondary links and primary links can share and the conditions of the propagation environment (attenuation and interference).

$$c'_u = Blog_2(1 + SINR_u) \tag{3}$$

$$c''_v = Blog_2(1 + SINR_v) \tag{4}$$

Based on the above discussion, the access control and channel assignment in the HetNet is formulated as the following optimization problem:

$$Max \sum_{u=1}^{Sl} c'_u x_u + \sum_{v=1}^{Pl} c''_v \tag{5}$$

subject to:

$$SINR_u \geq \alpha \tag{6}$$

$$SINR_v \geq \beta \tag{7}$$

$$c'_u > 0, u = 1, 2, ..., Sl \tag{8}$$

$$c''_v > 0, v = 1, 2, ..., Pl \tag{9}$$

$$c'_u, c''_v \in R^+ \tag{10}$$

$$x_u = \begin{cases} 1, \text{ if } \mathrm{SINR}_u \geq \alpha \text{ and } \mathrm{SINR}_v \geq \beta \\ 0, \text{ otherwise} \end{cases} \tag{11}$$

The objective function in (5) is to maximize the sum throughput in the HetNet. The SINR requirements of secondary links and primary links are posed in (6) and

Table 3 Parameters used for experiments

Parameters	Values
Number of secondary links Sl =	10, 15, 20, 25
Number of primary links Pl =	1
Number of primary channels to share PC =	1
Number of runs =	5,000
SINR threshold α, β (dB) =	4, 6, 8, 10, 12, 14
Primary channel bandwith B =	20 MHz

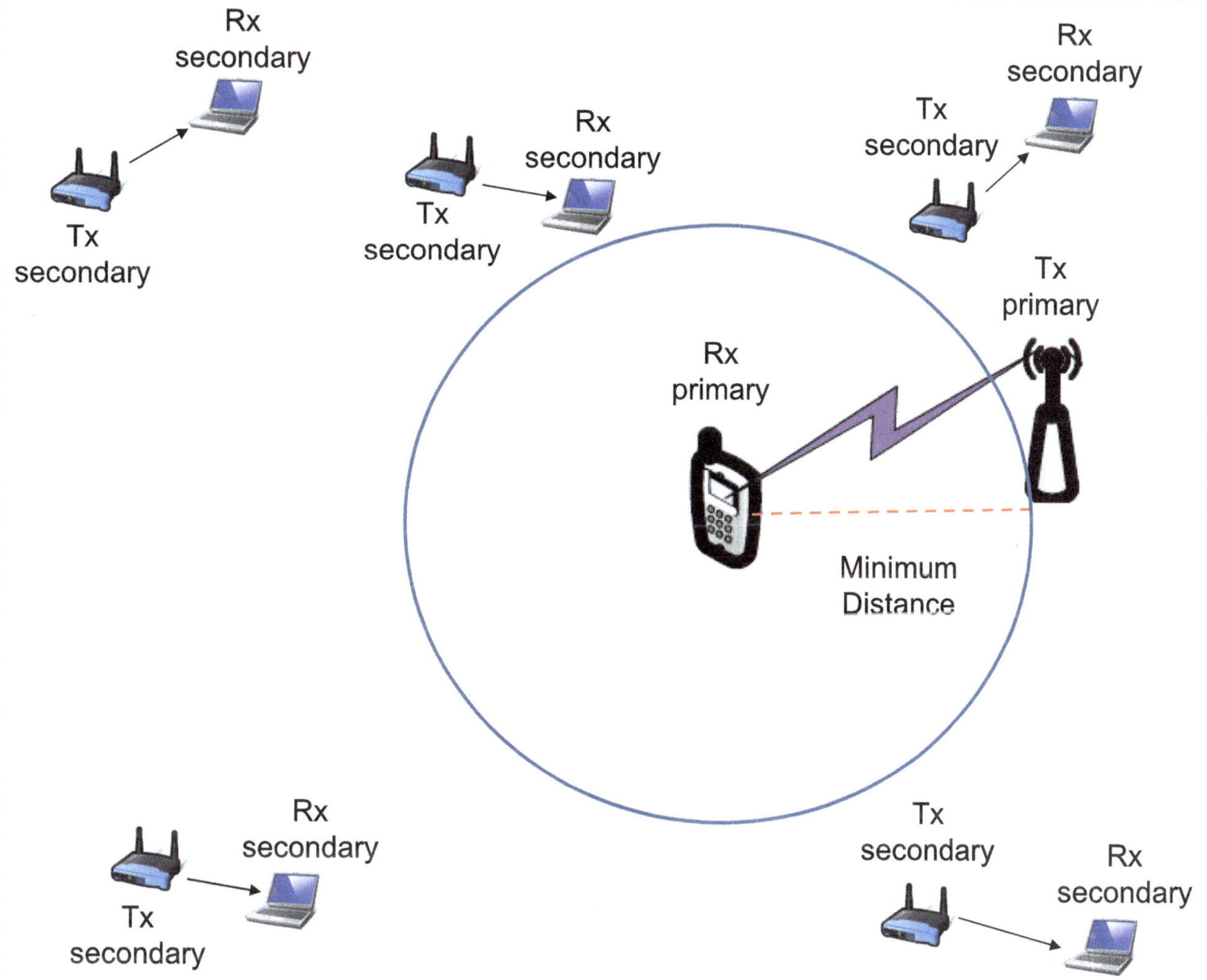

Figure 3 Primary exclusion zone.

(7) where α and β are the SINR thresholds for the secondary network and the primary network, respectively. Constraints from (8) to (10) are integrity restrictions. $x_u = 1$ if secondary link u is included in the solution and $x_u = 0$ if it remains out as indicated in (11).

As described in (5) to (11), the goal is to maximize the data rate of the HetNet without degrading the QoS of the primary and secondary networks. To decide which of the deployed secondary links can achieve this, a bio-inspired algorithm called SCPSO is used; that is, from a given scenario, the task of SCPSO is to evaluate different combinations of secondary users deployed in an area along with their primary channels assigned. A combination of secondary users is a candidate solution. In the context of SCPSO, a candidate solution represents a particle X_i; therefore, the number of particles is called the swarm S. Each particle has its own velocity (V_i) to fly through the search space. The particle is able to remember the best position ever visited in the search space since it has a memory (P_i); this position corresponds to the best fitness found by the particle, and it is known as *pbest*. On the other hand, the best position visited by the whole swarm is called *gbest*. At each iteration, the particle improves itself according to *pbest* (its own experience) and *gbest* (the experience of the best performer in the swarm) until a stop condition is met. Then, *gbest* is the solution of the problem, that is, the selected secondary users along with their primary channels allocated that achieve the goal.

The ith particle of the swarm S can be represented by a D-dimensional vector, $X_i = (x_{i1}, x_{i2}, \ldots, x_{iD})$, $x_{id} \in \{0,1\}$. The velocity of this particle is represented by another D-dimensional vector $V_i = (v_{i1}, v_{i2}, \ldots, v_{iD})$, $v_{ij} \in [-V_{\max}, V_{\max}]$ where $V_{\max}$ is the maximum velocity. The best previously visited position of the ith particle is denoted as $P_i = (p_{i1}, p_{i2}, \ldots, p_{iD})$, $p_{id} \in \{0,1\}$. Then, the behavior of the whole system emerges from the interaction of particles.

Algorithm 1 describes the access control and channel assignment algorithm using SCPSO as follows:

Algorithm 1

Initialization stage

Locate randomly *Sl* and *Pl* over the coverage area *A*

Initialize randomly candidate solution vector X_i, where $x_{id} \in \{0,1\}$

Initialize randomly velocity vector V_i, where $v_{id} \in [-V_{max}, V_{max}]$

Set $P_i = X_i$

Initialize randomly candidate channel allocation vector X'_i with values from *Pl*

Set $P'_i = X'_i$

Initialize randomly vector *Spectrum Status* with values from *Pl*

End Initialization stage

Loop

For $i = 1$ to number of particles

Perform mapping between X'_i and *Spectrum Status* vectors to calculate SINR levels at secondary and primary links

Perform mapping between P'_i and *Spectrum Status* vectors to calculate SINR levels at secondary and primary links

Calculate the fitness, $f(X_i)$ and $f(P_i)$

If $f(X_i) > f(P_i)$ then do

For d= 1 to D

$p_{id} = x_{id}$

$p'_{id} = x'_{id}$

Next d

End do

12: $g = i$ // g is referred as the index of the best particle in the swarm

13: For j = 1 to number of particles

14: Perform mapping between P'_j and *Spectrum Status* vectors to calculate SINR levels at secondary and primary links

15: Calculate the fitness, $f(P_j)$ and $f(P_g)$

16: If $f(P_j) > f(P_g)$ then $g = j$

17: Next j

18: For d = 1 to D

19: $v_{id} = w\, v_{id} + c_1 r_1(p_{id} - x_{id}) + c_2 r_2(p_{gd} - x_{id})$

20: $v_{id} = w^1\, v_{id} + c_3\,(gbest - pbest)$

21: $x_{id} = x_{id} + v_{id}$

22: $x_{id} = x_{id}\ mod\ (2)$

23: if $x_{id} = 1$ then allocate randomly a new channel to x'_{id}

24: Next d

25: Next i

26: Until stopping criterion met

27: Return (P_g, $f(P_g)$)

In Algorithm 1, besides the SCPSO vectors (X_i, P_i, and V_i), we include additional vectors related to the ith particle:

- P'_i which stores the best channel allocation for secondary links find so far.
- X'_i that represents a candidate channel allocation for secondary links.
- Spectrum status which has the channel allocation for primary links.

Fitness calculation includes the phases of updating *pbest* and *gbest*. In update *pbest* phase (from step 3 to step 11 at Algorithm 1), the particle compares $f(X_i) > f(P_i)$ and overwrites *pbest* if $f(X_i)$ is higher than $f(P_i)$. In contrast, in update *gbest* (from step 12 to step 17 at Algorithm 1), all *pbest* values will be compared with the current *gbest* value, so if there is a *pbest* which is higher than the current *gbest*, then *gbest* will be over-written. If the stop condition is not met, X_i and V_i are updated according to equations from step 19 to step 22 in Algorithm 1; where c_1 and c_2 are the learning factors, c_3 is the socio-cognitive scaling parameter, r_1 and r_2 are random numbers uniformly distributed in [0,1], and w and w^1 are the inertia weights.

Infeasible solutions, that is, candidate solutions which cannot achieve one or both SINR constraints presented in (6) and (7) are penalized by setting total particle's fitness to zero; therefore, they are not chosen during the process. On the other hand, when conditions (6) and (7) are satisfied, sum data rate is performed as indicated in objective function (5), being the total particle's fitness. Further details and the complete implementation of our solution procedure based on the SCPSO algorithm are provided in [6].

3 Simulation framework

Figure 1 shows a snapshot sample (the deployment of secondary links and a primary link in an instant of time) of a particular HetNet that will be analyzed by the access control and channel assignment algorithm using SCPSO to determinate the exclusion zone and the number of selected secondary links which maximize throughput in the HetNet. Due to the distance between different links affects the performance of the HetNet, we considered a stochastic model for the location of the links. The stochastic geometry [13] is a tool for analyzing the distribution of the interference and disruptions that have the links, so that it is possible to statistically analyze network performance. For this, the stochastic models focus on the analysis of a single receiver, or link, as shown in Figure 1, the primary receiver is placed permanently in the center of the coverage area and its corresponding transmitter is located at a distance of 1,000 m. The location of the secondary transmitters is based on the spatial model point Poisson process (PPP) in which each node took a random location represented by a pair of coordinates (a_i, b_i).

The metrics for evaluating the performance of the HetNet are directly related to the SINR on links. The SINR depends on many factors such as the distance between the desired transmitter and the desired receiver, and the set of interfering transmitters and their distances from the desired receiver [14]. The SINR is considered in this work by applying Equations (1) and (2) from Section 2, respectively.

During the simulation, several experiments are performed, each of them characterized by a different QoS requirement ($\alpha = \beta$), then a number of independent runs are taken by each experiment. A run represents a new snapshot with random locations for secondary nodes. Consequently, at each run, different combinations of secondary links coexisting with the primary link are evaluated to find out the best solution, *i.e.*, the combination of secondary links that maximizes throughput in the heterogeneous system. After a number of runs (or snapshots) are taken, the next step is to select the best snapshot; that is, the one with the highest throughput. Then, that snapshot is analyzed to provide the deployment guidelines. Figure 2 summarizes the aforementioned.

Some of the assumptions made for snapshots are:

- The secondary link is located in an area of 25 km^2 and its distance is limited to 1,000 m.
- Transmission powers of the primary and secondary are the same.
- The links require the same channel bandwidth since they use the same type of traffic (homogeneous traffic).
- Power attenuation due to multipath and shadowing phenomena is not considered for calculation of SINR.
- For simplicity, the SINR thresholds α and β coincide.

The parameter settings for the SCPSO algorithm are shown in Table 2.

The number of experiments depends on α and β; to analyze, in this work, six experiments are performed, each one representing a different QoS level in the system (α, β = 4, 6, 8, 10, 12, and 14 dB). At each experiment, the cases of 10, 15, 20, and 25 secondary links along with 1 primary link are evaluated. The configuration parameters for each experiment are shown in Table 3.

The following section shows deployment results from the simulation framework.

4 Deployment parameters of HetNets

Through the simulation framework described in the last section, we determine the design parameters that are required for achieving the maximum throughput in the HetNet. The results for the primary exclusion zones and the number of secondary users coexisting with the primary user are shown in the next subsections.

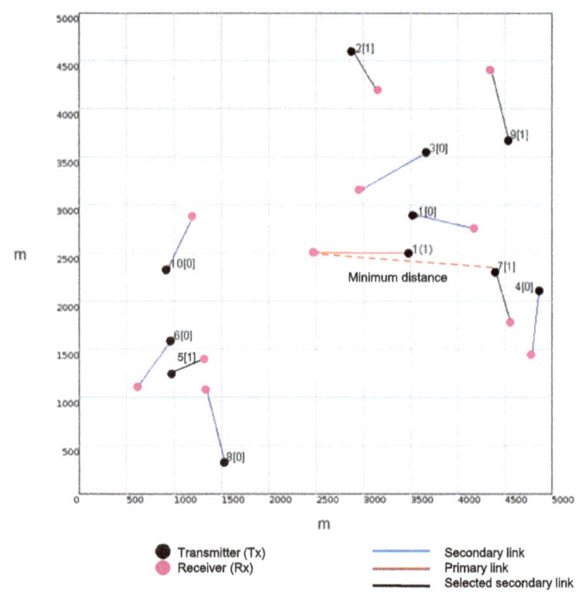

Figure 4 A snapshot sample after applying the access control and channel assignment algorithm based on SCPSO.

4.1 Primary exclusion zones

In the context of HetNets, interference caused by secondary transmitters to the primary receiver should be as low as possible. A solution proposed is to reduce the transmission power of the secondary users. However, this action has an impact on the capacity of the secondary network since the desired signal is affected, decreasing the received SINR. Another solution is to deny transmission to any secondary node when it is in close proximity to a primary receiver. The above is achieved by setting a primary exclusion zone which is the radius of protection for a primary receiver, *i.e.*, the minimum distance to place one or more secondary transmitters with respect to the primary receiver, to use simultaneously the frequency band (see Figure 3).

From Figure 3, it is assumed that each secondary link (transmitter-receiver pair) represents a secondary network or femtocell network. In this case, the secondary transmitter is able to assign a communication channel to its secondary receiver when it requests. Assuming that in an instant of time, the secondary transceiver does not have available channels to assign to its user, it can use its cognitive abilities to exploit any of the primary channels that are assigned to the primary network; so that a primary channel will be allocated to its requested secondary user. Under this situation, the secondary link can exploit concurrently with the primary link the channel, as long as a minimum distance is placed between them; so that the transmission of the secondary link does not interfere with the transmission of the primary link.

After performing the simulation framework described in Section 3, the best snapshot is analyzed by experiment and case to set the primary exclusion zone. Supposing that Figure 4 was the best snapshot, it shows the selected secondary links (black lines) after performing the access control and channel assignment algorithm

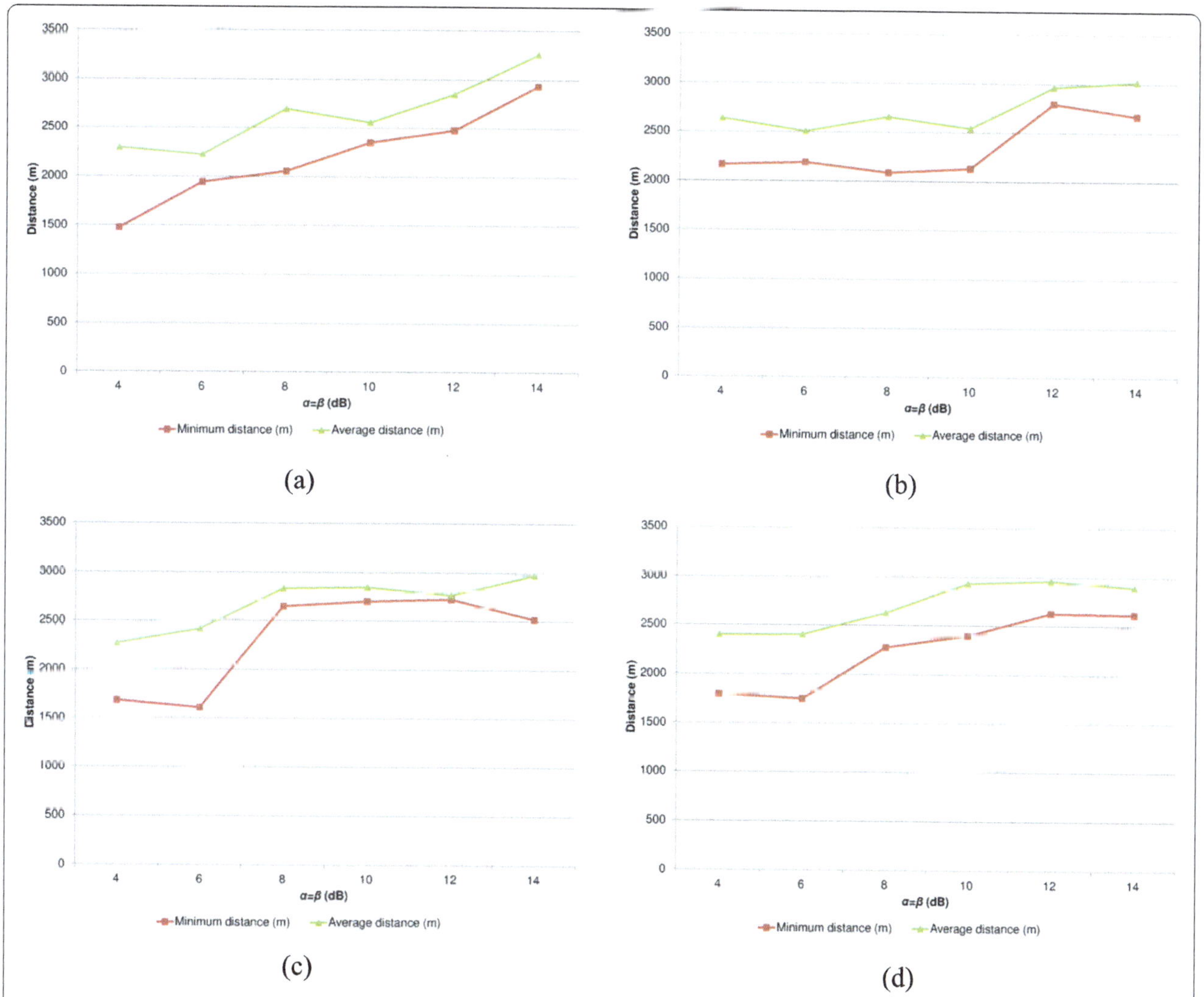

Figure 5 The minimum and average distances by experiment and case. (a) 10 secondary users, **(b)** 15 secondary users, **(c)** 20 secondary users, and **(d)** 25 secondary users.

using SCPSO. From the ten secondary users that are requesting for a primary channel, only four of them are selected (secondary links 2, 5, 7, 9). They maximize the throughput of the HetNet. The selected secondary links along with the primary link have the same channel (channel 1). That is indicated in square and rounded brackets for the selected secondary links and the primary link, respectively. Then, the distance between each selected secondary transmitter and the primary receiver is calculated. Among the calculated distances, the shortest distance is selected, and it is considered as the radius of the exclusion zone. Distances are calculated from the location coordinates (a_i, b_i) known of each node.

In this context, Figure 5a,b,c,d shows the values of the minimum (exclusion zone) and average distance that should exist between the secondary transmitters and the primary receiver for different cases (the number of secondary links) and conditions of SINR (the experiments). In this sense, the HetNet performance is not compromised due to mutual interference.

The results in Figure 5a,b,c,d suggest that the higher the QoS level in the HetNet, the longer the distance that should be deployed the secondary transmitters. However, secondary users with lower transmission powers can transmit at a shorter distance from the primary receiver. Those results represent the coverage area (primary exclusion zone) in which a primary receiver can decode the information sent by the primary transmitter without being affected by the interference from secondary links. Outside the coverage area, the secondary links can use the same primary channel to perform their communication, as shown in Figure 3. Then, the radius of the exclusion zone depends on the value of α and β.

4.2 Multiple secondary users

The main challenge in HetNets with multiple secondary users is on how they exploit concurrently a primary channel and take it dynamically. The limiting factor for spectrum reuse is interference; therefore, its control is essential to achieve maximum performance in the HetNet [15]. One of the strategies to achieve the aforementioned is to limit the number of secondary users in the HetNet.

Works [16] and [17] analyze the effect of scaling in wireless networks, that is, quantifying the increase of data rate in the network as the number of links grows. In [16], the authors present results in scaling a homogeneous ad hoc network, determining that the effective throughput of the network is increased by a factor of $\sqrt{n}$, where n represents the number of links in the network. We consider a similar experiment as presented in [17], but unlike that work, we evaluate HetNets which include primary and secondary users. Secondary transmitters are restricted by a distance of 1,000 m with respect

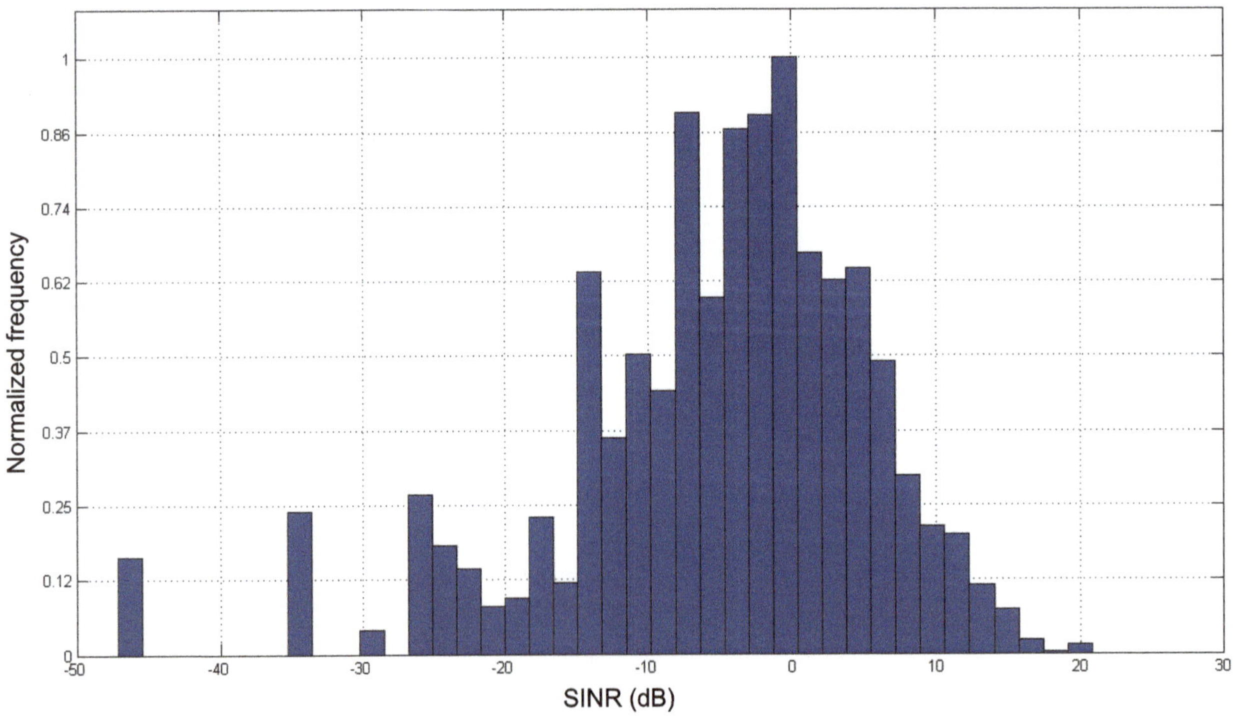

Figure 6 Distribution of SINR values obtained from the evaluated scenario.

to the secondary receiver. Additionally, our scenario differs from the reported in [16] in which one-hop is considered unlike multiple-hops as in an ad hoc network. Unlike the experiment presented in [17], our work does not evaluate the data rate of the network when the number of secondary links is gradually increased since the access control and channel allocation algorithm is able to select the set of secondary links that meet SINR restrictions.

Figure 6 shows the distribution of SINR values in the primary receiver obtained from our evaluation scenario. It is noted that SINR values are in the range from −46 to 22 dB. This graph shows that the access control and channel allocation algorithm based on SCPSO evaluates different combinations of secondary users to find the best solution.

On the other hand, Figure 7a,b,c,d shows the result of the number of selected secondary links and average number of selected secondary links that can share a channel with the primary user. The number of selected secondary links is derived from the best snapshot by experiment (SINR threshold) and case (number of secondary links); therefore, it corresponds to the minimum distances shown in Figure 5a,b,c,d. In contrast, the average number of selected secondary links is calculated taking into account the number of selected secondary links in each of the 5,000 runs (snapshots) by experiment and case.

From Figure 7a,b,c,d, the observation is as the SINR thresholds values increase, the number of secondary users that the HetNet admits to share a primary channel decreases. On the other hand, when the SINR thresholds values decrease, the HetNet can admit a greater number of secondary users but at the cost of sacrificing QoS. The above is also true, for the average number of selected secondary links, except for the cases of 20 secondary users and 25 secondary users.

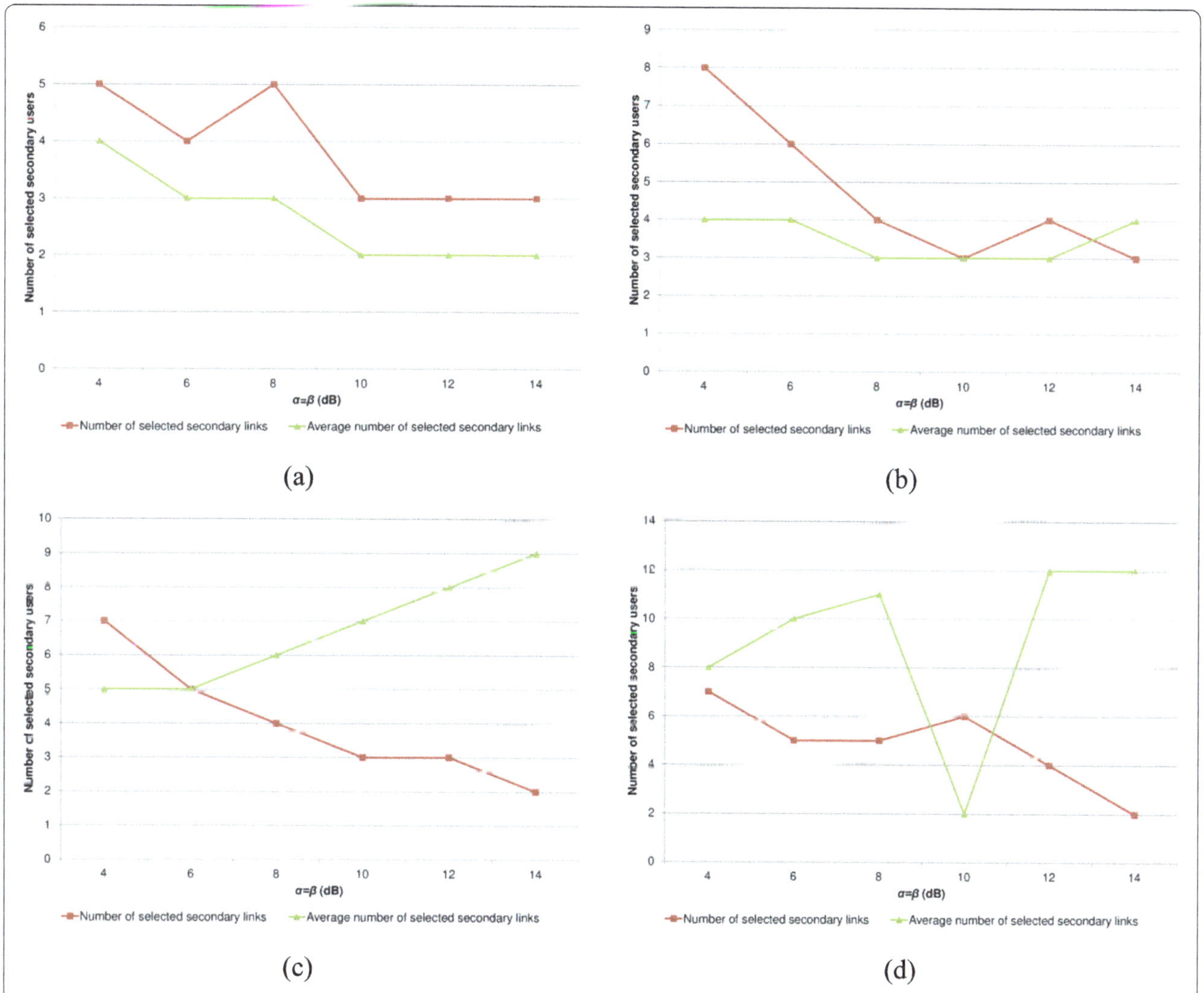

Figure 7 The number of selected secondary links by experiment and case. (a) 10 secondary users, **(b)** 15 secondary users, **(c)** 20 secondary users, and **(d)** 25 secondary users.

The benefit of the HetNet is the reuse of spectrum. However, the risks are equally important to consider, for example to control the number of admitted secondary users to coexist with the primary user. If the above is not controlled, it can lead to a 'tragedy of the commons' in which many users try to access the same spectral resource and neither of them is able to communicate due to the interference. A higher permissible density of secondary users also represents a more efficient reuse [18]. These facts show the importance for providers and regulators in the tradeoff between the permissible density of secondary users and the SINR threshold (QoS).

We will now present system performance results in terms of data rate. Figure 8a,b,c,d shows the maximum and average data rate that the system achieves by experiment and case. The maximum data rate of the system is the best found, *i.e.*, the run (or snapshot) with the highest data rate from the set of 5,000 runs. Those maximum data rates corresponds to the minimum distance and the number of selected secondary users from Figures 5a,b,c,d and 7a,b,c, d, respectively. In contrast, the average maximum data rate is obtained from the maximum data rates of each of the 5,000 runs. Results from Figure 8a,b,c,d suggest that the higher the SINR threshold values, the lower the data rate. This observation is consistent with results reported in [19]: the throughput is degraded significantly when the interference constraint is made more stringent; therefore, the performance of the underlay approach depends mainly on the interference constraint.

From those challenging scenarios of 20 and 25 secondary users (Figure 8c,d), the average maximum data rate tends to decrease dramatically, especially when the QoS requirements are higher. That implies that as the number of users increases, the interference also arises; making more challenging for the SCPSO to converge or to find a solution. To address that issue, a future work is to propose different

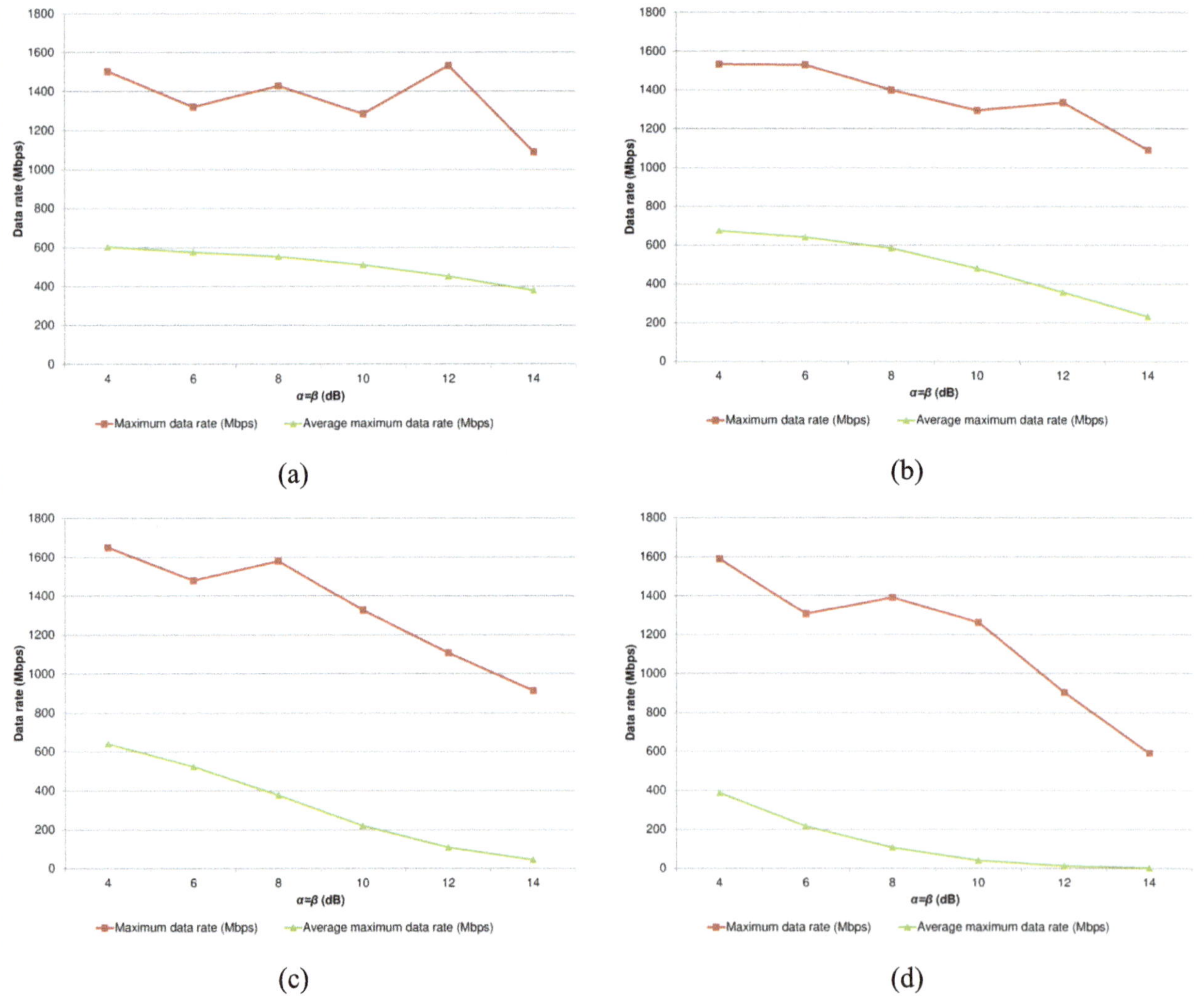

Figure 8 The maximum and average data rate by experiment and case. (a) 10 secondary users, **(b)** 15 secondary users, **(c)** 20 secondary users, and **(d)** 25 secondary users.

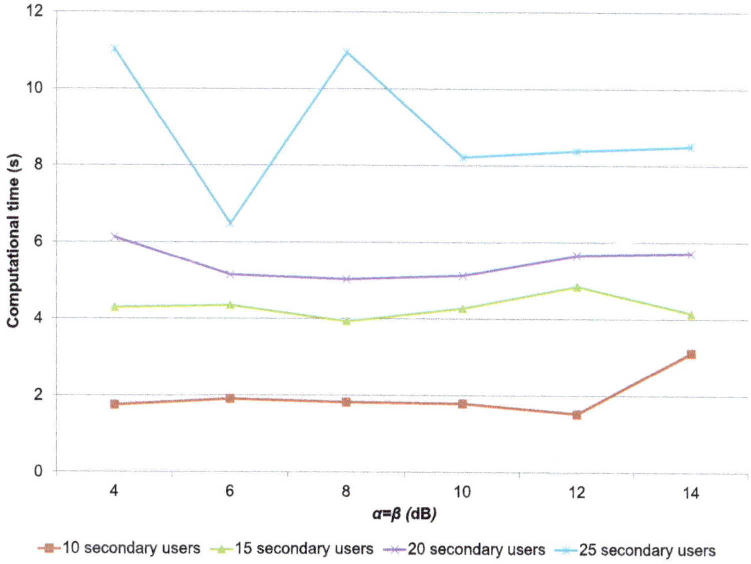

Figure 9 Computational time for the maximum data rates found.

parameters for the SCPSO by experiment and case, since keeping them fixed for all the experiments (the SINR thresholds) and cases (the number of secondary links) is not the best strategy to optimize its performance in challenging scenarios.

On the other hand, Figure 9 gives further insight into the computational time from those maximum data rates reported in Figure 8a,b,c,d. The running time is in various seconds, even in the most challenging scenarios such as in the cases of 20 and 25 secondary users. It increases with the number of secondary users deployed in the area and the SINR thresholds.

Finally, Figures 10 and 11 depict the convergence of average data rate through iterations, that is, considering the maximum data rates of each of the 5,000 runs. In Figure 10, the experiment of 4 dB for the cases of 10 and 25 secondary users is considered, which represents the less challenging scenario. In contrast, in Figure 11, the experiment of 14 dB for the cases of 10 and 25 secondary users, that is the most challenging scenario, is considered. From Figure 10, we observe that the access control and channel assignment algorithm using SCPSO takes more iteration to converge when there are more secondary users deployed in the area. That insight is

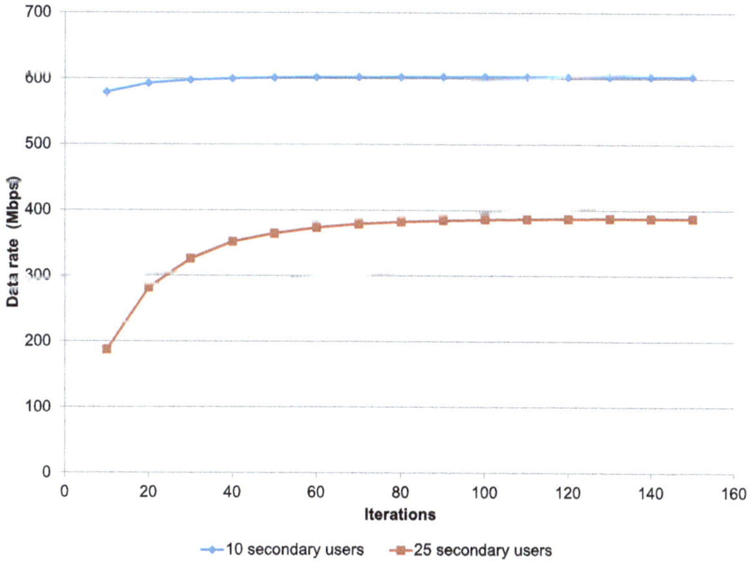

Figure 10 Convergence of average data rate (experiment = 4 dB, cases = 10, 25 secondary users).

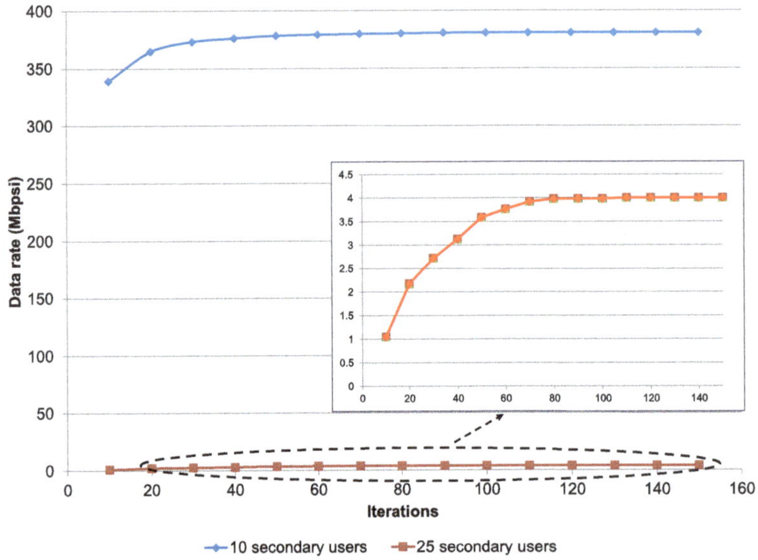

Figure 11 Convergence of average data rate (experiment = 14 dB, cases = 10, 25 secondary users).

also true for challenging scenarios such as that described in Figure 11; the average data rate is low in the case of 25 users. The above is consistent with the observation identified in Figure 8a,b,c,d, about to set different parameters for SCPSO by experiment and case to optimize its performance to find a solution.

5 Conclusions

In this work, we presented a framework to calculate the radius of the exclusion zone and the permissible density of secondary users to achieve the maximum throughput in the HetNet.

The interference from the secondary users to the primary user can be limited by defining a protection distance (primary exclusion zone). In this context, from simulation was observed that the higher the QoS level in the HetNet, the longer the distance that should be deployed in the secondary transmitters.

Simulation results also showed that from the number of secondary users wishing to operate concurrently with the primary user, a certain number of secondary users will be accepted depending on the QoS requirement.

Even when the computational time for a run was in seconds, a main issue was the parameter selection (tuning) on SCPSO algorithm. Future work will focus on identifying parameter values that improve the search ability of SCPSO to find a solution in challenging scenarios, since keeping them fixed for all the experiments and cases was not the best strategy to optimize its performance on the HetNet application.

Competing interests

The authors declare that they have no competing interests.

References

1. N Himayat, S Yeh, AY Panah, S Talwar, M Gerasimenko, S Andreev, Y Koucheryavy, *Proceedings of the IEEE International Conference on Computing, Networking and Communications (ICNC)*. Multi-radio heterogeneous networks: Architectures and performance (IEEE Honolulu, USA, 2014), pp. 252–258
2. SM Cheng, WC Ao, FM Tseng, KC Chen, Design and analysis of downlink spectrum sharing in two-tier cognitive femto networks. IEEE Trans. Veh. Technol. **61**(5), 2194–2207 (2012). doi:10.1109/TVT.2012.2187692
3. Y Liu, LX Cai, X Shen, H Luo, Deploying cognitive cellular networks under dynamic resource management. IEEE Wirel. Commun **20**, 82–88 (2013). doi:10.1109/MWC.2013.6507398
4. C Jiang, H Jiang, C Jiang, S Fan, Y Ren, C Chen, *Proceedings of the IEEE Military Communications Conference (MILCOM)*. Analysis on deployment parameters of secondary users in cognitive radio networks (IEEE San Jose, USA, 2010), pp. 743–748
5. MF Hanif, M Shafi, PJ Smith, P Dmochowski, *Proceedings of the IEEE International Conference on Communications (ICC)*. Interference and Deployment Issues for Cognitive Radio Systems in Shadowing Environments (IEEE Dresden, Germany, 2009), pp. 1–6
6. A Martínez-Vargas, ÁG Andrade, R Sepúlveda, O Montiel-Ross, in *Recent Advances on Hybrid Approaches for Designing Intelligent Systems*, ed. by O Castillo, P Melin, W Pedrycz, J. Kacprzyk, (Springer International Publishing Switzerland, 2014), p. 151
7. K Deep, JC Bansal, *Proceedings of the IEEE International Conference on Emerging Trends in Engineering and Technology (ICETET)*. A Socio-Cognitive Particle Swarm Optimization for Multi-Dimensional Knapsack Problem (IEEE Nagpur, India, 2008), pp. 355–360
8. MF Hanif, PJ Smith, PA Dmochowski, Statistical interference modelling and deployment issues for cognitive radio systems in shadow fading environments. IET Commun. **6**, 1920–1929 (2012)
9. O Durowoju, K Arshad, K Moessner, *Proceedings of the IEEE Vehicular Technology Conference (VTC Fall)*. Distributed Power Control and User Selection Algorithms for Cognitive Radios (IEEE Las Vegas, USA, 2013), pp. 1–5
10. R Dahama, KW Sowerby, GB Rowe, Estimating protection distances in spectrum sharing systems. IEEE Trans. Signal Process. **61**, 4284–4295 (2013)
11. P Tian, H Tian, L Gao, J Wang, X She, L Chen, *Proceedings of the IEEE International Symposium on Wireless Personal Multimedia Communications (WPMC)*. Deployment analysis and optimization of Macro-Pico heterogeneous networks in LTE-A system (IEEE Taipei, Taiwan, 2012), pp. 246–250

12. C Avin, Y Emek, E Kantor, Z Lotker, D Peleg, L Roditty, SINR diagrams: convexity and its applications in wireless networks. JACM **59**(18), 1–18 (2012). 34. doi:10.1145/2339123.2339125
13. RK Ganti, M Haenggi, Interference and outage in clustered wireless ad hoc networks. IEEE Trans Inf Theor **55**, 4067–4086 (2009)
14. JG Andrews, RK Ganti, M Haenggi, N Jindal, S Weber, A primer on spatial modeling and analysis in wireless networks. IEEE Commun. Mag. **48**, 156–163 (2010)
15. EZ Tragos, S Zeadally, AG Fragkiadakis, VA Siris, Spectrum assignment in cognitive radio networks: a comprehensive survey. IEEE Commun Surv Tutor **15**, 1108–1135 (2013)
16. P Gupta, PR Kumar, Towards an information theory of large networks: an achievable rate region. IEEE Trans. Inf. Theory **49**, 1877–1894 (2003)
17. M Vu, N Devroye, M Sharif, *V Tarokh, in Proceedings of the IEEE International Conference on Cognitive Radio Oriented Wireless Networks and Communications (CrownCom)* (USA, Scaling Laws of Cognitive Networks (IEEE Orlando, 2007). pp. 2–8
18. J Lee, JG Andrews, D Hong, Spectrum-sharing transmission capacity. IEEE Trans. Wirel. Commun. **10**, 3053–3063 (2011)
19. A Alsarhan, A Agarwal, I Obeidat, M Bsoul, A Khasawneh, Y Kilani, Optimal spectrum utilization in cognitive network using combined spectrum sharing approach: overlay, underlay and trading. Int J Bus Inform Syst **12**, 423–454 (2013)

2

Research on an improved parallel algorithm based on user's requirement in cognitive radar network

Bingxin Yan, Shubin Wang* and Yuanyuan Bao

Abstract

In cognitive radar network, a parallel algorithm can allocate idle spectrum to cognitive radars, but it often ignores the user's requirement. According to the graph theory, the topology of the network is divided into two different groups. In the process of distributing spectrum, each group summarizes the distributed information after completing a distribution. When a user's demand is satisfied, every group is informed to not allocate the spectrum for the user anymore and deletes it from the topology at the same time. As a result, by deleting the node which the user's requirement has already been met, the other nodes which are interfered with the node on the same channel can participate into spectrum allocation. The improved algorithm can ensure high spectrum utilization taking a little time that is spent on the spectrum allocation. The simulation result shows that the spectrum utilization of the improved algorithm is higher than that of the traditional parallel algorithm, and the user's satisfaction increases greatly. Therefore, this paper aims to attempt to explore an improved parallel algorithm to resolve the problem by considering user's requirement.

Keywords: Cognitive radar; Spectrum utilization; Graph theory; Parallel algorithm; User's requirement

1 Introduction

With increasing application of spectrum, it is becoming more and more scare [1]. Radar/communications sharing has been a recurring topic herein. The growing demand for 'beachfront' mobile spectrum to assure economic growth and provide for key wireless public safety systems require a reexamination of the large exclusive allocation for 'radiolocation' - radar in spectrum policy jargon.

Traditionally, radars were given large exclusive bands as primary allocations and the owner of the radar was able to make design tradeoffs between system cost/complexity and the amount of spectrum utilization they achieved. Since spectrum was a free cost to the designer there were few incentives to minimize spectrum footprint. However, with the growing interests on radar networks [2,3], spectrum allocation in radar network needs to be explored. However, the fixed spectrum management which allocates certain spectrum to the user leads to the low spectrum utilization rate. As a new kind of intelligent spectrum sharing technology, cognitive radar has emerged recently. In cognitive radar networks, spectrum allocation can be established based on the spectrum detection.

At present, there are three main spectrum allocation models for cognitive radar, including graph coloring model, game theory model, and auction bidding model, which could benefit our approach on radar network spectrum allocations. Among them, graph coloring model is commonly used. Based on graph coloring model, a lot of spectrum allocation algorithms have been studied. However, list coloring algorithm [1] does not distinguish the different channels. Coloring sensitive graph algorithm (hereafter referred to CSGC) [4,5] not only overcomes this weakness by checking the channel condition, but also puts forward several different label rules based on different target function. Nevertheless, the CSGC also has drawbacks that only one spectrum can be allocated in single distribution cycle. A parallel spectrum allocation algorithm distributes many spectrums in single distribution cycle improving the spectrum efficiency. Literature [2] and [3] refer to the

* Correspondence: wangshubin@imu.edu.cn
College of Electronic Information Engineering, Inner Mongolia University, Hohhot, China

parallel algorithm as well. In order to get higher channel multiplexing rate, the literature [6] improved the parallel algorithm which make other users share the same channel with the user that have the largest label. But neither these two algorithms consider the user's requirement, which leads to the waste of spectrum. To enhance the spectrum utilization ratio, aggregation distribution algorithm aggregates small spectrum fragment into a larger one [7-9]. The algorithm allocates the spectrum to users according to their demands, and considers the limitation of hardware equipment. Compared with the traditional way of continuous spectrum allocation, it significantly improves the spectrum utilization, but it is no doubt that the time is obviously increased. Bipartite matching [10] can also be used to distribute spectrum, but it is very complex. In order to increase the channel utilization, the literature [11] improves the CSGC in terms of user's priority. According to historical assignment information, literature [12] improves the CSGC, which can effectively avoid some users occupy certain spectrum exclusively. But the time that spends on allocation is not reduced yet. By using the SNR threshold and deleting the users whose requirement is wider than its available channel, power control is applied into the spectrum allocation [13-15]. The previous literature only consider one objective function, but the literature [16] proposes a new method that can consider both utilization and fairness of the network simultaneously through setting weight of network reward and fairness. The method solves the multiple functions problem in cognitive radar spectrum allocation. Chemical reaction [17] is also employed into spectrum allocation, but neither of them is simple. From the analysis above, we could see that the parallel algorithm has advantage of spending shorter time and the algorithm based on user's demand is able to greatly improve the utilization. In order to combine both advantages, the paper explores improved the parallel algorithm based on the user's requirement for radar networks.

2 Related content

2.1 Parallel algorithm

The parallel algorithm [6] is also called grouping algorithm. It divides the topology into many sub graphs, and then colors the sub graphs. Assume that all spectrums are orthogonal to each other. That is to say, coloring for a sub graph does not affect the others, and updating the topology does not influence others too. So, it can get the same allocation as the serial algorithm. To sum up, parallel algorithm is a good way to reduce the time of allocation because of coloring many sub graphs.

2.2 The algorithm of user's requirement

Variety of algorithms based on graph coloring model involves the factor of user's demand [8,9]. It is a kind of method that takes the user's demand for spectrum into consideration in the distribution of spectrum. When the user's requirement is met, the allocation of spectrum will stop offering to the user. This method not only can greatly improve the user satisfaction, but also can avoid spectrum is exclusively occupied by a certain user.

2.3 Label rule

The Label rule is firstly introduced by the CSGC algorithm [4], and it is proposed based on above graph-coloring algorithm whose main idea is considering the SINR of different users on every channel. The better the channel condition is, the greater the weight of channel will be. Every time channel with the greatest weight will be allocated to a corresponding user to maximize the system benefits. In order to achieve different target functions, the algorithm uses different label rules to meet the needs of the target function. When the goal maximizes the spectrum utilization in a collaborative mode, the label rule can be expressed as following:

$$\text{label}i = \max_{j \in li} {}^{\text{bi,j}} \big/ {}_{(Ri,j+1)} \tag{1}$$

$$\text{colori} = \arg\max_{j \in li} {}^{\text{bi,j}} \big/ {}_{(Ri,j+1)} \tag{2}$$

In the two equations above, ***Ri,j*** represents the number of node which can't be shared channel j with user i. ${}^{bi,j} \big/ {}_{(Ri,j+1)}$ represents the benefit that the user i bring to the system. l_i represents the number of available spectrum of user i.

2.4 Allocation model

We assume that the idle spectrum have been detected by some way. Then the graph coloring is used to establish the model to solve the spectrum allocation problem. According to the graph theory, spectrum allocation problem can be viewed as graph coloring problem by painting each secondary user into a vertex and each channel into a color. When we allocate channel m to user n, it is just similar to assign color m to vertex n. Then the spectrum allocation problem can be described by different matrices, which are listed in the following paragraph.

Available channel $\mathrm{L} = \{\mathrm{li,j}\}\mathrm{N} \times \mathrm{M}$. It is a matrix of N by M. N represents the number of secondary users, and M represents the numbers of spectrum. li, j consists of bandwidth of spectrum. If li, j is nonzero, it shows that the spectrum j is available to the user i. Otherwise, it means the channel j to the user i is not available.

Benefit $B = \{bi, j = \{0, 1\}\} N \times M$. The matrix represents the benefit to system which secondary user n applies the channel j.

Demand is the matrix 1 by N, indicating the need of spectrum for each user. It takes megahertz as its unit.

Interference constraint $= \{ci, k, m = \{0, 1\}\}N \times N \times M$. The matrix is a three-dimensional matrix, and it represents the number of interferences that are caused by secondary user i sharing the same channel m with secondary k.

Allocation $= \{ai, j = \{0, 1\}\}N \times M$. If $ai, j = 1$, it shows that the spectrum j is assigned to the user i.

The degree of interference $= \sum_{k=1, k \neq i}^{N} ci, k, j$. It represents the number of nodes which can't be shared channel j with user i.

In order to ensure that all users to make full use of the allocated spectrum, the distribution of spectrum should satisfy the following requirements:

1. Any user won't be able to communicate properly unless its requirement is met.

2. Due to the guard bandwidth between spectrums is very thin, it can be neglected.

3. The algorithm improves spectrum efficiency and fairness. As for the drawbacks of the parallel algorithm, it does not improve.

3 Improved parallel algorithm based on user's requirement

3.1 Algorithm principle

Assume that a cognitive radar network is composed of N1 primary users and N2 secondary users in the area of d×d. The whole available spectrum can be divided into m different channels which are orthogonal to each other. Each cognitive user can detect the available spectrum. The idle spectrum detection, spectrum allocation, and data transmission form one distribution cycle. In a cycle, we assume that the location of user and the available spectrum are fixed. The primary user (PU) or the secondary user (SU) has a transmission radius respectively in each channel. Only when the SU's transmission radius is not covered by the PU's, the SU can share the PU's spectrum.

In the study of graph coloring model, the topology is composed of vertexes and edges. Every vertex represents a cognitive radar user. What's more, it is associated with an available channel list, which is determined by the location of a user. Take Figure 1 as an example, among which P represents PU, S represents SU, and Ch represents channel. Set the number of primary user N1 as 4, the number of secondary user N2 is 4, and the available channel number Ch is 3. According to the graph theory, spectrum allocation problem can be regarded as graph coloring problem by painting each secondary user into certain vertex and each channel into a color. When channel m is allocated to user n, it is just similar to assign color m to vertex n. The details are expressed in Figure 2. Then the previous matrices can be used in part 2 to describe distribution of process.

The process of grouping algorithm is as follows:

1. Delete these nodes whose demand is larger than that of total number of available spectrum from topology. For one thing, even if all available spectrums are allocated to it, the user's demand will still not satisfied either. For the other thing, other nodes that interfere with this node on the channel could obtain the spectrum by deleting this node.
2. Divide the available topology applying user grouping and frequency grouping respectively. In frequency grouping, the topology can be divided according to the number of available spectrum. That is to say, if a topology consists of M available spectrum, it can be divided into M groups. According to Equation 4 and 5, we can calculate the values of Label and corresponding Color in each group. Select the maximum value of Label, and allocate the corresponding Color to it (If there are two or more largest Labels, we choose one randomly). Delete other nodes that can't simultaneously use the spectrum with it in the topology of the group. Besides, user grouping is another matter. The topology can be divided according to the number of users. Select the maximum Label that corresponds to the spectrum according to Equation 4 in each group. If there are same spectrums, and two or more users interfere with each other on the spectrum, the spectrum is assigned to the largest value of Label. And the other Labels correspond to the other users cannot take part in allocating spectrum in the current cycle. If there is no same spectrum, directly allocate spectrum to the user who has the largest value of Label.
3. Each group summarizes the distributed information after completing one allocation. Then the node is deleted from each sub topology. Because it can not only avoid wasting the spectrum resource, but also can enhance the fairness. Figure 3 is an example. Assume that the Figure3 (A) is an initial topology and the node 3 has got enough spectrums. If the improved parallel algorithm is applied based on user's requirement, then the node 3 is removed from topology of this group. As is shown in Figure (B). The nodes 1 and 5 may participate into the allocation of the spectrum after that. In other words, it enhances the probability of more users to use the channel.

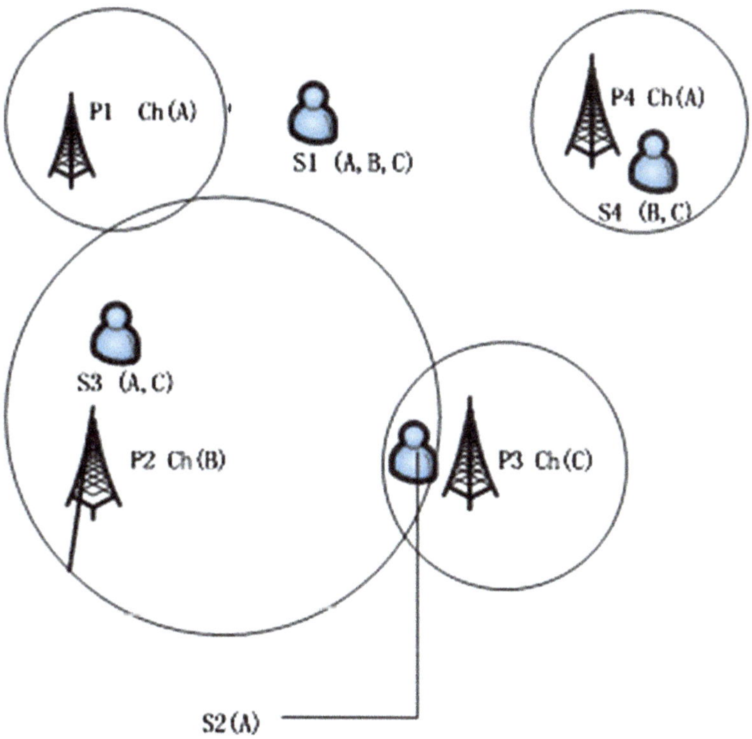

Figure 1 A network of cognitive radar. It is a distribution of different users in practical scenario. PU has authored channel, but SU does not have. SU only can use the idle channel and that do not cause any interference to PU.

4. The program terminates, when there is no available spectrum or all users' requirements have been met.

The procedure of the algorithm can be summarized as algorithm 1 and algorithm 2.

Algorithm 1 Frequency Grouping

(1) Initialize. According to the idle channel of PU and the distribution of SU, system produces the matrix of Frequency, Demand, Available Channel and Benefit randomly.
(2) Judge whether there is any user whose demand is larger than the total number of available channel. If so, delete it from the topology.
(3) Divide topology into different groups according to the number of available channels.
(4) Calculate the value of Label and Color.
(5) Allocate the corresponding Color to the maximum value of Label.
(6) Gather the spectrum that have been allocated in each group. If any user's demand has been met, delete it from each topology. Otherwise, go to the step 4.
(7) Unless all users' demands have been met or there is no available spectrum, the program would stop.

Algorithm 2 User Grouping

(1) Initialize. According to the idle channel of PU and the distribution of SU, system produces the matrix of Frequency, Demand, Available Channel and Benefit randomly.
(2) Delete the user whose demand is greater than the total number of available channel from the topology.
(3) Divide topology into different groups according to the number of users.
(4) Calculate the value of Label and Color.
(5) Allocate spectrum to the user whose Label is the largest. (If there are the same spectrum, and two or more users interference with each other on the spectrum, the spectrum is assigned to the node that has the largest value of Label.)
(6) Gather the spectrum that have been allocated in each group. If any user's demand has been met, delete it from each topology. Then update the topology. Otherwise, go to step 4.
(7) Unless all the user's demand had been met, or there is no available spectrum, the program would stop.

3.2 Objective function and labeling standards

Our objective function maximums the system spectrum utilization. In order to enhance the spectrum utilization, the assigned spectrum should be ensured to be larger than the users' requirement. Otherwise, the spectrum will not be allocated to the user, and be deleted from the topology. In this way, the objective function can be expressed as following:

$$f = max\sum_{i=1}^{N}\sum_{j=1}^{M} ai,j \times yi \times bi,j \tag{3}$$

In the equation above, $yi = \frac{Demandi}{\sum_{i=1}^{N} Ti} <= 1$, and $\sum_{i=1}^{N} Ti$ indicates the total bandwidth of available spectrum for user i. We take the MHz as its unit. In order to make sure of the $yi \leq 1$, the Node whose Demand is larger than $\sum_{i=1}^{N} Ti$ is deleted from the topology. After deleted, the node would not take part in allocation. So deleting a node does not affect the value of ai, j and objective function f. When the function is used during allocating spectrum, the labeling method (Collaborative-Max-

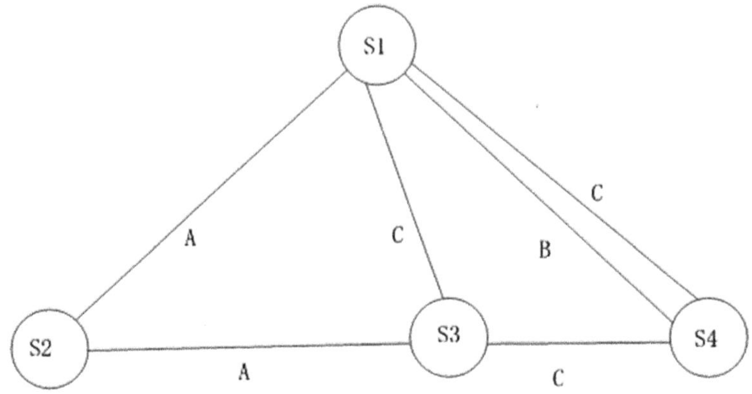

Figure 2 Topology of cognitive radar. According to the graph theory, user can be shown by node, and the constraint between them can be shown by edge. Then the topology is obtained, which consist of edges and nodes.

Proportional-Fair) will be introduced. As a result, the labeling standards change as follows:

$$Label i = \max \frac{yi \cdot \frac{bi,j}{Ri,j+1}}{\sum_{i=1}^{N} ai,j \cdot bi,j} \tag{4}$$

$$Color i = \arg\max \frac{yi \cdot \frac{bi,j}{Ri,j+1}}{\sum_{i=1}^{N} ai,j \cdot bi,j} \tag{5}$$

3.3 The time for parallel algorithm based on user's requirement

List coloring spectrum allocation algorithm and CSGC spectrum allocation algorithm are both based on the cognitive radar network topology of the static spectrum allocation. That is to say, topology of cognitive radar network always does not change. If there is a new node to get into the radar network, cognitive radar will rebuild topology. This process needs more time.

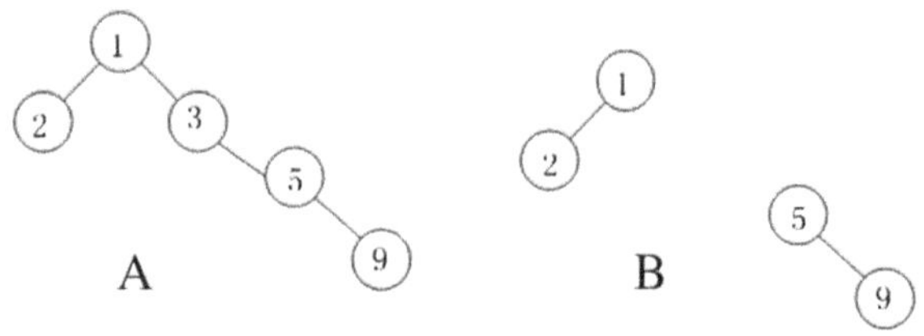

Figure 3 An illustration of the fairness. It is one of sub topologies. Compared with the traditional parallel algorithm, the improved method enhances the fairness of different secondary users to use the channel. Figure 3 is an example. Assume that the Figure3 **(A)** is an initial topology and the node 3 has got enough spectrums. If the improved parallel algorithm is applied based on user's requirement, then the node 3 is removed from topology of this group. As is shown in Figure **(B)**.

As the network the cognitive users and available channel increases, the time of list coloring algorithm and CSGC algorithm increases as well. It can be measured according to Equation (6):

$$T\ \text{cost} = \sum_{i=1}^{N}\sum_{j=1}^{M} ai,j \tag{6}$$

In Equation 6, an,m represents the allocation matrix. The traditional parallel algorithm of spectrum allocation reduces the time of CSGC spectrum allocation algorithm. The detailed calculation of the time is listed as followed:

$$T\,cost = \max_{1\le j\le M} \sum_{i=1}^{N} ai,j \tag{7}$$

The time cost of the improved parallel algorithm is little longer than that of the traditional one. The time of frequency grouping is shown as Equation 8:

$$T\,cost = \max_{1\le j\le M} \sum_{i=1}^{M} ai,j + \max_{1\le j\le M}(\lambda j) \cdot treport \tag{8}$$

In the Equation 8, λj represents the number of allocation of each group, which is determined by the

Table 1 Sets parameters

The parameter name	Value
The number of primary user	10
The number of channel	3–20
The number of secondary user	20
The primary user's coverage radius	5 km
The secondary user's coverage radius	3.5 km
The user's demand of spectrum	randomly generated from 0 to a largest T

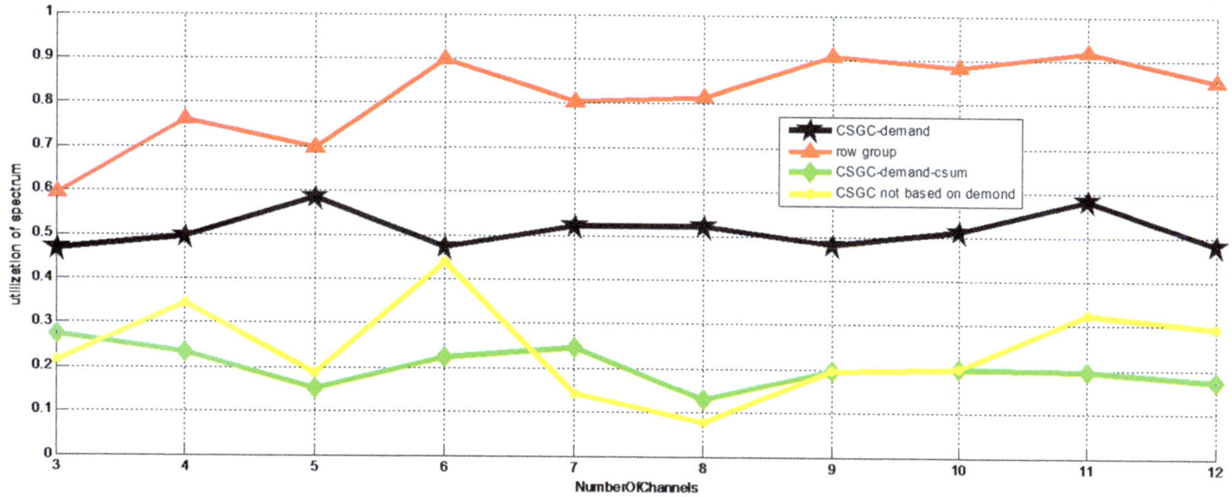

Figure 4 Utilization changes as available channels change. These four algorithms of allocation are compared. The four curves (red, green, yellow and black) indicate the utilization of spectrum changes as channels change. Among them, the red one is the improved frequency grouping method.

matrix of Available Channel. The allocation is carried on until the matrix Available Channel of each group is equal to zero. As the number of allocation in each group may not equal, $\max_{1\leq j\leq M}(\lambda j)$ represents the max number of allocation of groups. *treport* represents the time that each group spend to report their distribution to the **cognitive base station(CBS)** in order to judge whether the node has already reached the users' requirements. Just because of it, the time cost of the improved parallel algorithm is little longer than that of the traditional one.

The time of user's grouping is listed as Equation 9:

$$T\,\text{cost} = \max_{1\leq i\leq M} \sum_{j=1}^{M} ai,j + \max_{1\leq i\leq N} \eta i{\cdot}treport \tag{9}$$

In the equation above, ηi represents the number of allocation in each user grouping. The allocation is carried on, until the matrix of Available Channel of each group is zero. Similarly, for the number of allocation is not equal in each group, $\max_{1\leq i\leq N}(\eta i)$ represents the max number of allocation in each group.

4 Simulation and analysis

4.1 Simulation conditions

Assume that PU and SU are distributed in region of 1,000 × 1,000 randomly. Each user can use anyone of M channels. The bandwidth of available channel generates randomly from 1 to 3 MHz in TV spectrum. The demand of user i (Di) is an integer and is produced ran-

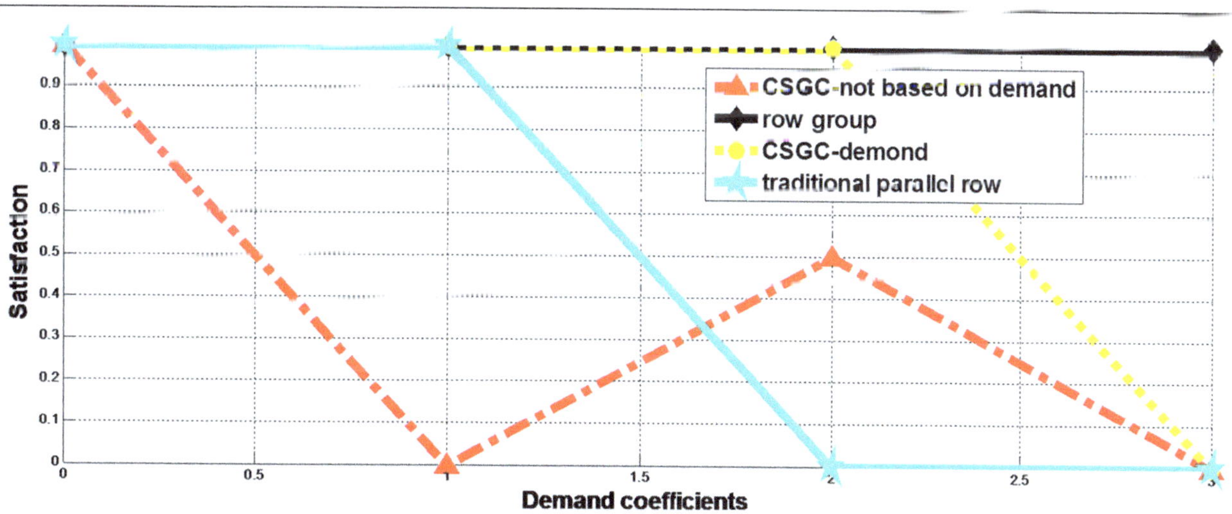

Figure 5 Satisfaction changes as Demand coefficients change. These four algorithms of allocation are compared. The four curves (red, yellow, blue, and black) indicate the satisfaction as changes over demand coefficient. Among them, the black one is user grouping method.

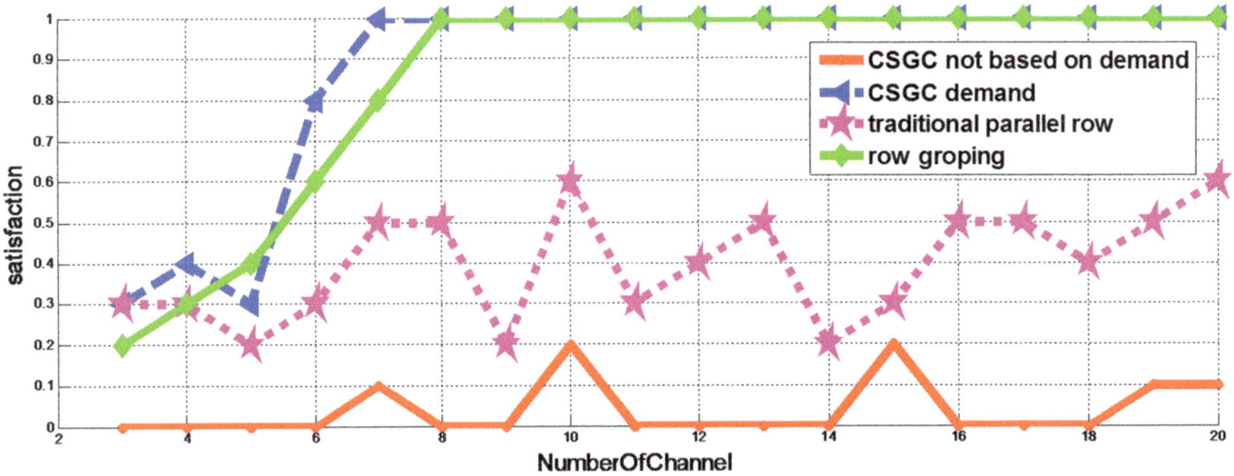

Figure 6 Satisfaction changes as number of channel change when the Demand matrix is fixed. These four algorithms of allocation are compared. The four curves (red, green, purple, and blue) indicate the satisfaction of spectrum as channels changes. Among them, the green one is the improved frequency grouping method.

domly. Di ∈ [0, α Ti], in which α is defined as the requirement coefficient and α ∈ [1,3]. The parameters are shown in Table 1.

4.2 Results and analysis

Satisfaction is one of the factors to measure fairness. Spectrum utilization rate is an important factor weighing whether the method is good or not. The definition of satisfaction is as follows:

$$\text{Satisfaction} = N_i/N_t \tag{10}$$

In Equation 10, N_i represents the number of nodes whose demand has been met, and N_t represents the total number of bandwidth requirements of all users.

$$\text{Utilization rate} = N_x/N_n \tag{11}$$

In the equation above, N_x represents the number of available spectrum, and N_n represents the number of total spectrum.

First, we change the number of available channels to compare the utilization of spectrum among the

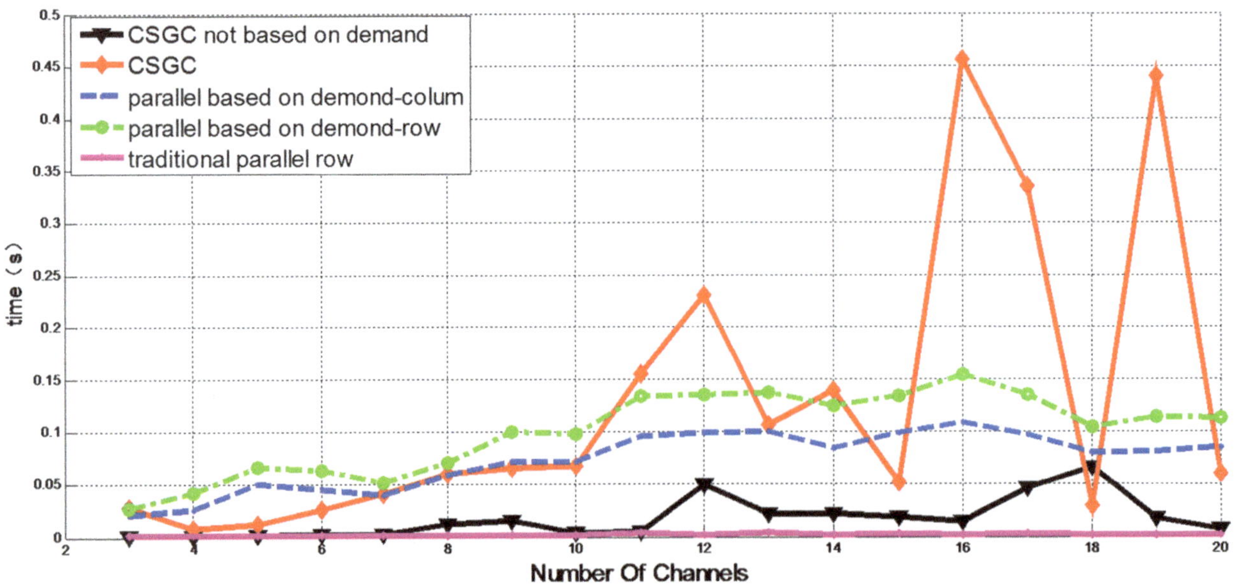

Figure 7 The cost of time changes as available channels. The time of six algorithms are compared with each other. Among them, the blue and purple curves represent the two algorithms of the two grouping improved algorithms.

algorithms, which is shown in Figure 4. Since the traditional parallel algorithm does not consider the different requirements of bandwidths, it may still allocate channels to the SU according to the Label rule that is given in Equation 4, though the SU's requirement has been satisfied. As a result, the total utilization rate is not very high. At the same time, those who have interference with the SU on the same channel cannot use the channel any more. This also decreases the total spectrum utilization. Even when available channels increases obviously, the traditional parallel's total spectrum utilization has little change. However, the improved parallel algorithm would not allocate extra channel to the SU whose requirement is satisfied. Therefore, the improved method has higher utilization rate than the traditional parallel algorithm.

Then, compare is made on these four algorithms from the perspective of the satisfaction rate, and the SU's demand is produced randomly, which is shown in Figure 5. When the demand coefficient is small, the user's satisfaction ratio is higher than that of normal parallel algorithm. Because the improved algorithm not only takes the different requirements of the bandwidths of SU's into account but also delete the users whose demands is much larger than the number of total available spectrum. Therefore, there are more users' requirements are satisfied. But as the demand coefficient increases, there are fewer users' needs can be satisfied, thus the satisfaction ratio gradually reduces. What 's more, according to the Figure 5, besides the improved parallel algorithm, CSGC-demand have high satisfaction. This is because it takes the user's requirement as its priority. It gets high satisfaction, but with much longer time cost. To be clear, the following figures are one of a certain results in many executions instead of statistical data.

In order to compare the satisfaction rate of algorithms clearly, we assume that the Demand matrix is fixed as [4, 2, 1, 4, 3, 5, 6, 3, 1, 7, 2, 8, 4, 1 ,6, 4, 4, 7, 6, 4] and frequency matrix is generated from [1, 3] randomly, which is shown in Figure 6. The result shows that satisfaction rate of the improved parallel algorithm tends to 1. However, the satisfaction rate of other algorithms that is not based on user's demand fluctuates greatly.

Next, we pay attention to the time of all algorithms. As is depicted in Figure 7, the improved grouping algorithm spends much shorter time than the CSGC algorithm on spectrum allocation. This is because the improved algorithm allocates many spectrums to many users simultaneously. Compared with traditional parallel algorithm, the improved algorithm spends much longer time on allocating spectrum because of summarizing allocated information in each cycle to CBS.

Above all, figures shows that the spectrum utilization of the improved algorithm is higher than the traditional parallel algorithm, and the user's satisfaction increases greatly, but the time cost of it is a little longer than others.

5 Conclusions

This paper explores an improved parallel algorithm based on user's requirement of spectrum. Two different methods are used to illustrate the improved parallel algorithm. Finally, we could safely draw the conclusion that the improved algorithm not only has much higher spectrum utilization rate, but also improves the fairness and customer's satisfaction with a little spending of time. However, compared with the traditional parallel algorithm, the improved parallel algorithm is a little more complex.

Competing interests

The authors declare that they have no competing interests.

Acknowledgements

Shubin Wang (wangshubin@imu.edu.cn) is the correspondent author and this work was supported by the National Natural Science Foundation of China (61261020), and the Natural Science Foundation of Inner Mongolia, China (2012MS0903), and PetroChina Innovation Foundation (2014D-5006-0603), and the Scientific Research Initial Fund for Higher Talents Program of Inner Mongolia University, China.

References

1. W Wang, X Liu, List coloring based channel allocation for open spectrum wireless networks, in *IEEE 62nd Vehicular Technology Conference* (Dallas, TX, USA, 2005)
2. L Qilian, Radar sensor wireless channel modeling in foliage environment: UWB versus narrowband. IEEE Sensors Journal. 1448–1457 (2011). doi:10.1109/JSEN.2010.2097586
3. L Jing, L Qilian, Design and analysis of distributed radar sensor networks. IEEE Trans on Parallel and Distributed Processing. 1926–1933 (2011). doi:10.1109/TPDS.2011.45
4. H Zheng, C Peng, Collaboration and Fairness in Opportunistic Spectrum Access, in *the 2005 IEEE International Conference on Communications* (Seoul, Korea, 2005)
5. C Peng, H Zheng, B Zhao, Utilization and fairness in spectrum assignment for opportunistic spectrum access. Mobile Netw. Appl. 555–576 (2006). doi:10.1007/s11036-006-7322-y
6. Y Liu, M Jiang, X Tan, Time Overhead and Fairness in Spectrum Allocation for Cognitive Radio, in *2011 the 6th International ICST Conference on Communications and Networking in China* (Harbin, China, 2011)
7. JD Poston, WD Horne, Discontinuous OFDM considerations for dynamic spectrum access in Idle TV Channels, in *the 1rd IEEE International Symposium on New Frontiers in Dynamic Spectrum Access Networks* (Baltimore, MD, USA, 2005)
8. H Lee, S Vahid, K Moessner, *A Survey of Radio Resource Management for Spectrum Aggregation in LTE- Advanced, IEEE Communications Surveys & utorials*, 2013, pp. 745–760. doi:10.1109/SURV.2013.101813. 00275
9. F Sheng, L Ma, X Tan, C Yin, Y Yu, Spectrum Allocation Algorithm Aware Spectrum Aggregation in Cognitive Radio Networks, in *2013 the Third International Conference on Instrumentation, Measurement, Computer, Communication and Control* (Shenyang, China, 2013)
10. C Zhao, M Zou, B Shen, B Kim, K Kwak, Cooperative spectrum allocation in centralized cognitive networks using bipartite matching, in *the IEEE Global Telecommunications Conference* (New Orleans, LO, USA, 2008)
11. S Zhang, P Ren, C Zhang, in *Software Engineering and Knowledge Engineering: Theory and Practice*, ed. by W Yanwen (Springer, Berlin Heidelberg, 2012), p. 1017
12. L Yang, X Xie, Y Zheng, A Historical-information-based algorithm in dynamic spectrum allocation, in *the IEEE International Conference on Communication Software and Networks* (Macau, China, 2009)

13. Y Wang, Z Wei, H Du, L Sang, D Yang, A spectrum allocation algorithm based on bandwidth matching and interference avoidance in cognitive radio networks, in *the IEEE 23rd International Symposium on Personal Indoor and Mobile Radio Communications* (Sydney, Australia, 2012)
14. C Wang, X Hong, H Chen, J Thompson, On capacity of cognitive radio networks with average interference power constraints, in *IEEE Transactions on Wireless Communications*, 2009, pp. 1620–1625. doi:10.1109/TWC.2009.071075
15. A Hoang, Y Liang, Maximizing spectrum utilization of cognitive radio networks using channel allocation and power control, in *the IEEE 64th Vehicular Technology Conference* (Montréal, Québec, Canada, 2006)
16. H Gao, J Cao, Non-dominated sorting quantum particle swarm optimization and its application in cognitive radio spectrum allocation. Journal of Central South University. 1878–1888 (2013). doi:10.1007/s11771-013-1686-5
17. AYS Lam, VOK Li, Chemical reaction optimization for cognitive radar spectrum allocation, in *the IEEE Global Telecommunications Conference* (Miami, Florida, USA, 2010)

3

Cooperative power control in OFDMA small cell networks

Gaofeng Nie[1], Hui Tian[1*], Jiazhi Ren[1], Jing Wang[2], Liu Liu[2] and Huiling Jiang[2]

Abstract

This paper aims at improving the system throughput of orthogonal frequency division multiple access small cell networks. Different with traditional schemes that neglect the cooperation among small cells, a scheme named as resource block exclusion-based power control (RBEBPC) is proposed by sharing the interference correlated information. RBEBPC consists of two steps that are iteratively conducted. First, based on current power allocation results, partial system resource blocks are excluded by playing the formulated cooperative coalition formation games. Second, the transmission power of each small cell is determined by solving a modified throughput maximization problem after the resource block exclusion. As the generated interference is constrained in the second step, part of the small cells transmit without full power. Thereby, the overall system interference keeps non-increasing after RBEBPC is adopted. Simulation results indicate that about 15% system throughput gain and 13% power saving gain are obtained compared to traditional iterative water filling scheme.

Keywords: Cooperative power control; Orthogonal frequency division multiple access; Small cell; Coalition game; Iterative water filling

Introduction

The wireless data traffic of cellular system is soaring. Based on the report of future mobile data forecast by Cisco, the compound annual growth rate of mobile data traffic from 2011 to 2016 will be 78% [1]. The wireless industry faces a big challenge: a 1,000-fold data traffic increase in a decade [2]. The emergence of small cell releases the cost pressure on the dense cell deployment, which makes it one of the most promising solutions to the explosive wireless traffic growth. Small cell is the node whose transmission power is lower than that of the macro base station [3]. Both picocell and femtocell are included in the concept of small cell. Small cell is cheap, plug and play, and self-configuring, which can be easily embedded in the existing cell deployment. The deployment of small cell shortens the distance between service providers and subscribers. As a consequence, subscribers can achieve the same quality of service with lower power consumption. The shorter distance also enables the small cell to use a higher carrier frequency that is not suitable for the macro cell.

However, when small cells share the same spectrum band with macro cells, strong interference may exist. There are two categories of interference in small cell networks. The first category, which is known as cross-tier interference, is the interference between small cells and macro cells. The second category, which is known as co-tier interference, is the interference among small cells. Strong interference drops the advantages of small cell deployment and becomes the bottleneck of throughput improvement. Some previous works on small cell interference mitigation focus on orthogonal resource reuse from time or frequency domain. Authors of [4-7] propose orthogonal time domain interference mitigation schemes, e.g., almost blank subframe and the enhanced versions. Authors of [8,9] provide frequency domain interference mitigation scheme, e.g., fractional frequency reuse. The orthogonal resource reuse schemes have the common drawback that the interference is mitigated at the cost of resource utilization. Authors of [10] analyze the development trend of small cell networks and point out that small cell deployment would become denser in the future. Marginal reward would be obtained in the orthogonal

*Correspondence: tianhui@bupt.edu.cn
[1] State Key Laboratory of Networking and Switching Technology, Beijing University of Posts and Telecommunications, Xitucheng Road, 100876 Beijing, China
Full list of author information is available at the end of the article

resource reuse schemes due to the large quantity of weak interference sources. In addition to orthogonal resource reuse, power control is also considered to mitigate the interference. Authors of [11] provide a price-based power control scheme to manage the first category of interference. But only one sub-channel case is considered. In [12], hierarchical competition schemes for downlink power allocation in orthogonal frequency division multiple access (OFDMA) femtocell networks are proposed. A Stackelberg game model is used to manage the cross-tier interference of Femtocell and macro cell. In [13], the potential game is used with a designed payoff function to manage both the co-tier and cross-tier interferences. Some works adopt the centralized manner to manage the power control issues in small cell networks. Authors of [14] provide centralized suboptimal algorithms to manage the downlink scheduling and power allocation issues in multicell OFDMA networks. Authors of [15] provide a binary power allocation scheme to mitigate the second category of interference but the research is limited to indoor scenario.

As described in [3], the scenario where spectrum bands used for small cells and macro cells are different is one of the scenarios remaining to be studied. Besides, the deployment of small cells are more random and flexible than conventional macro cells, which makes the interference mitigation issues among small cells rather challenging. In this paper, we resort to the cooperative game theory and convex optimization theory to deal with the interference issues in small cell networks. Our target is to establish an effective method to improve the system throughput. Considering the deployment trend [2,3], we only deal with the co-tier interference of the small cells. Based on the resource structure of OFDMA small cell networks, the proposed scheme, which is called resource block exclusion based power control (RBEBPC), conducts the following two steps iteratively. 1) Small cells cooperatively exclude the disadvantage resource blocks by using the coalition formation game theory. An interference constraint is obtained based on the channel exclusion results for each small cell. 2) The transmission power of small cell is determined by solving the modified throughput maximization problems. The system total interference is non-increasing during the conduction of RBEBPC. Besides, RBEBPC is also energy efficient, i.e., not all available power is used during the transmission.

System model and basic assumptions

In this paper, bold lower case letters denote vectors and bold upper case letters denote matrices. The element of the i-th row and the j-th column of $\mathbf{X}$ is denoted by $[\mathbf{X}]_{ij}$. The transpose of $\mathbf{X}$ is $\mathbf{X}^T$. $\mathbb{R}^{M\times N}$ represents the $M \times N$ real space and $\mathbb{R}$ represents the real line. Calligraphic upper case letters denote sets. The cardinal number of $\mathcal{S}$ is denoted by $|\mathcal{S}|$, and the power set of $\mathcal{S}$ is denoted by $\eta(\mathcal{S})$. Notation $\binom{n}{k}$ represents 'n choose k' and equals $\frac{n!}{k!(n-k)!}$. Notation $\Longleftrightarrow$ means 'if and only if'. Finally, 's.t.' is the abbreviation of 'subject to'.

As shown in Figure 1, the considered small cell network is under the coverage of a macro cell. The frequency bandwidth used by the macro cell is F1. The frequency bandwidth shared by small cells is F2. The used spectrum of F1 is orthogonal to that of F2, i.e., there is no cross-tier interference between the macro cell and small cells. Each small cell is assumed to have a backhaul link to the macro cell so that some mutual information can be exchanged. Small cells in the considered system model are denoted by $\mathcal{N} = \{1, 2, \ldots, N\}$. Without loss of generality, each small cell is assumed to serve only one user equipment (UE). In other words, in our system, each small cell first schedules a UE, then performs power allocation. The set of UEs is denoted by $\mathcal{M} = \{1, 2, \ldots, M\}$ and $M = N$. The i-th UE is assumed to be served by the i-th small cell. F2 equals W and is equally divided into L sub-channels. The available sub-channels for small cells in the system are denoted by $\mathcal{L} = \{1, 2, \ldots, L\}$.

Figure 2 shows the resource structure of the small cell network that is based on the above description. The system resource can be depicted from two dimensions: the small cell dimension and the frequency dimension. In the small cell dimension, each small cell owns at most L available sub-channels and all these sub-channels are able to share the total transmission power $P0$. However, in the frequency dimension, each sub-channel can be occupied by at most N small cells and these small cells *cannot* share the transmission power. The resource block of the l-th sub-channel occupied by the j-th small cell (denoted by RB_{lj}) can be uniquely determined by the tuple (l, j).

The wireless channel is assumed to vary in a slower pace compared to our power control scheme, i.e., the channel

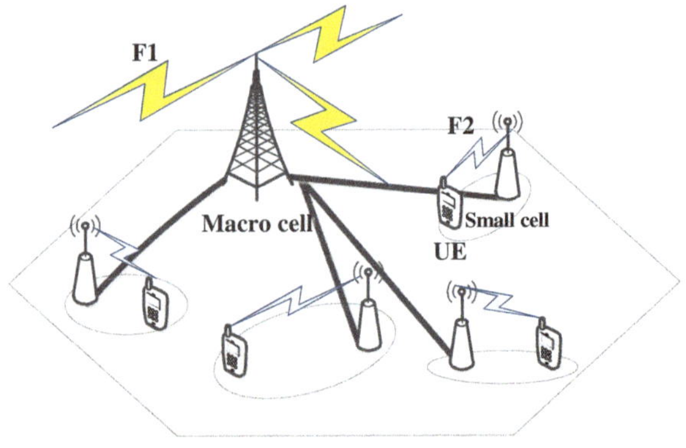

Figure 1 System model. The macro cell transmits data with frequency band F1. Small cells share frequency band F2. The used spectrum of F1 is orthogonal to that of F2.

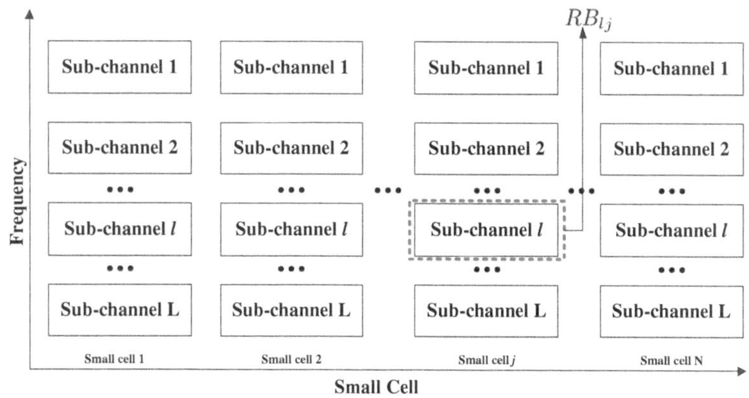

Figure 2 Resource structure of the small cell network.

gain during our analysis remains constant. The channel gain between the i-th UE and the j-th small cell on the l-th sub-channel is denoted by G_{ij}^{l}. The system channel gain matrix is denoted by $\mathbf{G} \in \mathbb{R}^{M\times N\times L}$. The channel gain matrix related to the i-th UE is defined as $\mathbf{G}_{i\cdot} \in \mathbb{R}^{N\times L}$, which contains all the channel gain between the i-th UE and all the small cells on all the sub-channels. Similarly, the channel gain related to the j-th small cell and the l-th sub-channel are defined as $\mathbf{G}_{\cdot j} \in \mathbb{R}^{M\times L}$ and $\mathbf{G}_{\cdot\cdot}^{l} \in \mathbb{R}^{M\times N}$ separately.

For the i-th small cell, the power allocation result is denoted by $\mathbf{p}_i = [p_i^1, p_i^2, \ldots, p_i^L]^T \in \mathbb{R}^{L\times 1}$, where p_i^l represents the allocated power level on the l-th sub-channel. The system power allocation result is denoted by $\mathbf{P} = [\mathbf{p}_1, \mathbf{p}_2, \ldots, \mathbf{p}_N] \in \mathbb{R}^{L\times N}$.

Based on above assumptions, the throughput of the j-th small cell is as follows:

$$R_j = \frac{W}{L}\sum_{l=1}^{L}\log_2\left(1+\frac{p_j^l G_{jj}^l}{\sigma^2+I_j^l}\right), \tag{1}$$

where σ^2 is the noise power of each sub-channel and

$$I_j^l = \sum_{i=1, i\neq j}^{N} p_i^l G_{ji}^l \tag{2}$$

is the suffered interference on the l-th sub channel. The transmission power of each small cell is usually constrained by:

$$\sum_{l=1}^{L} p_j^l \leq P0, \forall j \in \mathcal{N}, \tag{3}$$

where $P0$ is the maximum transmission power of each small cell and

$$p_j^l \geq 0, \forall j \in \mathcal{N}, \forall l \in \mathcal{L}. \tag{4}$$

The system throughput maximization problem of the small cell network is described as:

$$\mathbf{P1}: \quad \max \sum_{j=1}^{N} R_j \tag{5}$$

s.t. (Equation 3), (Equation 4).

Motivation of cooperative power control

Since **P1** is non-concave to its domain, it is difficult to obtain the global optimal solution. Some sub-optimal solutions concentrate on the non-cooperative manner such as the iterative water filling (IWF) to slove **P1**. Due to the neglect of possible mutual cooperation, the non-cooperative sub-optimal solutions to **P1** may not be efficient. To see this, a simple example of small cell network that motivates the cooperation is analyzed below.

Let us consider a simple small cell network with parameters $W = 2$, $L = 2$, $P0 = 1$, $\mathcal{N} = \{1,2\}$, $\mathcal{M} = \{1,2\}$, $\sigma^2 = 0.01$. The performance of IWF and equal power allocation (EPA) is discussed. The IWF scheme, which provides a non-cooperative sub optimal solution to **P1**, can be perfectly represented as a non-cooperative game $\mathcal{O} = \{\mathcal{N}, \mathcal{Y}, \mathcal{R}\}$. $\mathcal{N}$ is the player set. $\mathcal{Y}$ is the strategy space whose elements are system power allocation results. $\mathcal{R}$ is the payoff set, in which the j-th ($\forall j \in \mathcal{N}$) element is R_j. In IWF, the j-th small cell selfishly determines $\mathbf{p}_j$ to maximize R_j based on previous interference level. By conducting the power allocation for several rounds, the system can reach a stable state where each small cell has no incentive to change its power allocation results. The EPA scheme just equally allocates the available power to all sub-channels. Two scenarios with above system parameters are considered. Scenario 1 is a strong interference scenario, whose system channel matrix $\mathbf{G} \in \mathbb{R}^{2\times 2\times 2}$ is described as:

$$\mathbf{G}_{\cdot\cdot}^{1} = \begin{bmatrix} 0.7 & \mathbf{0.3} \\ \mathbf{0.3} & 0.5 \end{bmatrix}, \mathbf{G}_{\cdot\cdot}^{2} = \begin{bmatrix} 0.9 & \mathbf{0.5} \\ \mathbf{0.5} & 0.8 \end{bmatrix}. \tag{6}$$

Scenario 2 is a weak interference scenario whose system channel matrix $\mathbf{F} \in \mathbb{R}^{2\times 2\times 2}$ is as follows:

$$\mathbf{F}^1_{..} = \begin{bmatrix} 0.7 & \mathbf{0.03} \\ \mathbf{0.03} & 0.5 \end{bmatrix}, \mathbf{F}^2_{..} = \begin{bmatrix} 0.9 & \mathbf{0.05} \\ \mathbf{0.05} & 0.8 \end{bmatrix}. \quad (7)$$

The system throughput of IWF with above system configuration is shown in Figure 3. The start point of each sub-figure is the system throughput of EPA. We can see that the system throughput improves about 8% at the stable state in scenario 1. But in scenario 2, the throughput improvement of IWF can be neglected.

However, if small cells in scenario 1 follow some cooperation rules, the following power allocation can be obtained:

$$\hat{\mathbf{P}} = \begin{bmatrix} p^1_1 & p^1_2 \\ p^2_1 & p^2_2 \end{bmatrix} = \begin{bmatrix} 1 & 0 \\ 0 & 1 \end{bmatrix} \quad (8)$$

and the corresponding throughput is $(R_1, R_2) = (6.5850, 5.9307)$. Since both of the small cells improve the throughput by more than 70% compared to the stable state of IWF, $\hat{\mathbf{P}}$ is more favored. If the total transmission power is further reduced by 20%, the following solution is achieved:

$$\tilde{\mathbf{P}} = \begin{bmatrix} p^1_1 & p^1_2 \\ p^2_1 & p^2_2 \end{bmatrix} = \begin{bmatrix} 0.8 & 0 \\ 0 & 0.8 \end{bmatrix} \quad (9)$$

and the corresponding throughput is $(R_1, R_2) = (6.2668, 5.6147)$, which is comparable to the full transmission power allocation results $\hat{\mathbf{P}}$. The great improvements of $\hat{\mathbf{P}}$ and $\tilde{\mathbf{P}}$ motivate the cooperation of small cell networks where strong interference exists. Besides, for the non-cooperative solutions such as IWF, the convergence is not always guaranteed [16].

In this paper, we aim at achieving such cooperation gain. Some papers seek cooperation solution to the interference issues in small cells via coalition game. Authors of [17] use the recursive core approach to cluster the small cells. Small cells first form clusters by using the recursive core approach. To mitigate the interference, a time division duplex scheduling strategy is used in the formed clusters. The scheme proposed in [18] promotes the scheme in [17] by allowing a small cell to join more than one coalitions. There are obvious differences between the schemes in [17,18], and RBEBPC which will be introduced in the next section. First, both the schemes proposed in [17,18] are distributed schemes, while RBEBPC is a centralized solution. Second, no power control is used in the schemes of [17,18], while RBEBPC uses the power control method to deal with the interference.

Resource block exclusion-based power control

In RBEBPC, the goal is to improve the system throughput by excluding the strong interference resource blocks and by redistributing the system interference among small cells. To conduct RBEBPC, some mutual interference correlated information, such as **G** and previous **P**, should be exchanged among small cells. Considering the large number of small cells and the backhaul ability of each small cell, the centralized cooperation structure is adopted to guarantee the efficiency of information exchange. The macro cell, which covers all the small cells in the system, acts as the centralized node to collect and forward all the necessary information. The procedure of RBEBPC is described in Algorithm 1.

In order to obtain **G**, in the initialization state, all small cells transmit with equal power on all available sub-channels (line 3 in Algorithm 1). Then, the macro cell collects and forwards the necessary part of **G** to the small cells (line 4 and line 5). After information collection, small cells will conduct the cooperative power control for K rounds. In each round, small cells first report the used power levels to the macro cell (line 7). The macro cell, based on the reported power level, plays the

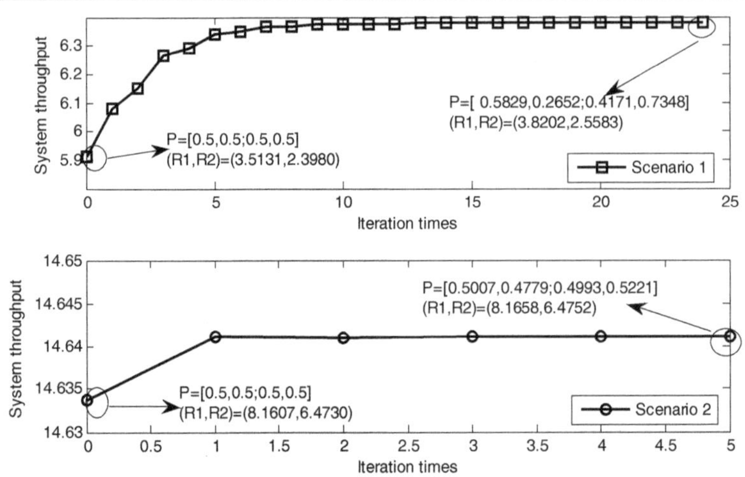

Figure 3 System throughput of IWF.

resource block exclusion game and calculates the interference constraint based on the resource block exclusion results for each small cell (*step 1*). Then each small cell solves a throughput maximization problem by following the received constraint from the macro cell (*step 2*).

Algorithm 1: Resource block exclusion-based power control.

```
begin
    Basic information collection
    Each small cell transmits with
    p = [P0/L, P0/L, ..., P0/L]^T ∈ R^{L×1};
    The i-th (∀i ∈ N) small cell collects and
    forwards G_{i·} to the macro cell;
    The macro cell delivers G_{·j} to the j-th (∀j ∈ N)
    small cell;
    Cooperation based power control
    for k = 1, k ≤ K, k++ do
        Small cells report last transmission power
        levels to the macro cell;
        Step 1- Resource block exclusion and
        interference calculation. The macro cell
        determines the excluded resource blocks
        and calculates interference constraint for
        each small cell;
        Step 2 - Power optimization. Each small
        cell solves a modified throughput
        maximization problem to determine the
        transmission power.
```

The necessary signaling overhead among the small cells, the serving UEs, and the macro cell in RBEBPC scheme can be estimated as follows. In the *Basic information collection* phase, if the quantification of a sub-channel gain of a certain small cell needs α bits, each UE should feedback αNL bits to the access small cell to indicate the interfering channel gain on all the sub-channels. Then, each small cell forwards the αNL bits to the macro cell via the backhaul link. Each small cell receives αNL bits information to indicate the channel gains correlated to the interfering UEs in the system. In the *Cooperation based power control* phase, each small cell forwards βL bits to indicate the previous power level in step 1, where β is the necessary bits to quantify the power level on a sub-channel. Each small cell receives NL bits that indicate the resource block exclusion results and receives $\gamma(L+1)$ bits that indicate the suffered interference on each sub-channel and the generated total interference, where γ is the necessary bits to quantify the interference level on a sub-channel. So in Algorithm 1, the signaling overhead between a UE and the serving small cell is αNL bits. The signaling overhead between a small cell and the macro cell is $2\alpha NL + K(\beta L + NL + \gamma(L+1))$ bits. As for the typical wireless channel, the time scale for change of path amplitude is several hundreds of millisecond [19]. In long-term evolution (LTE), the transmission time interval is 1 ms, which is much smaller than the time of path amplitude change. So in the extreme case, the backhaul link between the small cell and the macro cell should satisfy the data rate $\max\{\alpha NL, \beta L, NL + \gamma(L+1)\}$ kbps. For the typical configuration $\alpha = \beta = \gamma = 64$, $N = 10$, $L = 50$, the backhaul link data rate should satisfy 0.032 Gbps. If the backhaul links are implemented by using the standardized passive optical network (PON) [20] systems whose downstream is about 2.5 Gbps, the signaling overhead of RBEBPC is acceptable in practice.

In the following subsections, the resource block exclusion, interference calculation, and power optimization in RBEBPC are described in detail.

Resource block exclusion

Given the system power allocation result $\mathbf{P}$ and system channel gain matrix $\mathbf{G}$, a binary matrix $\mathbf{A} \in \{0,1\}^{L\times N}$ is defined to map the available sub-channels in the system:

$$[\mathbf{A}]_{lj} = \begin{cases} 1, & \text{if } [\mathbf{P}]_{lj} > 0, \\ 0, & \text{if } [\mathbf{P}]_{lj} \leq 0. \end{cases} \tag{10}$$

For the j-th small cell, the l-th sub-channel is available if and only if $[\mathbf{A}]_{lj} = 1$. Otherwise, $[\mathbf{A}]_{lj} = 0$ and the j-th small cell does not use the l-th sub-channel.

In Figure 2, the total available system resource blocks can be partitioned into $\mathcal{C} = \{\mathcal{N}_1, \mathcal{N}_2, \ldots, \mathcal{N}_L\}$, where $|\mathcal{C}| = L$ and $\mathcal{N}_l \subseteq \mathcal{N}$ denotes the set of small cells for which the l-th sub-channel is available (i.e., $\mathcal{N}_l = \{j \in \mathcal{N} | [\mathbf{A}]_{lj} = 1\}$). Given a sub-channel, the allocated power is not transferable, i.e., the power cannot be shared among small cells in an arbitrary ratio. For the sub-channels with strong interference, the system throughput will improve if some small cells mute the sub-channels. In this subsection, we resort to the coalition game to determine the sub-channels that need to be muted. Note that the system interference on the l-th sub-channel is only determined by the small cells in $\mathcal{N}_l$. Besides, the system interferences of different sub-channels are independent once $\mathbf{P}$ is given. So in the following analysis, we only consider the l-th sub-channel and the correlated small cells in $\mathcal{N}_l$.

Definition 1. Let $\mathcal{N} = \{1, 2, \ldots, N\}$ be a set of fixed players called the *grand coalition*. Non-empty subsets of $\mathcal{N}$ are called *coalitions*. A *collection* (in the grand coalition $\mathcal{N}$) is any family $\mathcal{D} = \{\mathcal{D}_1, \mathcal{D}_2, \ldots, \mathcal{D}_s\}$ of mutually disjoint coalition, and s is called its *size*. If additionally $\bigcup_{t=1}^{s} \mathcal{D}_t = \mathcal{N}$, the collection $\mathcal{D}$ is called a *partition* of $\mathcal{N}$.

Based on Definition 1, the throughput of a coalition $\mathcal{E} \subseteq \mathcal{N}_l$ is as follows:

$$R(\mathcal{E}) = \sum_{j\in\mathcal{E}} \frac{W}{L}\log_2\left(1+\frac{p_j^l G_{jj}^l}{\sigma^2 + I_j^l}\right). \quad (11)$$

If the small cells in the coalition $\mathcal{E}$ mute in a cooperative manner and only one small cell in $\mathcal{E}$ transmits, the achievable throughput is as follows:

$$\tilde{R}(\mathcal{E}) = \max_{j\in\mathcal{E}} \frac{W}{L}\log_2\left(1+\frac{p_j^l G_{jj}^l}{\sigma^2 + \sum_{i\in\mathcal{N}_l\backslash\mathcal{E}} p_i^l G_{ji}^l}\right). \quad (12)$$

The formulated coalition game of the l-th sub-channel is denoted by $\mathcal{G}^l = (\mathcal{N}_l, v)$, where $\mathcal{N}_l$ is the set of players and $v : \eta(\mathcal{N}_l) \mapsto \mathbb{R}$ is the payoff function of the coalitions [21]. The payoff function should take the interference and system throughput into consideration. A suitable payoff function for coalition $\mathcal{E}$ ($\forall\mathcal{E} \in \eta(\mathcal{N}_l)$) is as follows:

$$v(\mathcal{E}) = \begin{cases} \tilde{R}(\mathcal{E}), & \text{if } R(\mathcal{E}) < \tilde{R}(\mathcal{E}), \\ 0, & \text{if } R(\mathcal{E}) \geq \tilde{R}(\mathcal{E}). \end{cases} \quad (13)$$

Note that for $\emptyset$, we have $v(\emptyset) = 0$. Based on Equation 13, the establishment of coalition $\mathcal{E}$ ($|\mathcal{E}| \geq 2$) has two effects. The first effect is the reduction of transmission resource blocks for small cells in $\mathcal{E}$. Before the formulation of $\mathcal{E}$, small cells in $\mathcal{E}$ can have at most $|\mathcal{E}|$ transmission resource blocks (i.e., each small cell owns a transmission resource block) and at least two transmission resource blocks (i.e., $\mathcal{E}$ is established based on two existing coalitions). However, when $\mathcal{E}$ is formulated, small cells in $\mathcal{E}$ only have one transmission resource block. The second effect is the reduction of interference of the remaining transmission resource block. Since the small cells in $\mathcal{E}$ mute in a cooperative manner, the interference of small cells in $\mathcal{E}$ is canceled.

Some properties of $\mathcal{G}^l$ are summarized as follows.

Definition 2. The form of a coalition game $(\mathcal{N}, v)$ is the characteristic form, if and only if the value of a coalition $\mathcal{E} \in \eta(\mathcal{N})$ solely depends on the players in $\mathcal{E}$ and is in independent with how the players in $\mathcal{N}\backslash\mathcal{E}$ are structured.

Property 1. The form of game $\mathcal{G}^l$ is the characteristic form.

Proof. This property directly follows by the definition of payoff function (Equation 13). □

From *Property 1*, we can see that the value of the formulated payoff function is only sensitive to the players in the given coalition. In small cell networks, small cells are coupled with each other via the effect of interference. In (Equation 13), the effect of small cells beyond the given coalition is treated as a whole and is independent with their partition structure. By using Equation 13, we can obtain the value of a coalition once the coalition is given.

Given a partition $\mathcal{P} = \{\mathcal{P}_1, \mathcal{P}_2, \ldots, \mathcal{P}_s\}$ of $\mathcal{N}_l$, we can find a unique vector $\mathbf{v} = [\, v(\mathcal{P}_1), v(\mathcal{P}_2), \ldots, v(\mathcal{P}_s)]^T \in \mathbb{R}^{s\times 1}$ to represent the value of each coalition in $\mathcal{P}$. Besides, since the coalition value of $\mathcal{G}^l$ is the maximum achievable throughput by a single member small cell, game $\mathcal{G}^l$ has a transferable utility, i.e., the achievable throughput can be arbitrarily portioned among the small cells of a coalition (for example, via the proper choice of coding strategy [18]).

Definition 3. A transferable utility coalition game is supperadditive if and only if $\forall\mathcal{E}_1 \subseteq \mathcal{N}_l$, $\mathcal{E}_2 \subseteq \mathcal{N}_l$ and $\mathcal{E}_1 \cap \mathcal{E}_2 = \emptyset$:

$$v(\mathcal{E}_1 \cup \mathcal{E}_2) \geq v(\mathcal{E}_1) + v(\mathcal{E}_2). \quad (14)$$

For the transferable utility coalition game with supperadditivity, establishing a bigger coalition is always beneficial.

Property 2. Due to the implied tradeoff between interference and throughput in (Equation 13), the formulated game $\mathcal{G}^l$ is not supperadditive and the grand coalition is not always formed; thus, disjoint independent coalitions will form in the network.

The proof of *Property 2* can be found in the Appendix. The payoff function (Equation 13) shows that in the formulated coalition, only one player transmits and the rest in the coalition keep silent. Once a coalition is formulated, the available number of resource blocks reduces but the quality of the remaining resource block improves. Due to the random deployment of small cell and the fluctuation of the wireless channel quality, small cells may not be prone to form the grand coalition. In small-scale area where strong interference exists, small cells form a coalition with a higher possibility since the payoff of the coalition is probably higher if the interference is removed. The main obstacle of the grand coalition construction lies in (Equation 13) because the reduction of the available resource blocks is the cost of coalition construction. In practice, the construction of grand coalition needs harsh conditions. In the case that the interferences among small cells are so strong that if more than one small cell share the sub-channel the total throughput of the sub-channel reduces, the grand coalition will form.

Definition 4. A comparison relation $\triangleright$ is defined for two collections $\mathcal{D} = \{\mathcal{D}_1, \mathcal{D}_2, \ldots, \mathcal{D}_s\}$ and $\mathcal{F} = \{\mathcal{F}_1, \mathcal{F}_2, \ldots, \mathcal{F}_w\}$ that satisfy $\bigcup_{m=1}^{s} \mathcal{D}_m = \bigcup_{n=1}^{w} \mathcal{F}_n = \mathcal{H} \subseteq \mathcal{N}$. Thus, $\mathcal{D} \triangleright \mathcal{F}$ means that the way $\mathcal{D}$ partitions $\mathcal{H}$ is preferred to the way $\mathcal{F}$ partitions $\mathcal{H}$.

Definition 5. For collection $\mathcal{D}$ and $\mathcal{F}$ defined in *Definition 4*:

$$\mathcal{D} \triangleright \mathcal{F} \iff \sum_{m=1}^{s} v(\mathcal{D}_m) > \sum_{n=1}^{w} v(\mathcal{F}_n). \tag{15}$$

Definitions 4 and 5 provide a preference among partitions. Definition 5 indicates that 'social welfare' (the total throughput) is considered as the baseline. So the defined preference is coordinate with the target of system throughput improvement. We can use the exhaustive search method to obtain the maximum throughput partition structure of $\mathcal{G}^l$, in which a rather large search space should be considered. For the player set $\mathcal{N}_l$, the number of possible partition, which is given by a value known as Bell number, grows sharply with the number of players in $\mathcal{N}_l$. For example, the Bell number of $\mathcal{N}_l$ is 115,975 when $|\mathcal{N}_l| = 10$. However, by following some simple rules, we can obtain a stable practical partition structure of $\mathcal{G}^l$.

For any two collections $\mathcal{D} = \{\mathcal{D}_1, \mathcal{D}_2, \ldots, \mathcal{D}_s\}$, $\mathcal{F} = \{\mathcal{F}_1, \mathcal{F}_2, \ldots, \mathcal{F}_w\}$ and the grand coalition $\mathcal{N}$ that satisfy $(\bigcup_{m=1}^{s} \mathcal{D}_m) \cup (\bigcup_{n=1}^{w} \mathcal{F}_n) = \mathcal{N}$ and $(\bigcup_{m=1}^{s} \mathcal{D}_m) \cap (\bigcup_{n=1}^{w} \mathcal{F}_n) = \emptyset$, we define two operation rules to construct the stable partition structure.

Definition 6. Merge rule: If $\{\bigcup_{m=1}^{s} \mathcal{D}_m\} \cup \mathcal{F} \triangleright \mathcal{D} \cup \mathcal{F}$, merge $\mathcal{D}$ into $\{\bigcup_{m=1}^{s} \mathcal{D}_m\}$:

$$\text{Merge}(\mathcal{D} \cup \mathcal{F}) = \begin{cases} \{\bigcup_{m=1}^{s} \mathcal{D}_m\} \cup \mathcal{F}, & \text{if } \{\bigcup_{m=1}^{s} \mathcal{D}_m\} \cup \mathcal{F} \triangleright \mathcal{D} \cup \mathcal{F}, \\ \mathcal{D} \cup \mathcal{F}, & \text{if } \mathcal{D} \cup \mathcal{F} \triangleright \{\bigcup_{m=1}^{s} \mathcal{D}_m\} \cup \mathcal{F}. \end{cases} \tag{16}$$

Definition 7. Split rule: If $\mathcal{D} \cup \mathcal{F} \triangleright \{\bigcup_{m=1}^{s} \mathcal{D}_m\} \cup \mathcal{F}$, split $\{\bigcup_{m=1}^{s} \mathcal{D}_m\}$ into $\mathcal{D}$:

$$\text{Split}(\{\bigcup_{m=1}^{s} \mathcal{D}_m\} \cup \mathcal{F}) = \begin{cases} \{\bigcup_{m=1}^{s} \mathcal{D}_m\} \cup \mathcal{F}, & \text{if } \{\bigcup_{m=1}^{s} \mathcal{D}_m\} \cup \mathcal{F} \triangleright \mathcal{D} \cup \mathcal{F}, \\ \mathcal{D} \cup \mathcal{F}, & \text{if } \mathcal{D} \cup \mathcal{F} \triangleright \{\bigcup_{m=1}^{s} \mathcal{D}_m\} \cup \mathcal{F}. \end{cases} \tag{17}$$

Note that the above operation rules of Definitions 6 and 7 use $\triangleright$ 'locally', by focusing on the coalitions that take part and result from the merge operation and split operation. Algorithm 2 summarizes the procedure that uses the merge-split rule to obtain the stable partition structure of $\mathcal{G}^l$. Due to *Property 1*, the conduction of the split rule in Algorithm 2 can be transformed into the merge rule, if the coalition remains to be splitted is treated as a smaller grand coalition. The split operation is equivalent to the merge operation in the smaller grand coalition as long as we treat the effect of the players beyond the smaller grand coalition as a whole.

It is difficult to directly describe the complexity of Algorithm 2, because the termination condition of Algorithm 2 depends on the specific state of the small cell network. But we can estimate the complexity of Algorithm 2 in some extreme cases. Since the split operation can be treated as a special kind of merge operation, we only analyze the complexity of merge operation here. In the worst case where no coalition with size bigger than two is formed, the potential number of possible coalitions that should be considered is $\zeta_1 = 2^{|\mathcal{N}_l|} - |\mathcal{N}_l| - 1 = \sum_{k=2}^{|\mathcal{N}_l|} \binom{|\mathcal{N}_l|}{k}$. While in the best case where the grand coalition is formed, the potential number of possible coalitions that should be considered is $\zeta_2 = |\mathcal{N}_l| - 1$. So for each merge or split operation, the number of coalitions that should be considered is between ζ_1 and ζ_2.

Definition 8. A partition $\mathcal{P} = \{\mathcal{P}_1, \mathcal{P}_2, \ldots, \mathcal{P}_s\}$ of the grand coalition $\mathcal{N}$ is $\mathbb{D}_c$ stable, if no players in $\mathcal{N}$ are interested in leaving $\mathcal{P}$ through any operation (not necessary merge or split) to form a partition different with $\mathcal{P}$. A partition $\mathcal{Q} = \{\mathcal{Q}_1, \mathcal{Q}_2, \ldots, \mathcal{Q}_w\}$ of the grand coalition $\mathcal{N}$ is $\mathbb{D}_{hp}$ stable if no coalition has the incentive to split or merge.

Property 3. The partition structure of $\mathcal{G}^l$ obtained by Algorithm 2 is $\mathbb{D}_{hp}$ stable. The existence of $\mathbb{D}_c$ stable partition in game $\mathcal{G}^l$ is not always guaranteed. If the $\mathbb{D}_c$ stable partition of $\mathcal{G}^l$ exists, the partition structure obtained by Algorithm 2 is $\mathbb{D}_c$ stable.

Algorithm 2: Stable partition of $\mathcal{G}^l$ via the merge-split rule.

```
begin
    Initialization
    Initialize the partition of G^l by
    P = {{l_1}, {l_2}, ..., {l_|N_l|}}, where l_q is the q-th
    element of N_l;
    Q = P;
    Coalition formation with merge-split rule
    repeat
        P = Q;
        repeat
            P=Merge(P);
        until merge operation terminates;
        Q = P;
        repeat
            Q=Split(Q);
        until split operation terminates;
    until merge operation and split operation
    terminate.
```

The proof of *Property 3* can be found in the Appendix. Property 3 indicates that the final outcome of Algorithm 2 is stable. The outcome of Algorithm 2 can not be improved by any merge or split operation. Generally speaking, the optimal partition results can be found by using the exhaustive method. But the number of possible cases (the Bell number) that should be considered in the exhaustive method is too huge to manage. The outcome of Algorithm 2 may not be globally optimal but it is at least stable.

Denote the outcome of Algorithm 2 by $\mathcal{P}^* = \{\mathcal{P}_1^*, \mathcal{P}_2^*, \ldots, \mathcal{P}_x^*\}$. From Equation 13, we can find the set of players $\mathcal{U} = \{y_1, y_2, \ldots, y_x\}$, where $y_t \in \mathcal{P}_t^*$ is the player whose throughput on the sub-channel equals $\tilde{R}(\mathcal{P}_t^*)$, if other players in $\mathcal{P}_t^*$ mute in the sub-channel. So the resource blocks with strong interference are marked by modifying the available sub-channel mapping matrix:

$$[\mathbf{A}]_{lj} = \begin{cases} 1, & \text{for } j \in \mathcal{U}, \\ 0, & \text{for } j \in \mathcal{N}_l \backslash \mathcal{U}. \end{cases} \quad (18)$$

The resource blocks with strong interference are excluded by setting zero value to the corresponding elements in the available sub-channel mapping matrix. Figure 4 shows a schematic diagram of resource block exclusion. Part of the total system blocks are excluded by playing the formulated games. A detailed example of the resource exclusion procedure can be found in the simulation part.

Interference calculation

After the resource block exclusion, the suffered interference of the j-th UE on the l-th sub-channel is as follows:

$$\tilde{I}_j^l = \sum_{i=1, i \neq j}^{N} p_i^l G_{ji}^l [\mathbf{A}]_{li}. \quad (19)$$

The interference power level on the excluded resource block is not considered in $\tilde{I}_j$. The generated interference of small cell j is defined as:

$$\bar{I}_j = \sum_{l=1}^{L} \sum_{i=1, i \neq j}^{M} p_j^l G_{ij}^l [A]_{li}. \quad (20)$$

The generated interference depends on three factors: the previous allocated power vector, the interference channel gain and the channel availability. If the sub-channel is unavailable for some small cells, the interference to the small cell on this sub-channel has no contribution to the generated interference. Based on the system channel gain matrix $\mathbf{G}$, previous system power allocation results $\mathbf{P}$ and system current mapping matrix $\mathbf{A}$, the macro cell calculates the generated interference vector $\bar{\mathbf{I}} = [\bar{I}_1, \bar{I}_2, \ldots, \bar{I}_N]^T \in \mathbb{R}^{N \times 1}$ and the suffered interference vector $\tilde{\mathbf{I}}_j = [\tilde{I}_j^1, \tilde{I}_j^2, \ldots, \tilde{I}_j^L]^T \in \mathbb{R}^{L \times 1}$ $(\forall j \in \mathcal{M})$ after the sub-channel exclusion operation. The two interference correlated vectors are delivered to the small cells to be used in the power optimization procedure.

Power optimization

After obtaining the interference constraint $\bar{\mathbf{I}}$ and available sub-channel mapping matrix $\mathbf{A}$, each small cell will optimize the transmission power on each available sub-channel based on the last transmission power allocation result $\mathbf{P}$. Instead of solving **P1**, the j-th $(\forall j \in \mathcal{N})$ small cell will solve the following problem:

$$\textbf{P2:} \quad \max \sum_{l \in \mathcal{L}_j} \frac{W}{L} \log_2(1 + \frac{p_j^l G_{jj}^l}{\sigma^2 + \tilde{I}_j^l}) \quad (21)$$

$$s.t. \quad \sum_{i \in \mathcal{M} \backslash \{j\}} \sum_{l \in \mathcal{L}_j} p_j^l G_{ij}^l [A]_{li} \leq \bar{I}_j, \quad (22)$$

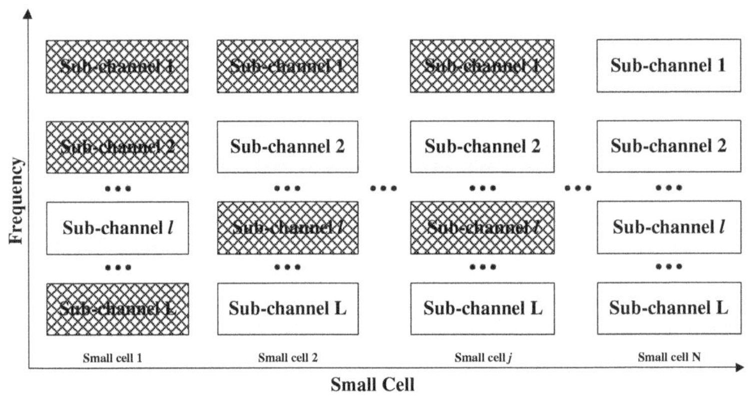

Figure 4 A schematic diagram of resource block exclusion. The grid resource blocks such as RB_{11} and RB_{l2} are excluded. From the small cell dimension, each small cell may own more than one sub-channel. From the frequency dimension, each sub-channel can be occupied by more than one small cell.

$$\sum_{l \in \mathcal{L}_j} p_j^l \leq P0, \tag{23}$$

$$p_j^l \geq 0, \forall l \in \mathcal{L}_j, \tag{24}$$

where $\mathcal{L}_j$ is the set of available sub-channels for the j-th small cell (i.e., $\mathcal{L}_j = \{l \in \mathcal{L} | [\mathbf{A}]_{lj} = 1\}$). The constraint of (Equation 22) is used to redistribute the total generated interference on all the available sub-channels. The formulated **P2** is a concave problem and the proof can be found in the Appendix.

Since **P2** is concave to the domain $\mathbf{p}_j \in \mathbb{R}^{|\mathcal{L}_j| \times 1}$, we can use the interior point method [22] to solve it. However, it is more convenient to solve the dual problem of **P2**, if we take the problem structure into consideration. First, due to concavity, the optimal solution to **P2** is equivalent to the optimal solution to the dual problem. Second, the number of variables in **P2** is $|\mathcal{L}_j|$. While in the dual problem, the number of variables reduces to two. Because constraints (Equation 24) are treated as slack conditions, only (Equation 22) and (Equation 23) are used when solving the dual problem. Third, in some special cases, the close form solution to the dual problem can be obtained, which greatly accelerates the computation. The details of solving the dual problem are represented as follows. The Lagrangian of **P2** is as follows:

$$\begin{aligned}
&L(\lambda, \mu, \theta, \mathbf{p}_j) = \\
&\sum_{l \in \mathcal{L}_j} \frac{W}{L} \log_2 \left(1 + \frac{p_j^l G_{jj}^l}{\sigma^2 + \tilde{I}_j^l}\right) + \lambda \Big(\bar{I}_j - \\
&\sum_{i \in \mathcal{M} \setminus \{j\}} \sum_{l \in \mathcal{L}_j} p_j^l G_{ij}^l [A]_{li} \Big) + \mu \left(P0 - \sum_{l \in \mathcal{L}_j} p_j^l\right) + \sum_{l \in \mathcal{L}_j} \theta_l p_j^l \\
&\equiv \sum_{l \in \mathcal{L}_j} [z \log_2 \left(b_j^l + p_j^l\right) + a_j^l] + \lambda \left(\bar{I}_j - \sum_{l \in \mathcal{L}_j} c_j^l p_j^l\right) + \\
&\mu \left(P0 - \sum_{l \in \mathcal{L}_j} p_j^l\right) + \sum_{l \in \mathcal{L}_j} \theta_l p_j^l,
\end{aligned} \tag{25}$$

where $z = \frac{W}{L}$, $a_j^l = \frac{W}{L} \log_2 \left(\frac{G_{jj}^l}{\sigma^2 + \tilde{I}_j^l}\right)$, $b_j^l = \frac{\sigma^2 + \tilde{I}_j^l}{G_{jj}^l}$, $c_j^l = \sum_{i \in \mathcal{M} \setminus \{j\}} G_{ij}^l [\mathbf{A}]_{li}$. The first-order partial derivatives of p_j^l is as follows:

$$\frac{\partial L(\lambda, \mu, \theta, \mathbf{p}_j)}{\partial p_j^l} = \frac{z}{\ln 2} \frac{1}{b_j^l + p_j^l} - \lambda c_j^l - \mu + \theta_l. \tag{26}$$

Based on the KKT condition of **P2**, we can see that θ is a slack variable vector that can be eliminated [22]. By solving $\frac{\partial L(\lambda, \mu, \mathbf{p}_j)}{\partial p_j^l} = 0$ and substituting the results to Equation 25, we can obtain the dual problem of **P2**:

$$\begin{aligned}
\mathbf{P3}: \quad & \min g(\lambda, \mu) \\
&= \min \max_{\mathbf{p}_j} L(\lambda, \mu, \mathbf{p}_j) \\
&= \min_{\lambda, \mu} \sum_{l \in \mathcal{L}_j} \left[-z \log_2(\lambda c_j^l + \mu) + (\lambda c_j^l + \mu) b_j^l\right] + \\
&\lambda \bar{I}_j + \mu P0 + \sum_{l \in \mathcal{L}_j} \left[z \log_2(\frac{z}{\ln 2}) + a_j^l - \frac{z}{\ln 2}\right]
\end{aligned} \tag{27}$$

$$s.t. \quad \lambda \geq 0, \mu \geq 0. \tag{28}$$

It is easy to verify that **P3** is a convex optimization problem. By using the KKT conditions, constraints (Equation 28) can be neglected, i.e., we can first solve **P3** without constraints (Equation 28) and then using (Equation 28) to examine the correctness of the solution. So **P3** can be efficiently solved by using the Newton Method, which is a method suitable for the convex optimization problem without constraints [22]. In some cases, we can obtain the close form optimal solution (λ^*, μ^*) to **P3**.

Case 1. $\lambda = 0$. If $\lambda = 0$, then $g(\lambda, \mu)$ degenerates into $g(0, \mu)$. By solving $\frac{\partial g(0,\mu)}{\partial \mu} = 0$, we can obtain the optimal μ:

$$\mu^* = \frac{z}{\ln 2} \frac{|\mathcal{L}_j|}{P0 + \sum_{l \in \mathcal{L}_j} b_j^l}. \tag{29}$$

Case 2. $\mu = 0$. If $\mu = 0$, then $g(\lambda, \mu)$ degenerates into $g(\lambda, 0)$. By solving $\frac{\partial g(\lambda,0)}{\partial \lambda} = 0$, we can obtain the optimal λ:

$$\lambda^* = \frac{z}{\ln 2} \frac{|\mathcal{L}_j|}{\bar{I}_j + \sum_{l \in \mathcal{L}_j} c_j^l b_j^l}. \tag{30}$$

When the optimal (λ^*, μ^*) is obtained, we can use Equation 26 and solve $\frac{\partial L(\lambda^*, \mu^*, \mathbf{p}_j)}{\partial p_j^l} = 0$ to obtain the optimal p_j^{l*}. Due to the elimination of θ, we must verify the correctness of the power results. If $p_j^{l*} \geq 0$ for all $l \in \mathcal{L}_j$, we can conclude that $\mathbf{p}_j^*$ is the solution to **P2**. If some elements of $\mathbf{p}_j^*$ are negative, we must remove these sub-channels with the minimum power and solve **P3** again until a solution $p_j^{l*} \geq 0$ for all $l \in \mathcal{L}_j$ is found.

The power optimization procedure is summarized in Algorithm 3. It is difficult to analyze the complexity of Algorithm 3 directly. But the number of iteration times of the Newton Method of Algorithm 3 can be estimated in some extreme cases. In the best cases where the optimal value can be achieved by (Equation 29) or (Equation 30), the iteration time of the Newton Method is one. In the worst case where the optimal value is obtained without

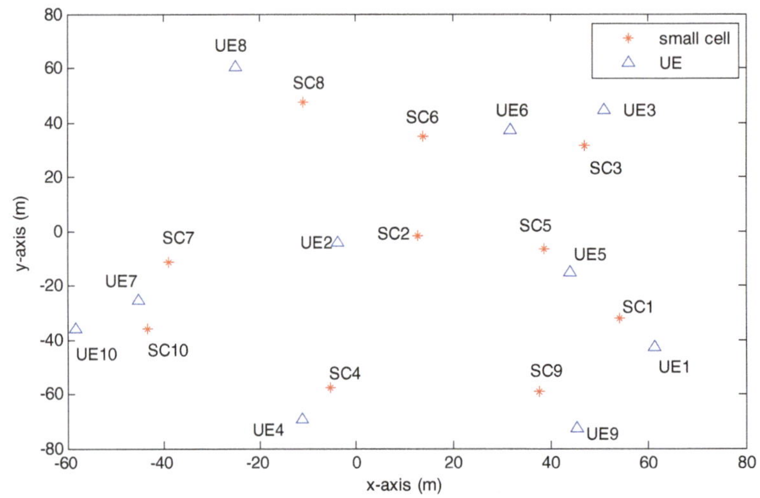

Figure 5 An example of the small cell network topology.

using these close form equations, the maximum number of iterations is bounded by $\frac{g(\lambda^0,\mu^0)-g(\lambda^*,\mu^*)}{\tau}+6$ [22], where λ^0 and μ^0 separately represent the initialized value of λ and μ, respectively, and τ is the maximum reduction value of function $g(\lambda, \mu)$ during the iterations.

Algorithm 3: Power optimization for the j-th small cell.

begin
- **Initialization**
  - $\lambda = 1, \mu = 1$, use (Equation 26) to obtain $\mathbf{p}_j$;
  - **if** $p_j^l \geq 0, \forall l \in \mathcal{L}_j$ **then**
    - $\mathbf{p}_j^* = \mathbf{p}_j$
  - **else**
    - **while** *$p_j^l \geq 0, \forall l \in \mathcal{L}_j$ is not true* **do**
      - solve **P3** with Newton Method (NM) without constraint (Equation 28);
      - **if** *$\lambda \leq 0$ in an iteration of NM* **then**
        - $\lambda^* = 0$ and solve (Equation 29) for μ^*;
      - **if** *$\mu \leq 0$ in an iteration of NM* **then**
        - $\mu^* = 0$ and solve (Equation 30) for λ^*;
      - solve (Equation 26) for $\mathbf{p}_j^l$
      - **if** *$p_j^l \geq 0, \forall l \in \mathcal{L}_j$ is not true* **then**
        - find $l_{\min}$ with minimum $p_j^{l_{\min}}$;
        - $p_j^{l_{\min}*} = 0, \mathcal{L}_j = \mathcal{L}_j \backslash \{l_{\min}\}$;
      - **else**
        - $p_j^{l*} = p_j^l, \forall l \in \mathcal{L}_j$;

Simulations and discussions

In this section, system throughput, power consumption, and interference variation are tracked and analyzed during the conduction of RBEBPC. Then, the average performance of the RBEBPC is evaluated . Based on the suggestion of [23], the deployment of small cells is based on the concept of cluster. A small cell cluster consists of a certain number of small cells that are randomly deployed in a certain area. The interference among different small cell clusters can be ignored because the total transmission power of a small cell is small and the distance between clusters is far. For simplicity, in our simulation, we use the square as the area of cluster and focus on the performance of single small cell cluster. The small cells are uniformly distributed in the square with side length d meters. UEs are uniformly distributed in the circles whose center is the serving small cells and radius equals R meters. Figure 5 shows a distribution example of small cells and UEs with parameters $d = 120, N = M = 10$. Besides, a minimum distance r is set between the small cell and the served UE.

Table 1 Simulation parameters

Parameters	Value
N	10
M	10
F2 (MHz)	1.25/5
L	6/25
d (m)	100 to 160
(R(m),r(m))	(15,10), (20,15), (25,20), (20,10)
System carrier frequency (GHz)	3.5
Noise power density (dBm/Hz)	-174
P0 (dBm)	14/20
Path loss model	Table A.2.1.1.2-3 in [23]

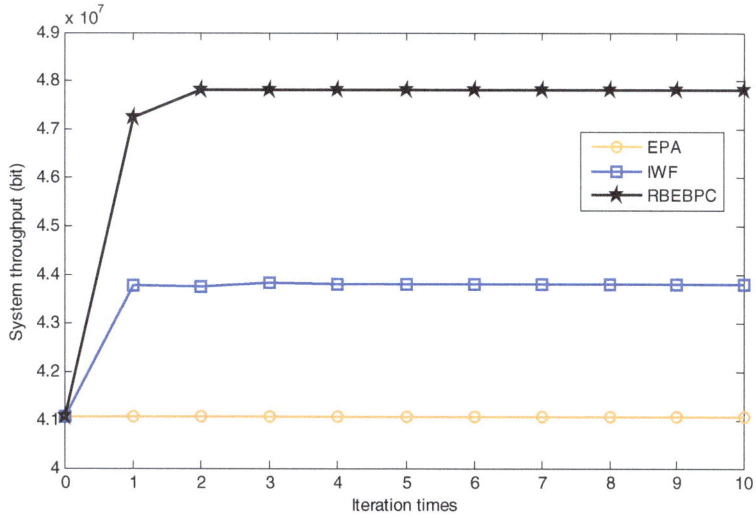

Figure 6 Track of the system throughput after each iteration.

The simulation parameters are summarized in Table 1. As a comparison, the performance of EPA and IWF is also tracked.

Track of RBEBPC

In this subsection, the system performance of RBEBPC based on the topology of Figure 5 is analyzed. For convenience, the value of K in Algorithm 1 equals 10 and the number of sub-channels is $L = 6$. The maximum power of each small cell is 14 dBm. The distance pair (R, r) equals (20, 10) and bandwidth F2 equals 1.25 M.

Figure 6 shows the system throughput variations of EPA, IWF, and RBEBPC. The initial points of Figure 6 are the system throughput of EPA. RBEBPC outperforms IWF, and IWF outperforms EPA. IWF arrives at the slight fluctuation state after the second iteration and RBEBPC arrives at the stable state after the third iteration.

The resource block exclusion results of the first round and the second round are shown in Tables 2 and 3 separately. After the second round ($k \geq 3$), no merge or split operation of the resource blocks happens. From the exclusion results, we can see that the merge or split operation happens frequently in the first or second iteration. So the value of K in Algorithm 1 can be set as a small positive integer. From Tables 2 and 3, we can see that coalitions containing two or less players are frequently formed. No grand coalition is formed.

Figure 7 shows the throughput difference between RBEBPC and IWF after each iteration. On the one hand, only small cell 6, small cell 7, and small cell 9 have negative throughput growth in RBEBPC compared to IWF. We can see from Table 2 that in the first round, small cell 6 joints the coalitions that has more than two players on all the available sub-channels. Only on sub-channel 4, small cell 6 is the player with advantage, which means that in the following iterations, small cell 6 only occupies the sub-channel 4. Small cell 7 is in the same condition as the small cell 6. Due to the interference constraint of Equation 22, small cell 9, will not greedily allocate the power to the sub-channel with the best channel gain. To see this, we can refer to the final power allocation results of IWF and RBEBPC in Tables 4 and 5 separately. From the topology of Figure 5, we can see that the possible coalition player of small cell 9 is small cell 1 and small cell 5. In the first round of resource block exclusion (Table 2), small cell 9

Table 2 Resource exclusion results, $k = 1$

l	$\mathcal{P}^*$	$\mathcal{U}$
1	{1}{2}{4}{5}{8}{9}{3, 6}{7, 10}	{1, 2, 4, 5, 8, 9, 3, 10}
2	{2} {4} {5} {8} {1,9} {3,6} {7,10}	{2,4,5,8,1,3,10}
3	{1} {2} {3} {4} {5} {9} {10} {6,7,8}	{1,2,3,4,5,9,10,8}
4	{5} {7} {8} {10} {1,9} {2,4} {3,6}	{5,7,8,10,1,4,6}
5	{4} {8} {9} {1,5} {3,6} {2} {7,10}	{4,8,9,1,3,2,10}
6	{1} {4} {5} {8} {9} {10} {2,7} {3,6}	{1,4,5,8,9,10,2,3}

Table 3 Resource exclusion results, $k = 2$

l	$\mathcal{P}^*$	$\mathcal{U}$
1	{1}{2}{3}{4}{5}{8}{9}{10}	{1, 2, 3, 4, 5, 8, 9, 10}
2	{1} {2} {3} {4} {5} {8} {10}	{1,2,3,4,5,8,10}
3	{1} {2} {3} {4} {5} {8} {9} {10}	{1,2,3,4,5,8,9,10}
4	{1} {4} {5} {6} {7} {8} {10}	{1,4,5,6,7,8,10}
5	{1} {2} {3} {4} {8} {9} {10}	{1,2,3,4,8,9,10}
6	{1} {3} {4} {5} {9} {10} {2,8}	{1,3,4,5,9,10,8}

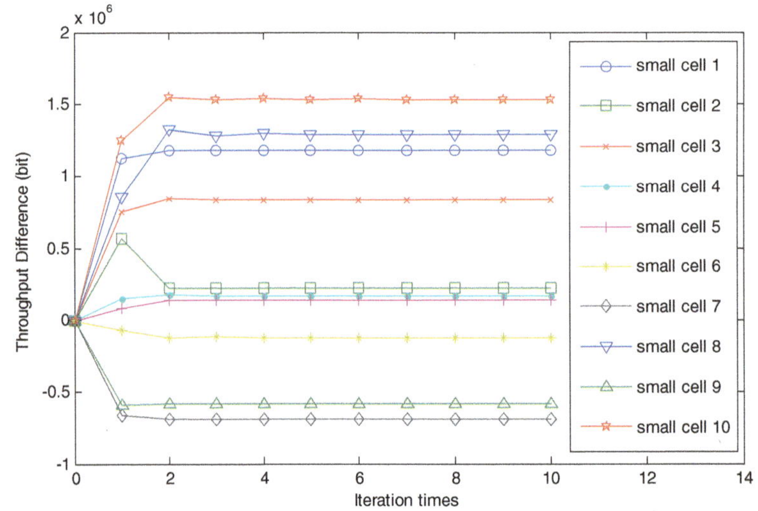

Figure 7 Track of throughput difference after each iteration.

abandons sub-channel 2 and sub-channel 4 compared to IWF. So the abandon of sub-channel and the interference constraint lead to the throughput reduction of small cell 9. On the other hand, the rest seven small cells benefit from RBEBPC compared to IWF.

Compare Table 4 to Table 5 and we can see that RBEBPC is energy-saving, i.e., not all small cells transmit with the total available power. The small cells can be divided into three groups based on the two tables. The first group (including small cells 2, 6, 7, and 9) consists of the small cells whose total transmit power and available resource blocks vary greatly in RBEBPC compared to IWF. The small cells belong to the first group follow the constraint (Equation 22) and transmit without full power. The second group (including small cell 3) consists of the small cells that use all the available power but different resource blocks compared to IWF. Small cells in the second group are influenced by the resource block exclusion operation. The disadvantage resource blocks are excluded. The third group (including small cell s1, 4, 5, 8, and 10) consists of the small cells that only have different power on all the sub-channels compared with IWF. In the third group of small cells, the power allocation difference among RBEBPC and IWF comes from the other small cells because all the small cells in the system are influenced by the interference. The power allocation variation of the other two groups leads to the power allocation result of the third group.

Figure 8 shows the system total interference under different iterations. The total system interference decreases in both IWF and RBEBPC. At the stable state, RBEBPC further reduces the interference by 56%. Compared to Figure 6, the reduction of interference transforms into the growth of system throughput.

Table 4 Power allocation results of IWF (Watt)

	$l = 1$	$l = 2$	$l = 3$	$l = 4$	$l = 5$	$l = 6$	Total
SC1	0.0043	0.0041	0.0046	0.0045	0.0042	0.0034	0.0251
SC2	0.0025	0.0065	0.0077	0	0.0034	0.0050	0.0251
SC3	0.0044	0.0033	0.0046	0.0037	0.0047	0.0044	0.0251
SC4	0.0042	0.0040	0.0041	0.0043	0.0043	0.0042	0.0251
SC5	0.0049	0.0049	0.0049	0.0052	0	0.0052	0.0251
SC6	0.0041	0.0092	0.0094	0.0025	0	0	0.0251
SC7	0.0041	0.0131	0	0.0080	0	0	0.0251
SC8	0.0050	0.0003	0.0050	0.0050	0.0048	0.0050	0.0251
SC9	0.0039	0.0043	0.0047	0.0029	0.0049	0.0044	0.0251
SC10	0.0043	0.0036	0.0044	0.0043	0.0043	0.0044	0.0251

Table 5 Power Allocation Results of RBEBPC (Watt)

	$l = 1$	$l = 2$	$l = 3$	$l = 4$	$l = 5$	$l = 6$	Total
SC1	0.0041	0.0044	0.0043	0.0045	0.0041	0.0038	0.0251
SC2	0.0026	0.0046	0.0028	0	0.0124	0	0.0224
SC3	0.0055	0.0026	0.0056	0	0.0069	0.0046	0.0251
SC4	0.0042	0.0040	0.0041	0.0043	0.0042	0.0043	0.0251
SC5	0.0032	0.0046	0.0077	0.0052	0	0.0044	0.0251
SC6	0	0	0	0.0042	0	0	0.0042
SC7	0	0	0	0.0042	0	0	0.0042
SC8	0.0042	0.003	0.0045	0.0040	0.0040	0.0047	0.0251
SC9	0.0110	0	0.0062	0	0.0042	0.0021	0.02344
SC10	0.0042	0.0041	0.0042	0.0042	0.0041	0.0042	0.0251

Average performance of RBEBPC

In this subsection, we will provide the average performance of RBEBPC. The used system parameters are $P0 = 20$ dBm, $F2 = 5$ MHz, $L = 25$. The proposed overlapping coalition formation (OCF) scheme in [18] is also simulated as the comparison scheme. Note that when we conduct the OCF, the power limitation of coalition formation is removed, i.e., all the small cells in the simulation areas have the potentials to form coalition.

Figure 9 shows the average system throughput of different values of K in Algorithm 2. The difference of average throughput can be neglected when K is bigger than 2. So in practice, the value of K can be set as a small positive number such as 2 or 3.

Figure 10 shows the average system throughput under different simulation areas. For each scheme, the average system throughput increases as the simulation area becomes bigger. The proposed RBEBPC scheme outperforms IWF by about 15% in dense case ($d = 100$) and by about 10% in sparse case ($d = 160$). IWF outperforms the OCF scheme. First, in OCF, the small cells are cooperatively scheduled rather than greedily transmit to achieve the maximum throughput. Second, no power control is used during the transmission in OCF. The gains of RBEBPC, IWF, and OCF over the EPA become weaker as d increases because the interference is not a key factor that influence the system throughput in the sparse scenario.

Figure 11 shows the cumulative distribution function (CDF) of the average system throughput. From the distribution, we can see that under these parameters, the OCF scheme improves the system throughput of EPA by about 30 Mbit, IWF improves the system throughput of OCF by about 20 Mbit and RBEBPC improves the system throughput of IWF by about 20 Mbit, which indicates

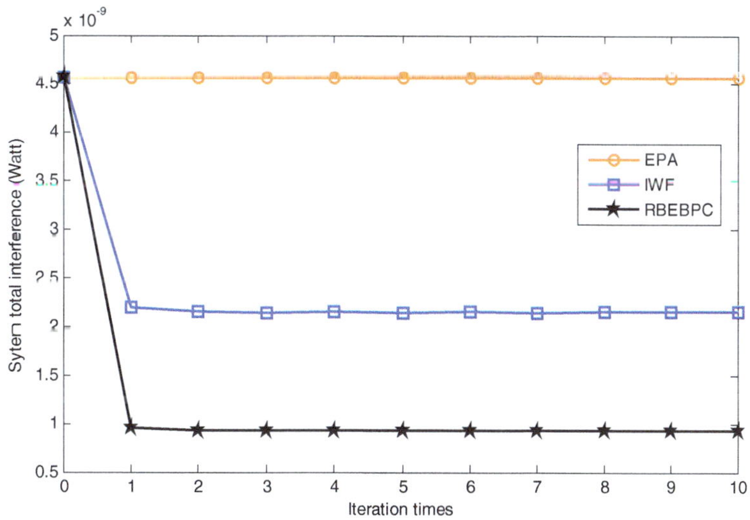

Figure 8 Track of system interference after each iteration.

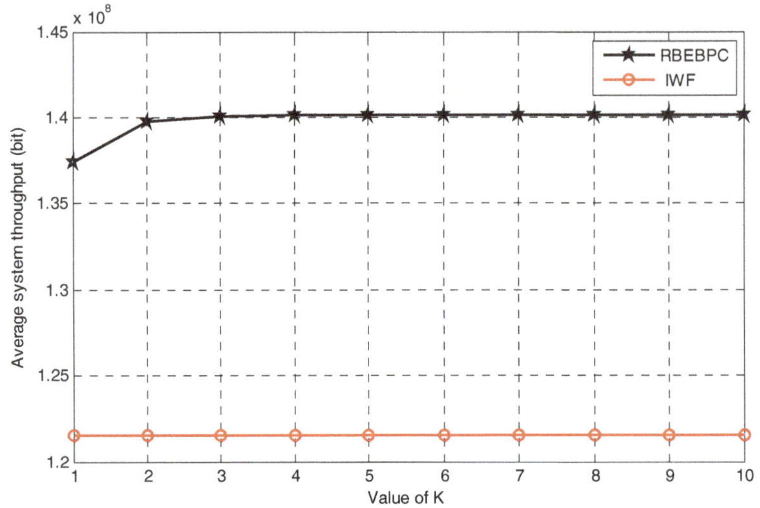

Figure 9 **Average system throughput of RBEBPC under different values of** K, $(R, r) = (20, 10)$, $d = 100$.

that RBEBPC is an effective method to improve the system throughput.

Figure 12 shows the CDF of small cell throughput growth in RBEBPC over the IWF scheme. About 36% of small cell suffers the loss of throughput (point A). Half of the throughput-decreasing small cells suffer a growth rate of −24% (point B). In the rest throughput-increasing small cells, half of them have a growth rate of 22% (point C). The extreme small cells, whose growth rate is more than 100% (point D), occupy about 5% of the total small cells. So the benefit of small cells in RBEBPC is bigger than the loss of the victim small cells.

Figure 13 shows the average system throughput of different link qualities. We consider three cases: the strong link, the moderate link, and the weak link. The elements of the triples represents the growth rate of IWF, OCF, and RBEBPC to the EPA separately. For all the schemes, the average system throughput grows when the link quality become stronger. The growth rates over EPA scheme become bigger when the link quality is weak because in this scenario, the interference has a stronger effect on the system throughput and the mitigation of interference achieves more significant payback in throughput.

Figure 14 shows the average system power consumption of different simulation areas. The IWF/EPA scheme will always use all the available power, while in RBEBPC, not all power is used. Compared with IWF/EPA, about

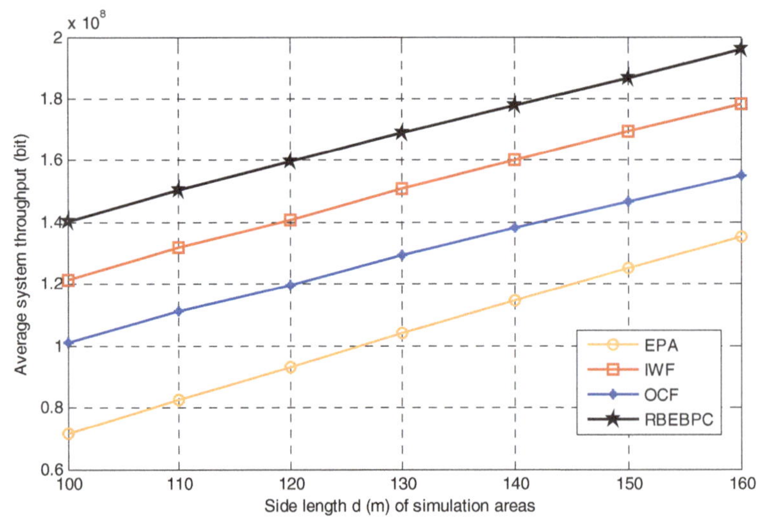

Figure 10 **Average system throughput under different** d, $(R, r) = (20, 10)$, $K = 10$.

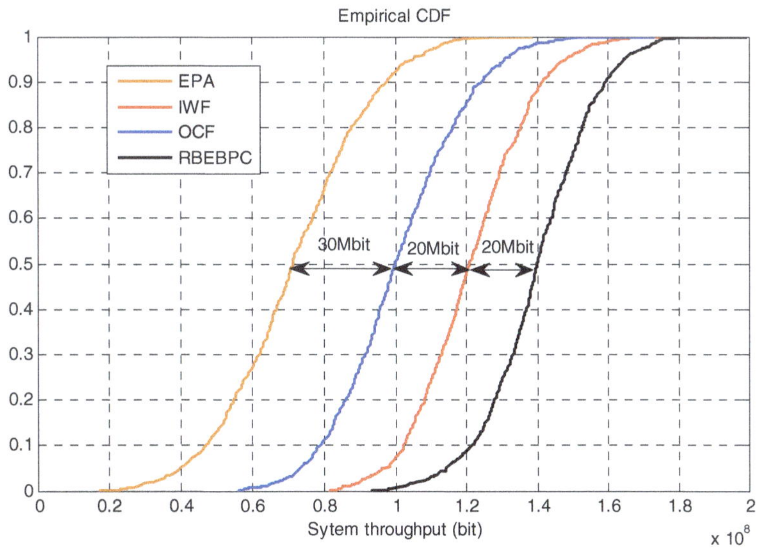

Figure 11 CDF of system throughput, $d = 100$, $(R, r) = (20, 10)$, $K = 10$.

13% power is saved in dense scenario ($d = 100$). In sparse deployment scenario, the small cells are prone to use more power. If the deployment of small cells is sparse enough, the interference among small cells will be neglected and the constraint in Equation 22 can be ignored and RBEBPC degenerates into IWF which will consume all the available power.

Conclusions

This paper focuses on the system throughput improvement via cooperative power control in OFDMA small cell networks. Based on the system resource structure, a cooperative power control scheme named as RBEBPC is proposed. Two steps are iteratively conducted in RBEBPC. First, the small cells play L independent coalition formation games to cooperatively determine the exclusion of resource blocks. Interference constraints are calculated based on the exclusion results for each small cell. Second, each small cell solves a modified throughput maximization problem to determine the power level on each sub-channel. By following the interference constraints, the system total interference is non-increasing and part of the small cells transmit without full power. Simulation results show that both system throughput and energy consumption are improved in RBEBPC compared to the traditional IWF scheme. However, the conduction of RBEBPC relies

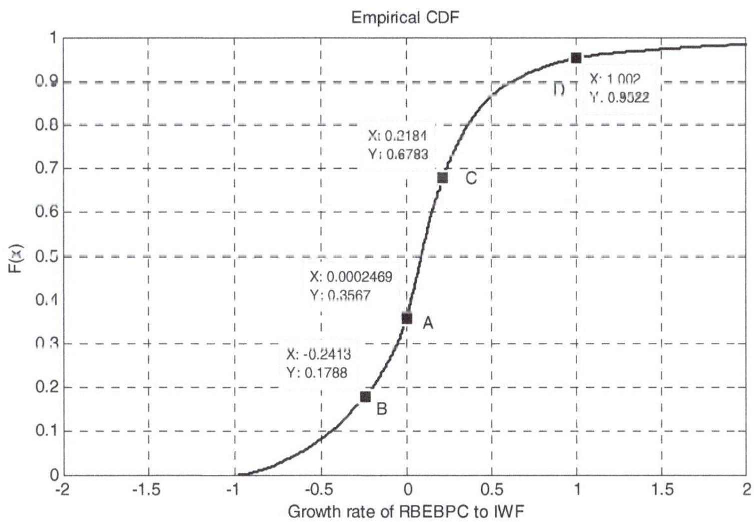

Figure 12 CDF of throughput growth rate, $d = 100$, $(R, r) = (20, 10)$, $K = 10$.

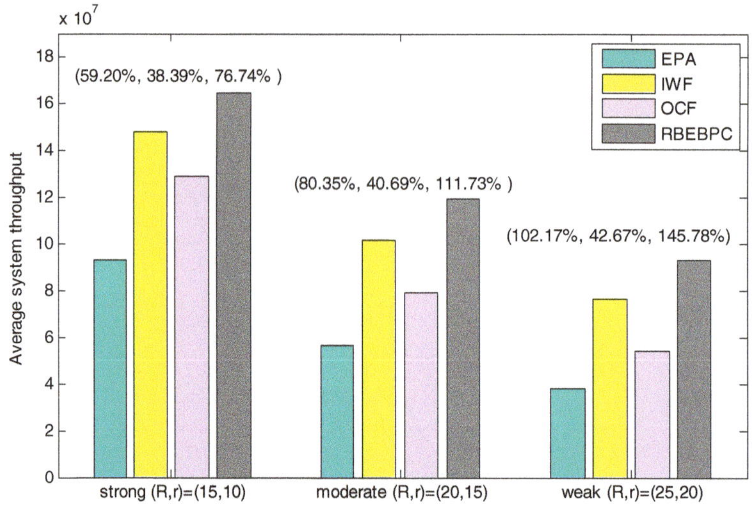

Figure 13 **Average system throughput under different link quality, $d = 100, K = 10$.**

on the backhaul link ability between the small cell and the macro cell. The performance improvement is based on the performance sacrifice of partial small cells, which may lead to the unfairness of the network. Both signaling overhead optimization and small cell fairness will be considered in our future study.

Appendix

Proof of Property 2

Proof. Consider two coalitions, $\mathcal{E}_1 \subseteq \mathcal{N}_l$, $\mathcal{E}_2 \subseteq \mathcal{N}_l$ and $\mathcal{E}_1 \cap \mathcal{E}_2 = \emptyset$, which satisfy *a)* slightly interference with each other such that $\tilde{R}(\mathcal{E}_1 \cup \mathcal{E}_2) \leq R(\mathcal{E}_1 \cup \mathcal{E}_2) = R(\mathcal{E}_1) + R(\mathcal{E}_2)$ and *b)* strong interference within each coalition such that $\tilde{R}(\mathcal{E}_1) > R(\mathcal{E}_1)$ and $\tilde{R}(\mathcal{E}_2) > R(\mathcal{E}_2)$. In this case, $v(\mathcal{E}_1 \cup \mathcal{E}_2) = 0 < \tilde{R}(\mathcal{E}_1) + \tilde{R}(\mathcal{E}_2) = v(\mathcal{E}_1) + v(\mathcal{E}_2)$, which means game $\mathcal{G}^l$ is not supperadditive. □

Proof of Property 3

Proof. The $\mathbb{D}_{hp}$ stability of partition structure obtained by Algorithm 2 follows by the termination condition of Algorithm 2. The necessary and sufficient conditions for $\mathbb{D}_c$ stability of a partition can be found in [24]. Since the partition structure of $\mathcal{G}^l$ is based on location and wireless channel state which are all random variables, these conditions are not always satisfied. But if $\mathbb{D}_c$ stable partition of $\mathcal{G}^l$ exists, the merge-split rule converges to this partition because the optimal $\mathbb{D}_c$ stable partition is the unique outcome of any arbitrary merge-split rule operation [24]. □

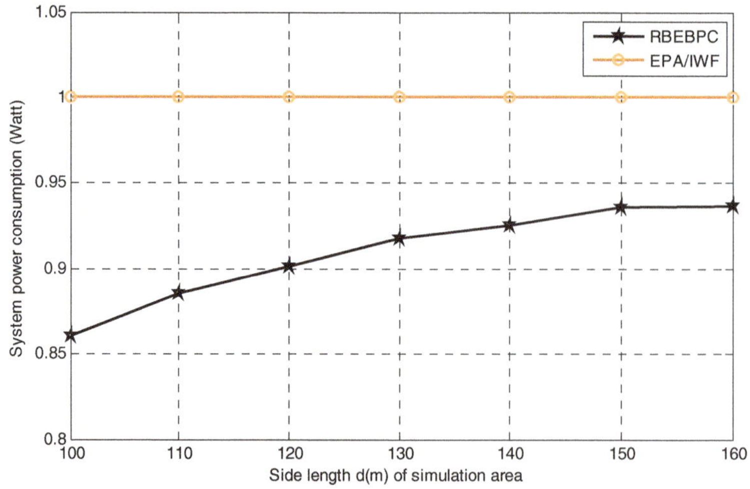

Figure 14 **Average system power consumption of different d, $(R, r) = (20, 10), K = 10$.**

Concavity Proof of P2

Proof. The domain of **P2** is non-empty because at least the solution $[0, 0, \ldots, 0]^T \in \mathbb{R}^{|\mathcal{L}_j| \times 1}$ satisfies all the constraints of **P2**. Since all the constraints of **P2** are linear, the domain formed by these constraints is a convex set. The Hessian of the target function is as follows:

$$\begin{bmatrix} \delta_j^1 & & & & & \\ & \delta_j^2 & & & & \\ & & \ddots & & & \\ & & & \delta_j^l & & \\ & & & & \ddots & \\ & & & & & \delta_j^{|\mathcal{L}_j|} \end{bmatrix} \in \mathbb{R}^{|\mathcal{L}_j| \times |\mathcal{L}_j|}, \quad (31)$$

where $\delta_j^l = \frac{W}{L \ln 2} \frac{-1}{(p_j^l + \frac{\sigma^2 + \tilde{I}_j^l}{G_{jj}^l})^2}, \forall l \in \mathcal{L}_j$. Note that when we calculate the Hessian, the interference on each subchannel is processed as a constant. Since the Hessian of the target function of **P2** is a diagonal matrix and each element in the diagonal is negative, the Hessian is negative semidefinite. Based on the second order conditions [22], we conclude that the target function of **P2** is a concave function. So **P2** is a concave optimization problem. □

Competing interests
The authors declare that they have no competing interests.

Acknowledgements
This work was supported by the 'Radio Resource Management for Small Cell Deployment' Project of DOCOMO Beijing Communications Laboratories Co., Ltd, the National Natural Science Foundation of China (Grant No.61471060) and the Funds for Creative Research Groups of China (Grant No.61421061).

Author details
[1] State Key Laboratory of Networking and Switching Technology, Beijing University of Posts and Telecommunications, Xitucheng Road, 100876 Beijing, China. [2] DOCOMO Beijing Communication Laboratories Co., Ltd.,Kexueyuan South Road, 100190 Beijing, China.

References
1. Cisco Visual Networking Index: Global Mobile Data Traffic Forecast Update, 2011–2016. http://www.cisco.com/en/US/solutions/collateral/ns341/ns525/ns537/ns705/ns827/white_paper_c11-520862.html
2. I Hwang, B Song, SS Soliman, A holistic view on hyper dense heterogeneous and small cell networks. IEEE Commun. Mag. **51**, 20–27 (2013)
3. TR 36.932, Scenarios and Requirements for Small Cell Enhancements for E UTRA and E-UTRAN (Release 12) (2012)
4. MI Kamel, KMF Elsayed, in *Global Telecommunications Conference*. Performance evaluation of a coordinated time-domain eICIC framework based on ABSF in heterogeneous LTE-advanced networks (IEEE Piscataway, New Jersey, USA, 2012), pp. 5326–5331
5. L Jiang, M Lei, in *Personal Indoor and Mobile Radio Communications*. Resource allocation for eICIC scheme in heterogeneous networks (IEEE Piscataway, New Jersey, USA, 2012), pp. 448–453
6. J Oh, Y Han, in *Personal Indoor and Mobile Radio Communications*. Cell selection for range expansion with almost blank subframe in heterogeneous networks (IEEE Piscataway, New Jersey, USA, 2012), pp. 653–657
7. J Pang, J Wang, D Wang, G Shen, Q Jiang, J Liu, in *Wireless Communications and Networking Conference*. Optimized time-domain resource partitioning for enhanced inter-cell interference coordination in heterogeneous networks (IEEE Piscataway, New Jersey, USA, 2012), pp. 1613–1617
8. S Uygungelen, G Auer, Z Bharucha, in *Vehicular Technology Conference*. Graph-based dynamic frequency reuse in femtocell networks (IEEE Piscataway, New Jersey, USA, 2011), pp. 1–6
9. W Noh, W Shin, C Shin, K Jang, Choi H-h, in *Wireless Communications and Networking Conference*. Distributed frequency resource control for intercell interference control in heterogeneous networks (IEEE Piscataway, New Jersey, USA, 2012), pp. 1794–1799
10. T Nakamura, S Nagata, A Benjebbour, Y Kishiyamaand, T Hai, S Xiaodong, Y Ning, L Nan, Trends in small cell enhancements in LTE advanced. IEEE Commun. Mag. **51**, 98–105 (2013)
11. P Li, Y Zhu, in *Personal Indoor and Mobile Radio Communications*. Price-based power control of femtocell networks: a Stackelberg game approach (IEEE Piscataway, New Jersey, USA, 2012), pp. 1185–1191
12. S Guruacharya, D Niyato, DI Kim, E Hossain, Hierarchical competition for downlink power allocation in OFDMA femtocell networks. IEEE Trans. Wireless Comm. **12**, 1543–1553 (2013)
13. L Giupponi, C Ibars, in *Personal Indoor and Mobile Radio Communications*. Distributed interference control in OFDMA-based femtocells (IEEE Piscataway, New Jersey, USA, 2010), pp. 1201–1206
14. L Venturino, N Prasad, X Wang, Coordinated scheduling and power allocation in downlink multicell OFDMA networks. IEEE Trans. Veh. Tech. **58**, 2835–2848 (2009)
15. J Kimm, D-H Cho, A joint power and subchannel allocation scheme maximizing system capacity in indoor dense mobile communication systems. IEEE Trans. Veh. Tech. **59**, 4340–4353 (2010)
16. KW Shum, K-K Leung, CW Sung, Convergence of iterative waterfilling algorithm for Gaussian interference channels. IEEE J. Sel. Area Comm. **25**, 1091–1100 (2007)
17. F Pantisano, M Bennis, W Saad, R Verdone, M Latva-aho, in *Wireless Communications and Networking Conference*. Coalition formation games for femtocell interference management: a recursive core approach (IEEE Piscataway, New Jersey, USA, 2011), pp. 1161–1166
18. Z Zhang, L Song, Z Han, W Saad, Z Lu, in *Wireless Communications and Networking Conference*. Overlapping coalition formation games for cooperative interference management in small cell networks (IEEE Piscataway, New Jersey, USA, 2013), pp. 643–648
19. D Tse, P Viswanath, *Fundamentals of Wireless Communication*. (Cambridge University Press, USA, 2004)
20. CS Ranaweera, PP Iannone, KN Oikonomou, KC Reichmann, RK Sinha, Design of cost-optimal passive optical networks for small cell backhaul using installed fibers. IEEE/OSA J. Opt. Commun. Netw. **5**, 230–239 (2013)
21. W Saad, Z Han, M Debbah, A Hjorungnes, T Basar, Coalitional game theory for communication networks. IEEE Signal Process. Mag. **26**, 77–97 (2009)
22. S Boyd, L Vandenberghe, *Convex Optimization*. (Cambridge University Press, USA, 2004)
23. 3GPP, TR 36.814, Further advancements for E-UTRA physical layer aspects (Release 9) (2010)
24. KR Apt, T Radzik, in *arXiv:cs/0605132v1 [cs.GT]*. Stable partitions in coalitional games (New York, USA, May, 2006)

4

Optimal power allocation on discrete energy harvesting model

Xiaolei Wang[1†], Jie Gong[1*†], Congshi Hu[2†], Sheng Zhou[1†] and Zhisheng Niu[1†]

Abstract

This paper studies the power allocation problem in energy harvesting systems with finite battery. We adopt the discretized energy arrival and power allocation model. Hence, the service process can be modeled as a finite state Markov chain. Based on the discretized model, we analyze the stationary distribution of the Markov chain and formulate the utility maximization problem, which is then reformed as a linear programming problem. By analyzing the linear programming problem, we provide some intuition on the structure of the optimal power allocation policy and find the condition in which the greedy power allocation is optimal. Numerical simulations show the influence of the energy arrival process on the optimal power allocation policy, and the results are consistent with our analysis.

Keywords: Energy harvesting; Markov chain; Power allocation

1 Introduction

With increasing CO_2 emissions in communication networks, how to realize green communications in the future has become an important and hot topic in the academic society. Besides the energy-efficient protocol design to reduce the energy consumption of the conventional wireless systems, utilizing renewable energy (e.g., solar or wind energy) to power on communication devices, namely, the *energy harvesting* technology, provides a new way for green by exploiting sustainable energy sources and hence is a promising solution to achieve environment-friendly communications. Recent developments in hardware have made the energy harvesting technology feasible for modern communication systems. For instance, a wind-powered tower for wireless base stations has been designed by Ericsson [1]. However, due to the randomness of the energy arrival process, how to optimally allocate the harvested energy is a challenging issue.

In recent years, a lot of research efforts have been focused on the energy harvesting systems. For the additive white Gaussian noise (AWGN) channel, the problem of minimizing the transmission completion time with infinite battery capacity in non-fading channel is studied in [2] for two scenarios, i.e., all packets are ready before transmission and packets arrive during transmission. Tutuncuoglu [3] finds the optimal transmission policy to maximize the short-term throughput with limited energy storage capacity, and exploits the relation between the throughput maximization and the transmission completion time minimization. For the fading channel, authors in [4] propose the directional water-filling (WF) algorithm which is proved throughput optimal for greedy source. Similar result is obtained in [5], which further considers the optimal solution with causal information. The algorithm is then extended to multiple antennas scenario in [6], where the spatial-temporal WF is proposed. Further, considering the dynamic data arrival with hybrid energy harvesting and power grid supplies, [7] proposes the optimal reverse multi-stage WF policy. Considering the circuit power consumption, a two-phase transmission policy is shown to be optimal [8]. In [9], the authors study the throughput maximization problem for the orthogonal relay channel with energy harvesting source and relay nodes under the deterministic model and show the structure of the optimal source and relay power allocation. Although the above algorithms give some insights about the optimal solution, they assume that all the energy arrival, the channel fading, and the data arrival must be explicitly known before transmission, which is called the *offline*

*Correspondence: gongj13@mail.tsinghua.edu.cn
†Equal contributors
[1]Tsinghua National Laboratory for Information Science and Technology, Department of Electronic Engineering, Tsinghua University, 100084 Beijing, People's Republic of China
Full list of author information is available at the end of the article

condition. Since the solutions based on the offline condition require accurate predictions for the system states, they are not always applicable in real communication systems.

Based on the *online* condition that only the past and current system states can be known, researchers have studied the optimal and sub-optimal power allocation policies in some special scenarios. Sharma [10] identifies throughput optimal and mean delay optimal energy management policies and shows a greedy policy to be optimal in low SNR regime with infinite battery capacity. And a throughput maximization algorithm in point-to-point communications with causal information based on Markov decision process (MDP) [11] approach is proposed in [12]. Recent work [13] studies the finite-horizon scheduling problem with discrete rates and proposes a low complexity threshold-based policy. However, the properties of the optimal solution can not be directly obtained via MDP approach. In addition, the MDP approach experiences very high computational complexity due to the curse of dimensionality, hence may not be applicable when the system state space grows large. From the information theory perspective, [14] studies the channel capacity of energy harvesting links with finite battery capacity and proves that the Markovian energy management policies are sufficient to achieve the capacity. Besides the throughput maximization problems, some other issues on the energy harvesting systems, such as the quality of service (QoS), the energy efficiency, and etc. are also studied. Huang [15] studies the utility optimization problem in energy harvesting networks under limited average network congestion constraint and develops a close-to-optimal algorithm using the Lyapunov optimization theory, which jointly manages the power allocation and the data access control. As the renewable energy is usually distributed asymmetrically in space domain, there are some papers considering the energy cooperation problem to balance the harvested energy in different places, including cellular network planning [16] and power grid energy saving [17], so that the overall system energy efficiency can be improved. But still, under the dynamic property of the energy harvesting process, how to allocate the energy to achieve the optimal system performance in general case is still an open question. It is desirable to explore the closed-form analytical solution for the online condition with some statistic characteristic of the energy harvesting process.

In this paper, we consider the power allocation problem in energy harvesting capacity to achieve the optimal system utility. Specifically, we study a single link with renewable energy transmitter, which only has the casual state information, including the distribution of the energy harvesting process, the past, and the current battery energy state. We model the energy arrival, storage, and usage as a discrete model and derive the optimal solution with closed-form expressions. The main contributions of this paper are presented as follows.

- We propose the discrete model for the energy harvesting system analysis. On one hand, the digital equipment has been widely used in modern communication systems, and it is feasible to give a discrete model for the energy harvesting process. On the other hand, the discrete model enable us to give a Markovian analysis and get some interesting closed-form analytical solution.
- For the independent identically distributed (i.i.d.) energy arrival process, we show the optimal solution can be obtained by solving a linear programming problem. Based on the linear programming formulation, we get some properties of the optimal power allocation policy and find the condition under which the greedy policy is optimal.
- Through extensive numerical simulations, we discuss the influence of the statistics of the energy arrival process on the optimal power allocation policy, which is shown consistent with our mathematical analysis.

The rest of the paper is organized as follows. Section 2 presents the system model. The problem is formulated and analyzed in Sections 3 and 4, respectively. Some numerical results are provided in Section 5 to evaluate the performance analysis. Finally, Section 6 concludes the paper.

Notations: Bold upper case and lower case letters denote matrices and vectors, respectively. $(\cdot)^{\mathrm{T}}$ denotes the transpose of a matrix or a vector. $\mathbf{0}_{n\times m}$ and $\mathbf{1}_{n\times m}$ represent the $n \times m$ matrices with all elements equal to 0 and 1, respectively. If $n = m$, they can be simplified as $\mathbf{0}_n$ and $\mathbf{1}_n$. $\mathbf{I}_n$ is the $n \times n$ unit matrix. $\mathbb{E}$ is the expectation operation.

2 System model

We consider a single link time-slotted wireless communication system with slot length T_f. The transmitter is powered by renewable energy, which is harvested from the environment and stored in a battery with finite capacity $B_{\max}$. The greedy data source is assumed to focus on the utility maximization with efficient harvested energy usage. The system model is illustrated in Figure 1. The utility is assumed to be a strictly concave and increasing function of allocated transmit power. As the slot length is fixed, the utility can be equivalently viewed as a function of used energy S_t, denoted by $u(S_t)$. While the transmit energy S_t depends on the system state, i.e., the amount of energy stored in the battery in slot t, denoted by E_t.

The state transition happens between time slots as shown in Figure 2. At the beginning of slot t, the power

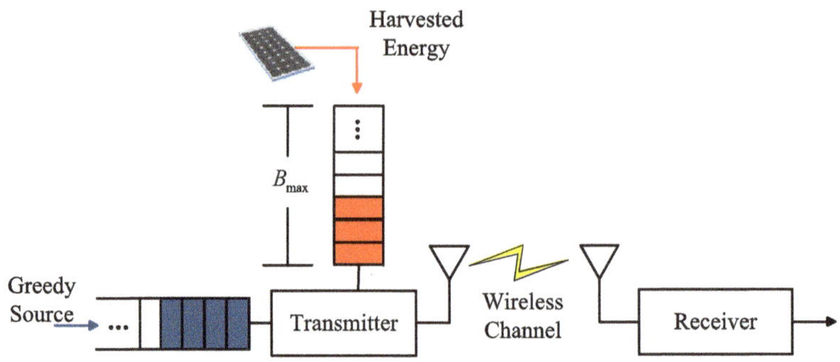

Figure 1 Single-link energy harvesting based wireless communication system.

allocation policy determines the amount of used energy S_t based on the system state E_t (red bar in the figure). Considering the energy causality constraint, the allocated energy cannot exceed the energy stored, i.e., $S_t \leq E_t$. Denote $\tilde{E}_t$ as the transition state after the power allocation process and before the energy harvesting process. According to the power allocation result, the transitional system state updates as $\tilde{E}_t = E_t - S_t$. Then, the amount of energy A_t is harvested during slot t is stored into the battery at the end of the slot. Note that the battery energy cannot exceed the battery capacity. Hence, the additional energy will be wasted if the battery is full. As a result, the system state is updated according to the following equation:

$$E_{t+1} = \min \{E_t - S_t + A_t, E_{\max}\}. \tag{1}$$

In this paper, we consider a discrete system model, i.e., the energy is discretized with unit ΔE. In the discrete model, the battery capacity can be expressed as $B_{\max} = N\Delta E$, where N is an integer. Hence, for the ease of description, we omit ΔE and denote the system state as $E_t = n, n = \{0, 1, \ldots, N\}$ which indicates that $n\Delta E$ amount of energy is stored in the battery. Besides, the energy arrival A_t and the power allocation S_t are also discretized with the unit energy ΔE, which will be detailed in the following subsections.

2.1 Energy arrival model

The amount of energy arrived in each time slot is assumed to be i.i.d. and takes non-negative integer values. The distribution of the energy arrival is expressed as:

$$\Pr[A_t = k] = h_k, \quad k = 0, 1, 2, \ldots, \tag{2}$$

where $h_k \geq 0, \sum_k h_k = 1$.

After harvesting the arrived energy, the system state transits from the transition state $\tilde{E}_t$ to the state of the next slot E_{t+1}. The state transition probability matrix due to energy arrival and harvesting process can be expressed as an $(N+1)\times(N+1)$ matrix $\mathbf{H}$, namely, *harvesting matrix*, with elements $H_{i,j}$ denoting the transition probability from $\tilde{E}_t = i$ to $E_{t+1} = j$ by harvesting $(j - i)\Delta E$ amount of energy. As the amount of harvested energy is non-negative, $\mathbf{H}$ is an upper triangular matrix. The elements of $\mathbf{H}$ can be calculated as:

$$H_{i,j} = \begin{cases} 0, & i > j, \\ h_{j-i}, & i \leq j < N, \\ 1 - \sum\limits_{k=0}^{N-i-1} h_k, & i < j = N, \\ 1, & i = j = N. \end{cases} \tag{3}$$

Note that $j = N$ represents that the battery is full. In this case, the amount of arrived energy larger than $N - i$ will tend to the same state $E_{t+1} = N$. Hence, the probability

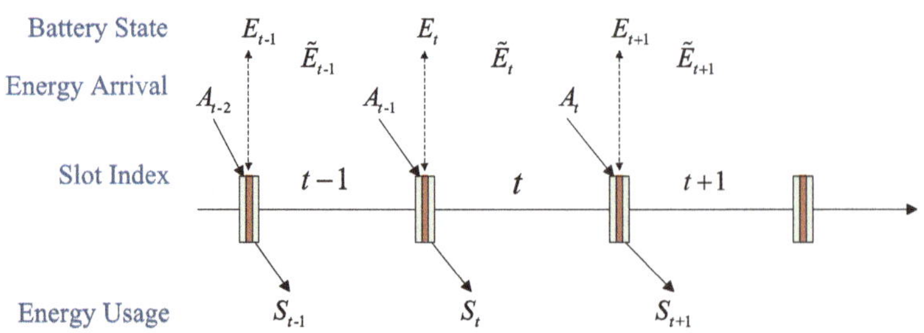

Figure 2 Discrete system state transition model. At the beginning of slot t, S_t is used to transmit. At the end of the slot, A_t is arrived and stored into the battery. Hence, the battery energy state at the beginning of slot $t+1$ is $E_{t+1} = \min\{E_t - S_t + A_t, E_{\max}\}$.

for $j = N$ is a summation of energy arrival distribution that satisfy $A_t \geq N - i$.

2.2 Power allocation policies

Recall that the power allocation policies only depend on the current system state. Similar as the energy arrival model, we express the state transition probability matrix between E_t and $\tilde{E}_t$ as a *policy matrix* **P** with elements $p_{i,j} \in \{0, 1\}, \forall i, j$ denoting the event of using $(i - j)$ unites of energy in state $E_t = i$. Then, the allocated power is calculated as $(i - j)\Delta E/T_f$. Since the allocated energy is non-negative, all the non-zero elements of the policy matrix **P** is on the left side of the diagonal elements, i.e.:

$$p_{i,j} = 0, \quad \forall i < j, \tag{4}$$

Besides, the deterministic policy tells us that only one action will be taken in each state. Hence, there is and only is one non-zero element in each row, which is as follows:

$$\sum\nolimits_{j=0}^{i} p_{i,j} = 1, \quad \forall i. \tag{5}$$

Note that $p_{i,j}$ is relaxed to take the value between 0 and 1 for the ease of mathematical formulation and theoretical analysis in the next section. However, a deterministic optimal policy is ultimately obtained through our solution, which means that the relaxed problem is equivalent with the original problem. We will discuss this issue in detail later.

2.3 Utility model

For any amount of allocated energy S_t, there is a corresponding utility $u(S_t)$. Since we consider the discrete energy model, the utility also takes value from a finite set $\{u_0, u_1, \ldots, u_N\}$. Specifically, u_k is the utility when the amount of $k\Delta E$ energy is allocated in a slot, i.e., $u_k = u(k\Delta E)$. As mentioned before, it is assumed to be increasing and strictly concave for $k > 0$ and $u_0 = 0$. For instance, if the optimal channel coding scheme with randomly generated codes is adopted, we can achieve the channel capacity given by:

$$u_k = \frac{1}{2}\log_2\left(1 + \frac{k\Delta E}{T_f\sigma^2}\right), \tag{6}$$

where σ^2 is the noise power. It is the well-known Shannon's equation [18]. Also, some other utility functions can be used, as long as they satisfy the monotonicity and the concavity properties.

3 Problem formulation

In this section, we formulate the utility maximization problem and transform the problem formulation into linear programming based on the Markovian property. We also simply discuss the linear programming problem formulation from the MDP point of view.

3.1 Utility maximization and Markov chain-based formulation

The objective of our problem is to maximize the average utility over a long time period, i.e.:

$$\max \lim_{T\to+\infty} \mathbb{E}\left[\frac{1}{T}\sum_{t=1}^{T} u(S_t)\middle| E_1 = k_0\right], \tag{7}$$

where k_0 is the initial battery energy, and the allocated energy S_t is determined by the battery energy state E_t. Notice that we can apply MDP approach [11] to solve the infinite horizon average utility maximization with finite number of states (the number of system states, i.e., the battery energy states is N). However, the MDP approach usually encounters the curse of dimensionality problem. In addition, the structure of the optimal policy is not clear as the MDP only outputs numerical results. To avoid the drawbacks of the MDP approach and analyze the optimal power allocation policy in detail, we consider to deal with the problem using linear programming. Specifically, for a given power allocation policy, the battery energy state forms a Markov chain. Firstly, we have the following lemma.

Lemma 1. *The problem (7) is irrelevant with the initial state k_0.*

Proof. The Markov chain with battery energy as state satisfies the weak accessibility ([11], Def. 4.2.2), as for a given energy arrival distribution $\{h_0, h_1, \ldots\}$ and for any states i, j, we can always find a stationary policy so that state i is accessible from state j. According to ([11], Prop. 4.2.3), the optimal average utility is the same for all initial states. □

Actually, the conclusion of Lemma 1 is easy to be understood as we consider the long-term average performance; the influence of the state at some specific time is negligible. Based on Lemma 1, we only need to consider the stationary behavior of the Markov chain. For a given power allocation policy **P**, there always exists a stationary system state distribution $\pi = [\pi_0, \pi_1, \ldots, \pi_N]^{\mathrm{T}}$, $\sum_{i=0}^{N} \pi_i = 1, \pi_i \geq 0, \forall i$ that satisfies:

$$\pi^{\mathrm{T}}\mathbf{PH} = \pi^{\mathrm{T}}, \tag{8}$$

where π_i is the probability that the battery energy is i and **PH** is the state transition probability matrix from state E_t

to state E_{t+1}. With the stationary distribution, the original problem (7) can be reformulated as:

$$\text{P-}\mathbf{1}: \max \sum_{i=0}^{N}\sum_{j=0}^{i} \pi_i p_{i,j} u_{i-j} \tag{9a}$$

$$\text{s.t.} \sum_{k=0}^{N} \pi_k \sum_{l=0}^{\min(k,i)} p_{k,l} H_{l,i} = \pi_i, \quad \forall i, \tag{9b}$$

$$\sum_{k=0}^{i} p_{i,k} = 1, \quad \forall i, \tag{9c}$$

$$\sum_{i=0}^{N} \pi_i = 1, \tag{9d}$$

$$p_{i,j} \in \{0,1\}, \quad \forall i \geq j, \tag{9e}$$

$$\pi_k \geq 0, \quad \forall k, \tag{9f}$$

where the harvesting matrix $\mathbf{H}$ and the utility function $u_k, k = 0, \ldots, N$ are predefined and (9b) is the expansion of (8). Note that different power allocation policies lead to different stationary distributions. On the contrary, if the state distribution varies, the optimal power allocation policy also changes. Hence, the unknown variables include both the power allocation policy $\{p_{i,j}, i \geq j\}$ and the stationary distribution $\{\pi_i\}$, which need to be jointly considered and optimized. As $p_{i,j} \in \{0,1\}$, it is a mixed optimization problem combining both integer programming and linear programming, which is difficult to be solved. To make it tractable, we transform the problem into a linear programming optimization problem in the rest of this section.

3.2 Problem reformulation with linear programming

Firstly, we relax the constraint (9e) and reformulate the problem as:

$$\text{P-}\mathbf{1R}: \quad \max \quad \sum_{i=0}^{N}\sum_{j=0}^{i} \pi_i p_{i,j} u_{i-j}$$

$$\text{s.t.} \quad \text{(9b) - (9d), and (9f)}$$

$$p_{i,j} \geq 0, \quad \forall i \geq j, \tag{10}$$

where $p_{i,j}$ becomes continuous variable, which can be considered as a probabilistic power allocation policy. In the next section, we will prove that the relaxation does not change the optimal solution. In other words, the optimal policy $p^*_{i,j}$ obtained by solving P-**1R** turns out to be of integer value.

Obviously, the problem P-**1R** is not a convex optimization problem since there is product $\pi_i p_{i,j}$ in the constraint (9b). To solve this problem, we transform it by multiplying π_i on both sides of the constraints (9c):

$$\sum_{k=0}^{i} \pi_i p_{i,k} = \pi_i, \forall i. \tag{11}$$

When $\pi_i \neq 0$, Equation (11) is the same with Equation (9c). On the other hand, if $\pi_i = 0$, it means that the stationary state i does not exist, leading to no influence on the total utility. Hence, by denoting $f_{i,j} = \pi_i p_{i,j}$, we can get the equivalent optimization problem of P-**1** as:

$$\text{P-}\mathbf{2}: \quad \max \quad \sum_{i=0}^{N}\sum_{j=0}^{i} f_{i,j} u_{i-j} \tag{12a}$$

$$\text{s.t.} \sum_{k=0}^{N}\sum_{l=0}^{\min(k,i)} f_{k,l} H_{l,i} = \pi_i, \quad \forall i, \tag{12b}$$

$$\sum_{k=0}^{i} f_{i,k} = \pi_i, \quad \forall i, \tag{12c}$$

$$\sum_{i=0}^{N} \pi_i = 1, \tag{12d}$$

$$0 \leq f_{i,j} \leq \pi_i, \quad \forall i \geq j, \tag{12e}$$

$$\pi_k \geq 0, \quad \forall k, \tag{12f}$$

We can see that the optimization problem P-**2** is a linear optimization as the objective function and the constraints are all linear functions. By solving the optimal $f_{i,j}$ and π_i, $p_{i,j}$ is also obtained by:

$$p_{i,j} = \frac{f_{i,j}}{\pi_i}, \pi_i > 0. \tag{13}$$

If $\pi_i = 0$, any $\left\{p_{i,j} : \sum_{j=0}^{N} p_{i,j} = 1,\ 0 \leq p_{i,j} \leq 1\right\}$ is optimal since the system state i has no influence on the total utility. Actually, $f_{i,j}$ is the probability that the system state is i and $j - i$ units of energy is used.

In the objective function of problem P-**2**, the variable π_i is not presented. And if $f_{i,j}$ is known, π_i can be calculated via (12b) or (12c). Substituting π_i by $\sum_{k=0}^{i} f_{i,k}$ in P-**2**, the optimal $\{f_{i,j}\}$ can be obtained by solving the following problem:

$$\text{P-}\mathbf{3}: \quad \max \quad \sum_{i=0}^{N}\sum_{j=0}^{i} f_{i,j} u_{i-j} \tag{14a}$$

$$\text{s.t.} \quad \sum_{k=0}^{i} f_{i,k} = \sum_{k=0}^{N}\sum_{l=0}^{\min(k,i)} f_{k,l} H_{l,i}, \forall i, \tag{14b}$$

$$\sum_{i=0}^{N}\sum_{k=0}^{i} f_{i,k} = 1, \tag{14c}$$

$$f_{i,j} \geq 0, \forall i \geq j. \tag{14d}$$

Note that the right-side inequality of (12e) and the constraint (12f) is omitted in P-**3** as they are guaranteed by (14c) and (14d). Further, the problem P-**3** can be expressed in a matrix form as:

$$\text{P-}\mathbf{3M}: \quad \max \quad \mathbf{u}^{\mathrm{T}}\mathbf{f} \tag{15a}$$
$$\text{s.t.} \quad \tilde{\mathbf{A}}\mathbf{f} = \left[\mathbf{0}_{1\times(N+1)}, 1\right]^{\mathrm{T}}, \tag{15b}$$
$$\mathbf{f} \geq \mathbf{0}_{\frac{1}{2}(N+1)(N+2)\times 1}, \tag{15c}$$

where $\mathbf{u} = \left[\mathbf{u}_0^{\mathrm{T}}, \mathbf{u}_1^{\mathrm{T}}, \ldots, \mathbf{u}_N^{\mathrm{T}}\right]^{\mathrm{T}}$ with $\mathbf{u}_i = \left[u_0, u_1, \ldots, u_{N-i}\right]^{\mathrm{T}}$, $\mathbf{f} = \left[\mathbf{f}_0^{\mathrm{T}}, \mathbf{f}_1^{\mathrm{T}}, \ldots, \mathbf{f}_N^{\mathrm{T}}\right]^{\mathrm{T}}$ with $\mathbf{f}_i = \left[f_{i,i}, f_{i+1,i}, \ldots, f_{N,i}\right]^{\mathrm{T}}$, and $\tilde{\mathbf{A}}$ is expressed as:

$$\tilde{\mathbf{A}} = \left[\tilde{\mathbf{A}}_0, \tilde{\mathbf{A}}_1, \ldots, \tilde{\mathbf{A}}_N\right], \tag{16}$$

where

$$\tilde{\mathbf{A}}_i = \begin{pmatrix} \mathbf{0}_{i\times(N-i+1)} \\ \tilde{\mathbf{H}}_i - \mathbf{I}_{(N-i+1)} \\ \mathbf{1}_{1\times(N-i+1)} \end{pmatrix}, \tag{17}$$

with

$$\tilde{\mathbf{H}}_i = \left[H_{i,i}, H_{i,i+1}, \ldots, H_{i,N}\right]^{\mathrm{T}} \cdot \mathbf{1}_{1\times(N-i+1)}. \tag{18}$$

Remark 1. The size of $\mathbf{f}_i$ and $\mathbf{u}_i$ are both $(N-i+1)\times 1$. As a result, the size of $\mathbf{f}$ and $\mathbf{u}$ are both $\frac{1}{2}(N+1)(N+2)\times 1$.

Remark 2. The size of $\tilde{\mathbf{A}}$ is $(N+2)\times\frac{1}{2}(N+1)(N+2)$.

The problem can be further simplified based on the following lemma.

Lemma 2. *The constraint (15b) can be equivalently written as:*

$$\hat{\mathbf{A}}\mathbf{f} = \mathbf{h}_0, \tag{19}$$

where $\mathbf{h}_0 = \left[H_{0,0}, H_{0,1}, \ldots, H_{0,N}\right]^{\mathrm{T}}$, *and* $\hat{\mathbf{A}} = \left[\mathbf{I}_{N+1}, -\hat{\mathbf{A}}_1, \ldots, -\hat{\mathbf{A}}_N\right]$ *with*

$$\hat{\mathbf{A}}_i = \begin{pmatrix} \mathbf{0}_{i\times(N-i+1)} \\ \tilde{\mathbf{H}}_i - \mathbf{I}_{(N-i+1)} \end{pmatrix} - \mathbf{h}_0\cdot\mathbf{1}_{1\times(N-i+1)}, \quad i = 1, \ldots, N. \tag{20}$$

Proof. Multiplying the constraint (15b) by an $(N+2)\times(N+2)$ matrix:

$$\mathbf{B} = \begin{pmatrix} \mathbf{I}_{N+1} & -\mathbf{h}_0 \\ \mathbf{0}_{1\times(N+1)} & 1 \end{pmatrix}, \tag{21}$$

i.e., $\mathbf{B}\tilde{\mathbf{A}}\mathbf{f} = \mathbf{B}\left[\mathbf{0}_{1\times(N+1)}, 1\right]^{\mathrm{T}}$, we have:

$$\begin{pmatrix} \hat{\mathbf{A}} \\ \mathbf{1}_{1\times\frac{1}{2}(N+1)(N+2)} \end{pmatrix}\mathbf{f} = \left[\mathbf{h}_0^{\mathrm{T}}, 1\right]^{\mathrm{T}}. \tag{22}$$

According to the fact that $\sum_{j=i}^{N} H_{i,j} = 1, \forall i$, we can easily prove that the sum of each column of $\hat{\mathbf{A}}$ equals to 1. Consequently, we have $\mathbf{1}_{1\times\frac{1}{2}(N+1)(N+2)}\hat{\mathbf{A}}f = \mathbf{1}_{1\times\frac{1}{2}(N+1)(N+2)}\mathbf{f}$, which means that the last constraint in (22) can be obtained by the summation of the first $(N+1)$ equality constraints, hence naturally holds as long as the other constraints are satisfied. As a result, (19) is obtained by removing the last constraint in (22). As the matrix $\mathbf{B}$ is invertible, the constraint (19) is equivalent with (15b). □

Based on Lemma 2, the number of the equality constraints is reduced from $(N+2)$ to $(N+1)$. Note that the rank of $\hat{\mathbf{A}}$ is $(N+1)$ since the first $(N+1)$ columns form an $(N+1)$ unit matrix, which means that the equality constraints are irreducible. We re-write the problem as:

$$\text{P-}\mathbf{4}: \quad \max \quad \mathbf{u}^{\mathrm{T}}\mathbf{f} \tag{23a}$$
$$\text{s.t.} \quad \hat{\mathbf{A}}\mathbf{f} = \mathbf{h}_0, \tag{23b}$$
$$\mathbf{f} \geq \mathbf{0}_{\frac{1}{2}(N+1)(N+2)\times 1}, \tag{23c}$$

In the rest of the paper, we focus on the solution for the problem P-**4**. As long as it is solved, the original problem P-**1R** is also solved. Specifically, the stationary distribution π can be calculated by (12b) or (12c), and the power allocation policy can be obtained by (13). In addition, the optimality proof of deterministic solution given in the next section guarantees the equivalence between problem P-**4** and problem P-**1**. Hence, problem P-**1** is also solved.

Remark 3. The optimization problem P-**4** can also be derived based on the MDP theory [11]. Specifically, for a given time index t, the system state $x_t \in \{0, \ldots, N\}$ is the battery energy state, the action $a_t(x_t) \in \{0, \ldots, x_t\}$ is the allocated energy. For the finite state problems, there exists an optimal stationary policy. Hence, we can omit the time index of a_t. Then, the cost function $g(x_t, a(x_t)) = -u_{a(x_t)}$ is the negative utility, and the state transition is calculated as $p_{ij}(a(i)) = H_{i-a(i),j}$. We re-write the average utility maximization problem as an average cost per-slot minimization problem:

$$\min \limsup_{T\to+\infty} \frac{1}{T}\mathbb{E}\left\{\sum_{t=0}^{T} g(x_t, a(x_t))\right\}. \tag{24}$$

The optimal policy satisfies the Bellman's equation ([11], Prop. 4.2.1), which in the problem (24) takes the form:

$$\lambda + s(i) = \min_{a(i)\in\{0,\ldots,i\}} \left[g(i,a(i)) + \sum_{j=0}^{N} p_{ij}(a(i))s(j) \right], i = 0,\ldots,N. \tag{25}$$

with a scalar λ anda vector s. The optimal cost λ^* can be determined by solving the following linear programming problem ([11], Sec. 4.5):

$$\max \quad \lambda \tag{26a}$$

$$\text{s.t.} \quad \lambda + s(i) \leq g(i,a(i)) + \sum_{j=0}^{N} p_{ij}(a(i))s(j),$$
$$i = 0,\ldots,N, a(i) = 0,\ldots,i. \tag{26b}$$

Applying the duality theory of linear programming ([19], Chap. 5), we can exactly get the linear program formulation as P-**3**, and hence, P-**4** is formulated. However, with Markov chain analysis instead of MDP theory, we can get the clear physical meaning of the variable $f_{i,j}$.

Remark 4. When solving the original problem P-**1**, it requires an exhaustive search over all the possible values of the parameters $p_{i,j} \in \{0,1\}, \forall i \geq j$, which is of exponential complexity. While the complexity of solving the linear programming problem P-**4** depends on the numerical algorithms which we apply. The most popular algorithms for solving linear programming problem are the *simplex algorithm* and the *interior point algorithm* [19]. It has been analyzed that the simplex algorithm is of exponential complexity in the worst case [20], while the complexity of a specific problem is case by case. We will show in the following section that in some cases, the optimal solution can be found without iteration. On the other hand, the interior point algorithm is of polynomial complexity. In summary, the complexity of the proposed linear programming is lower than that of the exhaustive search algorithm.

4 Optimal solution analysis

It is difficult to give an analytical solution for general conditions of **H** and **u**. In this section, we try to find some properties on the structure of the optimal power allocation policy and derive the condition in which some simple policy is optimal. We firstly present some general results about the optimal policy.

4.1 General properties

With a quick observation, we can firstly get the following property.

Proposition 1. *(Feasibility) The optimal solution* $\mathbf{f}^*$ *for the problem P-***4** *exists.*

Proof. Firstly, the problem P-**4** is feasible as we can at least find one solution satisfying all the constraints. For instance:

$$\mathbf{f}_g = \left[\mathbf{h}_0^{\mathrm{T}}, 0, \ldots, 0\right]^{\mathrm{T}}, \tag{27}$$

is feasible for the problem.

Secondly, the feasible region for the linear programming problem is finite according to the constraints $\mathbf{f} \geq \mathbf{0}$ and $\mathbf{1}_{1\times\frac{1}{2}(N+1)(N+2)} \cdot \mathbf{f} = 1$. Hence, the optimal solution exists. □

Proposition 1 tells us that the problem is feasible, and the existence of the optimal solution is guaranteed by the finite state constraint.

Lemma 3. *The optimal solution* $\mathbf{f}^*$ *for the problem P-***4** *can be achieved by the vector with at most* $(N+1)$ *non-zero elements.*

Proof. The optimal solution for a linear programming can be achieved by its *basic feasible solution* [19], of which the number of non-zero elements is no more than the rank of $\hat{\mathbf{A}}$. Since the rank of $\hat{\mathbf{A}}$ is $(N+1)$, its basic feasible solution has at most $(N+1)$ non-zero elements. □

The geographic explanation of Lemma 3 is that for a linear programming problem, the optimal solution can always be found at the vertex (corresponding to the basic feasible solution) of the convex polyhedron defined by the constraints (23b) and (23c) [19]. In this sense, we only need to focus on the basic feasible solutions with relatively small number of non-zero elements. However, it is not guaranteed that the optimal solution must have no more than $(N+1)$ non-zero elements. For instance, if there are two vertexes to be optimal, all the linear combinations of the two vertexes are also optimal.

Based on Lemma 3, we can guarantee the deterministic optimal policy can be obtained.

Proposition 2. ***(Deterministic optimal solution)*** *The optimal solution of P-***4** *can be achieved by a deterministic power allocation policy, i.e.,* $p_{i,j}^* \in \{0,1\}$.

Proof. Firstly, consider the case that $\pi_i^* > 0, \forall i$. That is, all the $(N+1)$ elements of π are non-zero. As we can find an optimal solution so that $\mathbf{f}^*$ has at most $(N+1)$ non-zero elements, based on (12c), there is and only is one non-zero element of $\{f_{i,0}^*, \ldots, f_{i,i}^*\}$ for a given i. Then, according to (13), we have $p_{i,j}^* \in \{0,1\}$.

If for some $k, \pi_k^* = 0$, the state k is a transient state. We can erase all the elements related with π_k^* in the problem formulation without changing the optimal solution. Specifically, in the original problem, by removing the k-th row and k-th column of the matrix **PH** in (8), the k-th constraint in (9b), and the elements related with π_k in (9a) and (9d), the optimal solution is not changed. Through the same deduction of the previous section, we can conclude the similar lemma that the optimal solution can be achieved with at most N non-zero elements. Similarly, we can prove $p_{i,j}^* \in \{0, 1\}, \forall i \neq k$. Then, by setting $p_{i,0}^* = 1$, a deterministic optimal policy is also obtained. The proof can also be extended to the case that for several ks, $\pi_k^* = 0$. □

Proposition 2 guarantees that a deterministic optimal power allocation policy can always be found. In fact, the deterministic optimal policy corresponds to the optimal basic feasible solution. Hence, in the sense of deterministic policy, the equivalence between P-**4** and P-1, or in other words, the equivalence between P-**1R** and P-**1** is guaranteed. By finding the optimal basic feasible solution of P-**4**, we can obtain the deterministic optimal power allocation policy for the original problem P-**1**. Again as discussed in the paragraph below Lemma 3, if there are two basic feasible solutions achieving optimal, the linear combination of the two solutions correspond to the probabilistic optimal policy. In practice, the deterministic optimal policy is desirable and also sufficient.

Theorem 1. ***(Structure of optimal policy)*** *For the optimal power allocation policy* $\mathbf{P}^*$, *if* $p_{i,j}^* > 0$ *and* $\pi_i^* > 0$, *for any m that satisfies* $\pi_m^* > 0$ *and* $m > i$, *we have:*

$$p_{m,n}^* = 0, \forall n < j. \tag{28}$$

Proof. See Appendix 1. □

Theorem 1 demonstrates that for the optimal policy, if k units of energy is allocated in state i, the allocated amount of energy k' in the next state $i+1$ will not increase drastically, i.e., $k' - k \leq 1$. The theorem can also be equivalently described as $p_{m,n}^* = 0, \forall n > j$, where $\pi_m^* > 0$ and $m < i$. Notice that the condition $\pi_i^* > 0$ indicates that state i is not a transient state. For a transient state, any power allocation policy is applicable as it does not change the objective. Hence, the conclusion does not hold for transient states.

4.2 Optimality of greedy policy

According to the feasible solution example (27), we can get that $p_{i,0} = 1, \forall i$, which turns out to be the *greedy policy*, i.e., in each slot, all the available energy in the battery is used up. In general, the greedy policy is not optimal. However, the following theorem provides the condition on which it is optimal.

Theorem 2. ***(Greedy optimal policy)*** *If*

$$\sum_{i=0}^{N-1} h_i (u_i - u_{i+1}) + u_N - u_{N-1} \geq 0, \tag{29}$$

the optimal solution for P-4 is $\mathbf{f}^* = \mathbf{f}_g$. *The optimal power allocation is the greedy algorithm, which means that we use up all the available energy in every slot. The stationary distribution of the system state is* $\pi = \mathbf{h}_0$.

Proof. See Appendix 2. □

For the ease of understanding the greedy optimal condition, we rewrite (29) as:

$$\sum_{i=0}^{N-1} h_i \Delta u_i \leq 1, \tag{30}$$

where

$$\Delta u_i = \frac{u_{i+1} - u_i}{u_N - u_{N-1}}, i = 0, \ldots, N-1 \tag{31}$$

is the relative utility gain by increasing one unit energy in power allocation state i. Thus, given the utility function, the condition can be viewed as the case that the weighted sum of energy arrival distribution up to $N-1$ weighted by relative utility gain is no more than 1. Since $\Delta u_i \geq 1$ and is decreasing function of i, the condition holds when the tail probability of energy arrival $1 - \sum_{i=0}^{N-1} h_i$ is large enough, and the value of h_i is relatively small when i is small. Intuitively, such kind of energy arrival distribution shows the property that the amount of energy arrival is quite large relative to the battery capacity. Hence, the optimal policy tends to use up all the available energy in the battery in every time slot (i.e., greedy policy). In this case, the battery can store as many energy as it can, and hence, the energy wasted due to the battery overflow is reduced.

On the other hand, with a fixed energy arrival distribution $\{h_i\}$, we can find the influence of the utility function on the optimality of greedy policy. Specifically, we adopt the Shannon's capacity as the utility function as expressed in (6). At low signal-to-noise ratio (SNR) regime, i.e., $\frac{k\Delta E}{T_f \sigma^2}$ is very small, we have the following approximation:

$$u_i \approx \frac{i \Delta E}{2 T_f \sigma^2}, \tag{32}$$

which results in $\Delta u_i \approx 1, \forall i$. Since $\{h_i\}$ is a probability distribution, (30) naturally holds in this low SNR case. It can be explained as when the channel condition is poor, the capacity gain is linearly proportional to the power allocation. Consequently, the greedy policy is optimal as it not only can obtain the same capacity with the same available energy, but also reduces the amount of wasted energy due to battery overflow.

A quick conclusion based on the Theorem 2 can be obtained as follows:

Corollary 1. *If the energy arrival is uniformly distributed between 0 and $2\bar{h}$ with average arrival rate $\bar{h}$, the greedy policy is optimal when:*

$$\bar{h} \geq \frac{u_N - u_0}{2(u_N - u_{N-1})}. \tag{33}$$

Proof. The energy arrival probability of uniform distribution is expressed as:

$$h_i = \frac{1}{2\bar{h}}, i = 0, \ldots, 2\bar{h}. \tag{34}$$

By substituting h_i in (29) with (34) and some derivation, we can get the condition on the average arrival rate. □

5 Numerical results

We calculate some numerical examples of the optimal power allocation to demonstrate the structure and the properties of the optimal policies. We define:

$$\gamma = \frac{\Delta E}{T_f \sigma^2} \tag{35}$$

as the reference SNR with one unit of energy, and adopt the Shannon's equation to calculate the utility, i.e.:

$$u_k = \frac{1}{2} \log_2(1 + k\gamma). \tag{36}$$

Set $N = 10$ and $\gamma = 1$ as an example, we examine the influence of the random energy arrival process on the optimal power allocation. Specifically, we consider some widely used distributions, including discrete uniform distribution, geometric distribution, Poisson distribution, and binomial distribution. Denote $Unif(\bar{h})$ as the uniform distribution that takes values in $\{0, 1, \ldots, 2\bar{h}\}$ with mean $\bar{h}$. Denote $Poiss(\bar{h})$ as the Poisson distribution with mean $\bar{h}$. Denote $Geom(p)$ as the geometric distribution with parameter p; thus, we have:

$$h_i = (1 - p)^i p, i \geq 0 \tag{37}$$

with mean $\bar{h} = (1-p)/p$. And denote $Bin(n, p)$ as the binomial distribution with parameters n, p. The probability density function is as follows:

$$h_i = \frac{n!}{i!\,(n-i)!} p^i (1-p)^{n-i}, 0 \leq i \leq n, \tag{38}$$

and the mean value is $\bar{h} = np$. Since $0 < p < 1$, we have $n > \bar{h}$. Notice that except for binomial distribution, all the studied distributions have only one parameter.

Tables 1, 2, and 3 provides the optimal power allocation policies for uniform distribution, Poisson distribution, and geometric distribution with different mean values. It can be seen that the minimum average arrival rates with which the greedy policy is optimal are different for different distributions ($\bar{h} = 13$ for uniform distribution, $\bar{h} = 8$ for Poisson distribution, $\bar{h} = 23$ for geometric distribution). The result is consistent with our observation. Specifically, if h_i is an increasing function of i, it is easier for the greedy policy to be optimal than the case that h_i is decreasing. As Poisson distribution increases before the average value, while geometric distribution is strictly decreasing, consequently higher average energy arrival rate is needed so that greedy policy is optimal for geometric distribution.

Another observation from these results is that different from infinite battery capacity case where the optimal power will not exceed the average energy arrival rate [[10] Theorem 1], the optimal power exceeds the average energy arrival rate at some cases. As the battery capacity is finite, the harvested energy may be wasted when the battery is full. Hence, more energy will be used when the battery energy is close to its limit in order to reduce the wastage of energy.

As the binomial distribution has two parameters, we provide the detailed numerical results in Tables 4, 5, and 6. It can be seen that when the average arrival rate $\bar{h} \leq 7$, the greedy policy will never be optimal. While for $\bar{h} = 8$, the greedy policy is shown optimal when $n \geq 11$. In addition, when $\bar{h} \geq 9$, the greedy policy is always optimal for any feasible values of $n > \bar{h}$. Since in the binomial distribution, h_i also follows the same feature of Poisson distribution that it firstly increases and then decreases, the threshold of $\bar{h}$ for the greedy policy to be optimal is relatively low.

Table 1 Optimal power allocation policy for uniform distribution with parameter $\gamma = 1$

Battery state	$\bar{h} = 4$	$\bar{h} = 8$	$\bar{h} = 12$	$\bar{h} = 13$
0	0	0	0	0
1	1	1	1	1
2	2	2	2	2
3	2	3	3	3
4	3	4	4	4
5	3	5	5	5
6	4	6	6	6
7	4	6	7	7
8	5	7	8	8
9	5	8	9	9
10	6	9	9	10

The numbers in the table indicate the amount of energy units used in corresponding state. The same meaning holds in the rest tables.

Table 2 Optimal power allocation policy for Poisson distribution with parameter $\gamma = 1$

Battery state	$\bar{h} = 4$	$\bar{h} = 5$	$\bar{h} = 7$	$\bar{h} = 8$
0	0	0	0	0
1	1	1	1	1
2	2	2	2	2
3	3	3	3	3
4	3	4	4	4
5	4	4	5	5
6	4	5	6	6
7	4	5	7	7
8	5	6	7	8
9	5	6	8	9
10	6	7	9	10

Table 4 Optimal power allocation policy for binomial distribution with parameters $\gamma = 1, \bar{h} = 4$

Battery state	$n = 5$	$n = 6$	$n = 8$	$n = +\infty$
0	0	0	0	0
1	1	1	1	1
2	2	2	2	2
3	3	3	3	3
4	3	3	3	3
5	4	4	4	4
6	4	4	4	4
7	4	4	4	4
8	4	4	5	5
9	4	5	5	5
10	5	5	6	6

Finally, Tables 7 and 8 compare the optimal policies for the considered distributions at high and low SNR regimes, respectively. It can be seen that at low SNR regime, the optimal policy for any distribution turns out to be the greedy policy. On the other hand, when the SNR is high, it is not true. Besides, the policies for Poisson distribution and binomial distribution are closer to the greedy policy at high SNR regime, which is consistent with the previous analysis.

Another interesting property can be seen from all the results, i.e., the optimal power allocation policy is a non-decreasing function of the battery energy state. Besides, the increasing step is no more than 1, which coincides to Theorem 1. The non-decreasing property can be explained as follows. The more the available energy is, the more allocated power is used in order to not only achieve higher data rate but also reduce the energy wastage due to battery overflow. As the battery capacity tends to infinity, the non-decreasing property still holds as shown in [10].

We also run some simulations to evaluate the performance of the optimal power allocation policy. For comparison, we consider the following two policies as baselines:

(1) Greedy power allocation policy, which allocates all the available energy to transmit data.

(2) Constant power allocation policy, which allocates the amount of energy equal to the average energy arrival rate. When the required constant energy is not available, the transmitter will allocate all the available energy.

The simulation results are shown in Figures 3, 4, 5, and 6. Specifically, under the uniform distributed energy arrival, it can be seen in Figure 3 that the constant power policy performs close to the optimal at two extreme cases, i.e., $\bar{h} = 1$ and $\bar{h} \geq 8$. While as was shown before, the greedy policy performs close to the optimal when $\bar{h} \geq 13$. In addition, the constant policy always performs

Table 3 Optimal power allocation policy for geometric distribution with parameter $\gamma = 1$

Battery state	$\bar{h} = 4$	$\bar{h} = 13$	$\bar{h} = 22$	$\bar{h} = 23$
0	0	0	0	0
1	1	1	1	1
2	2	2	2	2
3	2	3	3	3
4	2	4	4	4
5	3	5	5	5
6	3	5	6	6
7	3	6	7	7
8	4	7	8	8
9	4	8	9	9
10	5	8	9	10

Table 5 Optimal power allocation policy for binomial distribution with parameters $\gamma = 1, \bar{h} = 7$

Battery state	$n = 8$	$n = 9$	$n = 10$	$n = +\infty$
0	0	0	0	0
1	1	1	1	1
2	2	2	2	2
3	3	3	3	3
4	4	4	4	4
5	5	5	5	5
6	6	6	6	6
7	7	7	7	7
8	7	7	7	7
9	7	8	8	8
10	8	8	9	9

Table 6 Optimal power allocation policy for binomial distribution with parameters $\gamma = 1, \bar{h} = 8$ or $\bar{h} = 9$

Battery state	$\bar{h} = 8, n = 9$	$\bar{h} = 8, n = 10$	$\bar{h} = 8, n = 11$	$\bar{h} = 9, n = 10$
0	0	0	0	0
1	1	1	1	1
2	2	2	2	2
3	3	3	3	3
4	4	4	4	4
5	5	5	5	5
6	6	6	6	6
7	7	7	7	7
8	8	8	8	8
9	8	9	9	9
10	9	9	10	10

Table 8 Policy comparison under different energy arrival distribution in low SNR regime with parameters $\gamma = 0.01, \bar{h} = 6, n = 15$

Battery state	Unif.	Poiss.	Geom.	Bin.
0	0	0	0	0
1	1	1	1	1
2	2	2	2	2
3	3	3	3	3
4	4	4	4	4
5	5	5	5	5
6	6	6	6	6
7	7	7	7	7
8	8	8	8	8
9	9	9	9	9
10	10	10	10	10

better than the greedy policy. For the Poisson distributed energy arrival (Figure 4), the greedy policy performs worse than the constant policy at low energy arrival rate regime ($\bar{h} < 6$), but better at high energy arrival rate regime ($6 \leq \bar{h} < 10$). The performance converge to the optimal when $\bar{h} \geq 10$. In Figure 5, the behavior of these policies under Geometric distributed energy arrival is similar with the uniform distributed scenario. However, the two baseline policies in this scenario converge to the optimal much slower than in the uniform distributed scenario. At last, the performance under the binomial distributed energy arrival is illustrated in Figure 6, which is similar with the Poisson distributed case. To sum up, all the simulations show that the constant policy shows higher performance than the greedy policy in low energy arrival rate regime. However, there are still gaps to the optimal solution (maximum of 10% gap under geometric distribution and maximum of 2% to 3% under the rest distributions).

6 Conclusions

In this paper, we analyzed the optimal power allocation policy under a discrete system model using the Markov chain analysis. We proved that the problem can be solved via linear programming approach and analyzed the properties of the optimal policy. And for the greedy power allocation policy, we found the condition to guarantee its optimality. Numerical results show that under finite battery capacity condition, the optimal policy is quite different from that under infinite battery capacity condition. Specifically, different from the infinite battery case, the energy allocated in each slot might be larger

Table 7 Policy comparison under different energy arrival distribution in high SNR regime with parameters $\gamma = 10, \bar{h} = 6, n = 15$

Battery state	Unif.	Poiss.	Geom.	Bin.
0	0	0	0	0
1	1	1	1	1
2	2	2	1	2
3	3	3	2	3
4	3	4	3	4
5	4	5	3	5
6	5	5	3	5
7	5	6	4	6
8	6	6	5	6
9	7	7	5	7
10	7	8	5	8

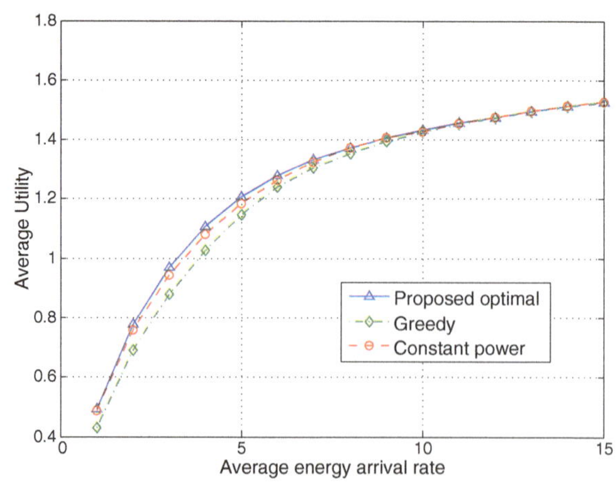

Figure 3 **The performance of the proposed optimal power allocation policy under uniform distributed energy arrival with parameter $\gamma = 1$.**

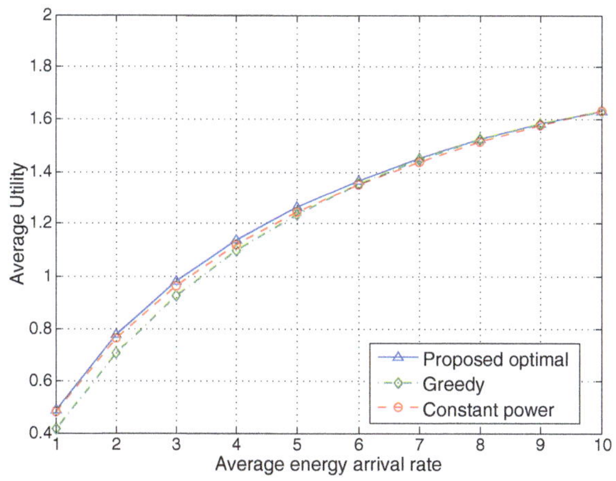

Figure 4 **The performance of the proposed optimal power allocation policy under Poisson distributed energy arrival with parameter $\gamma = 1$.**

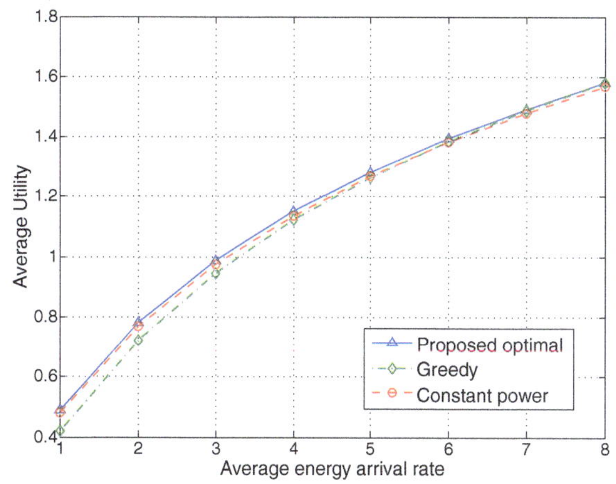

Figure 6 **The performance of the proposed optimal power allocation policy under binomial distributed energy arrival with parameter $\gamma = 1, n = 10$.**

than the average arrival rate. Also, extensive simulations showed the consistent results with the analysis for the greedy optimal condition. Based on the intuition provided through the analysis of discrete model, future work will extend our results to the continuous energy model scenario and the extensive wireless systems with multiple antennas/subcarriers.

Appendices

Appendix 1

Proof of Theorem 1

We prove the theorem by contradiction. Notice that $\pi_i^* > 0$ means that state i is not a transient state. Suppose for

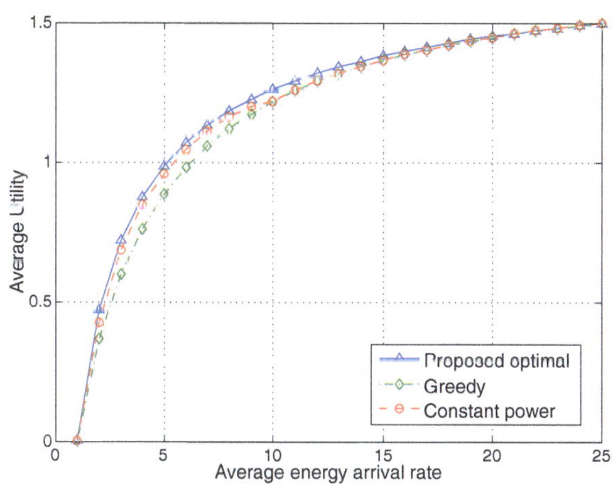

Figure 5 **The performance of the proposed optimal power allocation policy under geometric distributed energy arrival with parameter $\gamma = 1$.**

non-transient states $i, m, p^*_{i,j} > 0, p^*_{m,n} > 0$, where $m > i, n < j$. We provide the proof in the following two cases separately.

1) If:

$$\frac{p^*_{m,n}}{p^*_{i,j}} \geq \frac{\pi_i^*}{\pi_m^*}, \tag{39}$$

we construct a power allocation policy matrix $\mathbf{P}'$ by setting:

$$p'_{i,n} = p^*_{i,n} + p^*_{i,j}, \tag{40}$$

$$p'_{i,j} = 0, \tag{41}$$

$$p'_{m,n} = p^*_{m,n} - \frac{\pi_i^*}{\pi_m^*} p^*_{i,j}, \tag{42}$$

$$p'_{m,j} = p^*_{m,j} + \frac{\pi_i^*}{\pi_m^*} p^*_{i,j}, \tag{43}$$

and $p'_{k,l} = p^*_{k,l}$ for the other elements. We have:

$$\left(\pi^*\right)^{\mathrm{T}} \mathbf{P}' = \left(\pi^*\right)^{\mathrm{T}} \mathbf{P}^*, \tag{44}$$

$$\sum_{k=0}^{i} p'_{i,k} = 1, \quad \forall i \tag{45}$$

$$p'_{i,j} \geq 0, \quad \forall i \geq j \tag{46}$$

i.e., the power allocation policy $\mathbf{P}'$ still satisfies all the constraints (9b), (9c), and (10), and the stationary distribution π^* does not change under the new policy. Hence, all the constraints are satisfied with the new power allocation policy. Then, we compare the objective functions. The difference is only related with elements of $\mathbf{P}$ with indexes

$(i,n),(i,j),(m,n),(m,j)$. For the clarity of comparison, we only calculate:

$$D=\pi_i^* p_{i,n}^* u_{i-n}+\pi_i^* p_{i,j}^* u_{i-j}+\pi_m^* p_{m,n}^* u_{m-n}+\pi_m^* p_{m,j}^* u_{m-j}, \tag{47}$$

and

$$D'=\pi_i^* p_{i,n}' u_{i-n}+\pi_i^* p_{i,j}' u_{i-j}+\pi_m^* p_{m,n}' u_{m-n}+\pi_m^* p_{m,j}' u_{m-j}. \tag{48}$$

Joint with (40)-(43), we can get:

$$D-D'=\pi_i^* p_{i,j}^*\left[(u_{m-n}-u_{m-j})-(u_{i-n}-u_{i-j})\right]\overset{(a)}{<}0, \tag{49}$$

where (a) is due to the concavity of utility function. As a result, with the revised policy $\mathbf{P}'$, a larger average utility is obtained, which contradicts the optimality of $\mathbf{P}^*$.

2) If:

$$\frac{p_{m,n}^*}{p_{i,j}^*}<\frac{\pi_i^*}{\pi_m^*}, \tag{50}$$

similarly by setting:

$$p_{i,n}'=p_{i,n}^*+\frac{\pi_m^*}{\pi_i^*}p_{m,n}^*, \tag{51}$$

$$p_{i,j}'=p_{i,j}^*-\frac{\pi_m^*}{\pi_i^*}p_{m,n}^*, \tag{52}$$

$$p_{m,n}'=0, \tag{53}$$

$$p_{m,j}'=p_{m,j}^*+p_{m,n}^*, \tag{54}$$

and following the same deduction procedure of case 1), we can obtain the same contradiction.

To sum up, with the assumption that for non-transient states i,m, $p_{i,j}^*>0$ and $p_{m,n}^*>0, m>i, n<j$, it ends up that $\mathbf{P}^*$ is not optimal. Hence, the theorem is proved with the contradiction.

Appendix 2

Proof of Theorem 2

The theorem can be proved according to the *simplex method* [19]. Specifically, $\mathbf{f}_g=\left[\mathbf{h}_0^{\mathrm{T}},0,\ldots,0\right]^{\mathrm{T}}$ is a basic feasible solution, and the resulting objective function is $\mathbf{u}_0^{\mathrm{T}}\mathbf{h}_0$.

Consider a feasible solution $\mathbf{f}=\left[\mathbf{f}_0^{\mathrm{T}},\mathbf{f}_1^{\mathrm{T}},\ldots,\mathbf{f}_N^{\mathrm{T}}\right]^{\mathrm{T}}$ as initially defined in P-**3M**. From (19), we can get:

$$\mathbf{f}_0=\mathbf{h}_0+\sum_{i=1}^{N}\hat{\mathbf{A}}_i\mathbf{f}_i. \tag{55}$$

Then, the objective value can be written as:

$$\begin{aligned}\mathbf{u}^{\mathrm{T}}\mathbf{f}&=\mathbf{u}_0^{\mathrm{T}}\mathbf{f}_0+\sum_{i=1}^{N}\mathbf{u}_i^{\mathrm{T}}\mathbf{f}_i\\&=\mathbf{u}_0^{\mathrm{T}}\left(\mathbf{h}_0+\sum_{i=1}^{N}\hat{\mathbf{A}}_i\mathbf{f}_i\right)+\sum_{i=1}^{N}\mathbf{u}_i^{\mathrm{T}}\mathbf{f}_i\\&=\mathbf{u}_0^{\mathrm{T}}\mathbf{h}_0+\sum_{i=1}^{N}\left(\mathbf{u}_0^{\mathrm{T}}\hat{\mathbf{A}}_i+\mathbf{u}_i^{\mathrm{T}}\right)\mathbf{f}_i\end{aligned} \tag{56}$$

Denote $\mathbf{c}_i=\mathbf{u}_0^{\mathrm{T}}\hat{\mathbf{A}}_i+\mathbf{u}_i^{\mathrm{T}}$. To guarantee the optimality of $\mathbf{u}_0^{\mathrm{T}}\mathbf{h}_0$, as $\mathbf{f}_i\geq\mathbf{0}$, all the elements of $\mathbf{c}_i, i=1,\ldots,N$ should be no more than zero. The k-th element of $\mathbf{c}_i$ is as follows:

$$\begin{aligned}c_{i,k}&=\sum_{j=i}^{N}u_jH_{i,j}-\sum_{j=0}^{N}u_jH_{0,j}-u_{k+i}+u_k\\&=-\sum_{j=0}^{i-1}u_jh_j-\sum_{j=i}^{N-1}u_j(h_j-h_{j-i})+u_N\sum_{j=N-i}^{N-1}h_j-(u_{k+i}-u_k),\end{aligned} \tag{57}$$

where $i=1,\ldots,N, k=0,\ldots,N-i$. Due to the concavity of the utility function u_j, we have:

$$c_{i,k}-c_{i,k-1}=(u_k-u_{k-1})-(u_{k+i}-u_{k+i-1})\geq 0,\quad \forall k. \tag{58}$$

As a result, $c_{i,N-i}$ is the largest element for a given i. In addition, we have:

$$\begin{aligned}c_{i,N-i}-c_{i-1,N-i+1}&=\sum_{j=0}^{N-i}h_j\left(u_{j+i}-u_{j+i-1}\right)-\left(u_{N-i+1}-u_{N-i}\right)\\&\overset{(a)}{\leq}\sum_{j=0}^{N-i}h_j\left(u_{j+1}-u_j\right)-\left(u_N-u_{N-1}\right)\\&\overset{(b)}{\leq}\sum_{j=N-i+1}^{N-1}h_j\left(u_j-u_{j+1}\right)\\&\overset{(c)}{\leq}0,\quad i=2,\ldots,N,\end{aligned} \tag{59}$$

where inequality (a) holds due to the concavity of u_j, (b) is derived from (29), and (c) holds since $h_j\geq 0$ and u_j is increasing. Combining (58) and (59), we can conclude that $c_{1,N-1}$ is the largest element. Since $c_{1,N-1}=\sum_{i=0}^{N-1}h_i\left(u_{i+1}-u_i\right)-u_N-u_{N-1}\leq 0$, all the elements of $\mathbf{c}_i, i=1,\ldots,N$ is non-positive, and hence the optimality of the greedy solution is proved.

Competing interests
The authors declare that they have no competing interests.

Authors' contributions
The authors of the paper are of equal contribution to this paper. All authors read and approved the final manuscript.

Acknowledgements
This work is sponsored in part by the National Basic Research Program of China (No. 2012CB316001), and the Nature Science Foundation of China (No. 61201191 and 61401250), the Creative Research Groups of NSFC (No. 61321061), the Sino-Finnish Joint Research Program of NSFC (No. 61461136004), and Hitachi R&D Headquarter.

Author details
[1]Tsinghua National Laboratory for Information Science and Technology, Department of Electronic Engineering, Tsinghua University, 100084 Beijing, People's Republic of China. [2]China Mobile Communication Corporation (CMCC), No. 29, Financial Street, Xicheng District, Beijing 100033, People's Republic of China.

References
1. Ericsson, *Sustainable energy use in mobile communications*. (Ericsson Inc. white paper, 2007)
2. J Yang, S Ulukus, Optimal packet scheduling in an energy harvesting communication system. IEEE Trans. Commun. **60**(1), 220–230 (2012)
3. K Tutuncuoglu, A Yener, Optimum transmission policies for battery limited energy harvesting nodes. IEEE Trans. Wireless Commun. **11**(3), 1180–1189 (2012)
4. O Ozel, K Tutuncuoglu, J Yang, S Ulukus, A Yener, Transmission with energy harvesting nodes in fading wireless channels: optimal policies. IEEE J. Selected Areas Commun. **29**(8), 1732–1743 (2011)
5. CK Ho, R Zhang, Optimal energy allocation for wireless communications with energy harvesting constraints. IEEE Trans. Signal Process. **60**(9), 4808–4818 (2012)
6. C Hu, J Gong, X Wang, S Zhou, Z Niu, in *IEEE/CIC International Conference on Communications in China (ICCC)*. Spatial-temporal water-filling power allocation in MIMO systems with harvested energy, (2013), pp. 588–593
7. J Gong, S Zhou, Z Niu, Optimal power allocation for energy harvesting and power grid coexisting wireless communication systems. IEEE Trans. Commun. **61**(7), 3040–3049 (2013)
8. J Xu, R Zhang, Throughput optimal policies for energy harvesting wireless transmitters with non-ideal circuit power. IEEE J. Selected Areas Commun. **32**(2), 322–332 (2014)
9. C Huang, R Zhang, S Cui, Throughput maximization for the gaussian relay channel with energy harvesting constraints. IEEE J. Selected Areas Commun. **31**(8), 1469–1479 (2013)
10. V Sharma, U Mukherji, V Joseph, S Gupta, Optimal energy management policies for energy harvesting sensor nodes. IEEE Trans. Wireless Commun. **9**(4), 1326–1336 (2010)
11. DP Bertsekas, *Dynamic Programming and Optimal Control, Volume II*, 3rd edn. (Athena Scientific, Massachusetts, 2005)
12. CK Ho, R Zhang, in *IEEE International Symposium on Information Theory Proceedings (ISIT)*. Optimal energy allocation for wireless communications powered by energy harvesters, (2010), pp. 2368–2372
13. BT Bacinoglu, E Uysal-Biyikoglu, Finite-horizon online transmission scheduling on an energy harvesting communication link with a discrete set of rates. J. Commun. Netw. **16**(3), 293–300 (2014)
14. PK Deekshith, V Sharma, R Rajesh, AWGN channel capacity of energy harvesting transmitters with a finite energy buffer. arXiv preprint, available online: http://arxiv.org/abs/1307.4505
15. L Huang, MJ Neely, Utility optimal scheduling in energy-harvesting networks. IEEE/ACM Trans. Netw. (TON). **21**(4), 1117–1130 (2013)
16. M Zheng, P Pawełczak, S Stańczak, H Yu, Planning of cellular networks enhanced by energy harvesting. arXiv preprint, available online: http://arxiv.org/abs/1304.5088
17. Y-K Chia, S Sun, R Zhang, in *IEEE Wireless Communications and Networking Conference (WCNC)*. Energy cooperation in cellular networks with renewable powered base stations, (2013), pp. 2542–2547
18. TM Cover, JA Thomas, *Elements of Information Theory*. (John Wiley & Sons, New York, USA, 2012)
19. RJ Vanderbei, *Linear Programming: Foundations and Extensions*, 3nd edn. (Springer, Berlin, Germany, 2010)
20. V Klee, GJ Minty, *How good is the simplex algorithm?* (Academic Press, New York-London, 1972), pp. 159–175

5

Transmission priority scheme with adaptive backoff technique in fiber-wireless networks

Wan Hafiza Wan Hassan[1,2*], Horace King[1], Shabbir Ahmed[1] and Mike Faulkner[1]

Abstract

This paper considers the realization of fiber-wireless (Fi-Wi) networks using the gigabits passive optical network (GPON) and the infrastructure-based wireless local area network (WLAN). The bottleneck of such hybrid system is the WLAN where interference limits the performance. As such, we focus on enhancing the WLAN performance by analytically deriving the optimum contention window (CW) sizes of access points (APs) and wireless users (WUs), respectively. An adjustable transmission priority factor is introduced to allow uplink-downlink transmission fairness. Further, an adaptive backoff technique using information from monitoring the GPON and WLAN networks is proposed. Simulations show the CW sizes of all WUs are maintained within a standard deviation of 1.5% at a cost of a 3% loss in throughput due to the effect of convergence and other estimates.

Keywords: Fi-Wi networks; WLAN; GPON; Contention window (CW); Adaptive backoff; Access point (AP); Wireless user (WU)

Introduction

Fiber-wireless (Fi-Wi) networks combine the capacity of optical fiber networks with the ubiquity and mobility of wireless networks [1-7]. The use of energy-efficient and stable passive optical splitters makes them the favored choice among other fiber-to-the-home (FTTH) technologies. Gigabits passive optical network (GPON) and Ethernet passive optical network (EPON) are the major standards in PONs. In comparison to EPON, the current GPON offers higher capacity per user and allows more users per PON [8]. This makes GPON more attractive, and hence, it is adopted by major telecommunications service providers (e.g., National Broadband Network (NBN) in Australia and Ultra-Fast Broadband (UFB) network in New Zealand [9]).

In this work, we focus on the integration of GPON with a 'closed' WLAN system as shown in Figure 1. In this situation, the operator has a dedicated spectrum allocation which is shared by all access points. The integration is employed at the end-user terminal where an optical network terminal (ONT) of GPON and an access point (AP) of WLAN are incorporated into a single device, forming a Fi-Wi hybrid system (also known as radio and fiber (R&F) system) [3,10]. The integrated ONT-AP device acts as a gateway translating media access control (MAC) frames from the optical network to the wireless network and vice versa. Two distinct separate MAC protocols are used to access GPON and WLAN, respectively. Thus, the wireless MAC frames only traverse the WLAN and do not have to travel along the optical fiber to be processed at the central office. In contrast, radio over fiber (RoF), another possible enabling technology for the Fi-Wi system, transmits radio frequency (RF) signals from the central office to the remote access units (RAUs) as illustrated in Figure 2. The central office in RoF controls both optical and wireless media compared to just optical in R&F. This difference makes R&F more practical compared to RoF because an optical distribution use of wireless signals imposes additional fiber propagation delays which might exceed certain wireless MAC protocol timeouts and further worsens the overall network performance [10,11].

However, the increasing number of wireless terminals within households, especially in dense urban and suburban areas, results in dominant interference or congestion issues in WLANs. Such congestions occur when a WLAN channel is shared by many users. Hence, the scope of this paper is fully focused on the WLAN performance

*Correspondence: wanhafiza.wanhassan@live.vu.edu.au
[1]College of Engineering and Science, Victoria University, Ballarat Road, 8001 Melbourne, Australia
[2]School of Ocean Engineering, Universiti Malaysia Terengganu 21030 Kuala Terengganu, Malaysia

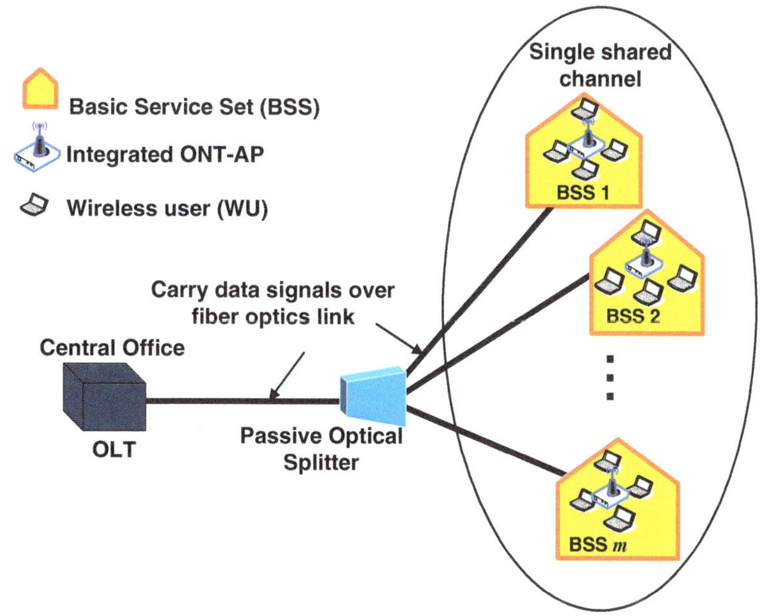

Figure 1 The Fi-Wi network scenario using the R&F technology.

where multiple APs are connected through GPON at close proximity to each other and share one single channel to serve their respective wireless users (WUs).

WLAN adopts the IEEE 802.11 standard which specifies the distributed coordination function (DCF) as its main medium access control protocol. DCF is a random access scheme, based on the carrier sense multiple access with collision avoidance (CSMA/CA) [12]. This protocol only allows a station to initiate a transmission after it senses the channel is unoccupied for a period defined as DCF interframe space (DIFS). If the channel is sensed busy either immediately or within the DIFS period, the station keeps on listening on the channel until it is sensed idle for a DIFS period. Consequently, the station generates a backoff interval which is randomly chosen from the backoff window (known as contention window (CW) size)

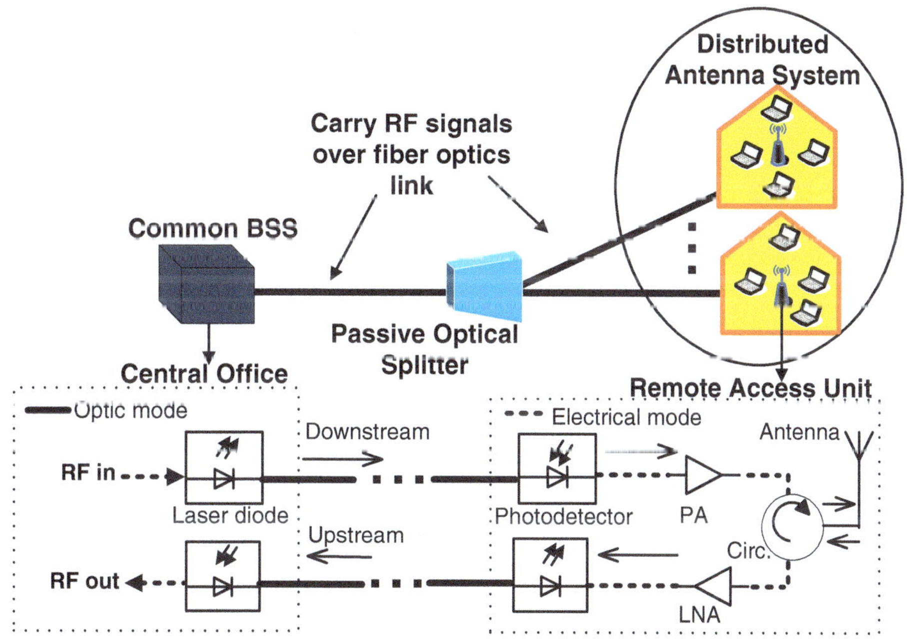

Figure 2 The Fi-Wi network scenario using the RoF technology [39].

[13]. The window size begins with a minimum CW size, and it is doubled at each retransmission up to a maximum CW size. Retransmission takes place whenever there is a packet collision, indicated by the absence of acknowledgment frame (ACK) from the receiver. This backoff algorithm is known as binary exponential backoff (BEB) technique [12].

It is extensively agreed that the BEB algorithm adopted in the MAC protocol for the IEEE 802.11 standard is the key factor to the WLAN performance degradation [13-22]. There are two major drawbacks found in BEB that cause the network degradation as discussed in [17]. First, the contention window is increased upon transmission failure regardless of the cause of failure. Second, after a successful packet transmission, the contention window is reset to the minimum size, thus forgetting its knowledge of the current congestion level in the network and increasing its collision chances.

Numerous modifications have been proposed in literature to improve the legacy BEB algorithm. In [14], the authors proposed to exponentially increase the CW size when there is a collision and exponentially decrease the CW size when there is a successful transmission, known as the EIED (exponentially increases exponentially decreases) algorithm. The exponential factors are then optimized to get maximum throughput. The proponents of [15] use a more conservative approach; they linearly decrease the CW size when there is a successful transmission. Similarly, [20] proposes to halve the CW size (unlike BEB where the CW size is reset to minimum) to increase the overall throughput. The IEEE 802.11e standard changes the minimum and maximum CW limits in the BEB algorithm to define different quality of service access categories [23,24]. In common, all of the above techniques steadily decrease the CW after a successful transmission to retain the overall network state information. Though this approach may have a positive correlation, it usually falls short of predicting the overall network status.

Previous works in [18] and [19] have shown that if exact knowledge (e.g., number of active stations and packet size) of the network is known, the CW size can be tuned to achieve a protocol capacity very close to its theoretical bound. In reality, it is almost impossible for stations to obtain an exact knowledge of the network, only estimates are likely. Listening to the channel to obtain the average packet length and slot transmission probability (alternatively described as slot utility or number of free slots) is the usual starting point [25].

The adaptive window algorithm (AWA) proposed by Bianchi et al. [22] used the number of active stations in the network to control the optimum CW size. A similar approach is employed in [18] and [19] to estimate the number of active stations which is then used in a complex p-persistent MAC protocol to select the optimum backoff interval. In some cases, it is possible to avoid the intermediate step of estimating the number of users [21,25,26].

In common, the above studies focused on the enhancement of the IEEE 802.11 protocol with all stations having an equal chance to transmit. In infrastructure networks, the AP requires more transmission opportunities to give fairness to the uplink/downlink performance [27]. Priority access for an individual user can be obtained by scaling the CW [26], changing the $CW_{\min}$ [28-32], utilizing reduced inter-frame spacings (PIFS and SIFS instead of DIFS) [33,34], and adjusting transmission opportunity (TXOP) limits [31,35]. Such techniques have been proposed for QoS enhancements and to provide uplink/downlink fairness in a single BSS infrastructure network. They are less suitable for an infrastructure WLAN network where there is more than one AP sharing the same spectrum and each serving its own basic service set (BSS) of associated WUs.

In this paper, we employ a novel approach known as transmission priority (TxPriority) scheme by introducing a priority factor that can be adjusted to allow AP transmission priorities as required within an infrastructure WLAN network. Mathematical analysis is carried out, and expressions are derived for optimum AP and WU contention window sizes. The derived formulations show that the optimum CW sizes depend on the number of active APs and WUs; unknown quantities. We capitalize on information from both GPON and WLAN networks to provide appropriate estimates and make the system adaptive. A new convergence function is introduced to improve the reliability of the adaptive system. Our mathematical formulations are verified using network simulator software OPNET 16.1 [36].

The remaining sections of this paper are organized as follows: 'Transmission priority scheme' section describes the transmission priority scheme and derives optimum contention window sizes. 'GPON network indicator' section outlines the GPON architecture and defines an indicator to estimate the number of active APs. 'Adaptive backoff technique' section estimates the number of active WUs and proposes an adaptive system. 'The convergence function' section presents a new convergence function. Finally, 'Conclusions' section concludes the paper and provides directions for future work.

Transmission priority scheme

This section derives optimum contention window sizes for the two types of contending stations, an AP and a WU in the WLAN network. The analysis considers the worst case scenario in terms of frequency availability in densely populated urban areas. All BSSs are within close proximity, i.e., all wireless stations, APs and WUs, share one single channel and can hear each other, albeit being potential interferers. The spectrum is considered 'closed', that is a single 20-MHz channel entirely dedicated to this Fi-Wi network. When a station wants to contend the channel, the backoff counter is randomly selected within the range $(0, \mathrm{CW}-1)$ and then decremented at each subsequent idle slot.

Consider a Fi-Wi network scenario comprising of m integrated ONT-APs and n associate WUs, contending the WLAN channel with contention window sizes of CW_{ap} and CW_{wu}, respectively. The probability that there is at least one transmission in a slot is given by,

$$P_{\mathrm{tr}} = 1-(1-p_{\mathrm{ap}})^m(1-p_{\mathrm{wu}})^n \tag{1}$$

where p_{ap} and p_{wu} are the probability of an AP and a WU transmitting. See Table 1 for a list of notations. The probabilities p_{ap} and p_{wu} are independent to each other and are given by the following, based on Bianchi's model in [22],

$$p_{\mathrm{ap}} = \frac{2}{(\mathrm{CW}_{\mathrm{ap}}+1)} \tag{2}$$

and,

$$p_{\mathrm{wu}} = \frac{2}{(\mathrm{CW}_{\mathrm{wu}}+1)}. \tag{3}$$

Further, the average number of consecutive idle slots per transmission $E[I]$ can be defined as

$$E[I] = \frac{1-P_{\mathrm{tr}}}{P_{\mathrm{tr}}}. \tag{4}$$

The probability that a transmission is a successful AP transmission P_s^{ap} is given by the probability that only one AP transmits on the channel provided that there is a transmission,

$$P_s^{\mathrm{ap}} = \frac{mp_{\mathrm{ap}}(1-p_{\mathrm{ap}})^{m-1}(1-p_{\mathrm{wu}})^n}{P_{\mathrm{tr}}}. \tag{5}$$

Table 1 List of notations

Notation	Description
p_{ap}	Probability of an AP transmitting
p_{wu}	Probability of a WU transmitting
m	Total number of APs (or BSS's)
n	Total number of WUs
CW_{ap}	Current contention window size for an AP
CW_{wu}	Current contention window size for a WU
P_s^{ap}	Probability that a transmission is a successful AP transmission
P_s^{wu}	Probability that a transmission is a successful WU transmission

Similarly, the probability that a transmission is a successful WU transmission is given by

$$P_s^{\mathrm{wu}} = \frac{np_{\mathrm{wu}}(1-p_{\mathrm{ap}})^m(1-p_{\mathrm{wu}})^{n-1}}{P_{\mathrm{tr}}}. \tag{6}$$

A transmission priority factor k can now be introduced. It is defined as the number of successful transmissions of WUs with respect to APs,

$$k = \frac{P_s^{\mathrm{wu}}}{P_s^{\mathrm{ap}}} = \frac{np_{\mathrm{wu}}(1-p_{\mathrm{ap}})}{mp_{\mathrm{ap}}(1-p_{\mathrm{wu}})}. \tag{7}$$

A value of $k < 1$ implies higher downlink throughput; while a value of $k > 1$ represents higher uplink throughput.

Note, successful transmissions from an AP and a WU are mutually exclusive events. Thus, the probability that a transmission is successful is given by

$$P_s = P_s^{\mathrm{ap}} + P_s^{\mathrm{wu}} = (1+k)P_s^{\mathrm{ap}}. \tag{8}$$

The system throughput S is derived from the effective channel utilization. It is defined as the ratio of time taken to successfully transmit payload bits to the sum of the packet transmission time T and the idle slot time $E[I]$,

$$S = P_s\frac{\gamma T}{T+E[I]} \tag{9}$$

where T is measured in slot times,

$$T = T_{\mathrm{Header}} + T_{\mathrm{Payload}} + \mathrm{SIFS} + \mathrm{ACK} + \mathrm{DIFS} \tag{10}$$

and γ is the fraction of T used for payload transmission,

$$\gamma = \frac{T_{\mathrm{Payload}}}{T}. \tag{11}$$

In addition, the uplink throughput S_{wu} and the downlink throughput S_{ap} are expressed as,

$$S_{\mathrm{wu}} = P_s^{\mathrm{wu}}\frac{\gamma T}{T+E[I]} \tag{12}$$

and

$$S_{\mathrm{ap}} = P_s^{\mathrm{ap}}\frac{\gamma T}{T+E[I]}. \tag{13}$$

Using Equation 7, p_{wu} can be substituted in terms of p_{ap}; thus, the system throughput in Equation 9 can be evaluated as follows,

$$S = \frac{(1+k)m\gamma p_{\mathrm{ap}}(1-p_{\mathrm{ap}})^{m+n-1}}{(1+p_{\mathrm{ap}}(\frac{\mathrm{km}}{n}-1))^n + \frac{1-T}{T}(1-p_{\mathrm{ap}})^{m+n}}. \tag{14}$$

Equation 14 corroborates that S depends on p_{ap} which in terms of a function CW_{ap} by Equation 2. An optimum value of CW_{ap} is obtained when S is maximized using the following equation,

$$\frac{dS}{d\mathrm{CW}_{\mathrm{ap}}} = \frac{dS}{dp_{\mathrm{ap}}}\frac{dp_{\mathrm{ap}}}{d\mathrm{CW}_{\mathrm{ap}}} = 0. \tag{15}$$

However, Equation 2 verifies that $\frac{dp_{ap}}{dCW_{ap}} \neq 0$ (except for the lower bound condition $CW_{ap} \rightarrow \infty$). Hence, the solution for Equation 15 is given by $\frac{dS}{dp_{ap}} = 0$; this equates to,

$$0 = \left(\left(1 + \left(\frac{\text{km}}{n} - 1\right) p_{ap}\right)^n + \frac{1-T}{T}(1-p_{ap})^{m+n}\right)(1-(m+n)p_{ap}) - p_{ap}(1-p_{ap})\left(\begin{array}{c} n\left(\frac{\text{km}}{n}-1\right)\left(1+\left(\frac{\text{km}}{n}-1\right)p_{ap}\right)^{n-1} \\ -\frac{1-T}{T}(m+n)(1-p_{ap})^{m+n-1} \end{array}\right). \quad (16)$$

Considering $m + n > n \gg 1$, the following approximation holds for small p_{ap},

$$(1-p_{ap})^{m+n} \approx 1-(m+n)p_{ap} + \frac{(m+n)(m+n-1)}{2} p_{ap}^2. \quad (17)$$

Thus, Equation 16 can be evaluated to give the following solution,

$$p_{ap} \approx \frac{\sqrt{(m+n)^2 + 2Q} - (m+n)}{Q} \quad (18)$$

where

$$Q = \frac{n-1}{n}(\text{km}-n)^2 T + (T-1)(m+n)(m+n-1) + 2T(\text{km}-n)(m+n-1). \quad (19)$$

Q is a negative number, and so Equation 18 becomes invalid when the term under the square root becomes negative, which (after some simplification) occurs when the larger of the two conditions holds

$$n > \max\left[\left((k+1)\sqrt{2T}-1\right)m, 2(T-m)\right]. \quad (20)$$

Now, the optimum CW_{ap} can be approximated from Equation 2,

$$\text{CW}_{ap}^{opt} \approx \frac{2Q}{\sqrt{(m+n)^2 + 2Q} - (m+n)}. \quad (21)$$

Further, Equations 7 and 21 can be used to approximate the optimum value of CW_{wu} as,

$$\text{CW}_{wu}^{opt} \approx \frac{n(\text{CW}_{ap}-1)}{\text{km}} + 2. \quad (22)$$

The derived formulation is verified in Appendix A.

Our proposed scheme aims to keep the CW sizes as close as possible to the optimum values at all the time. Hence, after a successful transmission, the CW sizes of the stations remain constant; they are not reset to a minimum value as in the BEB scheme.

Figure 3 illustrates the system throughput S with respect to the contention window size CW_{wu} for a network with $m = 15$, $n = 60$, $T = 30$, and $\gamma = 0.56$. The latter applies to the simulation parameters of Table 1. The figure further verifies that there exists an optimum contention window size that maximizes throughput. It demonstrates that a higher value of $k (k = 2)$ increases the probability of a WU transmission by reducing the optimum CW_{wu}^{opt} size, hence, increasing the overall uplink throughput. On the other hand, a lower value of $k (k = 0.5)$ reduces the probability of a WU transmission by increasing the optimum CW_{wu}^{opt} size. The figure also shows that the formulated optimum CW_{wu}^{opt} (Equation 22) is almost 20% higher than the actual CW_{wu}^{opt} (for $k = 1$) based on brute force evaluations of Equations 3, 7, and 14. However, the difference has a negligible effect on the throughput with an error of 0.3%.

The system simulations are carried out in OPNET 16.1[a] software using the IEEE 802.11a standard. The simulation parameters are summarized in Table 2.

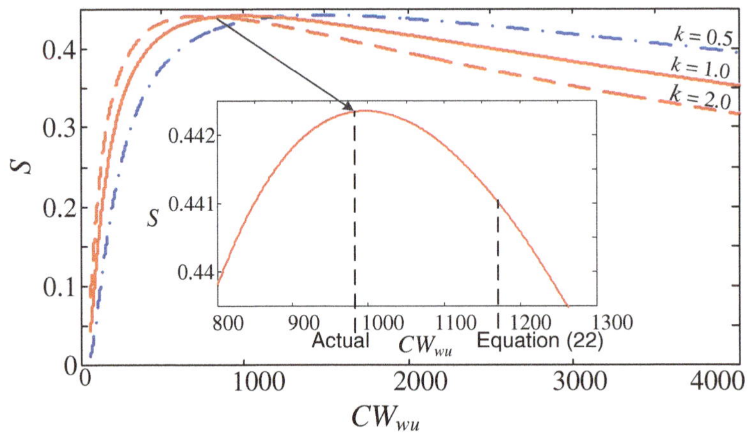

Figure 3 Normalized throughput, S vs CW_{wu} from Equations 3, 7, and 14 with transmission priority factor $k = [0.5, 1, 2]$.

Table 2 OPNET simulation parameters

Parameter	Value
Packet size	8,184 bits
MAC header	224 bits
PHY header	20 μs
ACK length	134 bits/control rate + PHY header
Data rate	54 Mbps
Control rate	6 Mbps
Channel bandwidth	20 MHz
Slot time	9 μs
SIFS time	16 μs
DIFS time	34 μs
ACK timeout	70 μs
Packet transmission time (T)	30 slots
Transmission priority factor (k)	1

Figure 4 compares the uplink and downlink saturation throughputs for the BEB [12] and AWA [22] schemes with the proposed TxPriority scheme having transmission priority factor $k = 1$. The normalized throughput is measured as the total bits/second successfully received by each station over the channel capacity (data rate = 54 Mbps). The traffic load is increased by increasing the number BSSs ranging from 1 to 30; each BSS comprises of one integrated ONT-AP and four associate WUs.

As expected, lower DL throughputs are observed in comparison to UL throughputs for the BEB and AWA schemes since they are both designed for *ad hoc* networks. The plots show that at a high traffic load (i.e., 30 BSS), the DL throughputs for BEB and AWA schemes are 0.06 and 0.09 with the corresponding UL throughputs of 0.25 and 0.35, respectively. These resulted in the ratio between DL and UL throughputs for both schemes to be approximately 0.25, which reflects the simulation scenario of having one AP serving every four WUs. On the other hand, our TxPriority scheme with $k = 1$ has an equal DL and UL throughputs of 0.22. An imbalance between DL and UL throughputs is noted for low BSS numbers due to the binomial approximations used in Equation 17 to derive optimum CW size (Equations 21 and 22). Apart from this imbalance, the TxPriority scheme attains throughputs close to the theoretical values derived in Equations 12 and 13 for uplink and downlink throughputs, respectively. Although the TxPriority scheme brings a balance between the uplink and downlink throughputs, it does not compromise the overall throughput. The throughput matches the theoretical results in Equation 9 and that of the AWA scheme. In both schemes, the throughput remained constant over increasing number of BSSs since the CW sizes are assigned on the number of active stations. The BEB scheme does not consider the number of contending stations; it simply doubles its CW size at every unsuccessful transmission and results in deteriorating throughput performance as the number of BSSs increases.

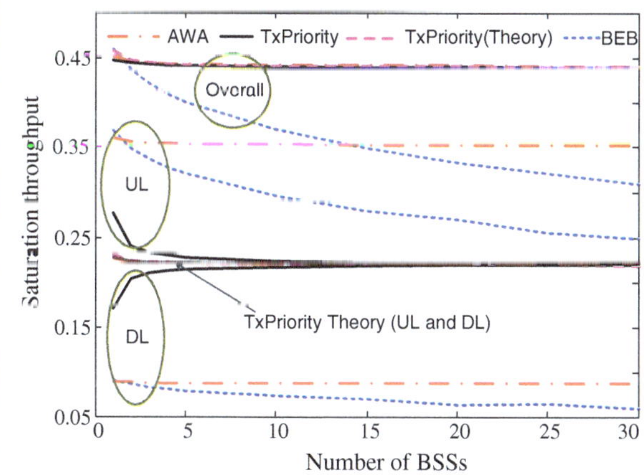

Figure 4 Normalized saturation throughput vs. number of BSSs for BEB, AWA, and TxPriority (simulation and theory) schemes.

Larger CW sizes also result in higher media access delays. Figure 5 shows that the BEB scheme has longer delays than the AWA and TxPriority schemes. The media access delay is measured as the average time that a packet has to wait before its transmission (i.e., the backoff counter decrements to zero). Both BEB and AWA schemes have their DL access delays equal to their UL access delays. In contrast, DL delays in the TxPriority scheme are always less than UL delays because the assigned CW_{ap} size is less than the CW_{wu} size. Nonetheless, the average overall delay of the TxPriority scheme is equal to that of the AWA scheme and considerably better than the exponentially increasing delay of the BEB. The TxPriority scheme aims to maximize the overall throughput and achieve the desired UL/DL ratio k by setting the optimum CW sizes of APs and WUs. The optimum CW sizes depend on four parameters k, T, m, and n. The priority factor k is set globally in accordance to the network demand. The packet transmission time T for each station is known from Equation 10. However, the values of m and n vary with the number of active users in the network. Hence, the scheme could be made adaptive if m and n were periodically estimated and each station updated their optimum CW sizes accordingly. The following section discusses how the estimate for m can be gained with information from the GPON network.

GPON network indicator

GPON employs a time division multiplexing (TDM) for downstream (1,480 to 1,500 nm) and time division multiplexing access (TDMA) for upstream (1,260 to 1,360 nm) transmissions. The downstream traffic is broadcast

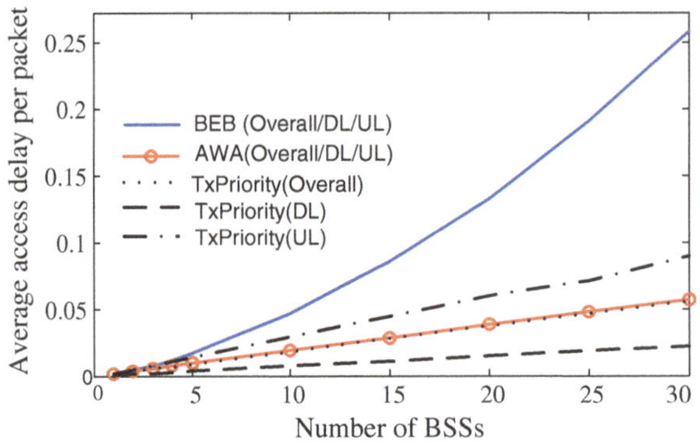

Figure 5 Media access delay per packet transmission vs. number of BSSs for BEB, AWA, and TxPriority schemes.

to all ONTs from OLT, and each frame is labeled with the address of its target ONT.

Figure 6 illustrates the GPON upstream and downstream frame format known as GPON encapsulation method (GEM) as specified in the ITU Standard G 984.3 [37]. The broadcast nature of downstream transmission allows for ONTs to retrieve network traffic information of all ONTs on the same PON. Each Port-ID within the GEM headers is unique to an ONT and is checked by all to identify their corresponding payload. Further, the upstream bandwidth mapping (US BW Map) subfield within the downstream control frame schedules upstream time slot for active ONTs. Hence, the total number of active integrated ONT-APs m can be estimated at the air interface by combining the estimate of active ONTs on the upstream line and downstream line, respectively. It is assumed that the GPON network is primarily used for high speed data services, and an active ONT-AP implies that it is serving at least one associate WU. Since the estimate m is only available to APs and not WUs, each

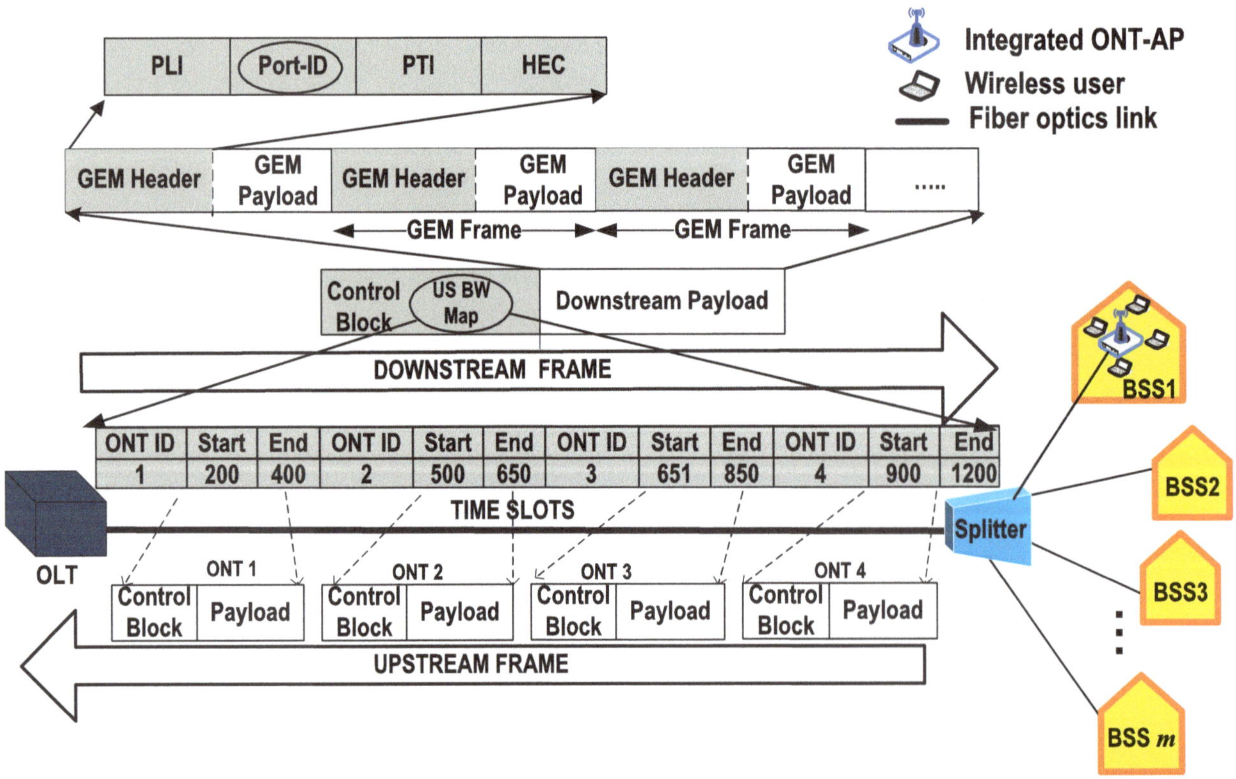

Figure 6 GPON frame format [37,40]. The shaded frames are non-encrypted and can be accessed by all ONTs. The circled subfields are used as indicators in estimating the number of active APs.

AP will periodically broadcast the information to their associate WUs.

The next section proposes a technique for estimating the number of active WUs n, thus, allowing each station to set optimum CW size.

Adaptive backoff technique

The value of n can only be estimated by observing the activity of the channel. The probability that there is a transmission on the channel can be approximated as,

$$\hat{P}_{\mathrm{tr}} = \frac{B}{B+I} \tag{23}$$

where B is the number of times the channel is busy within an observation period (i.e., starting from the time a station first contends the channel until it completes transmission) and I is the total number of idle slots.

Figure 7 illustrates an example of the estimate for station Y. Station Y with a packet to transmit has randomly chosen a backoff counter of 10. The backoff counter is decremented while the channel is sensed idle. When the backoff counter is decremented to 5, the channel is sensed busy due to station X's transmission. As a result, the backoff counter is frozen at 5 until the transmission is complete (i.e., station X completes transmission) and the channel is sensed idle. Station Y waits a further DIFS period before decrementing the backoff counter to 4. Subsequently, the backoff counter is decremented until it reaches zero and station Y transmits. As such, the value of B observed by station Y is 2 (i.e., one time busy due to station X's transmission and the second due to its own transmission) and the value of I is observed to be 9. Hence, the $\hat{P}_{\mathrm{tr}}$ is evaluated to be $2/11$. The stochastic process of selecting the backoff counter makes the estimate $\hat{P}_{\mathrm{tr}}$ prone to error when measured over a single observation time. To reduce the error, the estimate is averaged over ten observation periods and is given by,

$$\bar{P}_{\mathrm{tr}} = \frac{\sum_{i=1}^{10} B_i}{\sum_{i=1}^{10} B_i + \sum_{i=1}^{10} I_i}. \tag{24}$$

Figure 8 compares the $\bar{P}_{\mathrm{tr}}$ with the actual value of P_{tr} from Equation 1 for a scenario of $m = 15$ and $n = 60$. A standard deviation error of 4.4% is noted.

The raw estimate $\hat{n}$ is approximated by employing first-order binomial approximation into Equation 1,

$$P_{\mathrm{tr}} = np_{\mathrm{wu}} + mp_{\mathrm{ap}} - mnp_{\mathrm{ap}}p_{\mathrm{wu}} \tag{25}$$

and substituting using (2) and (3),

$$P_{\mathrm{tr}} = \frac{2n}{\mathrm{CW}_{\mathrm{wu}}+1} + \frac{2m}{\mathrm{CW}_{\mathrm{ap}}+1} - \frac{4mn}{(\mathrm{CW}_{\mathrm{wu}}+1)(\mathrm{CW}_{\mathrm{ap}}+1)}. \tag{26}$$

The algorithm starts here. At time (t), the smooth estimate $\bar{P}_{\mathrm{tr}}$ from Equation 24 is applied to Equation 26 to estimate the number of users, $\hat{n}$, (Equation 27).

$$\hat{n}(t) = \frac{(\overline{\mathrm{CW}}_{\mathrm{wu}}+1)((\overline{\mathrm{CW}}_{\mathrm{ap}})+1)\bar{P}_{\mathrm{tr}}(t)) - 2m)}{2(\overline{\mathrm{CW}}_{\mathrm{ap}}+1-2m)}. \tag{27}$$

Furthermore, a smoother behaving estimate $\bar{n}$ is obtained by exploiting a first-order recursive filter of $\hat{n}$,

$$\bar{n}(t) = \alpha\bar{n}(t-1) + (1-\alpha)\hat{n}(t) \tag{28}$$

where α is a smoothing factor. A value of $\alpha = 0.8$ is chosen since it has been shown in literature [18,22] and [19] as a good compromise between accuracy and precision.

The updated $\bar{n}(t)$ evaluates the new estimate of the $\overline{\mathrm{CW}}_{\mathrm{ap}}(t)$ and $\overline{\mathrm{CW}}_{\mathrm{wu}}(t)$, formulated in Equations 21 and 22 as follows,

$$\overline{\mathrm{CW}}_{\mathrm{ap}}(t) \approx \frac{2\overline{Q}}{\sqrt{(m+\bar{n}(t))^2 + 2\overline{Q}} - (m+\bar{n}(t))} \tag{29}$$

$$\overline{\mathrm{CW}}_{\mathrm{wu}}(t) \approx \frac{\bar{n}(t)(\overline{\mathrm{CW}}_{\mathrm{ap}}(t)-1)}{\mathrm{km}} + 2 \tag{30}$$

where the $\overline{Q}$ is calculated from Equation 19 using $\bar{n}(t)$. The process (Equation 24 to Equation 30) is then repeated with the new $\overline{\mathrm{CW}}(t)$'s for the next time period.

$$\overline{\mathrm{CW}}_{vwu} = \overline{\mathrm{CW}}_{\mathrm{wu}}(t) \tag{31}$$

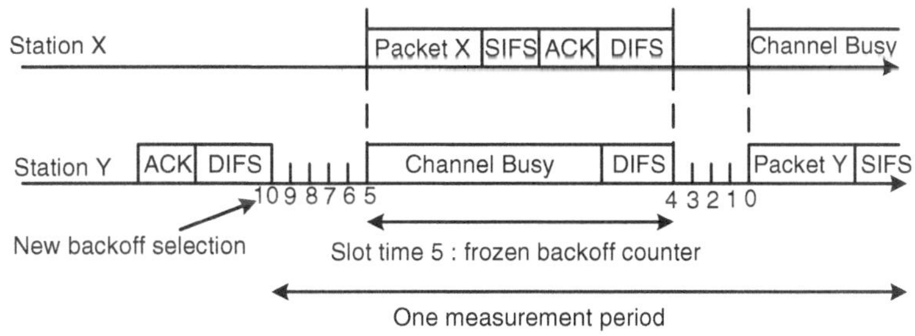

Figure 7 Estimating $\hat{P}_{\mathrm{tr}}$ over one observation period as measured by station Y.

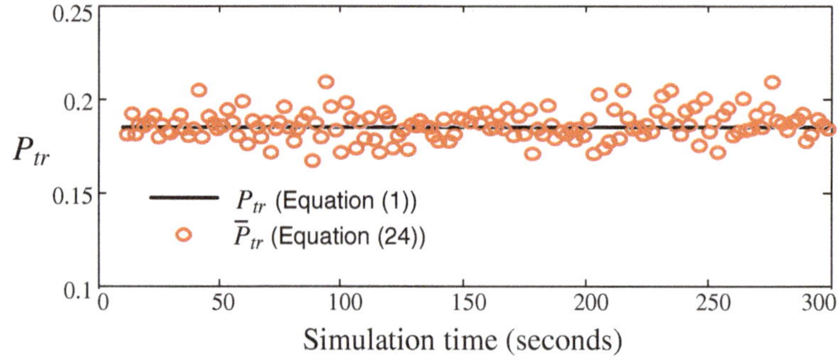

Figure 8 Average P_{tr} values as measured by one user (Equation 24) compared to theoretical P_{tr} values (Equation 1).

and,

$$\overline{\mathrm{CW}}_{\mathrm{ap}} = \overline{\mathrm{CW}}_{\mathrm{ap}}(t). \tag{32}$$

As it stands, the proposed adaptive technique does not converge to the optimum CW due to $\hat{n}$ not reaching the correct value. Figure 9 shows the effect of adaption on one station with all the other stations using the correct value of $n = 60$ and inhibiting their correction algorithms. When the adapting station starts off with an initial estimate above the correct value, ($\bar{n}(t = 0) = 70$), it further diverges until it reaches a stable value of $\bar{n}(t \to \infty) = 200$, far above the optimum value and resulting in a CW much larger than optimum. Similarly when starting with a lower estimate ($\bar{n}(t = 0) = 50$), the estimate drops to close on zero, forcing a small CW. Both situations cause the measured P_{tr} to change and affect the estimates of the other stations. In particular, the low CW causes the adapting station to hog the traffic which gives an impression to other stations that a higher number of users are accessing the channel; this is indicated by their estimates going over the actual value or $n = 60$.

The convergence function

In this section, a new convergence function is added into the computation of CW(t). The convergence function acts

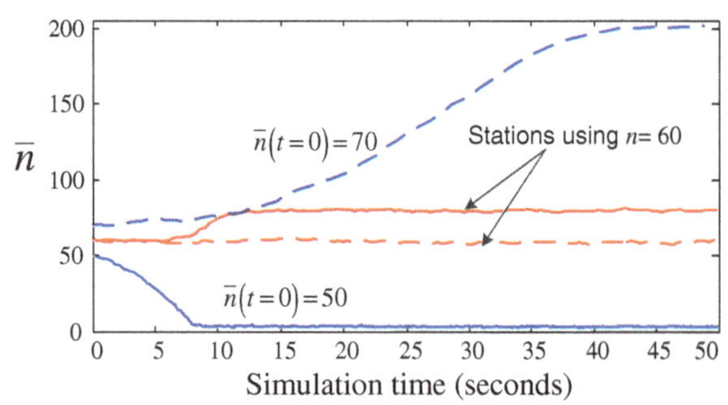

Figure 9 The convergence of $\bar{n}$ for one station. Other stations use ideal ($n = 60$). Initial condition: below $\bar{n}(t = 0) = 50$, solid; above $\bar{n}(t = 0) = 70$, dashed.

as a correcting factor in estimating the $\hat{n}$. In what follows, an analysis is carried out to show that the convergence function for an *ad hoc* network described in [22] is not appropriate for an infrastructure multiple AP scenario.

Equations 31 and 32 are replaced by:

$$\overline{\mathrm{CW}}_{\mathrm{wu}} = c(\bar{n})\overline{\mathrm{CW}}_{\mathrm{wu}}(t) \tag{33}$$

$$\overline{\mathrm{CW}}_{\mathrm{ap}} = c(\bar{n})\overline{\mathrm{CW}}_{\mathrm{ap}}(t) \tag{34}$$

where the convergence function,

$$c(\bar{n}) = \left(1 + \frac{h}{\sqrt{\bar{n}(t)}}\right). \tag{35}$$

Under conditions $CW_{\mathrm{ap}} \gg 1$ and $CW_{\mathrm{wu}} \gg 1$, Equation 27 can be simplified and rewritten as,

$$\hat{n}(t) = \frac{(\overline{\mathrm{CW}}_{\mathrm{wu}})((\overline{\mathrm{CW}}_{\mathrm{ap}})P_{\mathrm{tr}} - 2m)}{2(\overline{\mathrm{CW}}_{\mathrm{ap}} - 2m)}. \tag{36}$$

The $(\bar{\cdot})$ represents the estimate by the adapting station while P_{tr} is the actual value dominated by other stations which we assume operating close to the optimum CW size in which case the assumption that they all have the same p_{ap} or p_{wu} is justified. Substituting P_{tr} in Equation 36 with Equation 26 gives

$$\hat{n} = \frac{\overline{\mathrm{CW}}_{\mathrm{ap}}\overline{\mathrm{CW}}_{\mathrm{wu}}}{\overline{\mathrm{CW}}_{\mathrm{ap}} - 2m}\left(\frac{n}{\mathrm{CW}_{\mathrm{wu}}} + \frac{m}{\mathrm{CW}_{\mathrm{ap}}} - \frac{2mn}{\mathrm{CW}_{\mathrm{ap}}\mathrm{CW}_{\mathrm{wu}}} - \frac{m}{\mathrm{CW}_{\mathrm{ap}}}\right). \tag{37}$$

where n and m are the actual number of wireless users and access points, all (except for the adapting station) operating with contention windows of CW_{wu} and CW_{ap}, respectively. The convergence improves with the positive integer convergence factor, h, in Equation 35. However, Figure 10 indicates that the system throughput gradually decreases as the convergence factor increases. This observation is justified by Equation 33 to 35; increasing the value of h will further increase the CW size beyond

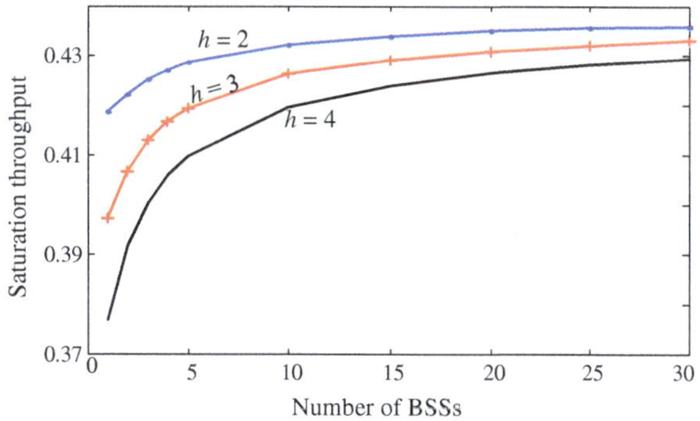

Figure 10 Effect of the convergence factor to the system throughput as the number of BSSs increases.

the optimum value and result in throughput degradation. Therefore, lower values of h are preferred.

An analysis is carried out to investigate the compatibility of the convergence function with the formulated CW size. Figure 11 plots $\hat{n} - \bar{n}$ against $\bar{n}$ as it deviates from the actual value of n for different sizes of m and varying values of h. To achieve effective convergence, $\hat{n} - \bar{n} > 0$ (i.e., $\hat{n} > \bar{n}$) is required when $\bar{n} < n$ and $\hat{n} - \bar{n} < 0$ (i.e., $\hat{n} < \bar{n}$) is required when $\bar{n} > n$. The convergence region is graphically represented by the shaded areas in Figure 11; while the non-shaded areas represent divergence. A stable value for $\bar{n}$ exists when the curve crosses the zero correction line with a negative slope. The steeper the curve the stronger the stability. For example, with no correction factor, $c(\bar{n}) = 1$, the curve is almost completely in the divergence region. If $\bar{n} < n$, then $\bar{n}$ will keep reducing until it gets to 0; or if it is greater that n, it will keep increasing until the curve re-enters the convergence region at a stable but incorrect value of $\bar{n}$. In Figure 11b, the incorrect value is $\bar{n} = 200$, way in excess of the actual value of $n = 60$ but in agreement with the simulations of Figure 9 which use the same values of n, m, and k.

Figure 11a illustrates the results for $m = 5$ having $n = 20$. The line $h = 1$ falls outside the shaded region and will not drive $\bar{n}$ toward the actual value of n. Nonetheless, the lines $h = 2$, $h = 3$, and $h = 4$ all fall within the shaded region and will allow convergence, $h = 2$ being the smallest results in the least throughput degradation and is considered to be the optimum choice. However, in Figure11b,c, for $m = 15$ and $m = 30$, both with the same $n/m = 4$ ratio, the optimum value of h increases to $h = 3$ and $h = 4$ respectively, indicating that the proposed scheme has to adaptively choose the optimum h for a particular size of m, thus, increasing its complexity. This is because the convergence equation defined in Equation 35 does not consider the number of APs (m). Hence, a modified convergence function $c(m, \bar{n})$ is proposed,

$$c(m, \bar{n}) = \left(1 + \frac{h + 2\log_{10}m}{\sqrt{\bar{n}(t)}}\right). \quad (38)$$

Figure 12a,c shows the results of the modified convergence function for the more extreme cases $m = 5$ and $m = 30$. All lines now fall within the shaded regions for

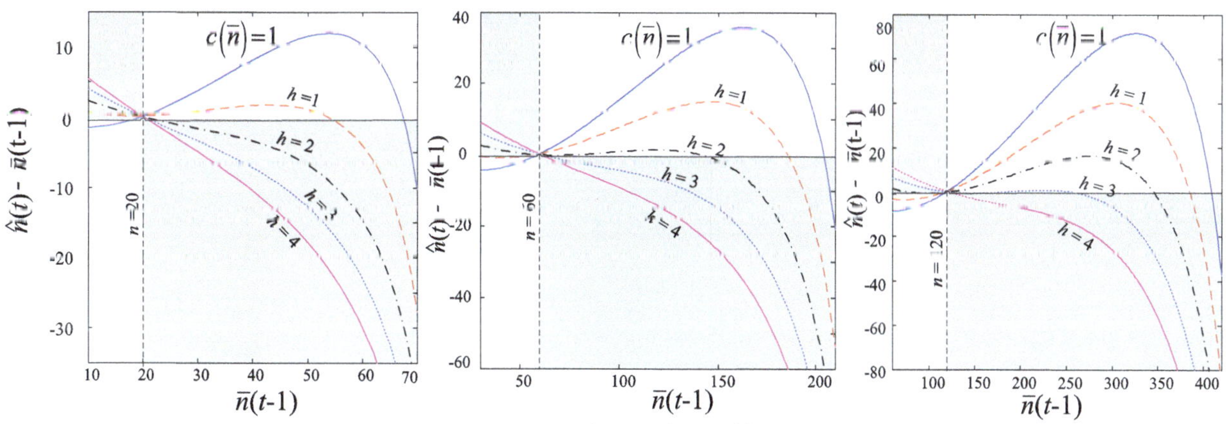

Figure 11 The change in $\bar{n}$ per iteration vs. $\bar{n}$ for different values of h, $\left(c(\bar{n}) = \left(1 + \frac{h}{\sqrt{\bar{n}}}\right)\right)$.

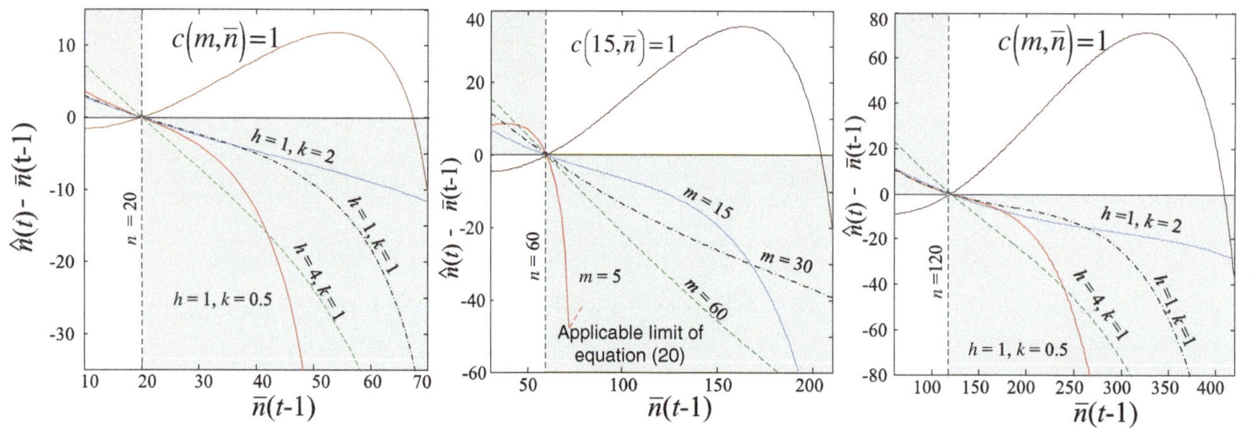

Figure 12 The change in $\bar{n}$ per iteration vs. $\bar{n}$ for different values of h, k, and m, $\left(c(m,\bar{n}) = \left(1 + \frac{h+2\log_{10}m}{\sqrt{\bar{n}}}\right)\right)$.

k between 0.5 and 2. As a result, the constant $h = 1$ can be used as the optimum choice irrespective of the size of m. Figure 12b shows the curves stay in the stable zones as the ratio n/m changes while holding the number of users constant at $n = 60$. The truncation of the $m = 5$ curve for large $\bar{n}$ is due to the invalidity condition (Equation 20). The maximum number of APs simulated was 60, but since the curves appear to be moving into more stable regions of the graph as m increases, the upper value of m for stability is likely to be much higher than this. However, the situation is quite different for n, which is limited by condition (Equation 20). For large m, the maximum value of n is set by the ratio $n/m < ((k+1)\sqrt{2T}-1)$. Using typical values of $k = 1$ and $T = 30$, $n < 14.4m$ contending users, a figure that is unlikely to be exceeded in a GPON residential environment. For small m, the n/m limit improves (increases) since the second term of Equation 20 now dominates.

Finally, the new convergence function can be incorporated into Equations 29 and 30 to give the CW sizes as follows,

$$\overline{\mathrm{CW}}_{\mathrm{ap}} = \left(1 + \frac{1+2\log_{10}m}{\sqrt{\bar{n}}}\right) \frac{2\overline{Q}}{\sqrt{(m+\bar{n})^2 + 2\overline{Q}} - (m+\bar{n})} \tag{39}$$

and,

$$\overline{\mathrm{CW}}_{\mathrm{wu}} = \left(1 + \frac{1+2\log_{10}m}{\sqrt{\bar{n}}}\right) \left(\frac{\bar{n}(\overline{\mathrm{CW}}_{\mathrm{ap}}-1)}{\mathrm{km}} + 2\right). \tag{40}$$

Figure 13 illustrates the $\overline{\mathrm{CW}}_{\mathrm{wu}}$ of one particular station for simulation scenarios of 5 BSSs, 15 BSSs, and 30 BSSs contending the channel. It is observed that all $\overline{\mathrm{CW}}_{\mathrm{wu}}$ converge within 15 s. The $\overline{\mathrm{CW}}_{\mathrm{wu}}$ for 5 BSSs converge almost instantaneously after the simulation starts.

The converged values of $\overline{\mathrm{CW}}_{\mathrm{wu}}$ are larger than the $\mathrm{CW}_{\mathrm{wu}}^{\mathrm{opt}}$ values by, 34%, 28%, and 25% for 5 BSSs, 15 BSSs, and 30 BSSs, respectively; a combined effect of the convergence function and other approximations. The error percentages decrease as the size of BSSs increases due to the fact that the defined convergence function (Equation 38) is a decreasing one. However, as seen in Figure 14, the overall system throughput of the proposed adaptive scheme, called the adaptive TxPriority (ATxPriority), only degrades by 3% compared to the TxPriority scheme.

Further, Figure 15 plots the cumulative distribution function (cdf) of all the WU stations' $\overline{\mathrm{CW}}_{\mathrm{wu}}$ for three network sizes after the networks have converged. In common, the slopes of the obtained cumulative distribution functions are steep, corroborating the small standard deviations, $\sigma_{\overline{\mathrm{CW}}_{\mathrm{wu}}}$, of : 2.3%, 1.1%, and 1.3% (relative to the measured means $\mu_{\overline{\mathrm{CW}}_{\mathrm{wu}}}$) for 5 BSSs, 15 BSSs, and 30

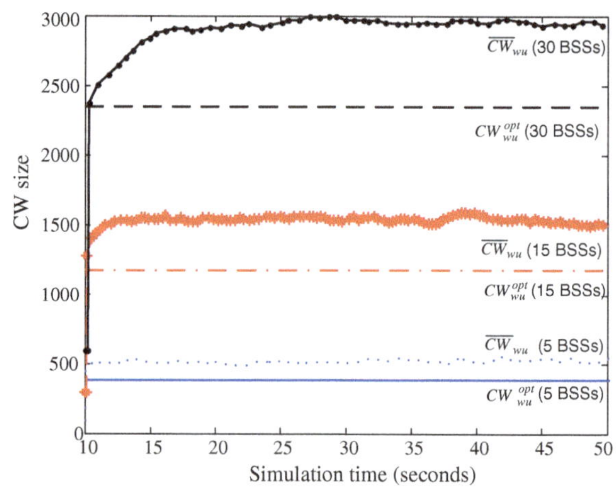

Figure 13 The convergence of $\overline{\mathrm{CW}}_{\mathrm{wu}}$ toward $\mathrm{CW}_{\mathrm{wu}}^{\mathrm{opt}}$ of one station for three sizes of BSS.

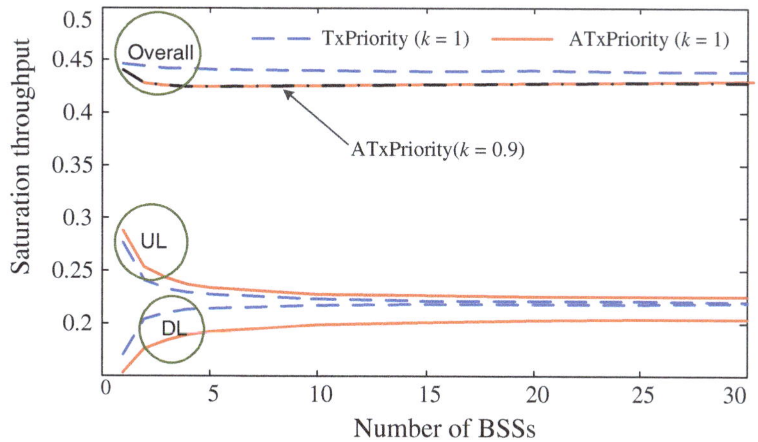

Figure 14 Comparison of normalized saturation throughput vs. number of BSSs between TxPriority and ATxPriority schemes.

BSSs, respectively. This leads to the conclusion that most WU stations within a network have an equal chance of transmission, thus, offering comparable fairness among all WUs.

Figure 16 compares the measured transmission priority factor k_{mea} (UL/DL) of the ATxPriority scheme to the TxPriority scheme. A 10% reduction is observed. This is due to a higher degradation of the downlink throughput experienced by the ATxPriority scheme than its uplink throughput, as shown in Figure 14. Equation 7 shows that the balance between the downlink throughput and the uplink throughput can be recovered by reducing the value of k. As illustrated in Figure 16, a value of $k = 0.9$ brings the measured transmission priority factor $k_{mea} = 1$. Nonetheless, the overall throughput of the ATxPriority remains unaltered (Figure 14, $k = 0.9$).

Conclusions

The paper proposes a transmission priority scheme with an adaptive backoff technique to enhance the WLAN performance of a Fi-Wi (GPON-WLAN) hybrid.The scheme introduces a transmission priority factor k to control the UL/DL fairness. The optimum CW sizes for AP and WU are derived. Performance evaluations show that the scheme is comparable to the adaptive window algorithm (AWA) scheme in terms of overall throughput and delay. Nonetheless, it outperforms the legacy BEB scheme with 40% overall throughput improvement and a maximum 80% of delay reduction. The adaptive backoff technique is introduced. This requires estimates of the number of APs (m) and the number of WUs (n). The first is directly estimated from GPON frame information, and the second is estimated by measuring the activity on the WLAN channel. However, the studied behavior of $\bar{n}$ (estimate of n) indicates that it does not converge to the correct value. Thus, a new convergence function, $c(m, \bar{n})$, is added. Simulations show its robustness for practical values of k, m, and n; but there is a small 3% reduction in overall throughput and a slight offset in k. The latter can be corrected by pre-compensation. The scheme shows

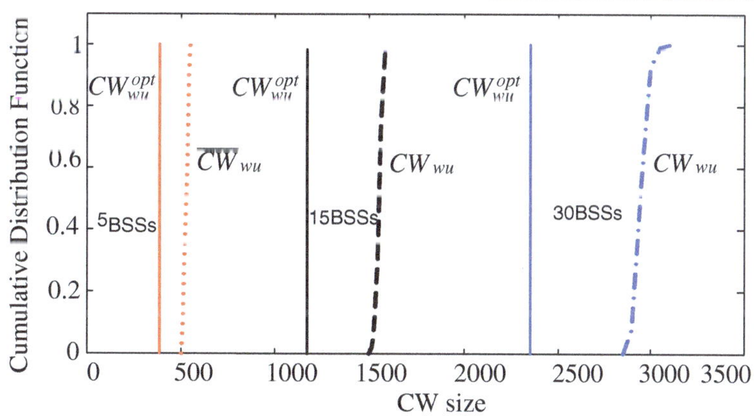

Figure 15 The cumulative distribution function of $\overline{CW}_{WU}$ of contending WUs for three sizes of BSS.

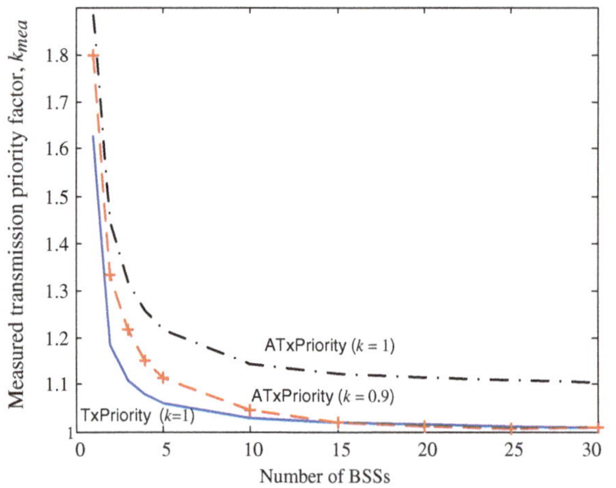

Figure 16 Comparison of the measured transmission ratio factor between TxPriority and ATxPriority schemes for different sizes of BSSs.

uniform convergence among all WUs with a measured standard deviation of their CW sizes being less than 1.5%. It is worth noting that the proposed algorithm is compatible with the whole range of WLAN standards because it is aimed at improving the legacy BEB technique in the DCF protocol adopted by all 802.11 family members [24,38]. The scheme improves the integration of fiber-wireless networks as it ensures each station in the BSS is given a fair transmission opportunity so that the huge bandwidth capacity provided by the GPON (backhaul) can be fully utilized. Our future work will further improve the scheme by utilizing additional information from GPON and removing some of the constraints, for example, allowing neighboring GPON clusters and non-closed spectrum.

Endnote

[a]OPNET software was chosen over other simulation packages because of their vast library of built-in functions.

Appendix A : validation of the formulated optimum CW size

The following formulation extension is carried out to validate the derived formulations with the previous published work in [22]. The formulation can be extended to an *ad hoc* network by letting $p_{\text{ap}} = p_{\text{wu}} = p$, which implies both stations are having equal chance of transmitting. As such, Equation 7 becomes

$$k = \frac{n}{m}. \tag{41}$$

Then, Equation 18 becomes

$$p = \frac{\sqrt{(m+n)^2 + 2(T-1)(m+n)(m+n-1)} - (m+n)}{(T-1)(m+n)(m+n-1)}$$
$$\approx \frac{\sqrt{2}}{(m+n)\sqrt{T}}. \tag{42}$$

Let the total number of stations, $m+n$, equals to N and so the optimum CW can be approximated from Equation 3 as,

$$\mathrm{CW}^{\text{opt}} \approx \sqrt{2TN}. \tag{43}$$

The above derived equation is identical to the optimum contention window size formulated in [22].

Competing interests
The authors declare that they have no competing interests.

Acknowledgements
We acknowledge the support from OPNET Technologies Inc. for providing us the educational version of OPNET 16.1 software. We also acknowledge Prof. Giuseppe Bianchi, for his kindness in providing us further information on proving the stability of an estimate.

References

1. M Milosavljevic, P Kourtessis, A Gliwan, JM Senior, in *Proc. ICTON 2009*, paper Tu.A5.4. Advanced PON topologies with wireless connectivity (Azores, Portugal, 2009)
2. BO Obele, MH Kang, in *Proc. IEEE NGMAST*. Fixed mobile convergence: a self-aware QoS architecture for converging WiMAX and GEPON access networks (Wales, UK, 2008), pp. 411-418
3. G Shen, RS Tucker, CJ Chae, Fixed mobile convergence architectures for broadband access: integration of EPON and WIMAX. IEEE Commun. Mag. **45**(8), 44–50 (2007)
4. N Ghazisaidi, M Martin, M Reisslein, VMP: A MAC protocol for EPON-based video-dominated FiWi access networks. IEEE Trans. Broadcasting. **58.3**(2012), 440–453 (2012)
5. J Wang, K Wu, S Li, C Qiao, *Performance modeling and analysis of multi-path routing in integrated fiber-wireless networks*, (San Diego, USA, 2010), pp. 311-315
6. M Maier, M Herzog, STARGATE: The next evolutionary step toward unleashing the potential of WDM EPONs. IEEE Commun. Mag. **45**(5), 50–56 (2007)
7. F Aurzada, M Lèvesque, M Maier, FiWi access networks based on next-generation PON and gigabit-class WLAN technologies: a capacity and delay analysis. IEEE Trans. Netw. Mag. **22**(4), 1176–1189 (2014)
8. S Bindhaiq, ASM Supa, N Zulkifli, AB Mohammad, RQ Shaddad, MA Elmagzoub, A Faisal, Recent development on time and wavelength-division multiplexed passive optical network (TWDM-PON) for next-generation passive optical network stage 2 (NG-PON2). Optical Switching Netw. **15**(2015), 53–66 (2014)
9. F Beltràn, Fibre-to-the-home, high-speed and national broadband plans: tales from down under. Telecommun. Policy. **38**(8), 715–729 (2014)
10. M Maier, N Ghazisaidi, M Reisslein, in *Proc. AccessNets*. The audacity of fiber-wireless (FiWi) networks (Hong Kong, China, 2009), pp. 16-35
11. HT Win, AK Pathan, On the issues and challenges of fiber-wireless (Fi-Wi) networks. J. Eng. (2013). [Online] Available: http://dx.doi.org/10.1155/2013/645745
12. IEEE 802.11. standard for wireless LAN medium access control (MAC) and physical layer (PHY) specifications (2007)

13. G Bianchi, Performance analysis of the IEEE 802.11 distributed coordination function. IEEE J. Selected Areas Commun. **18**(3), 535–547 (2000)
14. N-O Song, B-J Kwak, J Song, LE Miller, in *Proc. IEEE VTC-Spring*. Enhancement of IEEE 802.11 distributed coordination function with exponential increase exponential decrease backoff algorithm, vol. 4 (Jeju, South Korea, 2003), pp. 2775-2778
15. V Bharghavan, A Demers, S Shenker, L Zhang, in *Proc. SIGCOMM94*. MAWAW: A media access protocol for wireless LANs (London, England, 1994), pp. 212-225
16. WHW Hassan, H King, M Faulkner, in *Proc. 23rd IEEE International Symposium on Personal, Indoor and Mobile Radio Communications (PIMRC 2012)*. Modified backoff technique in fiber-wireless networks (Sydney, Australia, 2012), pp. 431-435
17. I Mouhamad, S Alouf, in *NETWORKING 2006 (Networking Technologies, Services, and Protocols; Performance of Computer and Communication Networks; Mobile and Wireless Communications Systems)*. Design and analysis of an adaptive backoff algorithm for IEEE 802.11 DCF mechanism (Springer Berlin, Heidelberg, 2006), pp. 184-196
18. F Cali, M Conti, E Gregori, IEEE 802.11 protocol: design and performance evaluation of an adaptive backoff mechanism. IEEE J. Selected Areas Commun. **18**(9), 1774–1786 (2000)
19. F Cali, M Conti, E Gregori, Dynamic tuning of the IEEE 802.11 protocol to achieve a theoretical throughput limit. IEEE/ACM Trans. Netw. **8**(6), 785–790 (2000)
20. P Chatzimisions, V Vitasas, A Boucouvalas, M Tsoulfa, Achieving performance enhancement in IEEE 802.11 WLANs by using DIDD backoff mechanism. Int. J. Commun. Syst. **20**, 23–41 (2007)
21. K Hong, S Lee, K Kim, Y Kim, Channel condition based contention window adaptation in IEEE 802.11 WLANs. IEEE Trans. Commun. **60**(2), 469–478 (2012)
22. G Bianchi, L Fratta, M Oliveri, in *Proc. IEEE Personal, Indoor and Mobile Radio Communications (PIMRC'96)*. Performance evaluation and enhancement of the CSMA/CA MAC protocol for 802.11 wireless LANs (Taipei, Taiwan, 1996), pp. 392-396
23. G Cena, L Seno, A Valenzano, C Zunino, On the performance of IEEE 802.11e wireless infrastructures for soft-real-time industrial applications. IEEE Trans. Industrial Inform. **6**(3), 425–437 (2010)
24. S Mangold, S Choi, P May, O Klein, G Hiertz, L Stibor, in *Proc. European Wireless*. IEEE 802.11e wireless LAN for quality of service, vol. 2 (Florence, Italy, 2002), pp. 32-39
25. B Luciano, M Conti, E Gregori, Runtime optimization of IEEE 802.11 wireless LANs performance. IEEE Trans. Parallel Distributed Syst. **15**(1), 66–80 (2004)
26. E Lopez-Aguilera, M Heusse, Y Grunenberger, F Rousseau, A Duda, J Casademont, An asymmetric access point for solving the unfairness problem in WLANs. IEEE Trans. Mobile Comput. **7**(10), 1213–1227 (2008)
27. WHW Hassan, H King, S Ahmed, M Faulkner, in *Proc. IEEE African Continent Conference (AFRICON 2013)*. Modified backoff technique with priority transmission scheme in fiber-wireless networks (Mauritius, 2013)
28. M Al-Hubaishi, T Alahdal, R Alsaqour, A Berqia, M Abdelhaq, O Alsaqour, Enhanced binary exponential backoff algorithm for fair channel access in the IEEE 802.11 medium access control protocol. Int. J. Commun. Syst. **27**(12), 4166–4184 (2013)
29. BAHS Abeysekera, T Matsuda, T Takine, Dynamic contention window control mechanism to achieve fairness between uplink and downlink flows in IEEE 802.11 wireless LANs. IEEE Trans. Wireless Commun. **7**(9), 3517–3525 (2008)
30. Y Gao, L Dai, Optimal downlink/uplink throughput allocation for IEEE 802.11 DCF networks. IEEE Wireless Commun. Lett. **2**(6), 627–630 (2013)
31. WS Lim, DW Kim, YJ Suh, Achieving fairness between uplink and downlink flows in error-prone WLANs. IEEE Commun. Lett. **15**(8), 822–824 (2011)
32. C Wang, Achieving per-flow and weighted fairness for uplink and downlink in IEEE 802.11 WLANs. EURASIP J. Wireless Commun. Netw. **1**(2012), 1–9 (2012)
33. WS Kim, BS Kim, Y Fang, Downlink and uplink resource allocation in IEEE 802.11 wireless LANs. IEEE Trans. Vehic. Technol. **54**(1), 320–327 (2005)
34. F Micò, J Villalón, A Valverde, in *Wireless and Mobile Networking Conference (WMNC)*. Guaranteed access mode for downlink traffic over IEEE 802.11 WLANs (Toulouse, France, 2011)
35. G Min, J Hu, ME Woodward, Modeling and analysis of TXOP differentiation in infrastructure-based WLANs. ELSEVIER Comput. Netw. **55**(11), 2545–2557 (2011)
36. OPNET Modeler 16.1. http:www.opnet.com
37. ITU-T Recommendation G.984.3. Gigabit-capable Passive Optical Networks (GPON): transmission convergence layer specification (March 2008)
38. D Skordoulis, Q Ni, H-H Chen, AP Stephens, C Liu, A Jamalipour, IEEE 802.11n MAC frame aggregation mechanisms for next-generation high-throughput WLANs. IEEE Wireless Commun. **15**(1), 40–47 (2008)
39. J Johny, S Shashidharan, in *Proc. 4th IEEE International Congress on Ultra Modern Telecommunications and Control Systems and Workshops (ICUMT)*. Design and simulation of a Radio Over Fiber system and its performance analysis (St. Petersburg, Russia, 2012), pp. 636-639
40. Huawei Technologies, GPON fundamentals [Online] Available: http://www.huawei.com

6

On the CFO/Channel estimation technique for MIMO-OFDM systems without using a prior knowledge of channel length

Tsui-Tsai Lin[1] and Fuh-Hsin Hwang[2*]

Abstract

In this paper, we propose an estimator which jointly estimates the carrier frequency offset (CFO) and channel state without a prior knowledge of channel length (CL) for a multiple-input multiple-output orthogonal frequency-division multiplexing system. A preliminary CFO/channel estimate is first reached based on the maximum-likelihood criterion. To improve the over-determined problem, a CL estimation technique is further developed. The first-order Taylor series expansion approach is used to determine the trimmed CFO estimate such that the exhaustive point search can be avoided. Simulation results have been presented to demonstrate that the proposed scheme provides a performance comparable to that of the optimal estimation which assumes that CL is known in advance.

Keywords: Multiple-input multiple-output (MIMO); Orthogonal frequency division multiplexing (OFDM); Carrier frequency offset (CFO); Maximum likelihood (ML); Channel length

1 Introduction

The multiple-input multiple-output (MIMO) scheme [1,2] incorporating with the orthogonal frequency division multiplexing (OFDM) [3,4] has attracted much attention of research workers for a long while. MIMO-OFDM [5-7] becomes a key technique in wireless communications because it provides many advantages including high-bandwidth efficiency, superior link quality, simple channel equalization, and resistance to the frequency-selective multipath distortion. In spite of providing these benefits, MIMO-OFDM is liable to performance degradation that is caused by some undesirable effects such as carrier frequency offset (CFO) [8-11]. This is because CFO destroys the orthogonality between subcarriers and thus reduces the detectability of OFDM symbols. In addition, the bit error rate (BER) performance of a MIMO-OFDM system is also in relation to the channel estimation accuracy [9,10]. Consequently, it is an important issue how we can improve the estimation accuracy of the channel for a reliable transmission, especially while CFO exists. To this end, in this paper, we develop an interesting technique which can jointly estimate both CFO and channel state prior to data decoding for a wireless communication system. In the past, numerous algorithms have been proposed to deal with CFO and/or channel estimation for MIMO-OFDM systems. For instance, a number of studies have been presented based on the optimal pilot design and the optimization criteria of interest for channel estimation in MIMO-OFDM systems [12-14]. In [12], a power-efficient training preamble is employed to improve the estimate complexity while the effect of CFO is considered. The Cramer-Roa lower bound (CRLB) is known to provide a lower bound on the statistical variance of any unbias estimator. In [14], the optimal pilot design and placement are determined by minimizing the CRLB in channel estimation. Zeng [15] has proposed a semiblind method for simultaneously estimating the CFO and channel of an uplink multiuser MIMO-OFDM system where a great many data are needed to estimate the subspace. In [16], a technique of joint CFO and channel estimation for uplink MIMO orthogonal frequency division multiple access (MIMO-OFDMA) systems is investigated while time-varying channels are considered. By means of the training sequence, the maximum likelihood (ML) estimator, which asymptotically achieves the CRLB, is proposed

*Correspondence: fuhsin@nknucc.nknu.edu.tw
[2]Department of Optoelectronics and Communication Engineering, National Kaohsiung Normal University, Kaohsiung 82444, Taiwan
Full list of author information is available at the end of the article

in [17]. Another novel ML-based scheme that estimates the aggregate effects of the channel and CFO by using two successive OFDM preambles can be found in [18]. Although the above-mentioned studies [16-18] including the ML estimate approaches [17,18] provide superior performance while taking their predominance into account, all of these approaches require a prior knowledge of channel length (CL), but CL is difficult to be acquired especially when CFO is considered in MIMO-OFDM systems. When CL is not known in advance or under-determined, the ML estimator's performance will not asymptotically approach the CRLB any more because the degrees of freedom are insufficient. It reveals that the knowledge of CL plays a very important role in CFO/channel estimate.

For CL estimation, a number of techniques can be found in the literatures. For example, the Akaike information criterion (AIC) [19] employs numerous data samples to represent the statistical characteristic of the channel/noise for CL estimation. However, AIC encounters over-estimate when signal-to-noise ratio (SNR) is high. The frequency-domain channel length indicator (FCLI) scheme [20] requires only one snapshot channel estimate while a prior information of noise power is assumed to be known, but FCLI becomes impractical as CFO exists. The conditional model order estimation (CME) [21] is proposed for MIMO communication channels. However, CME assumes that the CIR remains unchanged during a data frame and the processing of the block of available observations is performed offline. In [22], a joint estimate approach for channel and instantaneous CL is proposed. Unfortunately, it fails to work properly in the presence of CFO. Based on information criterion, a joint channel and CL estimation is proposed for single-input single-output (SISO) OFDM systems in [23] where the Levinson recursion technique is employed without significant incremental cost. Another CL estimate technique found in [24] is introduced based on the minimization of a combination of cost functions for single-input multiple-output (SIMO) channels where the channel is also assumed to be unchanged during a symbol frame. To the best of the authors' knowledge so far, no work has been devoted to the development of a joint channel and CFO estimation where the information of CL is not available for MIMO-OFDM systems. In light of these works, it is essential to develop an approach to estimate CL especially for MIMO-OFDM systems as CFO exists.

In this paper, we aim to propose a CFO/channel estimation technique while achieving the estimate of CL for a MIMO-OFDM system over Rayleigh fading. The design concept of the proposed scheme is originated from the estimate technique based on the ML criterion. To determine the channel impulse response, a preliminary CFO estimate is first developed, and we define a cost function according to the ML rule by introducing the cyclic prefix (CP) length which is more than CL. By locating the minimum value of the well-defined cost function, the CFO estimate is reached. The frequency-compensated data is then fed into the least square detector so as to compute the channel estimate while the square error function is being minimized. Although the coarse CFO and channel estimation can be obtained, the over-determined CL leads to a performance degradation. To overcome this problem, a novel CL estimation technique is derived by means of truncating the counterfeit signal paths. A fine CFO estimate employing the first-order Taylor series expansion [25] with the trimmed CL is achieved such that we can avoid employing the exhaustive point search approach. Simulation results prove the efficiency of the proposed CFO/channel estimation. Without using prior knowledge of CL, the proposed scheme is shown to provide a performance comparable to that of the optimal ML estimator which assumes a perfect knowledge of CL.

This paper is organized as follows. The data model and brief preview of the ML estimation algorithm are declared in Section 2. Furthermore, the performance analysis in terms of the estimation error of CL is also included in the section. Section 3 introduces the proposed CFO/channel estimation for a MIMO-OFDM system without knowing the CL value. Section 4 presents the computational complexity of the proposed estimator and demonstrates the comparison of complexity between the proposed and ML schemes. In Section 5, simulation results are given to verify the efficiency of the proposed scheme. Finally, Section 6 concludes the paper.

The notations used in this paper are defined as follows. Vectors and matrices are typed with boldface lower and capital letters. $\mathbf{I}_n$ represents an $n \times n$ identity matrix. $\mathbf{O}_{m\times n}$ is an $m \times n$ zero matrix. $\mathbf{1}_n$ and $\mathbf{0}_n$ stand for all-one and all-zero vectors with length n, respectively. $\mathrm{diag}\{a_1, \cdots, a_n\}$ is a diagonal matrix with diagonal entries taking from $a_1, \cdots, a_n$. $(\cdot)^*$, $(\cdot)^T$, and $(\cdot)^H$ are, respectively, the complex conjugate, transpose, and Hermitian transpose of $(\cdot)$. $\otimes$, $\Re\{\cdot\}$, $E\{\cdot\}$, $|\cdot|$, and $\|\cdot\|$ denote the operations taking the Kronecker product, real part, ensemble average, absolute value, and the Frobenius norm, respectively.

2 Data model and ML estimation

2.1 Data model

Let us consider an N-subcarrier MIMO-OFDM system deployed with N_t transmit antennas and N_r receive antennas in a multipath fading channel with length of L. The transmitted data symbols are drawn from a specified constellation set such as quadrature amplitude modulation (QAM) and then are modulated by an OFDM modulator. A CP of length N_g is inserted into the data stream to prevent against the intersymbol interference (ISI) where $N_g \geq L$. At the receiving end, the received symbols are processed by the matched filtering, symbol-rate sampling, and CP removing in sequential order. Let a specific $N \times 1$ received data vector of the jth receive antenna be given by:

$$\begin{aligned}\mathbf{y}_j &= \mathbf{E}(\epsilon_0)\mathbf{F}^H \sum_{i=1}^{N_t} \text{diag}\{\mathbf{x}_i\}\mathbf{H}_{ij} + \mathbf{w}_j \\ &= \mathbf{E}(\epsilon_0)\mathbf{F}^H \left[\text{diag}\{\mathbf{x}_1\}, \cdots, \text{diag}\left\{\mathbf{x}_{N_t}\right\}\right] \left(\mathbf{I}_{N_t} \otimes \mathbf{F}_L\right) \\ &\quad \times \left[\mathbf{h}_{1j}^T, \cdots, \mathbf{h}_{N_t j}^T\right]^T + \mathbf{w}_j \\ &= \mathbf{E}(\epsilon_0)\mathbf{F}^H \mathbf{X}\left(\mathbf{I}_{N_t} \otimes \mathbf{F}_L\right)\mathbf{h}_j + \mathbf{w}_j \\ &= \mathbf{E}(\epsilon_0)\mathbf{A}(L)\mathbf{h}_j + \mathbf{w}_j,\end{aligned} \tag{1}$$

where:

$$\begin{aligned}\mathbf{E}(\epsilon_0) &= \text{diag}\left\{1, e^{j2\pi\epsilon_0/N}, \cdots, e^{j2\pi(N-1)\epsilon_0/N}\right\} \in \mathcal{C}^{N\times N}; \\ \mathbf{h}_j &= \left[\mathbf{h}_{1j}^T, \mathbf{h}_{2j}^T, \cdots, \mathbf{h}_{N_t j}^T\right]^T \in \mathcal{C}^{LN_t}; \\ \mathbf{X} &= \left[\text{diag}\{\mathbf{x}_1\}, \text{diag}\{\mathbf{x}_2\}, \cdots, \text{diag}\left\{\mathbf{x}_{N_t}\right\}\right] \in \mathcal{C}^{N\times NN_t}; \\ \mathbf{A}(L) &= \mathbf{F}^H \mathbf{X}\left(\mathbf{I}_{N_t} \otimes \mathbf{F}_L\right) \in \mathcal{C}^{N\times LN_t}.\end{aligned} \tag{2}$$

The diagonal matrix $\mathbf{E}(\epsilon_0)$ with (i,i)th entry being $e^{j2\pi\epsilon_0(i-1)/N}$ denotes the CFO effect where ϵ_0 is the corresponding frequency offset normalized with respect to the subcarrier spacing. In order to simplify the expressions in (1), the $N \times LN_t$ matrix $\mathbf{A}(L)$ is constructed by placing the $N \times L$ pre-processed data matrices $\mathbf{F}^H\text{diag}\{\mathbf{x}_i\}\mathbf{F}_L$ for $i = 1, 2, \cdots, N_t$ side by side. $\mathbf{x}_i$ is the frequency-domain-transmitted symbol from the ith transmit antenna. $\mathbf{H}_{ij} = \mathbf{F}_L\mathbf{h}_{ij} \in \mathcal{C}^N$ is the channel frequency response where $\mathbf{h}_{ij} \in \mathcal{C}^L$ is the channel response between the ith transmit antenna and jth receive antenna. For simplicity, the path number of $N_{tr} = N_tN_r$ channels is assumed to be the same. The $N \times L$ matrix $\mathbf{F}_L$ is the first L columns of the $N \times N$ discrete Fourier transform (DFT) matrix $\mathbf{F} \in \mathcal{C}^{N\times N}$ with the (m,n)th entry being $[\mathbf{F}]_{mn} = e^{j2\pi mn/N}/\sqrt{N}$. Note that an N-point DFT is expressed as an $N \times N$ matrix multiplication as $\mathbf{X} = \mathbf{F}\mathbf{x}$, where $\mathbf{x}$ is the input time-domain signal, and $\mathbf{X}$ is the DFT of the signal. $\mathbf{w}_j \in \mathcal{C}^N$ is the noise vector of the jth receive antenna. Without loss of generality, we assume that all the entries of $\mathbf{x}_i \in \mathcal{C}^N$ are identically and independently distributed random variables with zero mean and a common variance. The received array data $\mathbf{y} = \left[\mathbf{y}_1^T, \cdots, \mathbf{y}_{N_r}^T\right]^T \in \mathcal{C}^{NN_r}$ can be expressed by:

$$\begin{aligned}\mathbf{y} &= \left[\mathbf{y}_1^T, \mathbf{y}_2^T, \cdots, \mathbf{y}_{N_r}^T\right]^T \\ &= \left[\mathbf{I}_{N_r} \otimes \mathbf{E}(\epsilon_0)\right]\left[\mathbf{I}_{N_r} \otimes \mathbf{A}(L)\right]\left[\mathbf{h}_1^T, \cdots, \mathbf{h}_{N_r}^T\right]^T \\ &\quad + \left[\mathbf{w}_1^T, \cdots, \mathbf{w}_{N_r}^T\right]^T \\ &= \mathcal{E}(\epsilon_0)\mathcal{A}(L)\mathbf{h} + \mathbf{w},\end{aligned} \tag{3}$$

where:

$$\begin{aligned}\mathcal{A}(L) &= \mathbf{I}_{N_r} \otimes \mathbf{A}(L) \in \mathcal{C}^{NN_r\times LN_{tr}}; \\ \mathcal{E}(\epsilon_0) &= \mathbf{I}_{N_r} \otimes \mathbf{E}(\epsilon_0) \in \mathcal{C}^{NN_r\times NN_r}; \\ \mathbf{h} &= \left[\mathbf{h}_1^T, \mathbf{h}_2^T, \cdots, \mathbf{h}_{N_r}^T\right]^T \in \mathcal{C}^{LN_{tr}}; \\ \mathbf{w} &= \left[\mathbf{w}_1^T, \mathbf{w}_2^T, \cdots, \mathbf{w}_{N_r}^T\right]^T \in \mathcal{C}^{NN_r}.\end{aligned} \tag{4}$$

2.2 ML estimation

For the considered system given the true CL L, the ML-based CFO/channel estimation is designed in accordance with a criterion of minimizing the log-likelihood function in terms of the unknown parameters $\mathbf{h}$ and ϵ_0. Specifically, we can solve the following minimization problem to obtain the CFO/channel estimates [17]:

$$\left(\hat{\mathbf{h}}, \hat{\epsilon}_0\right) = \arg\min_{\mathbf{h},\epsilon} \left\|\mathbf{y} - \mathcal{E}(\epsilon)\mathcal{A}(L)\mathbf{h}\right\|^2. \tag{5}$$

Given an ϵ, optimize (5) with respect to $\mathbf{h}$, and we have the ML estimation for $\mathbf{h}$ expressed by:

$$\hat{\mathbf{h}} = \left[\mathcal{A}^H(L)\mathcal{A}(L)\right]^{-1} \mathcal{A}^H(L)\mathcal{E}^H(\epsilon)\mathbf{y} \in \mathcal{C}^{LN_{tr}}. \tag{6}$$

Substituting (6) into (5) and applying the linear algebraic manipulations, we have the cost function:

$$\begin{aligned}C(\epsilon) &= \left\|\mathbf{y} - \mathcal{E}(\epsilon)\mathcal{A}(L)\hat{\mathbf{h}}\right\|^2 \\ &= \mathbf{y}^H\mathcal{E}(\epsilon)\bar{\mathbf{P}}(L)\mathcal{E}^H(\epsilon)\mathbf{y},\end{aligned} \tag{7}$$

where the projection matrix $\bar{\mathbf{P}}(L)$ is given by:

$$\bar{\mathbf{P}}(L) = \mathbf{I}_{NN_r} - \mathcal{A}(L)[\mathcal{A}^H(L)\mathcal{A}(L)]^{-1} \mathcal{A}^H(L) \in \mathcal{C}^{NN_r\times NN_r}. \tag{8}$$

It can be found that the quadratic cost function with respect to ϵ in (7) reaches a minimum value while the condition of $\epsilon = \epsilon_0$ is satisfied [26]. In this way, the CFO estimate can be implemented by searching the minimum of the cost function.

Because the computation of the frequency-dependent matrix $\mathcal{E}(\epsilon)\bar{\mathbf{P}}(L)\mathcal{E}^H(\epsilon)$, which requires $N^2N_r^2$ complex multiplications for each ϵ, is very time-consuming, we rewrite the cost function in (7) as:

$$C(\epsilon) = \mathbf{a}^H(\epsilon)\mathbf{D}_\mathbf{y}^H\bar{\mathbf{P}}(L)\mathbf{D}_\mathbf{y}\mathbf{a}(\epsilon), \tag{9}$$

where $\mathbf{D}_\mathbf{y} = \text{diag}\{\mathbf{y}\}$, and:

$$\mathbf{a}(\epsilon) = \mathbf{1}_{N_r} \otimes \left[1, e^{j2\pi\epsilon/N}, \cdots, e^{j2(N-1)\pi\epsilon/N}\right]^T \in \mathcal{C}^{NN_r}. \tag{10}$$

Then, the CFO estimate is chosen by the following criterion:

$$\begin{aligned}\hat{\epsilon}_0 &= \arg\min_{\epsilon} \ C(\epsilon) \\ &= \arg\min_{\epsilon} \ \mathbf{a}^H(\epsilon)\mathbf{D}_\mathbf{y}^H\bar{\mathbf{P}}(L)\mathbf{D}_\mathbf{y}\mathbf{a}(\epsilon).\end{aligned} \tag{11}$$

This criterion only needs to compute $\mathbf{D}_{\mathbf{y}}^{H}\bar{\mathbf{P}}(L)\mathbf{D}_{\mathbf{y}} \in \mathcal{C}^{NN_r \times NN_r}$ once while executing the CFO estimation, and the implementation complexity is thus greatly reduced. Note that the estimator's performance of the ML-based scheme depends on L. When L is perfectly known, the estimator of (11) reaches the CRLB of the CFO [17].

2.3 Performance analysis

The ML-based estimator has been shown to achieve the CRLB when the channel length is known in advance. Inevitably, the ML approach suffers from serious performance degradation when the CL estimate is erroneous. It is interesting to evaluate the impact of CL estimation error on the MSE performance of the CFO. In this subsection, the characteristic of the estimate error in L is examined. The theoretical analysis of the MSE performance associated with the ML-based estimator will be derived in terms of the CFO. Substituting L with the estimate $\hat{L}$, the cost function in (7) can be rewritten by:

$$C(\epsilon) = \mathbf{y}^{H}\mathcal{E}(\epsilon)\bar{\mathbf{P}}(\hat{L})\mathcal{E}^{H}(\epsilon)\mathbf{y}. \tag{12}$$

The ML estimator reaches the CFO estimate while the cost function is minimized. That is, the CFO estimate is derived by finding the solution for the derivative $dC(\epsilon)/d\epsilon$ being zero. Taking advantage of this feature, we analyze the MSE performance of the CFO in terms of the CL estimate.

Denote the estimated CFO as $\hat{\epsilon}_0 = \epsilon_0 + \Delta$ where Δ is the CFO estimation error and $\Delta \ll 1$. The first derivative of the cost function is approximated by the first-order Taylor series expansion [25] written as:

$$\begin{aligned} \dot{C}(\hat{\epsilon}_0) &= \left.\frac{dC(\epsilon)}{d\epsilon}\right|_{\epsilon=\hat{\epsilon}_0} \\ &\approx \dot{C}(\epsilon_0) + \Delta\ddot{C}(\epsilon_0). \end{aligned} \tag{13}$$

Since $C(\epsilon)$ is expressed by (12), we further rewrite $\dot{C}(\epsilon_0)$ and $\ddot{C}(\epsilon_0)$ by:

$$\begin{aligned} \dot{C}(\epsilon_0) &= 2\Re\left\{\mathbf{y}^{H}\dot{\mathcal{E}}(\epsilon_0)\bar{\mathbf{P}}(\hat{L})\mathcal{E}^{H}(\epsilon_0)\mathbf{y}\right\}; \\ &= 2\Re\left\{\dot{\tilde{\mathbf{y}}}^{H}\bar{\mathbf{P}}(\hat{L})\tilde{\mathbf{y}}\right\}; \\ \ddot{C}(\epsilon_0) &= 2\Re\left\{\mathbf{y}^{H}\ddot{\mathcal{E}}(\epsilon_0)\bar{\mathbf{P}}(\hat{L})\mathcal{E}^{H}(\epsilon_0)\mathbf{y}\right\} \\ &\quad + 2\mathbf{y}^{H}\dot{\mathcal{E}}(\epsilon_0)\bar{\mathbf{P}}(\hat{L})\dot{\mathcal{E}}^{H}(\epsilon_0)\mathbf{y} \\ &= 2\Re\left\{\ddot{\tilde{\mathbf{y}}}^{H}\bar{\mathbf{P}}(\hat{L})\tilde{\mathbf{y}}\right\} + 2\dot{\tilde{\mathbf{y}}}^{H}\bar{\mathbf{P}}(\hat{L})\dot{\tilde{\mathbf{y}}}, \end{aligned} \tag{14}$$

where:

$$\begin{aligned} \tilde{\mathbf{y}} &= \mathcal{E}^{H}(\epsilon_0)\mathbf{y}; \\ \dot{\tilde{\mathbf{y}}} &= \dot{\mathcal{E}}^{H}(\epsilon_0)\mathbf{y}; \\ \ddot{\tilde{\mathbf{y}}} &= \ddot{\mathcal{E}}^{H}(\epsilon_0)\mathbf{y}. \end{aligned} \tag{15}$$

Recall that the CFO estimate $\hat{\epsilon}_0$ is reached while the cost function in (12) is minimized. By finding the solution of $\dot{C}(\hat{\epsilon}_0) = 0$, the estimation error is obtained and expressed by:

$$\begin{aligned} \Delta &= -\frac{\dot{C}(\epsilon_0)}{\ddot{C}(\epsilon_0)} \\ &= -\frac{\Re\left\{\dot{\tilde{\mathbf{y}}}^{H}\bar{\mathbf{P}}(\hat{L})\tilde{\mathbf{y}}\right\}}{\Re\left\{\ddot{\tilde{\mathbf{y}}}^{H}\bar{\mathbf{P}}(\hat{L})\tilde{\mathbf{y}}\right\} + \dot{\tilde{\mathbf{y}}}^{H}\bar{\mathbf{P}}(\hat{L})\dot{\tilde{\mathbf{y}}}}. \end{aligned} \tag{16}$$

To investigate the effect of CL estimation error, the received data are rewritten in terms of CL estimate $\hat{L}$:

$$\mathbf{y} = \mathcal{E}(\epsilon_0)\left[\mathcal{A}(\hat{L})\bar{\mathbf{h}} + \mathbf{u}\right] + \mathbf{w}, \tag{17}$$

where $\mathbf{u} = u\left[L - \hat{L}\right]\left[\mathcal{A}(L)\mathbf{h} - \mathcal{A}(\hat{L})\bar{\mathbf{h}}\right] \in \mathcal{C}^{NN_r}$ is the residual signal term, $u[n]$ is the unit step function, and:

$$\bar{\mathbf{h}} = \begin{cases} \left\{\mathbf{I}_{N_{tr}} \otimes \left[\mathbf{I}_{\hat{L}} \;\; \mathbf{O}_{\hat{L}\times(L-\hat{L})}\right]\right\}\mathbf{h}, & \hat{L} < L \\ \left\{\mathbf{I}_{N_{tr}} \otimes \left[\mathbf{I}_{L} \;\; \mathbf{O}_{L\times(\hat{L}-L)}\right]^{T}\right\}\mathbf{h}, & \hat{L} \geq L \end{cases}. \tag{18}$$

For the over-determined case ($\hat{L} > L$), the residual signal vector $\mathbf{u}$ vanishes ($\mathbf{u} = \mathbf{0}$) and the augmented entries in $\bar{\mathbf{h}}$ are all null. Using (17) in company with the identity $\bar{\mathbf{P}}(\hat{L})\mathcal{E}^{H}(\epsilon_0)\mathbf{y} = \bar{\mathbf{P}}(\hat{L})\mathbf{v}$, (16) results as:

$$\Delta = -\frac{\Re\left\{\left[\dot{\mathcal{E}}_0^{H}\mathcal{A}(\hat{L})\bar{\mathbf{h}} + \dot{\mathbf{v}}\right]^{H}\bar{\mathbf{P}}(\hat{L})\mathbf{v}\right\}}{\Re\left\{\left[\ddot{\mathcal{E}}_0^{H}\mathcal{A}(\hat{L})\bar{\mathbf{h}}+\ddot{\mathbf{v}}\right]^{H}\bar{\mathbf{P}}(\hat{L})\mathbf{v}\right\} + \left[\dot{\mathcal{E}}_0^{H}\mathcal{A}(\hat{L})\bar{\mathbf{h}}+\dot{\mathbf{v}}\right]^{H}\bar{\mathbf{P}}(\hat{L})\left[\dot{\mathcal{E}}_0^{H}\mathcal{A}(\hat{L})\bar{\mathbf{h}}+\dot{\mathbf{v}}\right]}, \tag{19}$$

where $\mathbf{v} = \mathbf{u} + \mathcal{E}^{H}(\epsilon_0)\mathbf{w} \in \mathcal{C}^{NN_r}$, $\dot{\mathbf{v}} = \dot{\mathcal{E}}_0^{H}\mathbf{w}$, and $\ddot{\mathbf{v}} = \ddot{\mathcal{E}}_0^{H}\mathbf{w}$. It is clear that the estimation error in (19) depends on the value of $\hat{L}$ as well as the SNR. For sufficiently high SNR, the noise term is negligible such that the estimation error Δ can be approximated by:

$$\begin{aligned} \Delta &\approx -\frac{\Re\left\{\left[\mathcal{A}(\hat{L})\bar{\mathbf{h}} + \mathbf{u} + \mathcal{E}^{H}(\epsilon_0)\mathbf{w}\right]^{H}\dot{\mathcal{E}}_0\bar{\mathbf{P}}(\hat{L})\left[\mathbf{u} + \mathcal{E}^{H}(\epsilon_0)\mathbf{w}\right]\right\}}{\Re\left\{\left[\mathcal{A}(\hat{L})\bar{\mathbf{h}} + \mathbf{u}\right]^{H}\ddot{\mathcal{E}}_0^{H}\bar{\mathbf{P}}(\hat{L})\mathbf{u}\right\} + \left[\mathcal{A}(\hat{L})\bar{\mathbf{h}} + \mathbf{u}\right]^{H}\dot{\mathcal{E}}_0\bar{\mathbf{P}}(\hat{L})\dot{\mathcal{E}}_0^{H}\left[\mathcal{A}(\hat{L})\bar{\mathbf{h}} + \mathbf{u}\right]} \\ &= -\frac{\Re\left\{\left[\mathcal{A}(\hat{L})\bar{\mathbf{h}} + \mathbf{u}\right]^{H}\dot{\mathcal{E}}_0\bar{\mathbf{P}}(\hat{L})\left[\mathbf{u} + \mathcal{E}^{H}(\epsilon_0)\mathbf{w}\right] + \mathbf{w}^{H}\mathcal{E}(\epsilon_0)\dot{\mathcal{E}}_0\bar{\mathbf{P}}(\hat{L})\left[\mathbf{u} + \mathcal{E}^{H}(\epsilon_0)\mathbf{w}\right]\right\}}{\Re\left\{\left[\mathcal{A}(\hat{L})\bar{\mathbf{h}} + \mathbf{u}\right]^{H}\ddot{\mathcal{E}}_0\bar{\mathbf{P}}(\hat{L})\mathbf{u}\right\} + \left[\mathcal{A}(\hat{L})\bar{\mathbf{h}} + \mathbf{u}\right]^{H}\dot{\mathcal{E}}_0\bar{\mathbf{P}}(\hat{L})\dot{\mathcal{E}}_0^{H}\left[\mathcal{A}(\hat{L})\bar{\mathbf{h}} + \mathbf{u}\right]} \\ &\approx -\frac{\Re\left\{\left[\mathcal{A}(\hat{L})\bar{\mathbf{h}} + \mathbf{u}\right]^{H}\dot{\mathcal{E}}_0\bar{\mathbf{P}}(\hat{L})\left[\mathbf{u} + \mathcal{E}^{H}(\epsilon_0)\mathbf{w}\right]\right\}}{\Re\left\{\left[\mathcal{A}(\hat{L})\bar{\mathbf{h}} + \mathbf{u}\right]^{H}\ddot{\mathcal{E}}_0\bar{\mathbf{P}}(\hat{L})\mathbf{u}\right\} + \left[\mathcal{A}(\hat{L})\bar{\mathbf{h}} + \mathbf{u}\right]^{H}\dot{\mathcal{E}}_0\bar{\mathbf{P}}(\hat{L})\dot{\mathcal{E}}_0^{H}\left[\mathcal{A}(\hat{L})\bar{\mathbf{h}} + \mathbf{u}\right]}. \end{aligned}$$

Here, the performance analysis is presented by considering two distinct conditions including the over-determined and under-determined cases.

2.3.1 Over-determined case

In the case of the correct or over-determined estimation, i.e., $\hat{L} \geq L$, since $\mathbf{u} = \mathbf{0}_{NN_r}$, Δ can be further simplified to the expressions as:

$$\begin{aligned}\Delta &\approx -\frac{\Re\left\{\bar{\mathbf{h}}^H\mathcal{A}^H(\hat{L})\dot{\mathcal{E}}_0\bar{\mathbf{P}}(\hat{L})\mathcal{E}^H(\epsilon_0)\mathbf{w}\right\}}{\bar{\mathbf{h}}^H\mathcal{A}^H(\hat{L})\dot{\mathcal{E}}_0\bar{\mathbf{P}}(\hat{L})\dot{\mathcal{E}}_0^H\mathcal{A}(\hat{L})\bar{\mathbf{h}}} \\ &= -\frac{\Re\left\{\mathbf{h}^H\mathcal{A}^H(L)\dot{\mathcal{E}}_0\bar{\mathbf{P}}(\hat{L})\mathcal{E}^H(\epsilon_0)\mathbf{w}\right\}}{\mathbf{h}^H\mathcal{A}^H(L)\dot{\mathcal{E}}_0\bar{\mathbf{P}}(\hat{L})\dot{\mathcal{E}}_0^H\mathcal{A}(L)\mathbf{h}}.\end{aligned} \tag{20}$$

The second equality holds owing to $\mathcal{A}(\hat{L})\bar{\mathbf{h}} = \mathcal{A}(L)\mathbf{h}$ because (18) guarantees that the augmented channel responses in $\bar{\mathbf{h}}$ are zero. From (20), the MSE performance of the CFO mainly depends on the channel impulse response $\mathbf{h}$ and noise. To examine the noise effect on the MSE, let $\mathbf{h}$ be given and the mean and mean square of the estimation error can be thus derived as:

$$\begin{aligned}E\{\Delta\} &\approx 0; \\ E\left\{|\Delta|^2\right\} &\approx \frac{E\left\{\left|\Re\left\{\mathbf{h}^H\mathcal{A}^H(L)\dot{\mathcal{E}}_0\bar{\mathbf{P}}(L)\mathcal{E}^H(\epsilon_0)\mathbf{w}\right\}\right|^2\right\}}{\left|\mathbf{h}^H\mathcal{A}^H(L)\dot{\mathcal{E}}_0\bar{\mathbf{P}}(\hat{L})\dot{\mathcal{E}}_0^H\mathcal{A}(L)\mathbf{h}\right|^2} \\ &= \frac{\sigma_n^2\mathbf{h}^H\mathcal{A}^H(L)\dot{\mathcal{E}}_0\bar{\mathbf{P}}(L)\dot{\mathcal{E}}_0^H\mathcal{A}(L)\mathbf{h}}{2\left|\mathbf{h}^H\mathcal{A}^H(L)\dot{\mathcal{E}}_0\bar{\mathbf{P}}(\hat{L})\dot{\mathcal{E}}_0^H\mathcal{A}(L)\mathbf{h}\right|^2} \\ &= \frac{\sigma_n^2}{2}\left[\mathbf{h}^H\mathcal{A}^H(L)\dot{\mathcal{E}}_0\bar{\mathbf{P}}(L)\dot{\mathcal{E}}_0^H\mathcal{A}(L)\mathbf{h}\right]^{-1},\end{aligned}$$

respectively. As expected, for $\hat{L} \geq L$, the ML-based scheme is an unbias estimator, and the MSE value of the ML estimation is increased by the noise power.

2.3.2 Under-determined case

If CL estimate is less than the value of L, i.e., $\hat{L} < L$, the residual signal $\mathbf{u}$ exists and cannot be omitted, especially for high SNR. For $\sigma_n^2 \ll 1$, since $\bar{\mathbf{P}}(\hat{L})\mathcal{A}(\hat{L}) = \mathbf{O}_{NN_r\times NN_r}$, the CFO estimation error is simplified to the following expression:

$$\begin{aligned}\Delta &\approx -\frac{\Re\left\{\left[\mathcal{A}(\hat{L})\bar{\mathbf{h}}+\mathbf{u}\right]^H\dot{\mathcal{E}}_0\bar{\mathbf{P}}(\hat{L})\mathbf{u}\right\}}{\Re\left\{\left[\mathcal{A}(\hat{L})\bar{\mathbf{h}}+\mathbf{u}\right]^H\ddot{\mathcal{E}}_0\bar{\mathbf{P}}(\hat{L})\mathbf{u}\right\}+\left[\mathcal{A}(\hat{L})\bar{\mathbf{h}}+\mathbf{u}\right]^H\dot{\mathcal{E}}_0\bar{\mathbf{P}}(\hat{L})\dot{\mathcal{E}}_0^H\left[\mathcal{A}(\hat{L})\bar{\mathbf{h}}+\mathbf{u}\right]} \\ &= -\frac{\Re\left\{\mathbf{h}^H\mathcal{A}^H(L)\dot{\mathcal{E}}_0\bar{\mathbf{P}}(\hat{L})\mathcal{A}(L)\mathbf{h}-\mathbf{h}^H\mathcal{A}^H(L)\dot{\mathcal{E}}_0\bar{\mathbf{P}}(\hat{L})\mathcal{A}(\hat{L})\bar{\mathbf{h}}\right\}}{\Re\left\{\mathbf{h}^H\mathcal{A}^H(L)\ddot{\mathcal{E}}_0\bar{\mathbf{P}}(\hat{L})\mathcal{A}(L)\mathbf{h}-\mathbf{h}^H\mathcal{A}^H(L)\ddot{\mathcal{E}}_0^H\bar{\mathbf{P}}(\hat{L})\mathcal{A}(\hat{L})\bar{\mathbf{h}}\right\}+\mathbf{h}^H\mathcal{A}^H(L)\dot{\mathcal{E}}_0\bar{\mathbf{P}}(\hat{L})\dot{\mathcal{E}}_0^H\mathcal{A}(L)\mathbf{h}} \\ &= -\frac{\Re\left\{\mathbf{h}^H\mathcal{A}^H(L)\dot{\mathcal{E}}_0\bar{\mathbf{P}}(\hat{L})\mathcal{A}(L)\mathbf{h}\right\}}{\Re\left\{\mathbf{h}^H\mathcal{A}^H(L)\ddot{\mathcal{E}}_0\bar{\mathbf{P}}(\hat{L})\mathcal{A}(L)\mathbf{h}\right\}+\mathbf{h}^H\mathcal{A}^H(L)\dot{\mathcal{E}}_0\bar{\mathbf{P}}(\hat{L})\dot{\mathcal{E}}_0^H\mathcal{A}(L)\mathbf{h}}.\end{aligned}$$

The above expression indicates that the estimation error depends on both $\mathbf{h}$ and $\hat{L}$. Given $\mathbf{h}$, the ML estimator becomes bias ($E\{\Delta\} \neq 0$). The ML estimator fails to achieve a reliable performance even when the SNR is increased. This implies that the ML estimation strongly depends on the accuracy of the CL estimation.

3 Proposed joint CFO and channel estimation

It is well known that the ML-based scheme requires the prior knowledge of L to construct $\mathcal{A}(L)$. In practice, the instantaneous value of L is difficult to be reached, and it should be pre-determined prior to the ML estimation. In the case of the over-determined CL, the redundant dimension of the subspace in $\mathbf{P}(L)$ leads to a large noise power, which slightly degrades the estimation performance. With regards to the under-determined CL, the ML-based estimator breaks down because of insufficiency of the degree of freedom for the channel estimation in (6). Therefore, a correct CL is essential for the ML-based estimation. In light of the aforementioned problems, it encourages us to develop a joint CFO and channel estimation, which does not require a prior knowledge of CL. Since the maximum path delay is less than the CP length, N_g, we can regard the CP length as the preliminary estimate of CL to avoid the break down of performance owing to an insufficient degree of freedom. Here, we replace N_g with L in (11), and the coarse CFO estimate is derived by:

$$\hat{\epsilon}_c = \arg\min_{\epsilon}\ \mathbf{a}^H(\epsilon)\mathbf{D}_{\mathbf{y}}^H\bar{\mathbf{P}}(N_g)\mathbf{D}_{\mathbf{y}}\mathbf{a}(\epsilon). \tag{21}$$

Although the coarse CFO can be obtained, the estimator suffers slight performance degradation due to a large noise term introduced by the redundant dimension of $\mathcal{A}(N_g) \in \mathcal{C}^{NN_r\times N_gN_{tr}}$. To alleviate the degradation, one may properly choose an L from the effective channel response, which is constructed by means of truncating the estimated channel response. Based on the coarse CFO estimate in (21), the frequency-compensated data is given by $\mathcal{E}^H(\hat{\epsilon}_c)\mathbf{y}$ and the coarse channel estimate can be determined by the least square error criterion:

$$\min_{\mathbf{h}}\ \left\|\mathcal{E}^H(\hat{\epsilon}_c)\mathbf{y}-\mathcal{A}(N_g)\mathbf{h}\right\|^2, \tag{22}$$

whose solution is derived:

$$\hat{\mathbf{h}} = \left[\mathcal{A}^H(N_g)\mathcal{A}(N_g)\right]^{-1} \mathcal{A}^H(N_g)\mathcal{E}^H(\hat{\epsilon}_c)\mathbf{y} \in \mathcal{C}^{N_g N_{tr}}. \tag{23}$$

Note that there are a few of counterfeit signal paths since $N_g > L$. This implies that a truncated scheme is required where the effective paths are reserved and the other paths are eliminated. Owing to the hardware implementation limitations, the receive antennas are usually spaced closely, and the channel response for each path between a specified transmit antenna and any receive antenna is assumed to be similar. To relieve the noise effect, the effective channel response is obtained by averaging all normalized channel responses of the paths from one specified transmit antenna to all receive antennas. The normalized channel response associated with the qth receive antenna is defined by:

$$\tilde{\mathbf{h}}_q = \left(\mathbf{e}_q^T \otimes \mathbf{I}_{N_g}\right) |\hat{\mathbf{h}}| / \max\left\{\left(\mathbf{e}_q^T \otimes \mathbf{I}_{N_r}\right) |\hat{\mathbf{h}}|\right\} \in \mathcal{C}^{N_g}, \tag{24}$$

where $\mathbf{e}_q$ denotes the qth column of $\mathbf{I}_{N_{tr}}$. The resulting effective channel response is expressed by:

$$\hat{\mathbf{h}}_{eff} = \sum_{q=1}^{N_{tr}} \tilde{\mathbf{h}}_q / \max\left\{\sum_{q=1}^{N_{tr}} \tilde{\mathbf{h}}_q\right\} \in \mathcal{C}^{N_g}. \tag{25}$$

According to (25), the estimated channel length, $\hat{L}$, can be obtained by:

$$\hat{L} = \arg\max_i \ \hat{h}_{eff}(i) > \gamma, \tag{26}$$

where $\hat{h}_{eff}(i)$ denotes the ith element of $\hat{\mathbf{h}}_{eff}$ and γ is a preset threshold. The chosen parameter should be small enough so as to decrease the miss probability and increase the probability of detection while avoiding a high false alarm probability. More detailed discussion about the selection of γ will be given shortly in Section 5.

Replacing N_g with $\hat{L}$ in (21) and using the exhaustive point search, we can obtain the trimming CFO estimate. The computational complexity of the search for CFO has relations with the number of subcarriers, paths, and transmit and receive antennas as well as the searching granularity. The smaller the granularity is, the better the estimation accuracy will be. However, decreasing the granularity will increase the computational complexity, especially when there are a great number of subcarriers. A possible solution is to utilize the first-order Taylor expansion technique associated with the preliminary CFO estimate in (21). In this way, the concave cost function can be transformed into a quadratic function with respect to the normalized CFO, in which the trimming CFO estimate achieves the minimum and can be easily obtained. Denote $\mathbf{a}_0 = \mathbf{a}(0)$ and $\dot{\mathbf{a}}_0 = d\mathbf{a}(\epsilon)/d\epsilon|_{\epsilon=0}$ and we have:

$$\begin{aligned} \mathbf{a}(\hat{\epsilon}_c + \epsilon) &\approx \mathbf{a}(\hat{\epsilon}_c) + \epsilon \times \left.\frac{d\mathbf{a}(\epsilon)}{d\epsilon}\right|_{\epsilon=\hat{\epsilon}_c} \\ &= \mathcal{E}(\hat{\epsilon}_c)\left(\mathbf{a}_0 + \epsilon\dot{\mathbf{a}}_0\right). \end{aligned} \tag{27}$$

Substituting (27) into (9), the cost function can be rewritten by:

$$C(\epsilon) \approx \left[\mathcal{E}(\hat{\epsilon}_c)(\mathbf{a}_0 + \epsilon\dot{\mathbf{a}}_0)\right]^H \mathbf{D}_{\mathbf{y}}^H \bar{\mathbf{P}}(\hat{L})\mathbf{D}_{\mathbf{y}} \times \left[\mathcal{E}(\hat{\epsilon}_c)(\mathbf{a}_0 + \epsilon\dot{\mathbf{a}}_0)\right], \tag{28}$$

which reaches the minimum as the first derivative of (28) equals to zero. Applying some linear algebraic manipulation techniques, the trimming CFO estimate is derived by:

$$\hat{\epsilon}_0 = \hat{\epsilon}_c - \frac{\Re\left\{\mathbf{a}_0^H \mathcal{E}(\hat{\epsilon}_c)\mathbf{D}_{\mathbf{y}}^H \mathbf{P}(\hat{L})\mathbf{D}_{\mathbf{y}}\mathcal{E}^H(\hat{\epsilon}_c)\dot{\mathbf{a}}_0\right\}}{\dot{\mathbf{a}}_0^H \mathcal{E}(\hat{\epsilon}_c)\mathbf{D}_{\mathbf{y}}^H \mathbf{P}(\hat{L})\mathbf{D}_{\mathbf{y}}\mathcal{E}^H(\hat{\epsilon}_c)\dot{\mathbf{a}}_0}. \tag{29}$$

Substituting $\hat{L}$ in (26) and $\hat{\epsilon}_0$ in (29) into (23), we have the trimming estimate of the channel response.

4 Computational complexity

Conveniently enough, the proposed estimator's computational complexity is evaluated by counting the approximate number of total arithmetic operations including addition and multiplication floating-point operations (flops) in the algorithm. Because the computational complexity of the proposed scheme is dominated by the processing in the coarse CFO estimation, the total complexity load can be approximated as the required number of flops for determining the solution of (21). The computation for finding the inverse of the $N_g N_{tr} \times N_g N_{tr}$ block diagonal matrix $\mathcal{A}^H(N_g)\mathcal{A}(N_g)$ in (21) involves [27]:

$$\frac{1}{2}NN_t(1+N_t)\left[1+N_g(1+2N_g)\right] + \frac{4}{3}N_g^3 N_t^3 \tag{30}$$

Table 1 Comparison for the computational complexity between the proposed and ML schemes

Algorithm	Numbers of flops					
	(N, N_g, L)					
	(64,16,1)	(64,16,5)	(128,32,1)	(128,32,5)	(256,64,1)	(256,64,5)
Proposed	1.240E8	1.240E8	4.963E8	4.963E8	1.996E9	1.996E9
ML	1.235E8	1.236E8	4.929E8	4.930E8	1.969E9	1.970E9

$(N_g = N/4, N_t = N_r = 2, \mu = 10^4)$.

flops. In addition, constructing $\bar{\mathbf{P}}(N_g)$ costs about $N\left(1+N_g^2N_t^2+NN_gN_t\right)$ flops, and the calculation of $\Psi = \mathbf{D}_\mathbf{y}^H\bar{\mathbf{P}}(N_g)\mathbf{D}_\mathbf{y}$ approximately requires N^2N_r flops. For every single search loop of ϵ, the computation of $\mathbf{a}^H(\epsilon)\Psi\mathbf{a}(\epsilon)$ demands $N(1+N+NN_r)$ flops. Consequently, the overall flops for the proposed scheme is approximated by:

$$\begin{aligned}\text{Flop}_{\text{Proposed}} \approx\ & \frac{1}{2}NN_t(1+N_t)\left[1+N_g(1+2N_g)\right]+\frac{4}{3}N_g^3N_t^3 \\ & +N\left[1+N_g^2N_t^2+N(N_gN_t+N_r)\right] \\ & +\mu N\left[1+N(1+N_r)\right],\end{aligned} \tag{31}$$

where μ denotes the number of search points. Similarly, the total flop count for the ML approach in (11) is approximated by:

$$\begin{aligned}\text{Flop}_{\text{ML}} \approx\ & \frac{1}{2}NN_t(1+N_t)\left[1+L(1+2L)\right]+\frac{4}{3}L^3N_t^3 \\ & +N\left[1+L^2N_t^2+N(LN_t+N_r)\right] \\ & +\mu N\left[1+N(1+N_r)\right].\end{aligned} \tag{32}$$

As compared with (32), (31) shows that the proposed estimation seems to involve much more computational complexity than that of the ML scheme since $N_g \gg L$. Fortunately, because calculation of $\bar{\mathbf{P}}(N_g)$ is independent of the received data as well as the channel parameters, we can prestore the calculated $\bar{\mathbf{P}}(N_g)$ in the memory so

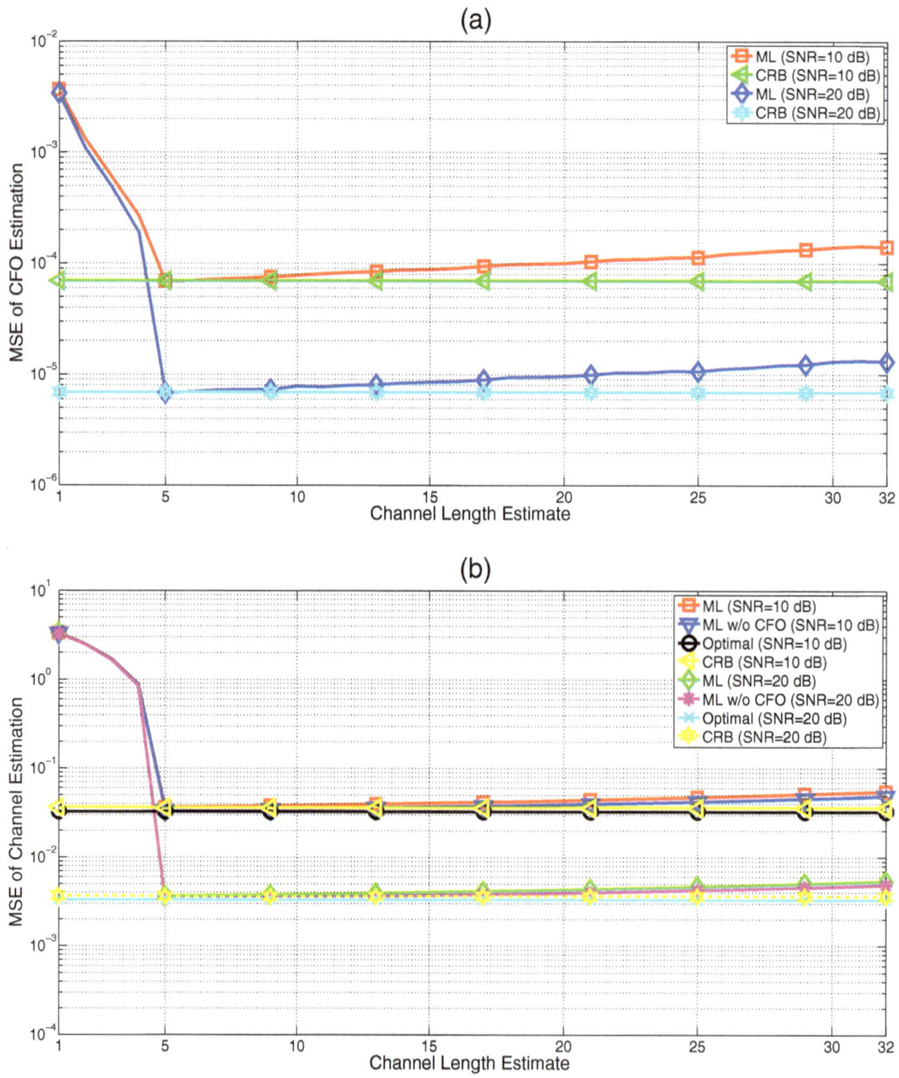

Figure 1 The MSE performance of the proposed estimation versus the channel length estimate $\hat{L}$ for the considered system. (a) MSE of CFO estimation; **(b)** MSE of the channel estimation.

as to avoid the copious calculations. That is, we simply have to consider the count of flops of the computation of Ψ and the search task in the coarse CFO estimation, and the computational complexity for the proposed scheme is thus reduced to:

$$\mathrm{Flop}_{\mathrm{Proposed}} \approx N^2 N_r + \mu N \left[1 + N(1 + N_r)\right]. \qquad (33)$$

Here, we give a numerical example to demonstrate the comparison of complexity between two schemes. Given $N_t = N_r = 2$ and $\mu = 10^4$, Table 1 lists the computational complexity of the proposed and ML approaches in terms of flops for different sets of combination of N, N_g, and L, respectively. For $(N, N_g, L) = (64, 16, 1)$, the flop counts for the proposed and ML schemes are 1.240×10^8 and 1.235×10^8, respectively, and the percentage of the difference in this case is only 0.34%. Furthermore, the maximum difference percentage in the table is under 1.5%. It shows that the proposed scheme operates at the expense of an acceptably increased complexity as compared with the ML approach.

5 Computer simulations

Some simulation results have been conducted to verify the efficacy of the proposed CFO/channel estimator for a 128-subcarrier MIMO-OFDM system ($N = 128$), which is deployed with two transmit/receive antennas in a Rayleigh fading channel with channel length of $L = 5$, where the channel impulse response is normalized such that $E\left\{\|\mathbf{h}_{ij}\|^2\right\} = 1$. Here, the QPSK constellation is used in the data transmission where a CP of length $N/4$ is inserted into the data stream. In the simulation mechanism, the CFO is uniformly and randomly generated from

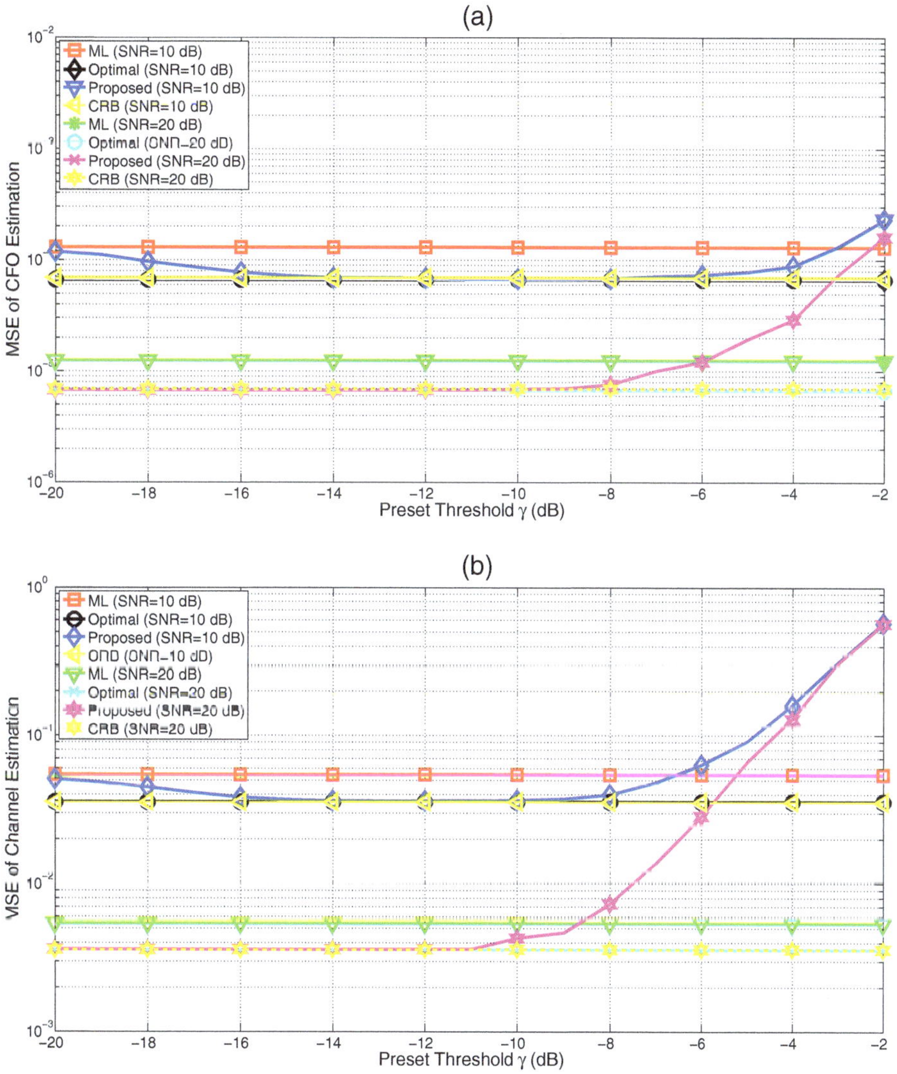

Figure 2 The MSE performance of the proposed estimation versus the preset threshold γ for the considered system. (a) MSE of CFO estimation; **(b)** MSE of the channel estimation.

the interval [−0.5, 0.5]. The SNR in decibel (dB) is defined as $\mathrm{SNR} = 10\log_{10}\left(E\left\{\|\mathcal{A}(L)\mathbf{h}\|^2\right\}/\sigma_n^2\right)$. To access the characteristic of the proposed estimator, the numerical results of the mean square errors (MSEs) of the CFO and channel estimation, which are defined by:

$$\mathrm{MSE}_\epsilon = \frac{1}{Q}\sum_{q=1}^{Q}\|\hat{\epsilon}_0(q) - \epsilon_0\|^2; \qquad \mathrm{MSE}_h = \frac{1}{Q}\sum_{q=1}^{Q}\sum_{j=1}^{N_r}\|\hat{\mathbf{h}}_j(q) - \mathbf{h}_j(q)\|^2, \tag{34}$$

are provided to evaluate the estimator's performance, where Q is the number of independent trials. For comparison, means of the Cramer-Rao bounds (CRBs), which depend on the channel impulse response $\mathbf{h}$, are also included [28,29]:

$$\mathrm{CRB}(\epsilon) = \frac{\sigma_n^2}{2}\left[\mathbf{h}^H\mathcal{A}^H(L)\dot{\mathcal{E}}_0^H\bar{\mathbf{P}}(L)\dot{\mathcal{E}}_0\mathcal{A}(L)\mathbf{h}\right]^{-1} = \gamma\sigma_n^2$$
$$\mathrm{CRB}(\mathbf{h}) = \sigma_n^2\left\{\left[\mathcal{A}^H(L)\mathcal{A}(L)\right]^{-1} + \gamma\mathbf{z}\mathbf{z}^H\right\}, \tag{35}$$

where:

$$\gamma = \frac{1}{2}\left[\mathbf{h}^H\mathcal{A}^H(L)\dot{\mathcal{E}}_0^H\bar{\mathbf{P}}(L)\dot{\mathcal{E}}_0\mathcal{A}(L)\mathbf{h}\right]^{-1}$$
$$\mathbf{z} = \left[\mathcal{A}^H(L)\mathcal{A}(L)\right]^{-1}\mathcal{A}^H(L)\dot{\mathcal{E}}_0\mathcal{A}(L)\mathbf{h}. \tag{36}$$

Throughout the simulations, the granularity of the point search is given by 10^{-4} for the ML-based estimation. Each simulation point is obtained by taking the averages of all

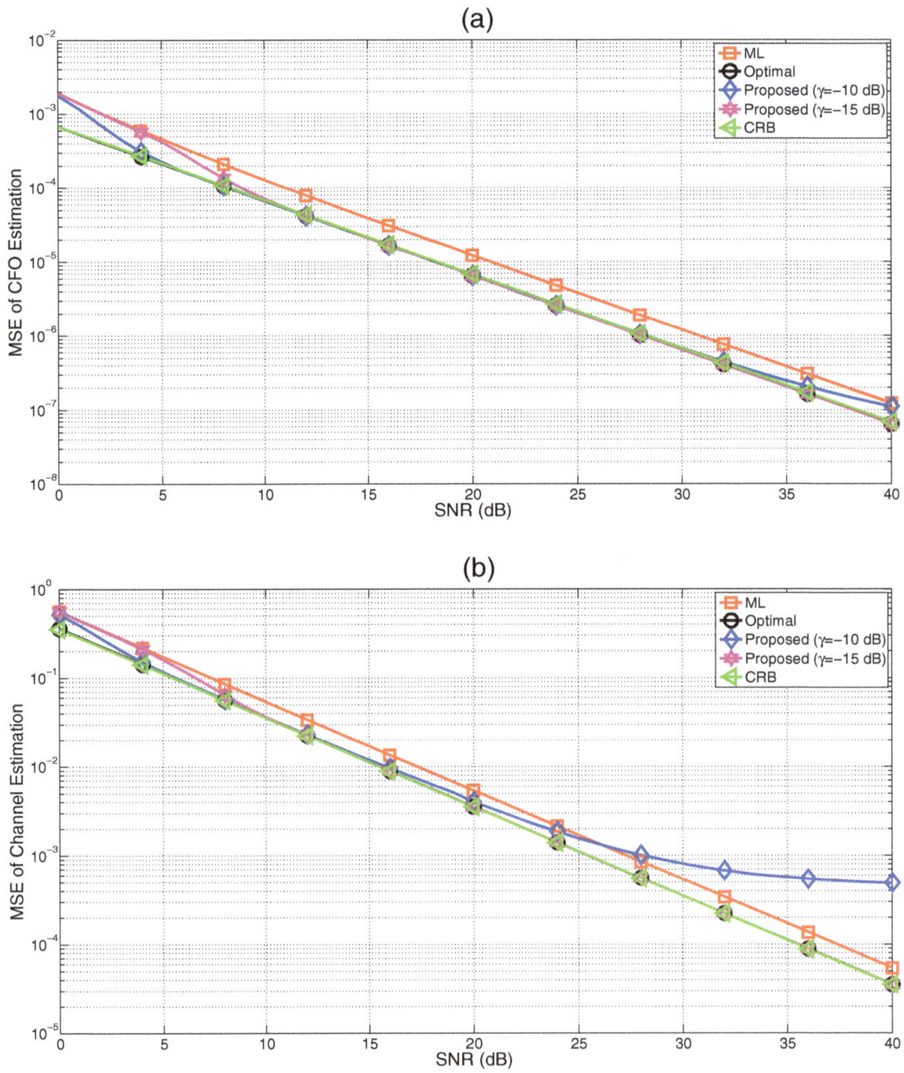

Figure 3 The MSE performance of the proposed estimation versus the input SNR for the considered system. (a) MSE of CFO estimation; **(b)** MSE of the channel estimation.

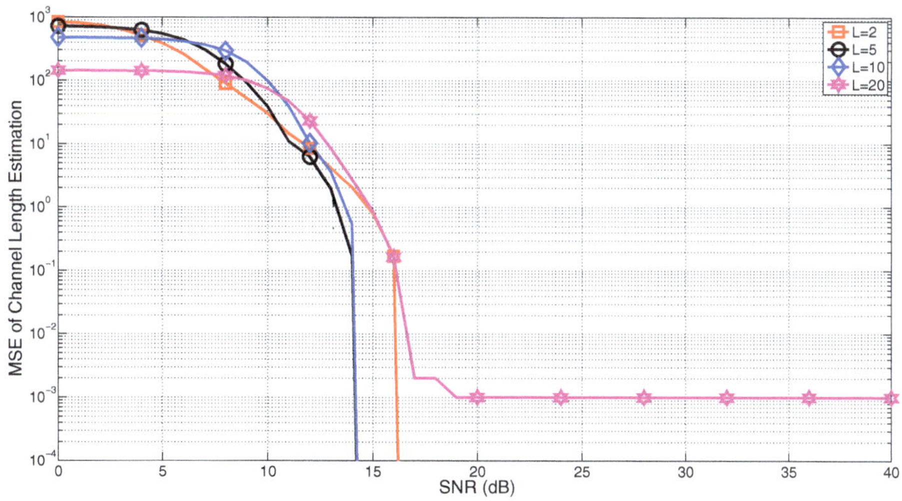

Figure 4 The MSE performance of the proposed CL estimation at different number of paths for the considered system.

independent trials, in which each trial uses a different set of the CFO, channel impulse response, and data/noise sequences.

In Figure 1, we investigate the effect of the estimation error in the channel length with different values of SNR. The MSE of the CFO versus CL estimate is plotted in Figure 1a. It shows that the MSE reaches a minimum value when channel length is correctly used ($\hat{L} = 5$). In regard to the under-determined case, the ML estimator in (11) is not able to provide a reliable performance. This is because the degree of freedom is not sufficiently large for reconstructing the received signal, which results in an erroneous cost function expressed by (7). A reliable CFO estimate cannot be obtained by searching the minimum of the erroneous cost function. For $\hat{L} > L$, the MSE in CFO increases by the value of $\Delta L = \hat{L} - L$. Furthermore, we examine the MSE performance of the channel estimation while varying CL estimate. For comparison, the simulation results associated with the ML estimator without CFO are also included. The corresponding results shown in Figure 1b and Figure 1a have a similar tendency. As expected, the MSE performance is slightly degraded owing to CFO.

In Figure 2, we examine the impact of the preset threshold γ on the estimate performance for SNR = 10 and 20 dB. For comparison, the results obtained with the ML estimator using the CP length and true CL are also included and represented by ML and optimal, respectively. This figure indicates that the proposed estimation approaches the optimal case when γ is sufficiently small

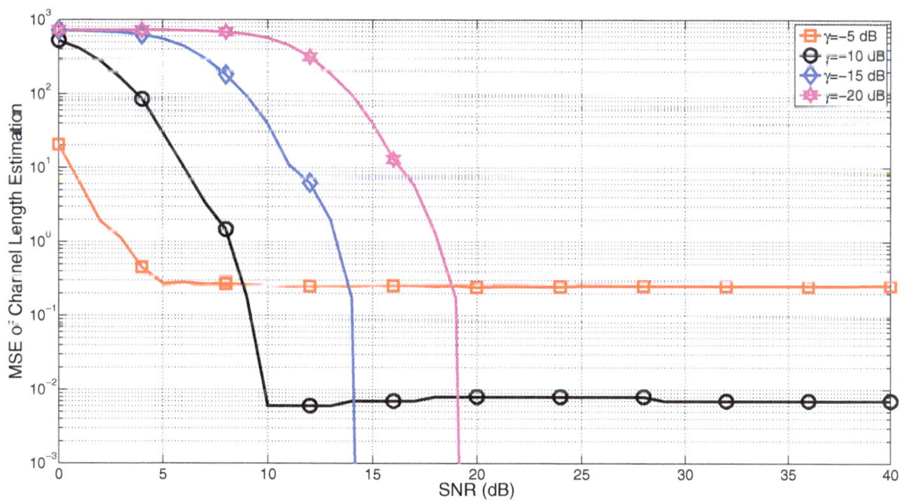

Figure 5 The MSE performance of the proposed CL estimation at different preset threshold, γ, for the considered system.

(-15 dB $< \gamma < -10$ dB). For a large preset threshold, the proposed scheme substantially degrades since the meaningful paths are not successfully detected. There is a large miss probability in the detection of the channel response, which prevents a reliable CFO and channel estimation. It is noteworthy that deep faded paths are filtered by a large preset threshold γ, especially for high SNR. On the contrary, for SNR = 10 dB, a performance degradation occurs as the preset threshold is very small ($\gamma = -20$ dB) owing to a large false alarm in detecting the channel response. It concludes that a proper choice of γ is essential in the presence of the CFO. Since the impact of the performance caused by a path signal which is much less than -10 dB is slight when compared with that caused by the dominant path, the acceptable range of SNR in this case is about $-20 < \gamma < -10$ dB.

In Figure 3, we investigate the impact of the SNR on the proposed CFO/channel estimator with $\gamma = -10$ and -15 dB. From Figure 3a, it can be shown that the MSE decreases as the value of SNR increases. In addition, using the preset threshold $\gamma = -15$ dB, the proposed scheme provides a performance similar to that of the ML estimator with a prior knowledge of channel length. The result verifies the efficiency of the proposed CFO estimation. For low SNR, a large false alarm introduced by noise drops the MSE performance in the CFO estimation. Figure 3b shows the MSE performance of the channel estimation. It is observed that the estimator's performance can be improved by increasing the SNR. The proposed scheme with CL estimation achieves a performance similar to that of the optimal estimation for $\gamma = -15$ dB. However, in the case of $\gamma = -10$ dB, the proposed estimator fails to offer a

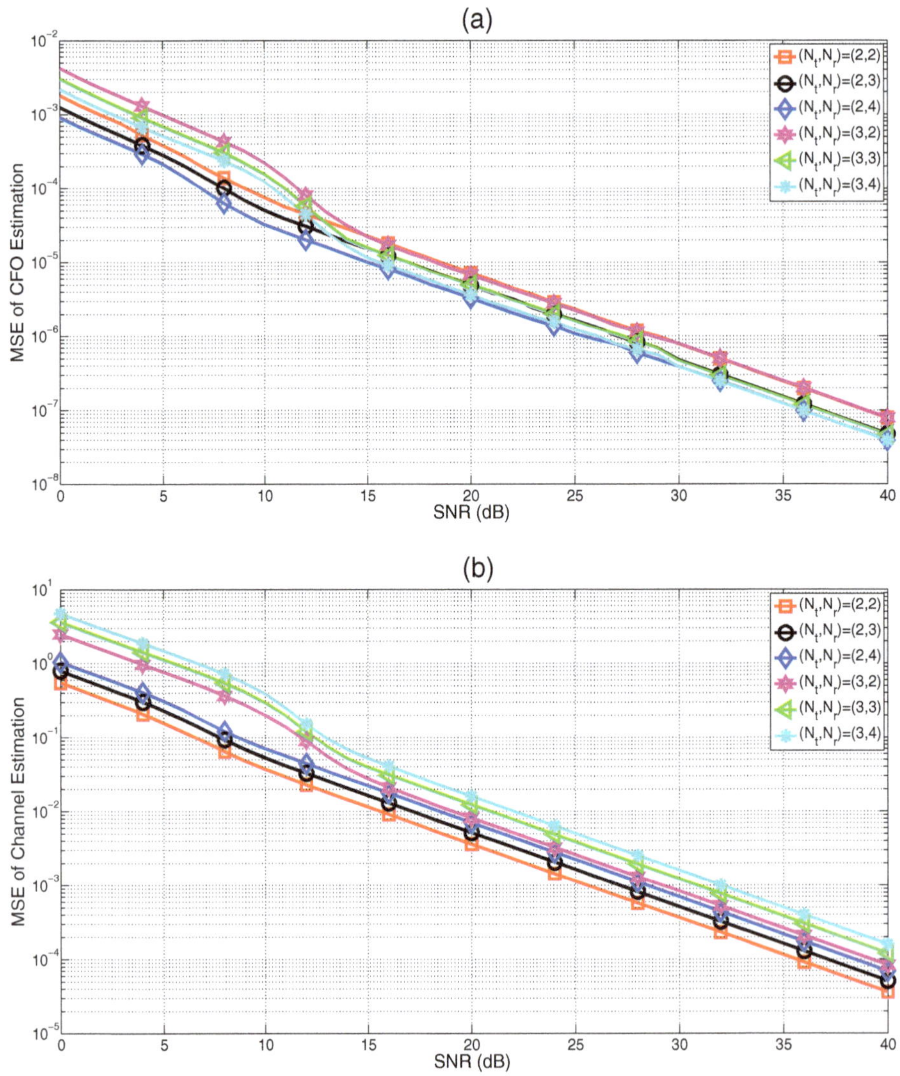

Figure 6 The MSE performance versus the input SNR for the considered system with different numbers of transmit/receive antennas. **(a)** MSE of CFO estimation; **(b)** MSE of the channel estimation.

performance comparable to that of the optimal case while the SNR is getting higher and higher. This is because the proposed CL estimator exhibits under-determining due to a deep fading.

In Figure 4, we present the MSE of CL estimation versus SNR for a set of different CL values while setting the preset threshold $\gamma = -15$ dB. It can be shown that in the figure the MSE of the proposed CL estimation suddenly approaches to zero while the SNR is greater than 15 dB. On the contrary, when the SNR is smaller than 10 dB, large noise power results in over-determination and thus significantly degrades the MSE performance of CL estimation. In a rich multipath environment ($L = 20$), the weak paths introduced by deep fading are removed because they can be treated as noise while the path number is being determined. This results in the under-determined problem.

In Figure 5, we illustrate the impact of the preset threshold γ on CL estimation. Figure 5 shows the MSE versus SNR for different values of γ. Given $\gamma = -15$ or -20 dB, the proposed CL estimator reaches CL perfectly when SNR is greater than 15 dB. On the contrary, when $\gamma > -10$ dB, the miss probability increases and the under-determination is brought about owing to the weak paths caused by deep fading being punctured. For the under-determination cases, there are error floors even when the SNR is increased.

Figure 6a plots the MSE curves of CFO versus SNR for several sets of (N_t, N_r). It shows that the MSE performance of CFO can be improved by increasing the receive diversity as N_t is given. When SNR is low, a little degradation in the MSE performance is induced because fewer degrees of freedom are provided for the minimization problem in (5). Figure 6b shows the MSE performance of the channel estimation. Given N_t, the MSE performance degrades as N_r increases because the larger the size of $\mathbf{h}$ is, the poorer the performance of channel estimation will be.

6 Conclusions

An algorithm for the joint CFO and channel estimation is presented for the MIMO OFDM scenario. As compared with the conventional ML estimate approach, the novelty of the proposed algorithm is that it does not require to know the channel length in advance. Furthermore, we derived a set of beneficial analyses for the MSE of channel length estimate, which can be further applied to the conventional ML estimator. Analytic and simulation results for the proposed estimate technique have been presented where the prior knowledge of channel length is not used because channel length estimate can be reached in our scheme simultaneously. It can be shown that the proposed scheme offers a performance comparable to that of the optimal ML-based estimator which operates well only when perfect channel length is assumed. The feasibility of implementation for the proposed approach is ensured by the presented complexity analyses.

Competing interests

The authors declare that they have no competing interests.

Acknowledgements

This works was supported in part by the Ministry of Science and Technology under Grand MOST 103-2221-E-239-013.

Author details

[1]Department of Electronic Engineering, National United University, Miaoli 36003, Taiwan. [2]Department of Optoelectronics and Communication Engineering, National Kaohsiung Normal University, Kaohsiung 82444, Taiwan.

References

1. O Besson, P Stoica, On parameter estimation of MIMO flat-fading channels with frequency offsets. IEEE Trans. Signal Process. **51**(3), 602–613 (2003)
2. H Sampath, S Talwar, J Tellado, V Erceg, A Paulraj, A fourth generation MIMO-OFDM broadband wireless system design, performance, and field trial results. IEEE Commun. Mag. **40**(9), 143–149 (2002)
3. BHu X Wang, A low-complexity ML estimator for carrier and sampling frequency offsets in OFDM systems. IEEE Commun. Lett. **18**(3), 503–506 (2014)
4. NL Hung, LN Tho, CC Ko, Turbo processing for joint channel estimation, synchronization, and decoding in coded MIMO-OFDM systems. EURASIP Wireless Commun. Netw (2009). doi:10.1155/2009/206524
5. H Minn, M Al-Dhahi, Y Li, Optimal training signals for MIMO OFDM channel estimation in the presence of frequency offset and phase noise systems. IEEE Trans. Commun. **54**(10), 1754–1759 (2006)
6. Y Jin, XG Xia, An interference nulling based channel independent precoding for MIMO-OFDM systems with insufficient cyclic prefix. IEEE Trans. Commun. **61**(1), 131–143 (2012)
7. X Ge, X Huang, Y Wang, M Chen, Q Li, T Han, C-X Wang, Energy-efficiency optimization for MIMO-OFDM mobile multimedia communication systems with QoS constraints. IEEE Trans. Veh. Technol. **63**(5), 2127–2138 (2014)
8. OH Salim, AA Nasir, H Mehrpouyan, W Xiang, S Durrani, RA Kennedy, Channel, phase noise, and frequency offset in OFDM systems: joint estimation, data detection, and hybrid Cramer-Rao lower bound. IEEE Trans. Commun. **62**(9), 3311–3325 (2014)
9. H Wang, XG Xia, Q Yin, Distributed space-frequency codes for cooperative communication systems with multiple carrier frequency offsets. IEEE Trans. Wireless Commun. **8**(2), 1045–1055 (2009)
10. S Rahimi, B Champagne, Joint channel and frequency offset estimation for oversampled perfect reconstruction filter bank transceivers. IEEE Trans. Commun. **62**(6), 2009–2021 (2014)
11. YCWu J Chen, S Ma, TS Ng, Joint CFO and channel estimation for multiuser MIMO-OFDM systems with optimal training sequences. IEEE Trans. Signal Process. **56**(8), 4008–4019 (2008)
12. M Ghogho, A Swami, Training design for multipath channel and frequency-offset estimation in MIMO systems. IEEE Trans. Signal Process. **54**(10), 3957–3965 (2006)
13. Z Zhang, W Zhang, C Tellambura, MIMO-OFDM channel estimation in the presence of frequency offset. IEEE Trans. Wireless Commun. **7**(6), 2329–2339 (2008)
14. M Dong, L Tong, Optimal design and placement of pilots symbols for channel estimation. IEEE Trans. Signal Process. **50**(12), 3055–3069 (2002)
15. Y Zeng, AA Leyman, TS Ng, Joint semiblind frequency offset and channel estimation for multiuser MIMO-OFDM uplink. IEEE Trans. Commun. **58**(9), 2697–2708 (2010)
16. KJ Kim, MO Pun, RA Iltis, Joint carrier frequency offset and channel estimation for uplink MIMO-OFDMA systems using parallel Schmidt Rao-Blackwellized particle filters. IEEE Trans. Commun. **55**(12), 2270–2278 (2007)
17. R Jose, KVS Hari, Maximum likelihood algorithms for joint estimation of synchronisation impairments and channel in multiple input multiple

output-orthogonal frequency division multiplexing system. IET Commun. **7**(15), 1567–1579 (2013)

18. E Manasseh, S Ohno, M Nakamoto, in *Proc. of Asia-Pacific Signal and Information Processing Association Annual Summit and Conference (APSIPA ASC 2011), Oct 18–21*. Preamble based joint channel and CFO estimation for MIMO-OFDM systems with null subcarriers (Xi'an, China, 2011)
19. H Akaike, A new look at the statistical model identification. IEEE Trans. Autom. Control. **AC19**(6), 716–723 (1973)
20. M Krondorf, TJ Liang, R Irmer, G Fettweis, in *Proc. Wireless Conf. April 2–5*. Improved channel estimation for complexity-reduced MIMO-OFDM receiver by estimation of channel impulse response length (Athens Greece, 2006), pp. 1–6
21. DKC So, RS Cheng, Iterative EM receiver for space-time coded systems in MIMO frequency-selective fading channels with channel gain and order estimation. IEEE Trans. Wireless Commun. **3**(6), 1928–1935 (2004)
22. J Siew, J Coon, A Nix, M Beach, S Armour, J McGeehan, in *Proc. Commun. Wireless Syst. Conf. Sept 20–22*. A joint channel and channel length estimation algorithm for MIMO-OFDM (Mauritius, 2004), pp. 285–289
23. A Tomasoni, G Gatti, S Bellini, M Ferrari, M Siti, Efficient OFDM channel estimation via an information criterion. IEEE Trans. Wireless Commun. **12**(3), 1352–1362 (2013)
24. J Via, I Santamaria, J Perez, Effective channel order estimation based on combined identification/equalization. IEEE Trans. Signal Process. **54**(9), 3518–3526 (2006)
25. M Greenberg, *Advanced Engineering Mathematics*, 2nd edn. (Prentice Hall, London, UK, 1997)
26. TC Chen, TT Lin, Blind carrier frequency offset mitigation in space-time block-coded multicarrier code division multiple access uplink transmission. Intl. J. Commun. Syst (2014). doi:10.1002/dac.2823
27. GH Golub, CF Van Loan, *Matrix Computation*, 3rd edn. (Johns Hopkins University Press, Baltimore, MD, 1996)
28. P Stoica, O Besson, Training sequence design for frequency offset and frequency-selective channel estimation. IEEE Trans. Commun. **51**(11), 1910–1917 (2003)
29. S Salari, Ardebilipour, M Ahmadian, Jion for multiple-input multiple-output-orthogonal frequency-division multiplexing systems. IET Commun. **2**(8), 1069–1076 (2008)

A distributed virtual MIMO coalition formation framework for energy efficient wireless networks

Rodrigo A Vaca Ramirez[1*], John S Thompson[2], Eitan Altman[3] and Victor Ramos[4]

Abstract

In this paper, we consider a distributed virtual multiple-input multiple-output (MIMO) coalition formation algorithm. Energy savings are obtained in the reverse link by forming multi-antenna virtual arrays for information transmission. Virtual arrays are formed by finding a stable match between two sets of single antenna devices such as mobile stations (MSs) and relay stations (RSs) based on a game theoretic approach derived from the concept of the college admissions problem. Thus, power savings are obtained through multi-antenna arrays by implementing the concepts of spatial diversity and spatial multiplexing for reverse link transmission. We focus on optimizing the overall consumed power rather than the transmitted power of MSs and RSs. Furthermore, it is shown analytically and by simulation that when the overall consumed power is considered, the energy efficiency of the single antennas devices is not always improved by forming a virtual MIMO array. Hence, single antenna devices may prefer to transmit on their own when channel conditions are favorable. In addition, the simulation results show that the framework we propose provides comparable energy savings and a lower implementation complexity when compared to a centralized exhaustive search approach.

Keywords: Game theory; Energy efficiency; Distributed decision making; MIMO

1 Introduction

Energy consumption has become a major research topic due to the growing energy costs which comes along with the global increase in the number of mobile subscribers. On one hand, the data volume of communication networks is expected to grow by a factor of 10 every 5 years, which brings a doubling of energy consumption over the same time period [1,2]. On the other hand, mobile stations' (MSs) capabilities and operation time are mostly constrained due to their limited battery resources. Most of the power expenditure of MSs in transmission mode is due to the power amplifiers and the signal processing module [3]. Thus, effective solutions allowing MSs to maximize their battery life while optimizing the overall power expenditure rather than only the transmit power are an open research field [1,4-6].

The use of multiple antennas in wireless links has emerged as an effective way to reduce the power consumption in the reverse link. It has been shown in [7] that multi-antenna systems require less transmitted power to achieve the same capacity requirements than single antenna devices. In Long-Term Evolution (LTE), a base station (BS) may support multiple antennas. However, MSs may not be equipped with more than one single antenna due to physical constraints [8,9]. Hence, implementing effective solutions allowing MSs to benefit from the advantages of multi-antenna systems without the extra burden of having multiple antennas physically present at the users' side has become a major issue for current communication systems.

Cooperative communications have recently attracted significant attention as an effective way to improve the performance of wireless networks [10,11]. By the use of cooperative techniques, wireless devices are allowed to share and use the network resources in a more efficient way [6,10,12-16]. As an example, the authors in [14] present a cooperative method to share the network resources and manage interference among femtocells in a distributed manner. Hence, femtocells form coalitions to improve their performance by sharing spectral resources

*Correspondence: rodrigo.vaca@cns.gob.mx
[1]Scientific Division, Federal Police, 01110, Constituyentes 947, Mexico City, Mexico
Full list of author information is available at the end of the article

and maximizing the spatial reuse. In [16], the authors consider the consequences that arise when two multi-antenna systems share the same spectrum band. They demonstrate that if cooperation between the two systems is possible, they may achieve a performance close to the maximum sum-rate.

An important application of cooperative techniques is the formation of virtual multi-antenna arrays. In this context, a number of single antenna devices may cooperate with each other by forming virtual multiple-input multiple-output (MIMO) transmitters or receivers to reap some of the benefits of multi-antenna systems [17]. The theoretical aspects of virtual MIMO have previously been covered in [17,18]. Work on virtual MIMO which considers energy efficiency as an optimization metric can be found in [8,12,19]. The authors in [12,19] illustrate the energy savings obtained when virtual MIMO techniques are used compared with non-cooperative approaches in wireless sensor networks. They argue that at certain ranges from the destination node, cooperative MIMO results in a more energy efficient solution that also reduces the total delay compared with the non-cooperative case. In [8], an approach to optimize the power allocation between transmitter and relay in order to minimize the overall energy per bit consumption is presented. Moreover, it is shown that by using an optimal power allocation, the virtual MIMO case achieves an energy efficiency performance close to the ideal MIMO system.

As mentioned previously, most of the current research in energy efficient virtual MIMO tackles the problem of 'why to cooperate.' Nevertheless, there are two questions that remain unanswered, namely 'when to cooperate' and 'with whom to cooperate.' In this work, we aim to provide an answer for both questions by providing a coalition formation framework that allows single antenna devices to decide with whom to cooperate in order to obtain energy savings in the *reverse link transmission*.

In addition, the implementation of cooperative solutions may face many challenges due to the large scale nature of wireless systems. Cooperation comes along with costs such as power expenditure that may limit or reduce the system's performance. Moreover, if cooperation between the users is regulated by a centralized entity, a significant amount of wireless signaling overhead is required between the users and the network. Furthermore, it is well known that the use of centralized techniques entails extra implementation costs and an increase in system's complexity [1,20,21]. Thus, the design of effective techniques that allow the single antenna devices to *autonomously* decide when and with whom to cooperate with the aim of obtaining power savings in the reverse link is a matter of vital importance in communication systems [22,23]. In this regard, game theory provides a powerful mathematical tool for the design of distributed solutions in cooperative communications [5,10,13,22]. Through the use of coalitional game theory, the authors in [22] propose a merge and split distributed algorithm to form multi-antenna coalitions among single antenna devices. The aim of their work is to maximize the users' rate while accounting for the cost of cooperation in terms of power. The main difference between the merge and split scheme and our proposal is that splitting may involve finding all the possible partitions of the set formed by the users in a coalition, which increases significantly the complexity of the method when compared to the scheme presented in this paper. The authors in [24] propose a relay selection method to improve the spatial diversity in the system by using the concept of amplify and forward. They model the cooperation in wireless networks as a Stackelberg game. Moreover, they consider as optimization parameters the transmitted power and the user rate. The proposal is quite novel and interesting; however, they take into account the transmitted power rather than the overall power consumption. In addition, the communication overhead and the complexity of the method due to the use of the Stackelberg game may add an extra limitation to the system. In [5], we propose an energy-efficient solution for virtual MIMO coalition formation, where cooperation is modeled as a game theoretical approach derived for the concept of stable marriage with incomplete lists. An optimal relay is selected to jointly minimize the reverse link power expenditure. Furthermore, we show that the communication overhead may be significantly reduced by using distributed techniques. Nevertheless, a major drawback of the framework proposed in [5] is that the number of elements able to join the coalition is constrained to a limited number, typically two.

The main contributions of this paper are as follows: (1) to provide a distributed low-complexity virtual MIMO coalition formation algorithm to reduce the energy consumption in the reverse link; (2) our solution can support any number of transmitters participating in the coalitions; (3) we focus on enhancing the MS performance by forming virtual coalitions with the relay stations (RSs). Moreover, we optimize the overall power consumption of MSs and RSs for reverse link transmission; (4) we analyze our proposal from both diversity and capacity perspectives; and (5) our proposal focuses on reducing the overall device consumed power rather than the transmitter radio frequency (RF) power, thus we take into account the power consumption of the RF components such as the power amplifiers and the baseband (BB) module.

The rest of the paper is structured as follows: Section 2 describes the problem scenario, Section 3 presents our power consumption model and performance measures. In

Section 4, our cooperative framework is shown. Moreover, in Section 5, we present a theoretical analysis of the consequences arising when optimizing the overall consumed power rather than the transmitted power when implementing spatial diversity and spatial multiplexing in multi-antenna systems. A summary of the comparison schemes and our simulation scenario is described in Section 6. Simulation results are presented in Section 7. Finally, Section 8 offers concluding remarks.

2 System scenario

In this section, the scenario adopted in this paper is described. We consider a system with N single antenna MSs transmitting data to a multi-antenna base station. In addition, R single antenna RSs are uniformly distributed through the cell. We assume $R \gg N$, since a bigger number of RSs than MSs provides better chances of having an increase in power savings due to the gains in capacity and diversity that can be obtained through the formation of virtual arrays. If the number of RSs is equal or less than the number of MSs, the performance of the system will get close to a single-input multiple-output (SIMO) system since the odds to find a suitable RS to cooperate may decrease. Moreover, previous work as the one presented in [25] suggests that RSs can be also solar powered, thus scenarios where a significant amount of RSs is deployed can be used as a suitable option to reduce power consumption.

Regarding power computations, we consider the overall power expenditure of MSs and RSs for coalition formation when forming virtual arrays. In order to improve the user's performance, single antenna devices (MSs and RS) are allowed to cooperate by forming $M_t \times M_r$ virtual MIMO coalitions, where M_r is the number of antennas at the BS, and M_t is the number of single antenna devices forming a virtual MIMO link. If cooperation is not feasible, MSs will prefer to transmit on their own to the BS in SIMO mode.

In Figure 1, the orthogonal frequency division multiple access (OFDMA) scenario is shown. The system bandwidth B (Hz) is divided into X resource blocks (RBs). In order to avoid mutual interference, an RB is assigned to each user independently. An RB defines the basic time-frequency unit with bandwidth $B_{RB} = B/X$ (Hz).

2.1 Virtual MIMO link

Figure 2 shows a virtual $M_t \times M_r$ MIMO link implementing spatial multiplexing. At the first time slot, the MS forwards the information vector $\mathbf{s}$ to its peers by using the cooperative link. In the subsequent slot, the MS and RSs will transmit the information vector $\mathbf{s}$ at the reverse link through the MIMO channel $\mathbf{H}$. We use decode-and-forward which is an obvious method for multiple transmitters cooperating to send data to a multiple antenna receiver as previously shown in [26]. In addition, to avoid mutual interference, the reverse and the cooperative link should be designed to be orthogonal to each other. When spatial multiplexing is implemented, we assume that the cooperative link has sufficient bandwidth for information transmission, thus cooperation is supported without any major interference issue. Hence, MSs can transmit their signal vector $\mathbf{s}$ to the cooperating peers and they can demultiplex it into independent information streams for simultaneous transmission in the next time slot. This adds an extra delay in the end-to-end transmission. Nevertheless, this is a common limitation of virtual MIMO cooperative approaches. A similar representation can be used for the spatial diversity concept by replacing the vector $\mathbf{s}$ with the information symbol s. Thereby, all antennas involved in the coalition transmit the same symbol s in the reverse link.

2.2 Cooperative link

For the single antenna devices (MSs and RSs) to cooperate among each other, the setup and maintenance of a cooperative link is required. The cooperative link is

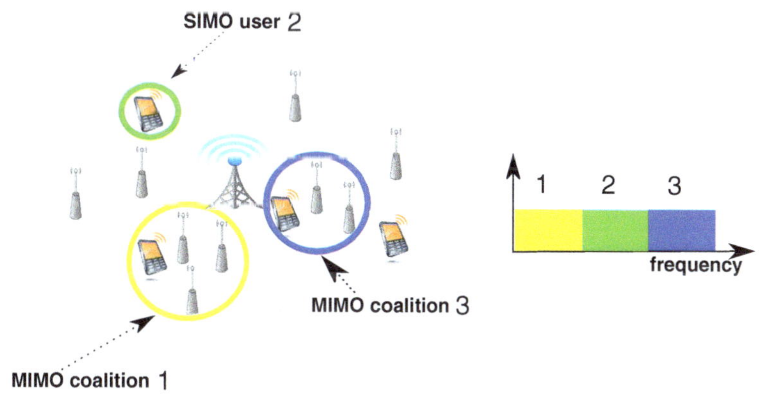

Figure 1 User cooperation example coalitions considering an OFDMA transmission model.

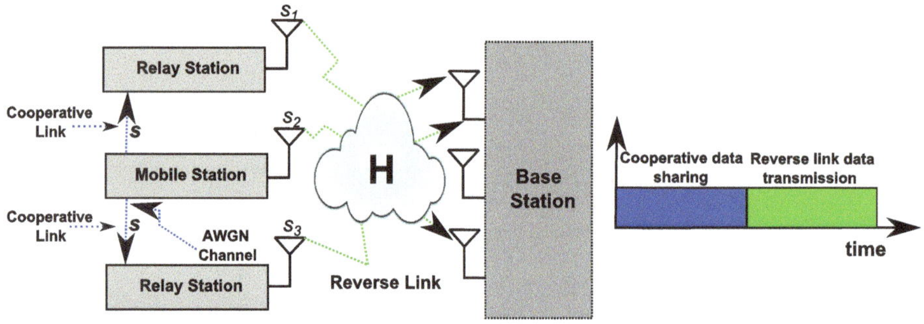

Figure 2 A virtual M_t x M_r MIMO link.

based on a short-range transmission, which is primarily used for information exchange between the transmitting peers. Thus, the channel between the MS-n and the RS-r can be modeled as a κth-power path loss (loss $\approx \frac{1}{l_{nr}^{\kappa}}$) with additive white Gaussian noise (AWGN). Accordingly, the received power P_{nr} at the RS-r, transmitted from the MS-n is given by:

$$P_{nr} = P_{tnr} \, l_{nr}^{-\kappa}, \tag{1}$$

where l_{nr} is the distance between the RS-r and the MS-n, and P_{tnr} is the transmitted power for cooperation. Hence, the signal-to-noise-ratio (SNR) at the RS side is represented by:

$$\eta_{nr} = \frac{P_{nr}}{N_0}, \tag{2}$$

where N_0 is the noise power. Moreover, due to the broadcast nature of the wireless channel, when the MS broadcasts its information to the farthest RS in the coalition, all other RSs can also receive and decode simultaneously this information. Thus, define $S'_n \in R$ as the subset of RSs which have formed a coalition with the MS-n. The cost of cooperation can be defined as the MS's maximum transmitted power to reach the farthest RS in the coalition. Thereby, define the set of distances between the MS-n and its S'_n subset of RSs as:

$$D_{nr}^{*} = \{l_{n(1)}, l_{n(2)}, \ldots, l_{n(\omega)}\}, \tag{3}$$
$$\text{s.t } l_{n(1)} \leq l_{n(2)} \leq \ldots \leq l_{n(\omega)},$$

where $\omega = |S'_n|$ and $|.|$ define the cardinality of the subset. Thus, by using Equations 1 and 2, the power spent for cooperation may be represented by:

$$P_{\text{tcop}} = \eta_{n(\omega)} l_{n(\omega)}^{\kappa} N_0. \tag{4}$$

In this work, we assume that there is sufficient bandwidth to support cooperative links without any major interference.

2.3 Reverse link channel model

The channel coefficient between a multi-antenna BS separated by a distance l_k from the kth MIMO coalition is determined by path loss, log-normal shadowing, and channel variations caused by frequency selective fading. In this work, a fading Rayleigh channel is considered, thus the fading coefficients for an $M_r \times M_t$ MIMO channel can be represented by a matrix:

$$\mathbf{H} = \begin{bmatrix} h_{1,1} & h_{1,2} & \cdots & h_{1,M_t} \\ h_{2,1} & h_{2,2} & \cdots & h_{2,M_t} \\ \vdots & \vdots & \ddots & \vdots \\ h_{M_r,1} & h_{M_r,2} & \cdots & h_{M_r,M_t} \end{bmatrix}, \tag{5}$$

where each matrix element defines a zero mean circular symmetric complex gaussian (ZMCSCG) random variable with unit variance [7]. If the MS prefers to transmit in SIMO mode, the channel can be defined by the following vector:

$$\mathbf{h} = [h_1, h_2, \ldots, h_{M_r}]^T. \tag{6}$$

Furthermore, path loss and shadowing are considered to attenuate the transmitted signal, thus, the received power P_r at the BS side is given by [27]:

$$P_r = P_t 10^{\frac{-L(l_k)+X_\sigma}{10}}, \tag{7}$$

where P_t represents the transmitted power, X_σ is the log-normal shadowing value (dB) with standard deviation σ, and $L(l_k)$ is the distance dependent path loss (dB) which is calculated as follows:

$$L(l_k) = a + b \log_{10}(l_k) \; [\text{dB}], \tag{8}$$

where $a = 15.3$ and $b = 37.6$ are path loss constants for a micro urban cell scenario. Moreover, since single antenna devices use a short-range transmitter for information exchange, a valid assumption is to consider that the elements involved in a MIMO coalition are sufficiently closely spaced to experience the same shadowing and path loss. Nevertheless, the channel fading experienced for the single antenna devices involved in a MIMO coalition can be considered as uncorrelated [28]. In addition, we assume the receiver and transmitter know the channel coefficients between them. State-of-the-art wireless standards such as LTE may implement closed-loop techniques to obtain current channel state information [29]. In this work, coalitions are formed to reduce the reverse link power consumption by using the concepts of spatial diversity and spatial multiplexing as shown below.

2.3.1 Spatial diversity

The received signal at the BS from the kth MIMO coalition is represented as:

$$\mathbf{y}_k = \sqrt{\frac{P_r}{M_t}}\mathbf{H}\mathbf{w}s + \mathbf{n}, \tag{9}$$

where M_t is the number of transmit antennas per *coalition*, s is the scalar information symbol with unit energy, $\mathbf{n}$ is the noise and, $\mathbf{w}$ is a complex weight vector that should satisfy $\|\mathbf{w}\|_F^2 = M_t$ to constrain the total average transmitted power, where $\|\cdot\|_F^2$ is the Frobenius norm.

Accordingly, the SNR for a MIMO coalition is given by [7]:

$$\eta_{k\_\text{mimo}} = \frac{\|\mathbf{g}^H\mathbf{H}\mathbf{w}\|_F^2 P_r}{M_t\|\mathbf{g}\|_F^2 N_0}, \tag{10}$$

where N_0 is the noise power and $\mathbf{g}$ is an $M_r \times 1$ complex weight vector which multiplies the received signal at the BS. Thus, maximizing the SNR at the receiver side is equivalent to maximizing the term $\|\mathbf{g}^H\mathbf{H}\mathbf{w}\|_F^2/\|\mathbf{g}\|_F^2$. The proper choices of $\mathbf{w}/\sqrt{M_t}$ and $\mathbf{g}$ that maximize the SNR are the input and output singular value vectors corresponding to the maximum singular value $\sigma_{\max}$ of $\mathbf{H}$ [7]. By the use of the singular value decomposition (SVD) the channel matrix can be represented as $\mathbf{H} = \mathbf{U}\Sigma\mathbf{V}^H$, where $\mathbf{V}^H$ represents the conjugate transpose of $\mathbf{V}$. Moreover, the columns of $\mathbf{V}$ and $\mathbf{U}$ are known as the input and output singular vectors, respectively. In addition, $\Sigma = \text{diag}\{\sigma_1, \sigma_2, \ldots, \sigma_J\}$ with $\sigma_i \geq 0$, where σ_i is the ith singular value of the channel, and J is the rank of $\mathbf{H}$. Thus, the received SNR at the BS side from the kth MIMO coalition may be expressed as follows:

$$\eta_{k\_\text{mimo}} = \frac{\sigma_{\max}^2 P_r}{N_0}, \tag{11}$$

In the case of a SIMO user, Equation 9 is re-written in the following way:

$$\mathbf{y}_k = \sqrt{P_r}\mathbf{h}s + \mathbf{n}. \tag{12}$$

Thereby, the received SNR at the BS may be represented by [7]:

$$\eta_{k\_\text{simo}} = \frac{\|\mathbf{h}\|_F^2 P_r}{N_0}, \tag{13}$$

2.3.2 Spatial multiplexing

When channel knowledge is assumed, the individual spatial channel modes may be accessed through linear processing at the transmitter and receiver side [7]. Hence, a signal vector $\mathbf{s}$ of dimension $J \times 1$ which is transmitted from the kth MIMO coalition through a rank J MIMO channel, $\mathbf{H}$, after linear processing at the BS side is represented by:

$$\begin{aligned}\tilde{y}_k &= \sqrt{\frac{P_r}{M_t}}\mathbf{U}^H\mathbf{H}\mathbf{V}\mathbf{s} + \mathbf{U}^H\mathbf{n},\\ &= \sqrt{\frac{P_r}{M_t}}\Sigma\mathbf{s} + \tilde{\mathrm{n}},\end{aligned} \tag{14}$$

where $\mathbf{V}$ represents the matrix with dimensions $M_t \times J$ that multiplies $\mathbf{s}$ at the transmitter side. Moreover, $\mathbf{U}^H$ represents the matrix with dimensions $M_r \times J$ that multiplies the signal at the receiver side. In addition, $\tilde{\mathbf{n}}$ is the ZMCSCG noise vector after processing, with dimensions $J \times 1$. The transmitted signal vector $\mathbf{s}$ must satisfy $\mathbf{E}\{\mathbf{ss}\}^H = M_t$ to constrain the total transmitted power. Furthermore, Figure 3 shows how $\mathbf{H}$ is decomposed into J parallel single-input single-output (SISO) channels under

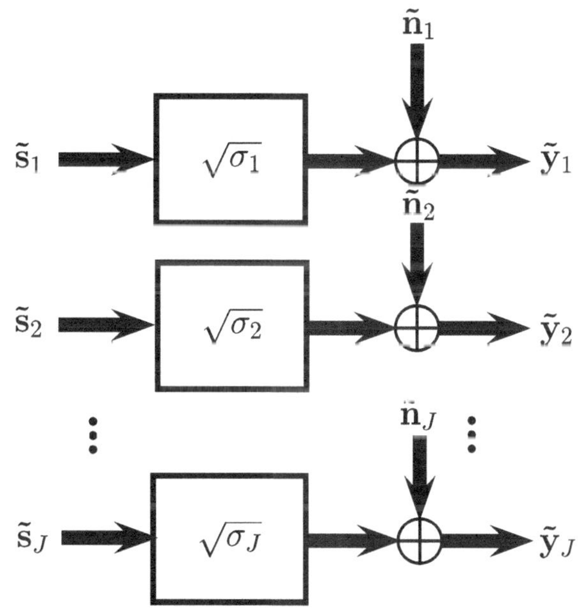

Figure 3 Modal decomposition of the channel.

the assumption of channel knowledge at the transmitter side, where each parallel sub-channel satisfies:

$$\tilde{y}_i = \sqrt{\frac{P_r}{M_t}}\sqrt{\sigma_i}s_i + \tilde{n}_i, \quad i = 1, 2, \ldots, J. \quad (15)$$

Hence, the total reverse link user throughput will become the sum of the individual parallel SISO channel capacities, where the SNR of the ith spatial channel (SC) is given by:

$$\eta_{i\_SC} = \frac{P_r \zeta_i \sigma_i}{M_t N_0}, \quad (16)$$

where $\zeta_i = \mathbf{E}\{\|\mathbf{s_i}\|^2\}$, $i = 1, 2, \ldots, J$, represents the transmitted power in the ith SISO parallel sub-channel and must satisfy $\sum_{i=1}^{J} \zeta_i = M_t$.

Moreover, since the transmitter may access multiple parallel SISO channels, the problem becomes how to allocate the power in a way that maximizes the mutual information. The optimal value of ζ_i is found iteratively through the use of the water-pouring method, which is explained in detail in [30].

When cooperation is not suitable, the MSs will transmit in SIMO mode, where the achievable SNR is defined by Equation 13.

3 Physical components power consumption model and performance metrics to optimize the overall consumed power

In this paper, we focus on optimizing the overall power consumption of the MS's components rather than only the transmitted power. For the MIMO user case, we consider the power expenditure in both the reverse and the cooperative link. When cooperation is not feasible, MSs would prefer to transmit in SIMO mode, hence only the reverse link power expenditure is taken into account. The reverse and cooperative link power consumption mainly depend on components such as the RF parts and the BB signal processing module [3]. The RF module incorporates the power expenditure of power amplifiers, and the BB module comprises the power consumption for channel coding/decoding and modulation/demodulation. For modeling the RF and BB module, we use the model previously presented in [3], where the authors make an analysis of the power expenditure for both modules in a LTE mobile station. Therefore, the overall consumed power in SIMO mode, P_{simo}, depends primarily on the transmitted power in the reverse link P_t.

$$P_{\text{simo}}(P_t) = P_{\text{circ}}(P_t). \quad (17)$$

Furthermore, the total consumed power to form a virtual MIMO link becomes a function of the transmitted power in the reverse link P_t, and how this is distributed between the mobile and the relay stations, which is defined by the weight vector, $\mathbf{w}$, when implementing spatial diversity and by the water filling coefficients ζ_i, $i = 1, 2, \ldots, J$, for the spatial multiplexing case. Thus, the total consumed power in the reverse link when implementing spatial diversity or spatial multiplexing respectively is obtained as follows:

$$P_{\text{mimo\_diversity}}(P_t) = \sum_{i=1}^{M_t} P_{\text{circ}}\left(\frac{P_t\|w_i\|^2}{M_t}\right), \quad (18)$$

$$P_{\text{mimo\_capacity}}(P_t) = \sum_{i=1}^{M_t} P_{\text{circ}}\left(\frac{P_t\zeta_i}{M_t}\right), \quad (19)$$

where P_{circ} is defined as the circuit power in the reverse link spent by each of the single antenna devices such as MSs or RSs forming the MIMO link. In addition, the power expenditure due to the cooperative link, P_{circop}, should be added to Equations 18 and 19. Thereby, the total power expenditure to form the virtual MIMO link when implementing spatial diversity or spatial multiplexing is given by:

$$P_{\text{mimo\_diversity\_total}}(P_t, P_{\text{tcop}}) = P_{\text{circop}}(P_{\text{tcop}}) + \sum_{i=1}^{M_t} P_{\text{circ}}\left(\frac{P_t\|w_i\|^2}{M_t}\right), \quad (20)$$

$$P_{\text{mimo\_capacity\_total}}(P_t, P_{\text{tcop}}) = P_{\text{circop}}(P_{\text{tcop}}) + \sum_{i=1}^{M_t} P_{\text{circ}}\left(\frac{P_t\zeta_i}{M_t}\right). \quad (21)$$

To model the circuit consumed power of the RF module, we consider a power amplifier array [3,31] which is based on four power amplifiers: a low-power amplifier (LPA) and three high-power amplifiers, HPA 1, HPA 2, and HPA 3, as presented in Figure 4. The power amplifier efficiency is assumed equal for both high-power amplifiers; however, HPA 1 and 2 are designed to transmit up to one fourth and to one half of the maximum transmitted power of HPA 3, respectively. Thus, the circuit power expenditure at the reverse link P_{circ} [W] is given by:

$$P_{\text{circ}}(P_t^{\star}) = \begin{cases} 2 + 0.005(P_t^{\star}) - A & 14 \geq P_t^{\star} \\ \frac{1.2+0.12(P_t^{\star})-(A-\frac{3}{4}P_{\text{BB}})}{4} & 17 \geq P_t^{\star} > 14 \\ \frac{1.2+0.12(P_t^{\star})-(A-P_{\text{BB}})}{2} & 20 \geq P_t^{\star} > 17 \\ 1.2 + 0.12(P_t^{\star}) - A & 24 \geq P_t^{\star} > 20 \end{cases} \quad (22)$$

where the $P_t^{\star}$ represents the transmitted power per antenna in [dBm], which is the input value converted to

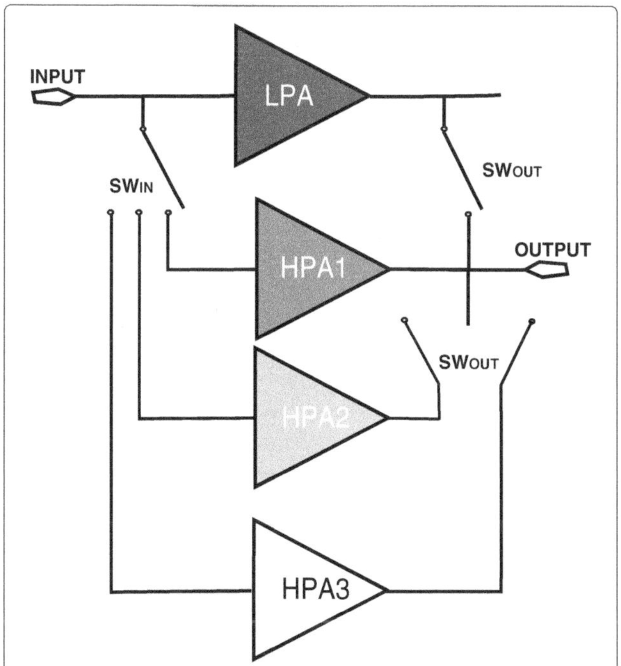

Figure 4 Internal model of the power amplifier for the RF module.

[dBm] of P_{circ} in Equations 17, 20, and 21, and A is a set of constant values defined as follows [3]:

$$A = P_{Tx} + P_{\text{con}} - P_{\text{BB}} \text{ [W]}. \tag{23}$$

The value P_{Tx} is the minimum power that the RF chain consumes in transmission mode, P_{con} is the MS's power consumption when connected to the BS, and P_{BB} is the power consumed by the BB module [3].

In addition, the cooperative link is constructed by using a short-range communication link, thus to model its circuit power expenditure P_{circop}, we use the LPA model shown below:

$$\begin{aligned} P_{\text{circop}}(P_{\text{tcop}}) &= 2 + 0.005(P_{\text{tcop}}) - A \text{ [W]} \\ 14 &\geq P_{\text{tcop}} \text{ [dBm]}. \end{aligned} \tag{24}$$

3.1 Performance metrics to optimize circuit consumed power

The achievable throughput on the link between the kth coalition and the BS when diversity is enhanced is calculated as [27]:

$$T_{k\_\text{diversity}}(\eta_k) = n_k^{RB} k_{sc} \varrho_s \varepsilon(\eta_k) \text{ [bits/s]}, \tag{25}$$

where n_k^{RB} is the number of resource blocks assigned to the kth coalition, k_{sc} is the number of subcarriers per resource block, ϱ_s is the symbol rate per subcarrier, and $\varepsilon(\eta_k)$ is the spectral efficiency for an LTE system [27]. Moreover, η_k in Equation 25 must be replaced by Equation 11 when the user transmits in MIMO or by Equation 13 when the user transmits in SIMO mode. In the case when spatial multiplexing is implemented, the throughput is given by:

$$T_{k\_\text{capacity}}(\eta_{i\_\text{SC}}) = n_k^{\text{RB}} k_{sc} \varrho_s \sum_{i=1}^{J} \varepsilon(\eta_{i\_\text{SC}}) \text{ [bits/s]}, \tag{26}$$

where η_{i_SC} is the SNR in the ith individual parallel SISO channel previously given in Equation 16 and J is defined as the rank of the channel. The user energy efficiency β_k measures the user throughput per unit of consumed energy.

$$\beta_k = T_k / P_{\text{total}\_k} \text{ [bits/J]}. \tag{27}$$

This is based on the total consumed power P_{total_k}, where P_{total_k} is equal to P_{simo} in Equation 17 when the coalition acts in SIMO mode, and to $P_{\text{mimo\_diversity\_total}}$ in Equation 20 or $P_{\text{mimo\_capacity\_total}}$ in Equation 21, when a virtual MIMO link is constructed to implement spatial diversity or spatial multiplexing, respectively. Moreover, T_k is replaced as required by $T_{k\_\text{diversity}}$ in Equation 25 or T_{k_capacity} in Equation 26. Additionally, the system energy efficiency β_{sys} is defined as the ratio between the total user throughput and the total power spent by all the users in the system:

$$\beta_{\text{sys}} = \frac{\sum_{k=1}^{N} T_k}{\sum_{k=1}^{N} P_{\text{total}\_k}} \text{ [bits/J]}. \tag{28}$$

4 College admission framework for distributed virtual MIMO coalition formation

In this paper, cooperation is modeled based on the college admissions problem from game theory which has been used for decades to model and find a *stable match* between medical school students and residency programs in the United States [32,33]. Moreover, we use the college admission framework (CAF) to find a *stable match* between two sets of elements (MSs and RSs). The CAF is a generalization of the stable marriage (SM) problem, thus the solution for both problems tends to be similar in practice [34]. However, coalitions are not limited only to two participants as in the SM case [5]. As described in [32], the CAF involves a set of colleges and a set of applicants. Each applicant lists in order of preference those institutions she/he aims to attend while each institution lists in order of preference those applicants it is willing to admit. Additionally, each institution has a limit in the number of applicants that is able to admit. Thus, the problem becomes to assign applicants to institutions by considering both, preferences and constraints. In our problem, the N MSs take the role of colleges and the R RSs become the applicants. Hence, RSs are assigned to MSs to form virtual MIMO coalitions with the aim of reducing the total energy consumption of the MSs. Notice that the purpose of this work differs from well-know antennas selection techniques previously proposed in the literature [35]. In this work, by using a distributed method, we aim to find the best match between two sets of N MSs and R RSs rather than only find the best antenna elements to transmit/receive over a well-constrained set as previously shown in [35]. Thus, by using the CAF each MS may have multiple RS but each RS belongs to one MS only.

Virtual MIMO coalitions implement spatial diversity or spatial multiplexing respectively to obtain power savings in the reverse link. An important property of the CAF is that it leads the system to a *stable solution* as described in [33,36]. *Stability* means that there are no RSs and MSs in the system such that both of the following assumptions are true:

- The RS is not included into any coalition or would prefer to form a virtual MIMO link with a different MS to the one that is currently matched with;
- The MS is able to include another RS into its MIMO coalition or would prefer to cooperate with a different RS to one of its current partner RSs.

Thus, as stated in [33], when in the CAF method the conditions above are met and no match variations are presented, it always produces a *stable* outcome.

A mapping M is a tuple of one MS with a subset of one or more RSs, such that each single antenna device (MS or RS) belongs exactly to one tuple. Hence, if $(n, S'_n) \in M$, we say that the subset S'_n of RSs is the cooperative partner set of MS-n in M and vice versa, where $S'_n \in R$. The distributed coalition formation algorithm is described as follows:

1. At the beginning of the algorithm, each MS in the system sends a broadcast message through the cooperative link to find the subset of RSs willing to cooperate and form a virtual MIMO link, which for the MS-n is denoted by $S_n \in R$.
2. Moreover, the RSs in the system exchange their channel statistics in the reverse link (fading coefficient, path loss, and shadowing) and the channel statistics in the cooperative link (path loss) with the subset of MSs willing to cooperate with them, which for the RS-r is denoted by $S_r \in N$. Thereafter, each mobile station has the means to rank its subset of suitable RSs, S_n, by using the following utility function, that in the diversity enhancement case is defined by:

$$U_{nr\_\text{diversity}}(\eta_{\text{target}}) = P_{\text{simo}}(\eta_{\text{target}}) - P_{\text{mimo\_diversity}}(\eta_{\text{target}}), \tag{29}$$

where $U_{nr\_\text{diversity}}$ represents the difference in power expenditure when the MS-n transmits on its own or forms a virtual MIMO link with the RS-r, and η_{target} is a fix target SNR that SIMO and MIMO coalitions aim to achieve. Thus, the higher is the value of the utility, the more MS-n will be willing to form a virtual link with RS-r. Moreover, a negative value of $U_{nr\_\text{diversity}}$ means that forming a coalition with the RS-r become less energy efficient, thus the MS will prefer to transmit in SIMO mode. In the case when implementing spatial multiplexing, each MS-n ranks each RS-r from its subset S_n by using the following utility function:

$$U_{nr\_\text{capacity}}(T_{\text{target}}) = P_{\text{simo}}(T_{\text{target}}) - P_{\text{mimo\_capacity}}(T_{\text{target}}), \tag{30}$$

where U_{nr_capacity} represents the difference in energy efficiency performance when the MS-n transmits on its own or forms a coalition with RS-r, and T_{target} represents a target transmission rate that both SIMO and MIMO users aim to achieve. Thus, as in the case where diversity is enhanced, the higher the value of the utility, the MS-n will be more willing to form a virtual MIMO link with the RS-r. The MS-n's preference list ι_n is formed by evaluating the utility for each RS in S_n with Equation 29 when diversity is enhanced or Equation 30 for the capacity enhancement case. Moreover, the RSs of the MS-n's preference list, ι_n, must be sorted in descending order as follows:

$$\iota_n = \{\text{RS}_{n(h)}, \text{RS}_{n(2)}, \ldots, \text{RS}_{n(1)}\}, \quad \text{s.t } U_{n(1)} \le U_{n(2)} \le \ldots \le U_{n(h)}, \tag{31}$$

where $U_{n(r)}$ represents the pairwise comparisons between the MS-n and the RS-r, and $h = |S_n|$, where $|.|$ is defined as the cardinality of the sub-set. These are the values obtained from Equation 29 or 30 whenever the preference list is designed to implement spatial diversity or spatial multiplexing, respectively. Notice that when $U_{n(r)}$ becomes negative, the MS-n will not consider the RS-r for coalition formation, thus RS-r will not be included in the MS-n ranking list, ι_n.
3. Mobiles are only required to exchange their channel statistics in the reverse link with the RSs willing to

cooperate with them. Based on this information, RSs are able to rank their subset of MSs, S_r, by using the following utility function when diversity is enhanced:

$$U_{rn\_\text{diversity}}(\eta_{\text{target}}) = P_{\text{circ}}\left(\frac{P_t(\eta_{\text{target}})\|w_{\text{rs}}\|^2}{M_t}\right) - P_{\text{circ}}\left(\frac{P_t(\eta_{\text{target}})\|w_{\text{ms}}\|^2}{M_t}\right). \quad (32)$$

For the capacity case, Equation 32 may be re-written as follows:

$$U_{rn\_\text{capacity}}(T_{\text{target}}) = P_{\text{circ}}\left(\frac{P_t(T_{\text{target}})\zeta_{\text{rs}}}{M_t}\right) - P_{\text{circ}}\left(\frac{P_t(T_{\text{target}})\zeta_{\text{ms}}}{M_t}\right). \quad (33)$$

Equations 32 and 33 represent the difference in power expenditure between the RS-r and the MS-n when forming a virtual MIMO link, respectively. Thus, the larger the value of the utility, the larger the power expenditure of the RS due to its better channel conditions in the reverse link when compared to the MS. Furthermore, the RS's preference list, ι_r, is obtained by evaluating each of the elements in the S_r subset by Equation 32 or 33 when using spatial diversity or spatial multiplexing, respectively. The elements of ι_r are also sorted in descending order as the ι_n case described previously in Equation 31.

4. Once the preference lists for MSs and RSs are obtained, Algorithm 1 from [33] can be performed. Algorithm 1 is implemented in a distributed way with the sole participation of MSs and RSs, thus a centralized entity such as a BS is not required for the algorithm implementation. Hence, RSs and MSs exchange signaling messages through the cooperative link with the corresponding members contained in their preference lists ι_r and ι_n, respectively. At the initial state, MSs transmit in SIMO mode, further, MSs and RSs aim to form virtual MIMO coalitions with the elements with the highest ranking in their preference lists. If that is not possible, they remove the element and propose or wait for a proposal of the next element in their corresponding list. When a new RS wants to join a coalition, the coalition checks to see if adding RS-r to the kth coalition will generate energy savings. The RS-r communicates its channel statistics in the reverse and cooperative links to the elements of the coalition. Hence, the kth coalition computes the difference in energy consumption by adding RS-r to the coalition which is calculated in an analogous way as presented in Equations 29 and 30 for the diversity and capacity case, respectively. If no power savings are obtained, the RS contacts the next MS on its list. The algorithm runs in an asynchronous way, thus the interaction between MSs and RSs can occur in an *ad hoc* fashion.

Algorithm 1: College admission framework (CAF), after [36]

```
Initialization: All MSs must be operating in SIMO mode and all the RSs are free;
while There is an MS-n wanting to form a MIMO link; do
    'r(h) is the highest ranked MS in the RS-r preference list, ι_r, to whom the RS-r has not proposed yet;
    if RS-r is contained in the MS_r(h)'s preference list; then
        if MS_r(h) is free; then
            the MS_r(h) and the RS-r become engaged;
        else
            MS_r(h) is already engaged with a subset of RSs, S̄_n ∈ R;
            if If adding the RS-r to the MS_r(h) current subset of RSs, S̄_n, provides energy savings; then
                RS-r becomes engaged;
            end if
            if If adding the RS-r to the MS_r(h) current subset of RSs, S̄_n, does not provides extra energy savings. Nevertheless, MS_r(h) prefers RS-r to the RS-t in its preference list, ι_r(h), where RS-t ∈ S̄_n; then
                RS-r becomes engaged;
                RS-t becomes free;
            else
                MS_r(h) is deleted from the list of the RS-r, ι_r;
            end if
        end if
    end if
end while
```

5 Analysis of the consequences in performance of MIMO systems when the overall consumed power is optimized

In this section, a theoretical analysis is provided of the consequences arising in terms of energy efficiency when the overall power consumption is considered as an optimization metric rather than the transmitted power. Hence, to show the effects on user performance, we analytically derive the statistics of the transmitted and overall consumed power when implementing spatial diversity or spatial multiplexing respectively. While these statistics can be obtained experimentally, we derive them in closed form.

5.1 Spatial diversity approach

From Equation 7, we know that the transmitted power of any signal, P_t, can be calculated by:

$$P_t = \frac{P_r}{10^{\frac{-L(l_k)+X_\sigma}{10}}}. \quad (34)$$

Moreover, if we combine Equation 11 with Equation 34, we obtain the transmitted power for a MIMO user:

$$P_{t\_mimo} = \frac{\eta_{k\_mimo} N_0}{\sigma^2_{max} 10^{\frac{-L(l_k)+X_\sigma}{10}}}. \quad (35)$$

Given that on average, $\mathbf{E}\{\sigma^2_{max}\} = M_t \times M_r$ and $X_\sigma = 0$ [7], this allows us to re-write the equation above as follows:

$$P_{t\_mimo} = \frac{\eta_{k\_mimo} N_0}{M_t \times M_r 10^{\frac{-L(l_k)}{10}}}. \quad (36)$$

To obtain the statistics for the transmitted power of a MIMO user, P_{t_mimo}, we assume that the MSs are uniformly distributed in the cell. Therefore, for a circular cell of radius R, it is known that the probability distribution function (PDF) of the distance of any point from the center is [12]:

$$f_{l_k}(l_k) = \frac{2l_k}{R^2} \quad l_k \in [0, R]. \quad (37)$$

In addition, from Equation 8, we observe that path loss is an element depending on distance, thus to derive its PDF, we use the transformation of random variables. Thereby, we obtain the inverse relationship of the distance as a function of path loss as follows:

$$l_k(L) = 10^{\left(\frac{L-a}{b}\right)}, \quad (38)$$

Hence, the path loss PDF $f_L(L)$ may be derived by:

$$f_L(L) = \left|\frac{dl_k}{dL}\right| f_{l_k}(l_k(L)) \quad L \in [-\infty, a + b\log_{10}(R)] \quad [dB], \quad (39)$$

$$f_L(L) = \frac{2\log(10)}{bR^2} 10^{\frac{2(L-a)}{b}}. \quad (40)$$

Once the statistics for the path loss are obtained, we proceed to derive the PDF of the transmitted power. From Equation 36, we are able to obtain the inverse relationship of the path loss as a function of the transmitted power for a MIMO user, P_{t_mimo}.

$$L(P_{t\_mimo}) = -10\log_{10}\left(\frac{\eta_{k\_mimo} N_0}{M_t M_r P_{t\_mimo}}\right). \quad (41)$$

Thus, the PDF of the transmitted power for a MIMO user can be obtained as follows:

$$f_{P_{t\_mimo}}(P_{t\_mimo}) = \left|\frac{dL}{dP_{t\_mimo}}\right| f_L(L(P_{t\_mimo}))$$
$$P_{t\_mimo} \in \left[0, \frac{\eta_{k\_mimo}\rho}{M_t M_r 10^{-\left(\frac{a+b\log_{10}(R)}{10}\right)}}\right], \quad (42)$$

$$f_{P_{t\_mimo}}(P_{t\_mimo}) = \frac{20}{bR^2 P_{t\_mimo}} 10^{\frac{-2}{b}\left(10\log_{10}\left(\frac{\eta_{k\_mimo}\rho}{M_t M_r P_{t\_mimo}}\right)+a\right)}. \quad (43)$$

From Equation 22, we see that the circuit consumed power, P_{circ}, depends of the transmitted power when converted to [dBm]. Thus, the inverse relationship of the transmitted power, P_{t_mimo}, in function of the transmitted power in [dBm], $P_{t\_mimo\_dBm}$, for a MIMO user case is given by:

$$P_{t\_mimo}(P_{t\_mimo\_dBm}) = 1e^{-3} \times 10^{\frac{P_{t\_mimo\_dBm}}{10}}. \quad (44)$$

Thereby, the PDF of the transmitted power in [dBm], $P_{t\_mimo\_dBm}$ is:

$$f_{P_{t\_mimo\_dBm}} = \left|\frac{dP_{t\_mimo}}{dP_{t\_mimo\_dBm}}\right| f_{P_{t\_mimo}}(P_{t\_mimo}(P_{t\_mimo\_dBm}))$$
$$P_{t\_mimo\_dBm} \in \left[-\infty, 10\log_{10}\left(\frac{1e^3 \times \eta_{k\_mimo} N_0}{M_t M_r 10^{-(a+b\log_{10}(R))}}\right)\right], \quad (45)$$

$$f_{P_{t\_mimo\_dBm}} = \frac{2\log(10)}{bR^2} 10^{\frac{-2}{b}\left(10\log_{10}\left(\frac{1e^3 \times \eta_{k\_mimo} N_0}{M_t M_r 10^{\frac{P_{t\_mimo\_dBm}}{10}}}\right)+a\right)}. \quad (46)$$

Moreover, we observe that the input of P_{circ} in Equation 18 is the transmitted power for each antenna in [dBm]. Thereby, assuming that the total transmitted power for a MIMO user, P_{t_mimo}, is divided evenly between each antenna as was proposed for this derivation, the PDF of the transmitted power per antenna in [dBm], P_t^*, is given by:

$$f_{P_t^*} = \frac{2\log(10)}{bR^2} 10^{\frac{-2}{b}\left(10\log_{10}\left(\frac{1e^3 \times \eta_{k\_mimo}\rho}{M_t^2 M_r 10^{\frac{P_t^*}{10}}}\right)+a\right)}$$
$$P_t^* \in \left[-\infty, 10\log_{10}\left(\frac{1e^3 \times \eta_{k\_mimo}\rho}{M_t^2 M_r 10^{-(a+b\log_{10}(R))}}\right)\right], \quad (47)$$

Finally, we derive the inverse relationship of the transmitted power per antenna in [dBm] as a function of the circuit consumed power in the reverse link by combining Equations 18 and 22 as follows:

$$P_t^*(P_{\text{mimo\_diversity}}) = \begin{cases} \frac{\gamma+A-2}{0.005} & 14 \geq P_t^*, \\ \frac{4\gamma+(A-\frac{3P_{\text{BB}}}{4})-1.2}{0.117} & 17 \geq P_t^* > 14, \\ \frac{2\gamma+(A-P_{\text{BB}})-1.2}{0.117}, & 20 \geq P_t^* > 17, \\ \frac{\gamma+A-1.2}{0.117} & 24 \geq P_t^* > 20. \end{cases} \quad (48)$$

where $\gamma = \frac{P_{\text{mimo\_diversity}}}{M_t}$. Hence, by using the transformation of random variables, the PDF of the circuit consumed power in the reverse link for a MIMO user is shown below:

$$f_{P_{\text{mimo\_diversity}}} = \begin{cases} \frac{363\log(10)}{bR^2M_t} 10^{\frac{-2}{b}\left(10\log_{10}\left(\frac{1e^3\times\eta_{k\_\text{mimo}}\rho}{W10^{\frac{\gamma+A-2}{0.05}}}\right)+a\right)} & P_{\text{mimo\_diversity}} \in [0, M_t(2+0.005W1-A)], \\ \frac{68\log(10)}{bR^2M_t} 10^{\frac{-2}{b}\left(10\log_{10}\left(\frac{1e^3\times\eta_{k\_\text{mimo}}\rho}{W10^{\frac{4\gamma+A-0.75P_{BB}-1.2}{1.17}}}\right)+a\right)} & P_{\text{mimo\_diversity}} \in \left[M_t(2+0.005W1-A), M_t\left(\frac{1.2+0.117W1-W2}{4}\right)\right], \\ \frac{34\log(10)}{bR^2M_t} 10^{\frac{-2}{b}\left(10\log_{10}\left(\frac{1e^3\times\eta_{k\_\text{mimo}}\rho}{W10^{\frac{2\gamma+A-P_{BB}-1.2}{1.17}}}\right)+a\right)} & P_{\text{mimo\_diversity}} \in \left[M_t\left(\frac{1.2+0.117W1-W2}{4}\right), M_t\left(\frac{1.2+0.117W1-W3}{2}\right)\right], \\ \frac{17\log(10)}{bR^2M_t} 10^{\frac{-2}{b}\left(10\log_{10}\left(\frac{1e^3\times\eta_{k\_\text{mimo}}\rho}{W10^{\frac{\gamma+A-1.2}{1.17}}}\right)+a\right)} & P_{\text{mimo\_diversity}} \in \left[M_t\left(\frac{1.2+0.117W1-W3}{2}\right), M_t(1.2+0.117W1-A)\right]. \end{cases} \quad (49)$$

where $W = M_t^2M_r$, $W1 = 10\log_{10}\left(\frac{1e^3\times\eta_{k\_\text{mimo}}N_0}{M_t^2M_r10^{-(a+b\log_{10}(R))}}\right)$, $W2 = A - (0.75 \times P_{\text{BB}})$, and $W3 = A - P_{\text{BB}}$.

Finally, by integrating the PDFs of the transmitted and circuit consumed power over their respective ranges, we obtain the cumulative distribution functions (CDFs) for transmitted and circuit consumed power, which are shown in Figures 5 and 6, respectively. In addition, we also find the CDFs by simulation to compare them with our theoretical derivations. Moreover, as an example, we consider SIMO and MIMO users with three and six antennas. Furthermore, we require the users to achieve the same target SNR, η_{target}, whether SIMO or MIMO is used, in order to make fair comparisons in terms of power expenditure. Notice that to obtain the statistics of the overall consumed power for the SIMO case, a similar procedure is followed as the one shown for the MIMO user case. For the required values to evaluate the statistics and perform

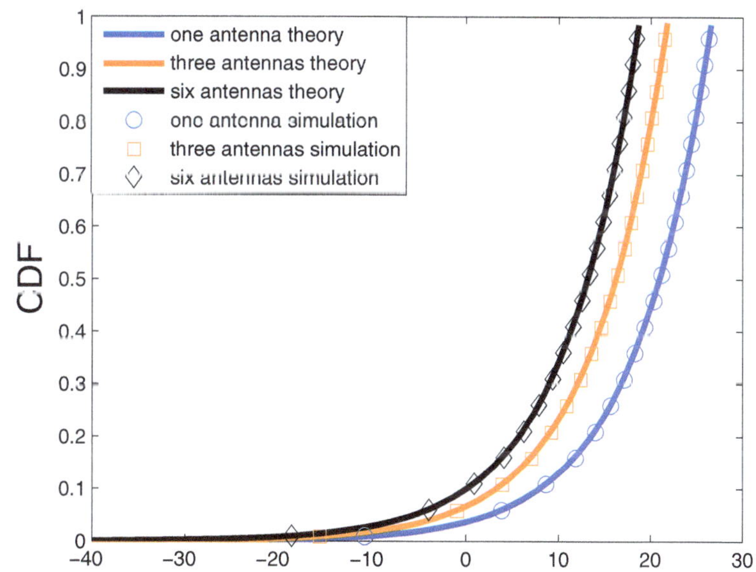

Figure 5 Transmitted power [dBm]. User performance differences, when enhancing diversity and optimizing transmitted power.

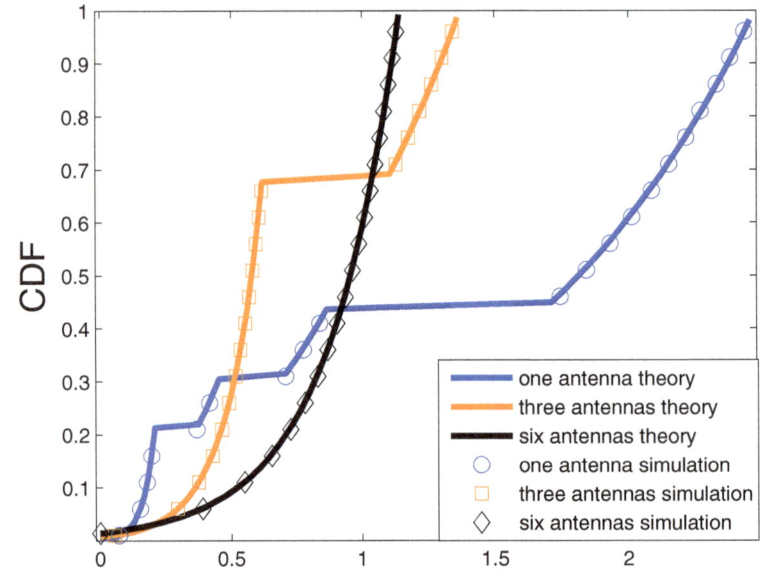

Figure 6 Overall consumed power (W). User performance differences, when enhancing diversity and optimizing overall consumed power.

the simulations, we consider the values shown in Table 1. These results are discussed further in subsection 5.3.

5.2 Spatial multiplexing approach

For the following derivations, we use Shannon's capacity formula for ease of analysis and without loss of generality. Thus, Equation 26 can be re-written as follows:

$$\begin{aligned} T_{k\_capacity} &= \sum_{i=1}^{J} \log_2\left(1+\eta_{i\_SC}\right) \\ &= \sum_{i=1}^{J} \log_2\left(1+\frac{P_r\zeta_i\sigma_i}{M_t N_0}\right). \end{aligned} \tag{50}$$

Moreover, if we assume equal gain conditions between the multiple parallel SISO channels $\zeta_i = 1$, $\mathbf{E}\{\|\mathbf{H}\|_F^2\} = M_r M_t = J\sigma_i$ [7], Equation 50 may be re-written as:

$$\begin{aligned} T_{k\_capacity} &= \sum_{i=1}^{J} \log_2(1+\eta_{i\_SC}) \\ &= J\log_2\left(1+\frac{P_r M_r}{J N_0}\right). \end{aligned} \tag{51}$$

Thus, by combining Equation 34 and Equation 51, we obtain the required transmitted power as:

$$P_t = \frac{\beta}{M_r 10^{\frac{-L(d_k)}{10}}}, \tag{52}$$

Table 1 Simulation parameters

Parameter	Value
MSs per macro-cell, N	20
RSs per macro-cell, R	95
Number of antennas at the receiver, M_r	6
Cell radius	150 m
Number of available RBs, X	20
Number of cells, D	1
Subcarriers per RB, k_{sc}	12
Symbol rate per subcarrier, ϱ_s	15 kbps
P_{Tx}	31.8 dBm
P_{con}	23.8 dBm
P_{BB}	11.7dBm
Maximum user transmit power	24 dBm
Shadowing, Std. Dev., σ	3 dB
η_{target}	17 dB
T_{target}	910 kbps
ε for 17 dB SNR	4.5 $\frac{\text{bits}}{\text{symbol}}$
κ	3.5
Path loss constant, a	15.3
Path loss constant, b	37.6

where $\beta = \left(2^{\frac{T_{k\_capacity}}{J}} - 1\right) JN_0$. To obtain the statistics of the transmitted power P_t, we assume that the MSs are uniformly distributed over the cell. Moreover, from Equation 52, we obtain the inverse relationship of the path loss in function of the transmitted power.

$$L(P_t) = -10\log_{10}\left(\frac{\beta}{M_r P_t}\right). \tag{53}$$

Thus, by using a similar approach as in Equation 39, the PDF of the transmitted power may be obtained as:

$$f_{P_t} = \frac{20}{R^2 b P_t} 10^{\frac{-2}{b}\left(a+10\log_{10}\left(\frac{\beta}{M_r P_t}\right)\right)} \quad P_t \in \left[0, \frac{\beta}{M_r 10^{-\left(\frac{a+b\log_{10}(R)}{10}\right)}}\right]. \tag{54}$$

Furthermore, we require to obtain the transmitted power in [dBm]. Thus, the inverse relationship of the transmitted power, P_t, as function of the transmitted power in [dBm], P_{t_dBm}, is given by:

$$P_t(P_{t\_dBm}) = 1e^{-3} 10^{\frac{P_{t\_dBm}}{10}}. \tag{55}$$

Thereby, by using a similar approach as in Equation 45, we derive the PDF of the transmitted power in [dBm] as follows:

$$f_{P_{t\_dBm}} = \frac{20\log(10)1^{-3}}{R^2 b} 10^{\frac{-2}{b}\left(a+10\log_{10}\left(\frac{\beta}{1e^{-3} M_r 10^{\frac{P_{t\_dBm}}{10}}}\right)\right)}$$
$$P_{t\_dBm} \in \left[-\infty, 10\log_{10}\left(\frac{1e^3 \times \beta}{M_r 10^{-\left(\frac{a+b\log_{10}(R)}{10}\right)}}\right)\right], \tag{56}$$

As in the diversity case, we should observe that in order to compute the circuit consumed power P_{circ}, Equation 22, we require the transmitted power per antenna. Thus, assuming that the transmitted power is divided evenly over all the antennas, the PDF of the transmitted power per antenna in [dBm] is:

$$f_{P_t^*} = \frac{20\log(10)}{R^2 b} 10^{\frac{-2}{b}\left(a+10\log_{10}\left(\frac{\beta}{M_r M_t 1e^{-3} 10^{\frac{P_t^*}{10}}}\right)\right)}$$
$$P_t^* \in \left[-\infty, 10\log_{10}\left(\frac{1e^3 \times \beta}{M_t M_r 10^{-\left(\frac{a+b\log_{10}(R)}{10}\right)}}\right)\right], \tag{57}$$

Finally, we derive the inverse relationship of the transmitted power per antenna in [dBm] as a function of the circuit consumed power in the reverse link by combining Equations 19 and 22 as shown:

$$P_t^*(P_{\text{mimo\_capacity}}) = \begin{cases} \frac{\gamma_1 + A - 2}{0.005} & 14 \geq P_t^*, \\ \frac{4\gamma_1 + (A - \frac{3P_{BB}}{4}) - 1.2}{0.117} & 17 \geq P_t^* > 14, \\ \frac{2\gamma_1 + (A - P_{BB}) - 1.2}{0.117}, & 20 \geq P_t^* > 17, \\ \frac{\gamma_1 + A - 1.2}{0.117} & 24 \geq P_t^* > 20. \end{cases} \tag{58}$$

where $\gamma_1 = \frac{P_{\text{mimo\_capacity}}}{M_t}$. Moreover, by using the transformation of random variables, the PDF of the circuit consumed power is shown below:

$$f_{P_{\text{mimo\_capacity}}} = \begin{cases} \frac{363\log(10)}{bR^2 M_t} 10^{\frac{-2}{b}\left(10\log_{10}\left(\frac{1e^3\beta}{M_t M_r 10^{\frac{\gamma_1+A-2}{0.05}}}\right)+a\right)} P_{\text{mimo\_capacity}} \in [0, M_t(2+0.005Z1-A)], \\ \frac{68\log(10)}{bR^2 M_t} 10^{\frac{-2}{b}\left(10\log_{10}\left(\frac{1e^3\beta}{M_t M_r 10^{\frac{4\gamma_1+A-0.75P_{BB}-1.2}{1.17}}}\right)+a\right)} P_{\text{mimo\_capacity}} \in \left[M_t(2+0.005Z1-A), M_t\left(\frac{1.2-0.117Z1-Z2}{4}\right)\right], \\ \frac{34\log(10)}{bR^2 M_t} 10^{\frac{-2}{b}\left(10\log_{10}\left(\frac{1e^3\beta}{M_t M_r 10^{\frac{2\gamma_1+A-P_{BB}-1.2}{1.17}}}\right)+a\right)} P_{\text{mimo\_capacity}} \in \left[M_t\left(\frac{1.2-0.117Z1-Z2}{4}\right), M_t\left(\frac{1.2+0.117Z1-Z3}{2}\right)\right], \\ \frac{17\log(10)}{bR^2 M_t} 10^{\frac{-2}{b}\left(10\log_{10}\left(\frac{1e^3\beta}{M_t M_r 10^{\frac{\gamma_1+A-1.2}{1.17}}}\right)+a\right)} P_{\text{mimo\_capacity}} \in \left(M_t\left(\frac{1.2+0.117Z1-Z3}{2}\right), M_t(1.2+0.117Z1-A)\right]. \end{cases} \tag{59}$$

where $Z1 = 10\log_{10}\left(\frac{1e^3\beta}{M_r M_t 10^{-(\frac{a+b\log_{10}(R)}{10})}}\right)$, $Z2 = A - (0.75 \times P_{\mathrm{BB}})$, and $Z3 = A - P_{\mathrm{BB}}$. Finally, as in the diversity case, by integrating the PDFs of the transmitted and circuit consumed power over their respective ranges, we obtain the CDFs for transmitted and circuit consumed power, which are shown in Figures 7 and 8, respectively. In addition, we also find the CDFs by simulation to compare them with our theoretical derivations. As an example, we consider SIMO and MIMO users carrying three and six antennas. Furthermore, we make the users independently of SIMO or MIMO to achieve the same transmission rate, T_{target}, in order to make fair comparisons in terms of power expenditure. To evaluate the statistics and perform the simulations, we consider the values shown in Table 1.

5.3 Analysis

From Figures 5 and 7, it is easy to see that increasing the number of antennas provides power savings at all percentiles of the CDF when only the transmitted power is optimized. However, this trend does not remain the same when optimizing circuit power consumption. In Figure 6, in the case of diversity, we see that the SIMO curve intersects the MIMO curves when transmitting with three and six antennas at the 30th and 45th percentile, respectively. Moreover, for the capacity case in Figure 8, we see that the SIMO curve intersects the MIMO curves when transmitting with three and six antennas at the 20th and 28th percentile, respectively. This intersection point represents that in the diversity case, SIMO is more power efficient for 30% and 45% of the users in the cell when compared to MIMO when transmitting with three and six antennas, respectively. The same relation holds for the spatial multiplexing case. This is because the MSs are able to experience better transmission conditions, when they are close to the BS. Thus, turning on the RF transmitter and the BB module of the relay stations is less power efficient than transmitting with only one antenna. Nevertheless, as the users get close to the cell edge increasing the number of transmit antennas tends to be an energy efficient solution when circuit power consumption is optimized. This fact can be seen from Figures 6 and 8, since as the number of antennas increases, it allows the three and six antennas curves to converge faster to the tail of the distribution. Our analysis in this section will be useful to understand the performance of our framework proposed in Section 7.

6 Comparison schemes and simulation scenario

To evaluate the performance of our proposal, we describe four distributed relay selection algorithms, which allow MSs and RSs to cooperate to form MIMO coalitions with the purpose of reducing the energy consumption in the reverse link. In addition, we present a baseline scheme where all MSs transmit on their own in SIMO mode. Finally, we present a centralized global optimum approach which is coordinated from the BS and based on an exhaustive search. For all the methods we describe, the communication between the MSs and RSs is made through the cooperative link. Thus, the subset of RSs willing to cooperate with the MS-n is limited by the range of the cooperative link, which naturally limits the complexity of the relay selection.

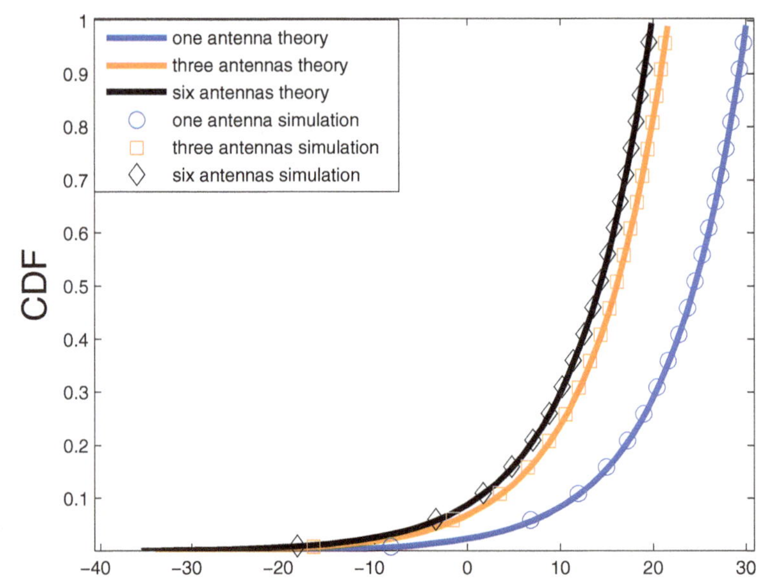

Figure 7 User performance differences, when implementing spatial multiplexing and optimizing transmitted power.

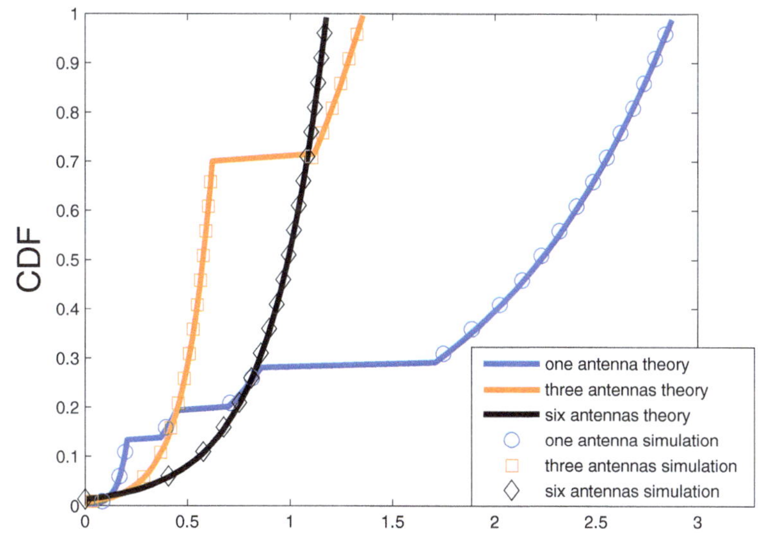

Figure 8 User performance differences, when implementing spatial multiplexing and optimizing overall consumed power.

6.1 Minimum relaying hop (MRH) path loss selection scheme

In [37], the authors propose a relay selection method as a function of path loss. Hence, the best RS for coalition formation is the one with the least path loss to the MS, this method always chooses the RS with the most energy-efficient cooperative link.

$$RS_c = \text{argmin}\{l^{\kappa}_{nr}\}. \tag{60}$$

From Equation 60, notice that to perform the RS selection, it is just required to know the channel statistics of the cooperative link.

6.2 Best worst (BW) channel selection scheme

The BW method considers the quality of the cooperative and the reverse link of each RS. This is because both links have a direct influence on the total consumed energy to form the virtual MIMO link. In [37], the best worst channel is used in which the relay whose worse channel is the best is selected:

$$\text{argmin}\left\{\left\| G_r, \frac{1}{l^{\kappa}_{nr}} \right\|\right\}, \tag{61}$$

where $G_r = \|\mathbf{h}_r\|_F^2 10^{\frac{-L(l_r)+X_\sigma}{10}}$ represents the channel path gain between the RS-r and the BS, and l_r defines the distance between the rth RS and the BS.

6.3 SM scheme

In [5], a distributed RS selection algorithm is presented which is based on the stable marriage process. This method, as in the BW channel selection scheme, requires the channel statistics from the RSs in the reverse and cooperative link plus the channel statistics of the MSs in the reverse link. Thereby, each MS and RS are able to rank their respective candidates for coalition formation. Noticed that the SM method has the same limitation as the MRH and BW methods in the sense that each MS is only able to select one RS.

6.4 SIMO transmission

We implement a baseline scheme, where all the MSs in the network transmit in SIMO mode.

6.5 College admission framework scheme

This scheme implements our RS selection method described in Section 4.

6.6 Centralized optimum scheme

We present a centralized global optimum scheme based on an exhaustive search approach. Thus, the BS collects the required channel statistics from RSs and MSs in order to form optimal coalitions. We implement this centralized approach with the aim of finding the price of anarchy for our proposed scheme. The *price of anarchy* is computed as the difference in performance between a centralized and a distributed approach [38].

6.7 Simulation scenario

We perform Monte Carlo simulations with Matlab® using the parameters presented in Table 1. This is done to compare the performance of our method with the schemes presented above. The simulation is comprised of a single cell with the MSs and RSs distributed uniformly over the cell area. The cell is served by an omnidirectional BS. Moreover, the system is noise limited, hence each coalition transmits in an independent RB to avoid co-channel interference. For the case when diversity is enhanced, we assume that all the *users* (SIMO or MIMO) independent of their distance to the BS try to achieve the same target SNR, η_{target}. In the case when spatial multiplexing is used, we assume that all the users in the network aim to achieve the same data rate, T_{target}.

7 Results

From the simulations, we generate the CDFs and the graphs that illustrate the performance in terms of *overall power expenditure* for the schemes presented in Section 6.

When diversity is enhanced in Figure 9, we show the overall consumed power at different distances from the BS, where all the users in the cell aim to achieve the same target SNR. We observe that the distributed and centralized approaches exhibit a similar performance when compared to the baseline method at close distances from the BS (up to 75 m). This is because as mentioned in the analysis presented in Section 5.1, MSs experience good transmission conditions close to the cell center. Thus, turning on the BB and the RF module of the RSs becomes less power efficient than transmitting with only one antenna. Conversely, when channel conditions are no longer so beneficial (e.g., after 75 m), we see that as the MSs move away from the BS, the increase from one to a higher number of transmit antennas allows the MS to obtain potential energy savings. Furthermore, from the analysis shown in Section 5.1 and the results presented in Figure 9, we can confirm that, when circuit power consumption is optimized and spatial diversity is enhanced, by increasing the number of antennas, the obtained power savings are more visible at the cell edge than at the cell center. In Figure 10, we evaluate the system energy efficiency, given by Equation 28. Notice that at the 50th percentile the CAF scheme is more energy efficient compared to the benchmark, the MRH path loss, the BW channel, and the SMI framework with improvements of 58%, 15%, 10%, and 5%, respectively. Nevertheless, the CAF scheme has losses of 10% compared to the centralized global optimum scheme. These losses are tolerable in practice due to the significant reductions in complexity for the CAF method compared to the centralized optimum scheme: this is discussed further at the end of this section. Moreover, the better performance in energy-efficiency terms for the CAF and the centralized optimum when compared to the other distributed approaches can be easily understood as a direct consequence of the bigger number of antenna elements than can be involved in the coalition.

When spatial multiplexing is used, we aim to obtain gains in energy efficiency by dividing the total data rate requirements between the elements forming the virtual MIMO link. Thereby, as in the diversity case, we can see from Figure 11 that most of the power savings due to coalition formation are observed at the cell border. This is because, it is more power efficient to deliver high transmission rates for SIMO users close to the BS than when close to the cell edge due to the improved propagation

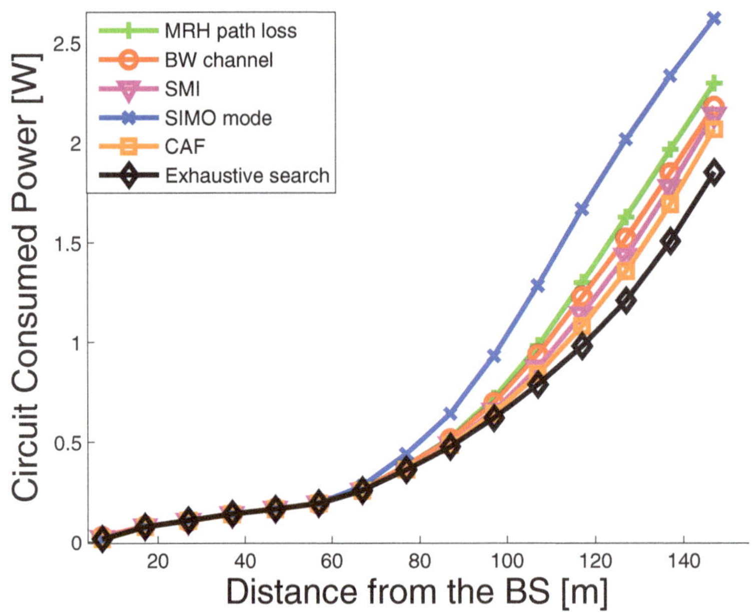

Figure 9 User circuit consumed power against distance from the BS for the SNR = 17 dB. Performance comparison when spatial diversity is implemented.

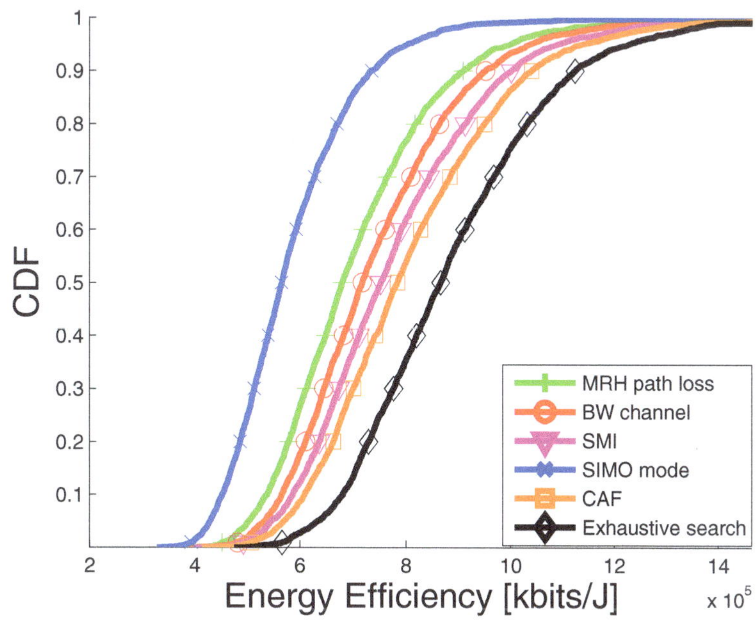

Figure 10 System CDF energy efficiency. Performance comparison when spatial diversity is implemented.

conditions. Thus, using a lower modulation order for transmitting from each antenna in a coalition when close to the cell edge becomes more power efficient than using a single transmitter. Hence, from the results obtained in Section 5.2 and Figure 11, it can be understood that increasing the number of transmit antennas to split the total rate requirement among the transmitters by implementing spatial multiplexing is more power efficient in terms of overall power consumption at the cell border than at the cell center.

Finally, in Figure 12, we show the performance in terms of energy efficiency for the approaches presented

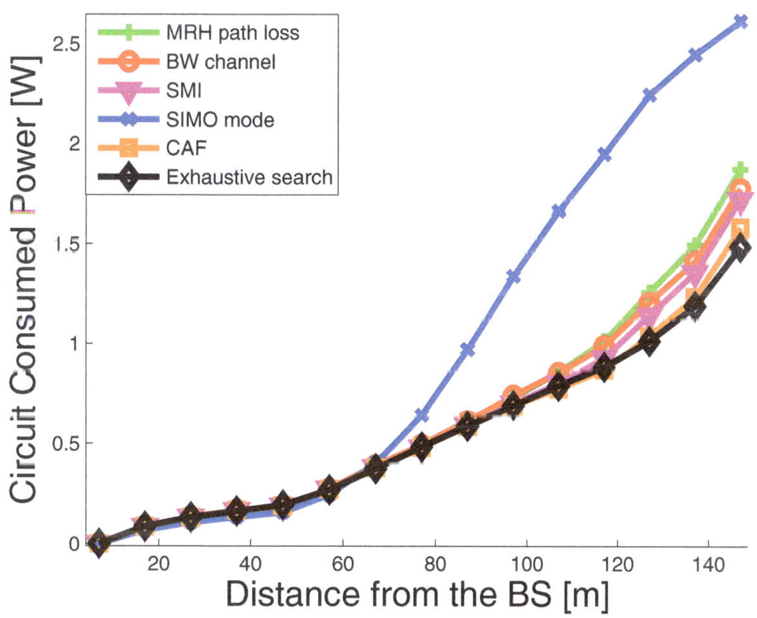

Figure 11 User circuit consumed power against distance from the BS for bit rate 920 kbps. Performance comparison when spatial multiplexing is implemented.

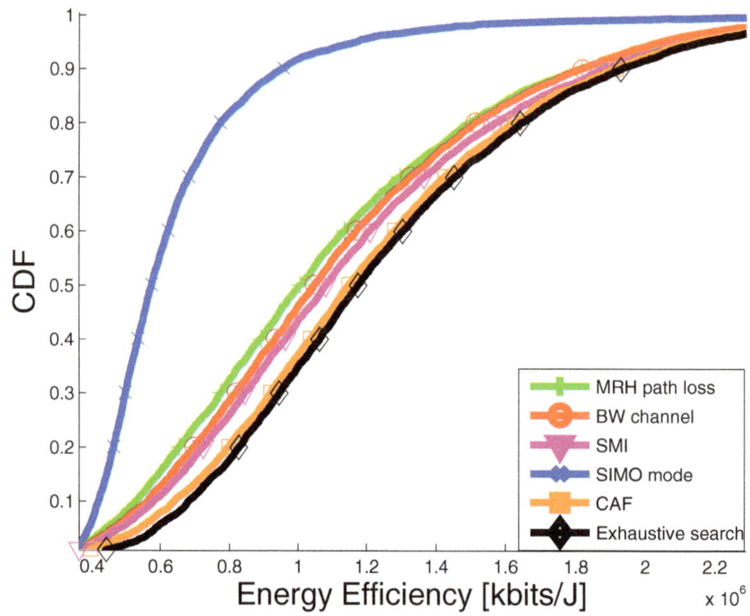

Figure 12 System energy efficiency when enhancing capacity. Performance comparison when spatial multiplexing is implemented.

in Section 6 when spatial multiplexing is implemented. We find that the centralized global optimum is 2% more energy efficient when contrasted to the CAF. Moreover, when comparing the CAF with the other distributed approaches, we see that the CAF method has improvements of 14%, 9%, 5%, and 91% over the MRH, the BW, the SMI, and the baseline SIMO mode, respectively. Thereby, we can confirm that increasing the number of antennas in order to use a lower modulation order results in an energy efficient solution for the network.

To conclude our comparison, the complexity of the centralized global optimum approach is compared with the proposed CAF method. On one hand, for the CAF method, each MS-n in the system has to evaluate each RS in its preferred subset of suitable candidates, S_n, by using Equation 29 or 30 depending on whether diversity or capacity are enhanced. Furthermore, each RS-r evaluates its preferred subset S_r of RSs by using Equation 32 or 33. Big $\mathcal{O}$ notation is used to describe the growth rate of the both schemes. Thus, the system performs arithmetic operations with a complexity of $\mathcal{O}(|S_n|^2)$ and $\mathcal{O}(|S_r|^2)$ when candidate MSs or RS are ranked, respectively, where $|.|$ is defined as the cardinality of the subset. If we assume that $R \gg N$, the complexity of the candidate ranking process is bounded by the number of RSs in the system rather than by the number of MSs. Thereby, this will allow us to upper bound the complexity of the candidate ranking by $\mathcal{O}(|S_n|^2)$ operations. Moreover, forming the MS's preference list, ι_n, Equation 31 requires a sorting operation which induces a complexity of $\mathcal{O}(|S_n|\log(|S_n|))$ operations. Finally, the complexity of the decision-making, Algorithm 1 can be upper bounded by a binary search operation which requires a complexity of $\mathcal{O}(\log(|S_n|))$ operations. Therefore, the dominant factor which determines the CAF scheme complexity will be the one with the largest exponent, thus the complexity of the method will be upper bounded by $\mathcal{O}(|S_n|^2)$ operations.

On the other hand, the centralized global optimum scheme is based on enumerating all possible alternatives for virtual MIMO coalition formation between the MS-n and its preferred subset of candidate RSs, S_n. This is done with the purpose of finding the optimal number of transmit antennas that would minimize the overall power consumption in the reverse link. Therefore, to guarantee that a given feasible solution is optimal, the solution should be compared with any other feasible solution. In general, an exhaustive search approach, where the number of elements is discrete, is considered $\mathcal{NP}$-complete [39]. A notable characteristic of $\mathcal{NP}$-complete problems is that the required time to solve the problem increases very quickly as the size of the problem grows [39]. To implement the exhaustive search scheme, each MS in the system will evaluate the total number of possible combinations in its preferred subset of candidate RSs, S_n. Hence, the total number of possible combinations is computed by $\sum_{k=1}^{|S_n|} \binom{|S_n|}{k}$, where $\binom{|S_n|}{k} = \frac{|S_n|!}{k!(|S_n|-k)!}$. Moreover, each combination is evaluated by Equation 20 or 21 depending if diversity or capacity are

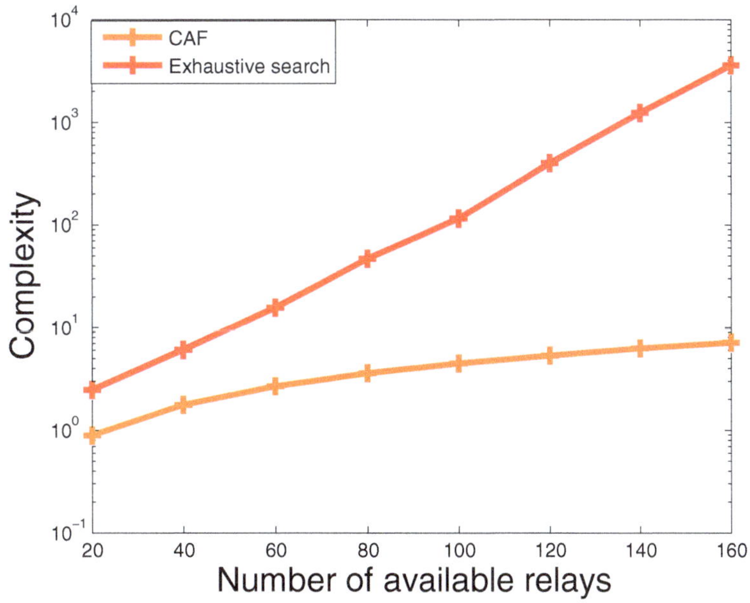

Figure 13 Complexity of the centralized optimum approach compared to the CAF method.

enhanced, respectively. Thus, this induces a complexity of $\mathcal{O}\left(\left(\sum_{k=1}^{|S_m|}\binom{|S_m|}{k}\right)^2\right)$ for the system. In addition, the complexity of both methods (exhaustive search and CAF) increases linearly with the number of MSs in the system, N. Hence, the exhaustive search method has a complexity of $\mathcal{O}\left(N \times \left(\sum_{k=1}^{|S_n|}\binom{|S_n|}{k}\right)^2\right)$ which is a higher order complexity when compared to the complexity of $\mathcal{O}(N \times |S_n|^2)$ for the CAF scheme. Furthermore, Figure 13 shows how the complexity of the system changes for both methods as the number of RSs increases in the system. It can be easily seen that as the number of RSs increases, the computational complexity of the exhaustive search increases exponentially, therefore it may not be a suitable solution to implement in real time systems.

8 Conclusions

In this paper, we considered a low complexity virtual MIMO coalition formation algorithm, which is based on game theory. The framework we proposed allows MSs to select the most suitable RSs providing the most power savings in the network. Thereby, we studied energy efficient coalition formation by using the concepts of diversity and spatial multiplexing, respectively. We have shown analytically and by simulation that increasing the number of transmit antennas is a more energy efficient solution for users close to the cell edge rather than for cell center users, when overall terminal power consumption is optimized. Furthermore, we have proven than the coalition formation algorithm we proposed is more energy efficient compared to the benchmark, the MRH path loss, the BW channel, and the SM framework with improvements of 58%, 15%, 10%, and 5% for the spatial diversity case. When implementing spatial multiplexing, the CAF method has improvements of 14%, 9%, 5%, and 91% over MRH, BW, SMI, and the baseline SIMO mode. It experiences only small performance losses of 10% and 2% when compared to an exhaustive search approach when implementing diversity and spatial multiplexing respectively. In addition, we presented a complexity analysis showing that the complexity of the method we proposed increases linearly as the number of RSs grows in the network. This is a much lower complexity when compared to the exponential growth of the exhaustive search scheme. Thus, the game-theory-based framework we proposed achieves a similar performance compared to a centralized scheme with a much lower complexity order. Hence, it may be a suitable energy-efficient solution for practical applications.

Competing interests
The authors declare that they have no competing interests.

Acknowledgements
Part of this work was done when Rodrigo Vaca was with the Institute of Digital Communications at the University of Edinburgh.
The work of Eitan Altman has been partially supported by the European Commission within the framework of the CONGAS project FP7-ICT-2011-8-317672.

Author details
[1]Scientific Division, Federal Police, 01110, Constituyentes 947, Mexico City, Mexico. [2]The University of Edinburgh, Institute for Digital Communications, King's Buildings, Mayfield Road, EH93JL Edinburgh, UK. [3]INRIA, BP93, 06902 Sophia Antipolis Cedex, France. [4]Department of Electrical Engineering, Universidad Autónoma Metropolitana (UAM), 09340 Iztapalapa, Mexico City, Mexico.

References

1. R Vaca, J Thompson, V Ramos, Non-cooperative uplink interference protection framework for fair and energy efficient orthogonal frequency division multiple access networks. IET Commun. **7**(18), 2015–2025 (2013). doi:10.1049/iet-com.2013.0125
2. C Han, T Harrold, S Armour, I Krikidis, S Videv, PM Grant, H Haas, JS Thompson, I Ku, C-X Wang, TA Le, MR Nakhai, J Zhang, L Hanzo, Green radio: radio techniques to enable energy-efficient wireless networks. IEEE Commun. Mag. **49**(6), 46–54 (2011). doi:10.1109/MCOM.2011.5783984
3. AR Jensen, M Lauridsen, P Mogensen, TB Sørensen, P Jensen, in *Proceedings of the IEEE 76th Vehicular Technology Conference (VTC-Fall)*. LTE UE power consumption model for system level energy and performance optimization (IEEE, Quebec City, Canada, 2012), pp. 1–6. doi:10.1109/VTCFall.2012.6399281
4. R Vaca, J Thompson, V Ramos, in *Proceedings of the IEEE 76th Vehicular Technology Conference (VTC-Fall)*. Uplink interference protection as a non-cooperative game over OFDMA networks (IEEE, Quebec City, Canada, 2012), pp. 1–5. doi:10.1109/VTCFall.2012.6399092
5. R Vaca, E Altman, J Thompson, V Ramos, in *Proceedings of the 24th Annual IEEE International Symposium on Personal Indoor and Mobile Radio Communications (PIMRC)*. A stable marriage framework for distributed virtual MIMO coalition formation (IEEE, London, UK, 2013), pp. 2722–2727. doi:10.1109/PIMRC.2013.6666606
6. R Vaca, J Thompson, E Altman, V Ramos, in *Proceedings of the IEEE INFOCOM Workshop on Green Cognitive Communications and Computing Networks*. A game theory framework for a distributed and energy efficient bandwidth expansion process (IEEE, Toronto, Canada, 2014), pp. 712–717. doi:10.1109/INFCOMW.2014.6849318
7. A Paulraj, R Nabar, D Gore, *Introduction to Space Time Wireless Communications*. (Cambridge University Press, Cambridge UK, 2003)
8. J Jiang, M Dianati, M Imran, Y Chen, in *Proceedings of the IEEE 76th Vehicular Technology Conference (VTC-Fall)*. Energy efficiency and optimal power allocation in virtual-MIMO systems (IEEE Quebec City, Canada, 2012), pp. 1–6. doi:10.1109/VTCFall.2012.6399002
9. S Sesia, I Toufik, M Bake, *LTE-the UMTS Long Term Evolution from Theory to Practice*. (Wiley, Chichester, USA, 2009)
10. Z Han, D Niyato, T Başar, A Hjørungnes, *Game Theory in Wireless and Communication Networks*. (Cambridge University Press, Cambridge, UK, 2012)
11. G Quer, F Librino, L Canzian, L Badia, M Zorzi, Inter-network cooperation exploiting game theory and bayesian networks. IEEE Trans. Commun. **61**(10), 4310–4321 (2013). doi:10.1109/TCOMM.2013.082813.120710
12. S Hussain, A Azim, J Park, Energy efficient virtual MIMO communication for wireless sensor networks. Telecommun. Syst. **42**(1–2), 139–149 (2009). doi:10.1007/s11235-009-9176-7
13. W Saad, Z Han, M Debbah, A Hjorungnes, T Başar, Coalitional game theory for communication networks. IEEE Signal Process. Mag. **26**(5), 77–97 (2009). doi:10.1109/MSP.2009.000000
14. F Pantisano, M Bennis, W Saad, R Verdone, M Latva-aho, in *Proceedings of the IEEE Wireless Communications and Networking Conference (WCNC)*. Coalition formation games for femtocell interference management: a recursive core approach (IEEE, Cancun, Mexico, 2011), pp. 1161–1166. doi:10.1109/WCNC.2011.5779295
15. AS Ibrahim, AK Sadek, W Su, KJR Liu, Cooperative communications with relay-selection: when to cooperate and whom to cooperate with? IEEE Trans. Wireless Commun. **7**(7), 2814–2827 (2008). doi:10.1109/TWC.2008.070176
16. EG Larsson, EA Jorswieck, Competition versus cooperation on the MISO interference channel. IEEE J. Selected Areas Commun. **26**(7), 1059–1069 (2008). doi:10.1109/JSAC.2008.080904
17. K Yazdi, E Gammal, P Schitner, in *Allerton Conference on Communications, Control and Computing*. On the design of cooperative transmission schemes, (2003)
18. N Jindal, U Mitra, A Goldsmith, in *International Symposium on Information Theory (ISIT)*. Capacity of ad-hoc networks with node cooperation (IEEE, 2004), p. 269. doi:10.1109/ISIT.2004.1365306
19. S Cui, A Goldsmith, A Bahai, Energy efficiency of MIMO and cooperative MIMO techniques in sensor networks. IEEE J. Selected Areas Commun. **22**(6), 1089–1098 (2004). doi:10.1109/JSAC.2004.830916
20. Z Han, D Niyato, W Saad, T Başar, A Hjørungnes, *Game Theory in Wireless and Communication Networks*. (Cambridge University Press, Cambridge, UK, 2012)
21. Z Han, R Liu, Fair multiuser channel allocation for OFDMA networks using Nash bargaining solutions and coalitions. IEEE Trans. Commun. **53**(8), 1366–1376 (2005). doi:10.1109/TCOMM.2005.852826
22. W Saad, Z Han, M Debbah, A distributed coalition formation framework for fair user cooperation in wireless networks. IEEE Trans. Wireless Commun. **8**(9), 4580–4593 (2009). doi:10.1109/TWC.2009.080522
23. H Burchardt, S Sinanovic, Z Bharucha, H Haas, Distributed and autonomous resource and power allocation for wireless networks. IEEE Trans. Commun. **61**(7), 2758–2771 (2013). doi:10.1109/TCOMM.2013.053013.120916
24. WA Prasetyo, H Lu, H Nikookar, in *Proceedings of the 41st European Microwave Conference (EuMC)*. Optimal relay selection and power allocation using game theory for cooperative wireless networks with interference. IEEE (IEEE, Manchester, UK, 2011), pp. 37–40
25. S Fujio, D Kimura, in *Proceedings of the IEEE 24th International Symposium on Personal Indoor and Mobile Radio Communications (PIMRC)*. Energy saving effect of solar powered repeaters for cellular mobile systems. IEEE (IEEE London, UK, 2013), pp. 2841–2845. doi:10.1109/PIMRC.2013.6666631
26. J Laneman, D Tse, G Wornell, Cooperative diversity in wireless networks: efficient protocols and outage behavior. IEEE Trans. Inf. Theory. **50**(12), 3062–3080 (2004). doi:10.1109/TIT.2004.838089
27. H Burchardt, Z Bharucha, G Auer, H Haas, Uplink interference protection and scheduling for energy efficient OFDMA networks. EURASIP J. Wireless Commun. Netw. **2012**(180) (2012). doi:10.1186/1687-1499-2012-180
28. H Dai, in *Proceedings of the IEEE International Conference on Acoustics, Speech and Signal Processing (ICASSP)*. Distributed versus co-located MIMO systems with correlated fading and shadowing, vol. 4 (IEEE, Toulouse, France, 2006), pp. 561–564. doi:10.1109/ICASSP.2006.1661030
29. Q Li, G Li, W Lee, M-I Lee, D Mazzarese, B Clerckx, Z Li, MIMO techniques in WiMAX and LTE: a feature overview. IEEE Commun. Mag. **48**(5), 86–92 (2010). doi:10.1109/MCOM.2010.5458368
30. T Cover, J Thomas, *Elements of Information Theory*. (Wiley-Interscience, Hoboken, NJ, USA, 1991)
31. K Bobae, K Cholho, L Jongsoo, A dual-mode power amplifier with on-chip switch bias control circuits for LTE handsets. IEEE Trans. Circuits Syst. **58**(12), 857–861 (2011). doi:10.1109/TCSII.2011.2172528
32. D Gale, M Sotomayor, Some remarks on the stable matching problem. Discrete Appl. Math. **11**(3), 223–232 (1985). doi:10.1016/0166-218X(85)90074-5
33. A Roth, E Peranson, The redesign of the matching market for American physicians: some engineering aspects of economic design. Am. Econ. Rev. **89**(4), 748–780 (1999). doi:10.1257/aer.89.4.748
34. K Iwama, S Miyazaki, in *Proceedings of the International Conference on Informatics Research for Development of Knowledge Society Infrastructure*. A survey of the stable marriage problem and its variants (IEEE, Kyoto, Japan, 2008), pp. 131–136. doi:10.1109/ICKS.2008.7
35. S Sanayei, A Nosratinia, Antenna selection in MIMO systems. IEEE Commun. Mag. **42**(10), 68–73 (2004). doi:10.1109/MCOM.2004.1341263
36. D Gale, LS Shapley, College admissions and the stability of marriage. Am. Math. Mon. **69**(1), 9–15 (1962)
37. V Sreng, H Yanikomeroglu, D Falconer, in *Proceedings of the IEEE 58th Vehicular Technology Conference (VTC-Fall)*. Relayer selection strategies in

cellular networks with peer-to-peer relaying (IEEE, 2003), pp. 1949–1953. doi:10.1109/VETECF.2003.1285365
38. EG Larsson, EA Jorswieck, J Lindblom, R Mochaourab, Game theory and the flat-fading gaussian interference channel. IEEE Signal Process. Mag. **26**(5), 18–27 (2009). doi:10.1109/MSP.2009.933370
39. Z Han, KJR Liu, *Resource Allocation for Wireless Networks.* (Cambridge University Press, Cambridge, UK, 2008)

A dynamic energy-saving strategy for green cellular railway communication network

Jiying Huang[1*], Zhangdui Zhong[2] and Hongwei Huo[3]

Abstract

Energy saving is seldom contemplated in cellular railway communication network because safety is always the main focus. However, we analyzed train schedules and determined that the cellular railway network had considerable potential for energy saving. We designed a simple train detection strategy using the handover procedure to initiate power saving. We also proposed a flexible and effective solution based on the mealy-type finite state machine to transit the cell state between an energy-saving mode and a full running mode. The key design principle of the proposed solution consists in forming an orthogonal group and cluster, based on network deployment and train schedules, to simplify the state transition. Our evaluations using a real network deployment led to a significant energy saving of up to 33% with different power-saving strategies and factors.

Keywords: Green cellular railway network; Energy saving; Finite state machine

1 Introduction

Mobile communication networks have experienced tremendous increases over the last 5 years in subscriber/device quantities, service variety, and traffic loads. This tendency leads directly to ever-increasing energy consumption and CO_2 emissions in mobile cellular networks. According to the Green radio report [1], 57% of the energy consumption of the information and communication technology (ICT) industry is attributed to users and network devices in mobile and wireless networks, the scale of which is still growing exponentially [2]. A quantitative analysis [3] has estimated that ICT contributed to about 2% of global CO_2 emissions in 2007, and the proportion for cellular networks will double by 2020. Therefore, compared with the traditional design of mobile wireless networks which mainly focuses on ubiquitous access and large capacity, the green cellular wireless network [4-6] has recently received the significant attention, with the aim of developing more energy-efficient systems to make a contribution to reducing the carbon footprint.

Because base stations (BSs) constitute the most significant component of wireless cellular networks, many green communication techniques focus on BSs activities. Recently, these green strategies have received increasingly substantial attention from both industry and academia, and several projects such as EARTH, Green Touch, OPERA-Net, Green Radio, and GSMAMEE [7] have been funded for research, deployment, and evaluation. As shown in Figure 1, a typical BS contains three main components of energy consumption: a DC power system, cooling, and a radio unit with RF conversion and signal processing control. According to a research study [8], 62% of energy consumption came from the radio unit, 25% of power was consumed by the cooling system, and 11% by the DC power system. In the literature [9,10], the radio base station system, which uses 10.3 kW of electricity to produce only 120 W of effective energy usage for signal transmitting and processing, has quite a low energy efficiency of 1.2%. Therefore, many green strategies [11,12] proposed to switch off the radio unit during its idle time.

In a railway system, the wireless communication platform is the vital control base that administers the operation of the trains and communications among the railway operational staff. With the rapid increase in high-speed railways, a dedicated communication system, currently GSM for railway (GSM-R) in China plays an increasingly important role. GSM-R is widely used in high-speed railways involving passenger transport, and in heavy haulage railways, to provide a communication

* Correspondence: EJIYING@126.COM
[1]School of Electronic and Information Engineering, Beijing Jiaotong University, No. 3 Shangyuancun, Beijing 100044, China
Full list of author information is available at the end of the article

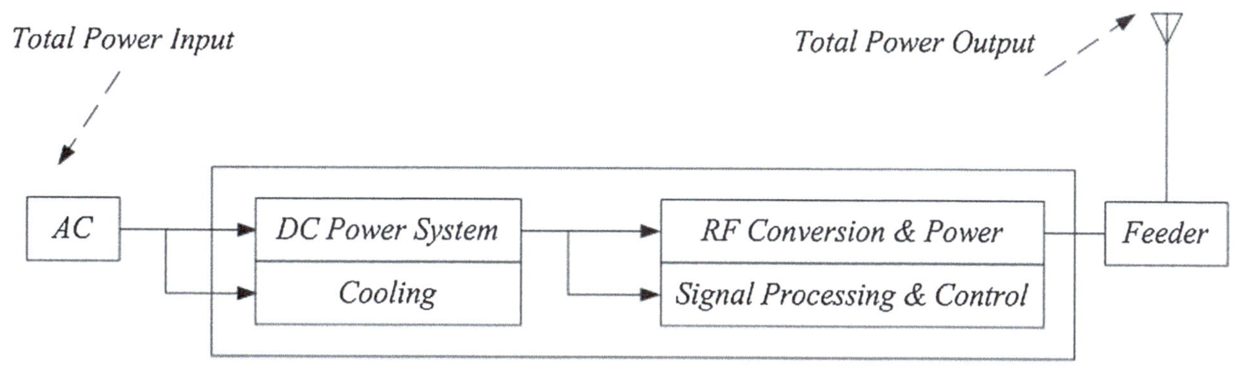

Figure 1 Typical base station architecture.

platform for control messages and application services. Twenty-one countries in the world apply GSM-R in railways, such as China, Germany, Italy, and France which have between 3,000 and 4,000 BSs [13]. A single S4/4/4 BS consumes 26,000 kWh per year [14]. Taking 3 km as an example for the coverage of 1 BS, 80 BSs in the 240-km railway system consume about 2,080,000 kWh per year. This is an enormous cost and causes large CO_2 emissions. From this viewpoint, it is important and urgent to apply power-saving strategies to cellular railway networks.

Compared with a mobile cellular network using a planar coverage scenario, mostly in hot spots, venues, malls, and cities, a dedicated wireless communication network GSM-R deployment has a specific chain structure, and cellular cells are aligned sequentially along railway tracks, as shown in Figure 2. The recent green communication techniques cannot be applied in a dedicated cellular railway network because base stations along the tracks are not allowed to power off due to high security requirements. These security requirements introduce greater challenges for energy saving in a cellular railway network.

In this study, our purpose is to make cellular railway networks more energy efficient. Shown in a real scenario, it is possible to apply the current power-saving strategy to a cell, especially at the idle time; there is less service requirement for railway communication. However, for cellular railway networks, the first question is what event should initiate power saving in a high safety environment. Note that it is impossible to obtain an accurate train schedule in advance. We designed a simple train detection strategy based on the handover process when establishing a communication. Another challenge was to design a mechanism to operate the cell state between an energy-saving mode and a full

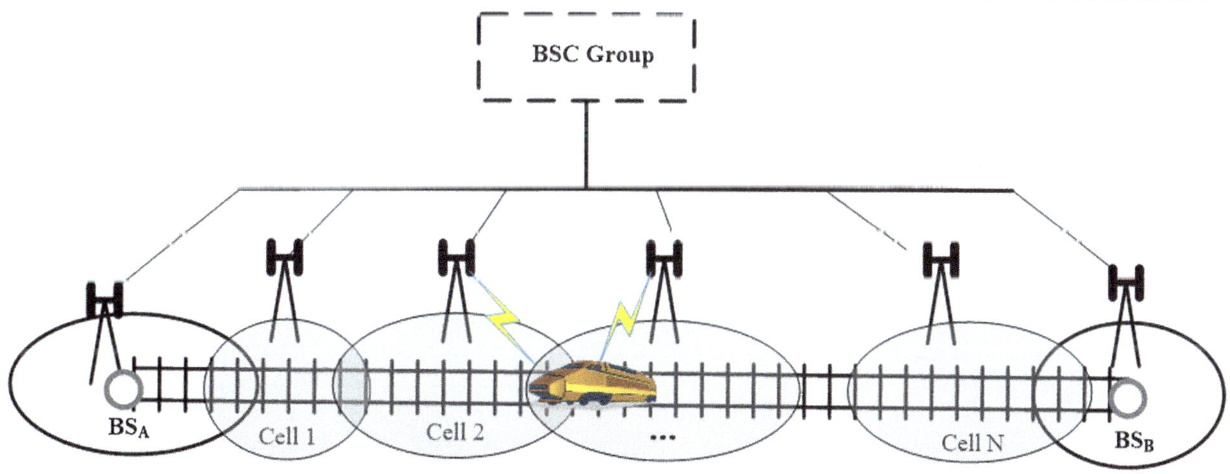

Figure 2 System model for cellular railway networks.

running mode. Indeed, the trains dynamically run in different cellular regions. We then proposed a flexible and effective solution based on the mealy-type finite state machine. It was shown that its evaluation, using a real network deployment, led to significant energy savings with different power-saving strategies and that the train schedules and speed factors were also verified for our solution. Compared with studies in the literature related to green cellular communication techniques, our contribution covers many aspects:

- To the best of our knowledge, this is the first paper to propose an energy-efficient solution for a green cellular railway network without breaking the rules in the 3GPP Standard.
- We present a new train detection strategy to trigger the power saving which monitors the HANDOVER message in a railway network, rather than obtaining advance the train schedules.
- We propose to take into consideration of the uncertainty of speed measures and bidirectional traffic of trains, and then propose a flexible and effective solution based on a mealy-type finite state machine.
- We provide a detailed algorithm applying a divide and conquer approach in our study, and BSs can easily transfer their state without complicated computation.
- We evaluated our method using the data from real scenarios of railway networks, to identify the many aspects of performance and the impact on the system configuration.

The remainder of this paper is organized as follows. Section 2 discusses related work, and Section 3 provides the model description of the system and proposes the solution and algorithm for two strategies, in detail. Section 4 evaluates the performance using simulation analysis, followed by concluding remarks in Section 5.

2 Related work

Because energy saving and environmental protection have become a global demand and an inevitable trend, the growing awareness of energy efficiency has paved the way for new technologies in designing green cellular networks. Energy efficiency has the advantages of maintaining profitability and reducing the overall impact on the environment. BS has attracted significant attention and has been considered in all stages of cellular networks, especially in the hardware and architecture design, and BS deployment and operation.

Hardware improvements such as using more energy-efficient power amplifiers, mean natural resources for cooling, and renewable energy sources can be considered a good approach to achieve energy efficiency. Ericsson Ψ (Psi) [15] and Flexenclosure E-site solutions [16] are examples of such efforts. A comprehensive survey on how to generate renewable energy is provided in [17], such as solar, wind, and fuel cells, for green base stations in 5G green cellular networks. However, these products and methods are inadequate in dealing with energy waste in a railway cellular network. In railways, it is possible that no trains pass through a specific area for long periods of time, but a cellular operation continues running, resulting in much energy wastage by BSs.

A solution with a new architecture was presented for energy saving, considering current cellular networks cannot meet the design requirements of energy efficiency. The studies [18,19] consider a heterogeneous architecture for network planning. Combining macro-cells with micro-cells and pico-cells, a theoretical relationship between energy efficiency and network deployment is derived and evaluated. Cognitive radio and cooperative relaying are also significant technologies to improve the throughput and coverage in future networks. A non-stationary spectrum sharing policy to improve the spectrum and energy efficiency is proposed in [20]. These new architectures and technologies cannot be directly used in railway cellular networks because they are complex and lack flexibility. Furthermore, cellular railway networks are not ideal because of the high security requirements.

Because the radio unit, as shown in Figure 1, consumes up to 62% of the entire energy usage, a dynamically energy-efficient BS operation based on an on/off switching operation can be one of the flexible solutions to reduce the power demands of base stations. Traffic loads in cellular networks have significant features in space and time because of a number of factors such as user mobility and behavior. The BSs are entirely shut down when experiencing low traffic load, and the corresponding traffic loads are transferred to neighboring cells during periods of low use. In [21], the authors suggest the possibility of energy saving using dynamic BS operation based on a simple analytic model from ideal networks. Some cells may be switched off to save energy during daylight for residential areas and at night for office districts. It shows energy savings of the order of 25 to 30% are possible with several regular cellular topologies. In [22], a profile-based approach was designed to profile base station traffic and approximate the network-wide energy proportionality using non-load-adaptive base stations, and power off of underused base stations under light traffic. This approach is quite suitable for a large, dense city, achieving 53% energy savings, while a medium-sized city could achieve 23% energy savings. A similar solution is proposed in [12] to save 50 to 80% energy in a metropolitan urban area, implementing a distributed approach using a newly introduced notion of network impact to take into account the additional load increments brought to its neighboring BSs.

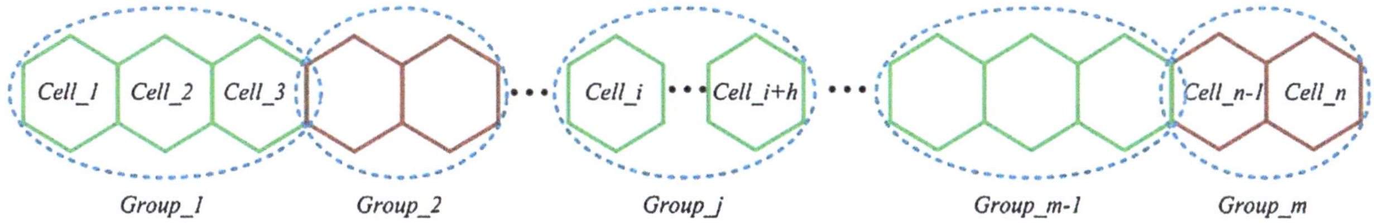

Figure 3 Cell and orthogonal group demonstration.

However, many of these studies only focused on ideal networks, such as a hexagonal model network. They suggested that energy could be saved using dynamic BS switching algorithms. Although other studies have considered actual deployed network scenarios, it was difficult to implement these algorithms in practice because of factors such as complexity, lack of flexibility, and diversity of practical network deployment scenarios. Finally, all these related studies take only regular wireless network deployment scenarios into consideration, instead of focusing on the special wireless network deployment scenarios along railway tracks in which the cell is based on a chain structure instead of a planar structure. In a railway-dedicated mobile network, the traffic load does not follow the daytime/nighttime regular pattern, and the user density is much lower than in public mobile communication networks. Therefore, a new energy efficiency model, based on the internal parameters of a railway cellular network in administering the operation of a network cell, is needed.

3 System description and the proposed algorithm

3.1 System model

We consider a wireless cellular railway network which contains a set of N cells, covering railway tracks from station BS_A to station BS_B, as shown in Figure 2. The cells in BS_A and BS_B, which monitor the trains' entry from BS_A to BS_B, will be active. They only carry out the energy-saving strategy some of the time. Our focus is on dynamic energy saving with active or not strategies for base stations when trains leave the cells controlled by these base stations.

The set of N cells is controlled by one or several base station controllers (BSCs), and the dynamic energy-saving strategies will be operated by these BSCs.

According to the system model, two main problems arise, for which we will provide corresponding strategies in the following subsections:

(1) How can the cell detect the entry or departure of trains in a particular interval without any prior knowledge of the train schedules?
(2) How can the cell change its energy-saving status given the time-varied distribution of trains on the railway track?

3.2 Train detection strategy

It is impossible to obtain the train schedules in advance because of the impact of many factors such as weather, freight requirements, and train status, and it is also not feasible to use the information to schedule the status of the cells. In a GSM-R network, there are two types of special terminals: a vehicle station with higher RF power capability (8 W) in trains and a normal handheld mobile station (2 W), to transmit and receive messages to control the train schedule.

When a train is scheduled in a typical railway, such as the Beijing-Guangzhou and Zhengzhou-Xi'an railways in China, a vehicle station will occupy the dedicated wireless channel from the train's start-off until its arrival at its destination. A connection should be initialized between the vehicle station and base station: the vehicle station sends out a Connect Management (CM) SERVICE REQUEST message to the network. The CM SERVICE REQUEST has a RF power-capability element to indicate the type of vehicle station and an International Mobile Subscriber Identity (IMSI). During the traffic channel assignment procedure, one dedicated traffic channel is

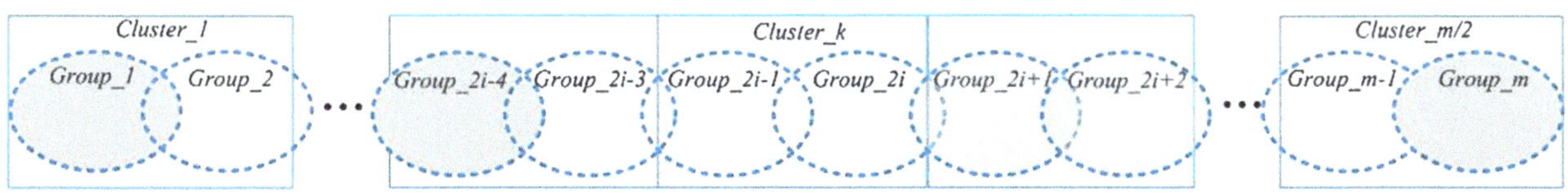

Figure 4 Two orthogonal groups in one cluster.

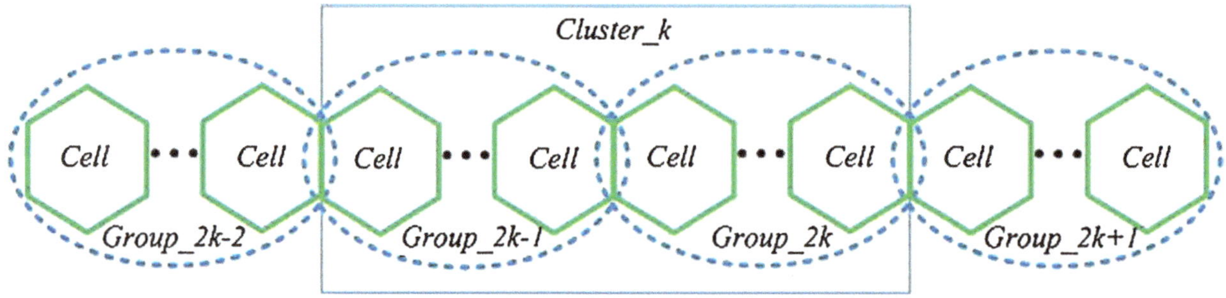

Figure 5 Details of *Cluster_k* in Figure 4.

then assigned to the vehicle station, and all the channel properties are contained in the message ASSIGNMENT COMMAND.

When a train enters into the next neighbor cell $i+1$ under base station $i+1$ from cell i, the vehicle station on the train (8 W), as shown in Figure 1, will send out the MEASURE REPORT message to the BSC, which decides whether the vehicle station should start the handover procedure using the message HANDOVER COMMAND. After finishing the successful handover procedure, the vehicle station is now connected to the new cell $i+1$, a HANDOVER COMPLETE message is sent from the vehicle station to the base station $i+1$ to indicate that establishing the new connection is completed. Besides, the old wireless channel in cell i is released by the message RF CHANNEL RELEASE. The message HANDOVER COMPLETE is the indicator that the train has entered the next cell and left the current one.

3.3 State transition strategy

The railway characteristics indicate that a cell's status is related to neighboring cells, and the finite state machine model is a powerful method for dealing with cell status transitions. The mealy-type state machine [23], a type of Finite state machine, is based on the present state of the machine, and the input conditions, and can effectively lead to a reduction in the number of states.

3.3.1 Cell state

Each cell along the railway tracks consists of two states, namely active and passive. Because there is a high risk to powering off the radio unit in a cellular railway network, the passive state uses a time slot power-saving strategy [24], which was designed to reduce the energy consumption on the idle time slot. The strategy is to adjust the power amplifier bias to a much lower level when no traffic is on the carrier. When the cell is controlled to be in an energy-saving mode by the BSC node, the corresponding cell enters into the passive state; otherwise, it stays in the active state, which is described as follows:

$$c_i = \begin{cases} 1\ (\text{Active}), & \text{BS is not in power saving mode} \\ 0\ (\text{Passive}), & \text{BS is in power saving mode} \end{cases} \tag{1}$$

where c_i is the state of cell i. The energy-saving mode can be achieved by activating different energy-saving strategies according to the base station's specific requirements.

3.3.2 Orthogonal group state

Based on the train detection strategy, first, all the cells along the railway are divided into a number of orthogonal groups without taking into account the overlapped area covered by two adjacent cells. The purpose of forming orthogonal groups is to hide the differences in several special cells that also cover railway stations or residential areas along the railway tracks. On the other hand, some cells have a relatively small radius, and at the same time, the train's speed may be very high, which may lead to insufficient time to perform cell state transition. Using the concept of an orthogonal group, we assume the whole unit performs a state transition. To sum up, the cells are firstly encapsulated to form an orthogonal group to simplify the problem to be solved.

Table 1 States of *Cluster_k* of Figure 5

Group_2*k* − 1	Group_2*k*	Cluster state
0 (passive)	0 (passive)	S_{00}
0 (passive)	1 (active)	S_{01}
1 (active)	0 (passive)	S_{10}
1 (active)	1 (active)	S_{11}

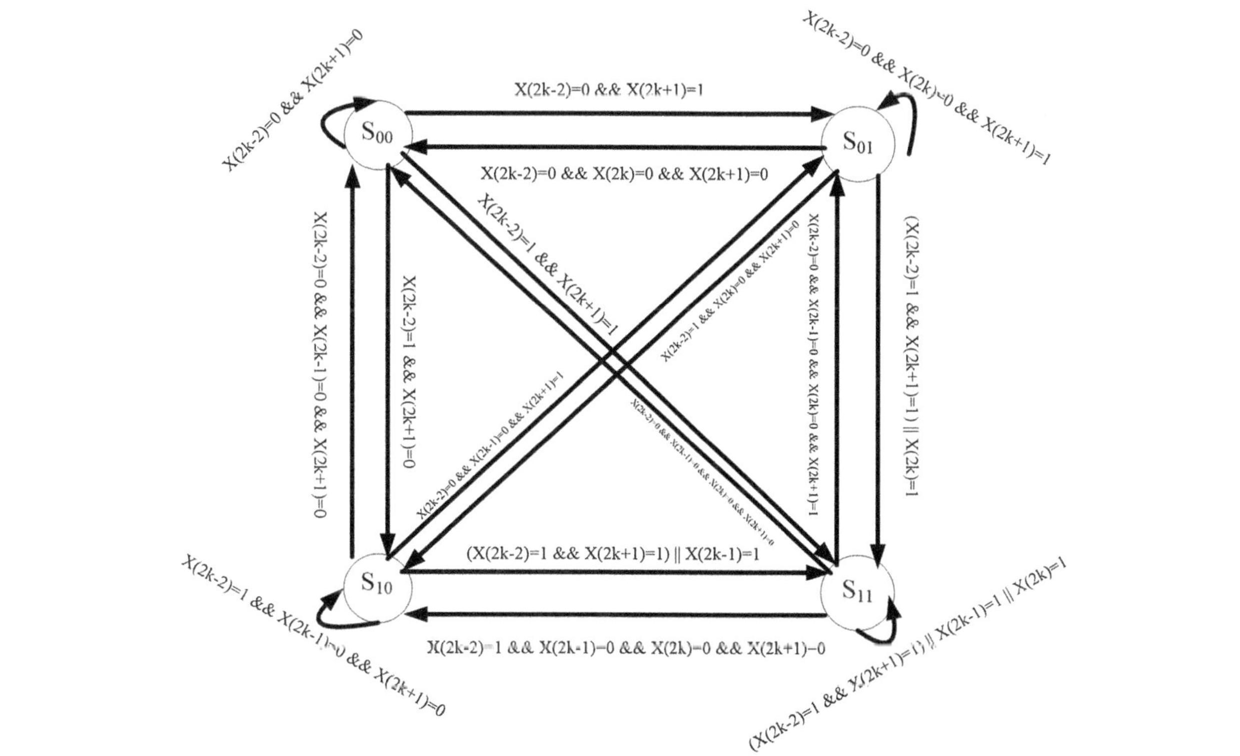

Figure 6 State transition diagram of *Cluster_k* of Figure 5.

The number of cells in each orthogonal group can be flexibly configured as depicted in Figure 3. We set the orthogonal group to a range for each group that makes less impact on trains' speeds. The state of the group has a close relationship with the state of the cells. If all the cells in the group are in a passive state, that is to say, all BSs are in power-saving mode, the whole orthogonal group is in a passive state. Correspondingly, if even one cell in the orthogonal group is active, the state of the whole group is set to active. g_i is the state of ith orthogonal group, h_j is the number of cells in this group. The state of orthogonal group is given by

$$g_i = \begin{cases} 1 \text{ (Active)}, & c_{h_1}||c_{h_2}||\ldots||c_{h_j} = 1 \\ 0 \text{ (Passive)}, & c_{h_1}||c_{h_2}||\ldots||c_{h_j} = 0 \end{cases} \quad (2)$$

3.3.3 State transition by dividing orthogonal groups into orthogonal clusters

Let m be the number of orthogonal groups. As a result, 2^m possible group states exist. The group state is transferred between different states depending on the present group state and the input event. Obviously, the state machine will become more complex with an increase in groups, because more groups mean more possible group states.

To solve the state transition problem in a more efficient manner, we use the divide and conquer method [25] in this paper. In computer science, divide and conquer is an important algorithm design paradigm based on multi-branched recursion. The divide and conquer algorithm works by recursively breaking down a problem into two or more sub-problems of the same or related types, until these become simple enough to be solved directly. The solutions to the sub-problems are then combined to solve the original problem.

In Algorithm 1, the procedure from Step 3 to Step 14 shows how to construct orthogonal groups using input information from Steps 1 and 2. With the divide and conquer design paradigm, we divide the orthogonal groups

Table 2 Scenario parameters

Parameter	Value
Number of BSs	43
BS core power	60 W
Cell radius average	3.9 km
Inter-cell distance standard deviation	2.02 km
Power-saving strategies	10%, 20%, 25%, 35%
Train scheduling time	30 min
Train speed	80 km/h
Time interval for state	10 s
$d_{\text{threshold}}$	3 km

Algorithm 1 The outline of the proposed method

Input: the number of cells n, the set of cells $C = \{c_i,\ldots, c_n\}$, the inter-cell distance $d_{i,i+1}$ ($i = 1,\ldots, n - 1$), the special cell indicator λ_i (i.e., c_i is a special cell if $\lambda_i = 1$), predefined threshold inter-cell distance $d_{\text{threshold}}$;

Group (set $G = \{g_1 = \emptyset,\ldots, g_j = \emptyset,\ldots\}$) formation based on the input information;

$j = 1$;

FOR $i = 1,\ldots, n$ **DO**

  IF $(d_{i,i+1} < d_{\text{threshold}}$ && $d_{i-1,i} \geq d_{\text{threshold}})\|(\lambda_i = 1)$ **THEN**

    c_i and c_{i+1} are added into g_j, $g_j = g_j \cup \{c_i,c_{i+1}\}$;

  ELSE IF $(d_{i,i+1} \geq d_{\text{threshold}}$ && $d_{i-1,i} \geq d_{\text{threshold}})\|(\lambda_i = 0)$ **THEN**

    c_i is added into g_j, $g_j = g_j \cup \{c_i\}$;

  ELSE IF $d_{i,i+1} < d_{\text{threshold}}$ && $d_{i-1,i} < d_{\text{threshold}}$ **THEN**

    c_{i+1} is added into g_{j-1}, $g_{j-1} = g_{j-1} \cup \{c_{i+1}\}$;

  END IF

  $j = j + 1$;

END FOR

The set of groups G is formed and the element of G is m (i.e., m groups);

Cluster formation based on divide and conquer method;

FOR $j = 1,\ldots, m$ **DO**

  g_j and g_{j+1} are formed into cluster_$(j + 1)/2$;

END FOR

$[m/2]$ clusters are obtained as a result;

State transition for BS power saving cluster by cluster;

WHILE (TRUE)

  In each time interval t, wireless access request in each group is measured;

  FOR $i = 1,\ldots, [m/2]$ **DO**

    Cluster_i state transition is performed based on the measurement results;

  END FOR

END WHILE

into several orthogonal clusters which are outlined between Step 15 and Step 19, with each orthogonal cluster containing the same number of groups. As a result, performing the state transition of orthogonal groups in one orthogonal cluster becomes a sub-problem shown in the left steps of Algorithm 1. In this way, the solutions to sub-problems can first be obtained, and then combined, to achieve the final solution to the original problem.

3.3.4 Detailed minimum orthogonal cluster solution

To decrease the complexity to exploit the relationship between adjacent groups, a large state machine with 2^m group states is divided into *m/*2 small state machines with four group states (00, 01, 10, 11) for each small state machine. We call the cluster with two groups the minimum orthogonal cluster, and the solution is to solve the sub-problem with the group state transition in one cluster. *Cluster_k* in Figure 4 is selected to further illustrate how a small mealy-type state machine works to solve the sub-problem. The details of *Cluster_k* are given in Figure 5.

Obviously, four possible states (00, 01, 10, 11) exist for *Cluster_k* as demonstrated in Table 1. State transitions are conducted between the four states. In each time interval lasting *t*, the wireless access request of each orthogonal group is measured. For instance, we measure the wireless access request of each orthogonal group in the time interval (*t*, 2 *t*) at time 2 *t*. The results of measurement are the events or inputs to the related state machines, which will have an effect on the state transition following the measurement. The events or inputs can be described as follows:

$$X(i) = \begin{cases} 1, \text{WirelesAccessRequest}_{\text{Group}_i} > 0 \\ 0, \text{WirelesAccessRequest}_{\text{Group}_i} >= 0 \end{cases} \quad (3)$$

Wireless_Access_Request_Group$_i$ represents the wireless access request of the group *i* in a certain time interval. *X*(*i*) is an indicator of whether there is a wireless access request in group *i*. In addition, *X*(*i*) plays the role of the input event to state machines in the proposed solution.

We propose the specific rules (i.e., the state transition function) for a state transition in each cluster. Figure 6 shows the state transition diagram for *Cluster_k* of Figure 5. A state transition is executed if events occur.

A state transition from S_{00} to S_{01} is used as an example to describe the procedures as shown in Algorithm 2. It is assumed that the present state of *Cluster_k* is S_{00}. The measurement module measures the wireless access request in each time interval, and measurement results are output to set the energy-saving mode. If no wireless access requests are detected in Group_2*k* – 2 (*X*(2 *k* – 2) = 0)

Algorithm 2 State transition of *Cluster_k* from S_{00} to S_{01}

1: Initial conditions: *Cluster_k* is in state S_{00} (i.e., all the non-special BSs in Group_2*k* – 1 and Group_2*k* are in power saving model)

2: **WHILE(TRUE)**

3: In each time interval *t*, wireless access request in each group is measured;

4: Measurement results are used for judgment;

5: **IF** *X*(2*K* – 1) = 0 && *X*(2*K* + 1) = 1 **THEN**

6: All the BSs belonging to Group_2*k* should get out of power saving model;

7: State transition is executed and *Cluster_k* enters into state S_{01};

8: **ELSE THEN**

9: Other judgments are performed further for state transition;

10: **END IF**

11: **END WHILE**

and there are wireless access requests in Group_$2k+1$ ($X(2k+1)=1$), all the BSs in Group_$2k$ are controlled to be out of the energy-saving mode. As a result, S_{00} is transferred to S_{01}.

The state transition for each cluster is performed simultaneously, and after the state transitions of all the clusters have been performed, the state of each orthogonal group is determined, which solves the original problem.

4 Evaluation analysis

In this section, we evaluate our energy-saving solution based on real GSM-R network data along a certain railway. Forty-three cells that cover the railway tracks between two BSs are selected to be optimized for BS energy consumption. More parameters can be found in Table 2 according to GSM-R network planning along the tracks. The algorithms and procedures for the simulations[a] are carried out using the Python language, which has functional and structured programming to handle railway simulations. First, we set up our simulation scenario for the power-saving statistics of each BS. In the time interval of state transition, trains can be scheduled to locations according to their speeds and scheduling times, and each BS calculates the power consumption using its cell state and power-saving strategy. Next, orthogonal groups are formed for these 43 cells. To our mind, it is important for the orthogonal group to make a low impact on the speed of the trains for the cells with a small cell radius. We use Figure 7 to show the inter-cell distances of the 43 cells. It shows some cells are close enough to assign them to one orthogonal group according to our solution. Here $d_{\text{threshold}}$ is highlighted for 3 km, and for simplicity, we select one direction to create the orthogonal group which is carried by the BSC group in Figure 2 and is fixed for trains in both directions. As a result, cells with BS_9 and BS_{10}, and cells with BS_{24} and BS_{25} in Figure 7 are respectively assigned to one group. In the next three subsections, the BS energy-saving ratio for the selected GSM-R network is investigated with the variation in BS power saving in the passive state, time interval for train scheduling, and train speed. In each of the three cases, the head and tail cells, which cover two stations, are always in the active state, that is to say, the two cells are not controlled for power saving.

4.1 Impact of the RBS power-saving model

Although it is not permissible for the system to entirely turn off some underused BSs in a low traffic period, because of the dedicated cellular railway network's safety requirements, there are also three power-saving strategies that can be implemented in a cellular railway network according to the BS architecture in Figure 1. A summary of the approximate energy-saving ratios is shown in Table 3. As previously mentioned, the BS should take action for power saving when a cell is in the passive state. To evaluate the impact of the amount of BS power saving on the energy-saving ratio for each BS within the selected GSM-R network, extensive simulations were conducted. We assume the train speed remained at 80 km/h, and the time interval for train scheduling was 30 min. Figure 8 shows the simulation result. "Power saving 10%" means the BS power can be reduced by 10% for power saving when the corresponding cell has been in the passive state. With the increase in the amount of BS power saving, higher BS energy-saving ratios can be obtained. The BS energy-saving ratio cannot reach the amount of BS power saving because the cells along the railway are not always in a passive state. Besides, more energy saving can be achieved in the case of shorter inter-group distances.

4.2 Impact of train schedule

The time interval for scheduling trains has a considerable effect on the BS energy saving. If more trains pass through the railway, it will lead to longer times for BSs staying in their active state to serve these trains. Time intervals of 15, 30, 60, and 120 min, for scheduling trains, were chosen to evaluate our energy-saving solution. Moreover, during simulations, when the train speed was 80 km/h, the BS power reduced by 35% when the corresponding cell was in the passive state. As demonstrated in Figure 9, a longer time interval for scheduling trains results in a higher BS energy-saving ratio, because more BSs are stayed in the passive state. We also noticed that when the time interval for scheduling trains was short enough, some BSs belonging to groups with long inter-group distances did not enter into the power-saving model. A shorter time interval for scheduling trains means more trains running on the railway, and a longer inter-group distance means that trains need more time to pass through these groups, both of which lead to the phenomenon.

Table 3 Energy saving ratio with power-saving strategies

Power-saving strategy	Feature	Energy saving ratio (%)
Activation on radio unit	BTS power saving/time slot power saving [24,26]	20
Activation on cooling system	Passive cooling [27]	10
Activation on radio unit and DC power unit	DC power saving related on radio unit	25
Activation on all three parts	All	35

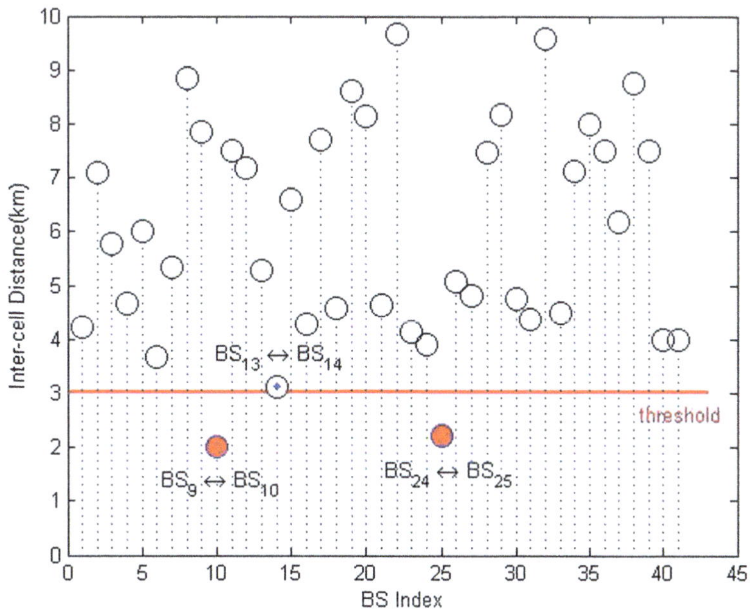

Figure 7 Inter-cell distances of the 43 cells.

4.3 Impact of train speed

The speed of trains was also investigated to evaluate the proposed solution for BS energy saving. The variation in train speed had an effect on the time taken by a train to pass through a group. In the simulations, the time interval for scheduling trains was 30 min, and the BS power reduced by 35% for power saving when the corresponding cell was in the passive state. Figure 10 shows that a higher train speed brings about a higher BS energy-saving ratio, because trains passed through a group in a shorter period of time. Thus, BSs could remain in the power-saving mode longer. In addition, Figure 10 demonstrates that the BS energy-saving ratio was much closer when the speed of the trains was great enough, because trains passed through a group in a very short time when at high speed.

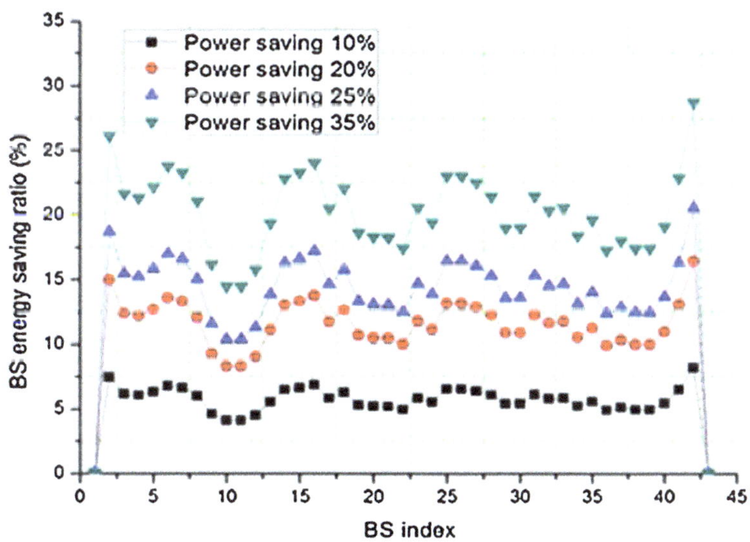

Figure 8 BS energy-saving ratio with variation of BS power saving in the passive state.

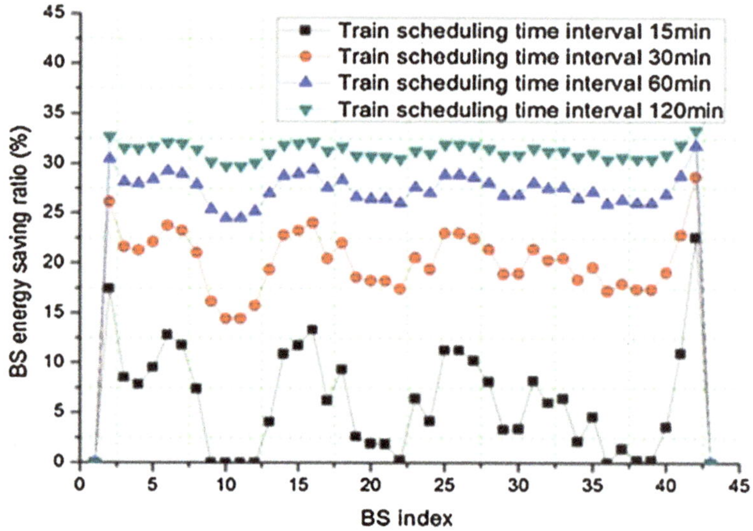

Figure 9 BS energy-saving ratio with variation of time interval for scheduling trains.

5 Conclusions

In this study, we focused on the issue of a dynamic energy-saving strategy in a cellular railway network. According to the practical analysis, we concluded that energy saving was necessary for cells that did not serve as a vehicle station. Taking into account the difficulty of collecting the train schedules in advance, we proposed a train detection strategy based on the handover process. Furthermore, a mealy-type finite state machine-based solution for BS energy saving along railway tracks was proposed, which flexibly and effectively transferred the cell's state and can be applied to power-saving models. Finally, we built our simulating environment based on a real cellular railway network, and the numerical results showed that energy saving can be achieved using different power-saving models, train scheduling, and speed, providing a promising solution for dynamically reducing energy consumption for BSs along railway tracks. Future study will involve more cellular networks which can provide communication services for both railway and the habitants along the tracks.

6 Endnotes

[a]In the evaluation part, we only implemented the power-saving strategy using the Python language on the top layer. We did not touch any implementation of

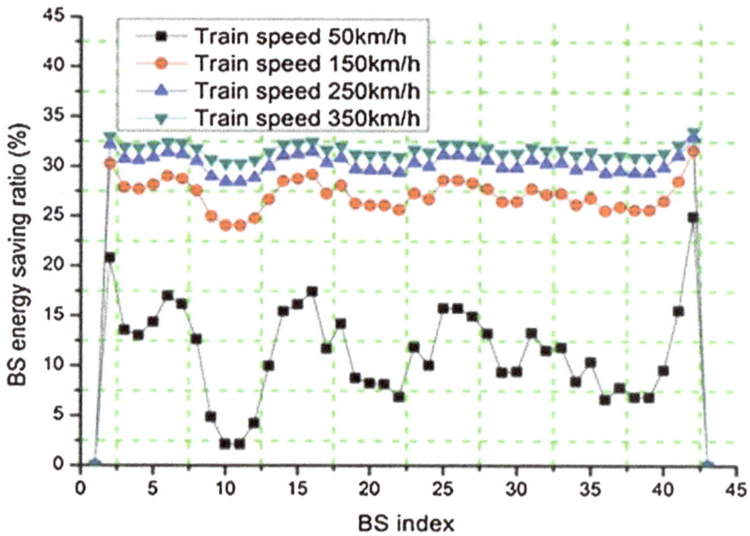

Figure 10 BS energy-saving ratio with variation in train speed.

physical (PHY) layer. We believe that this is sufficient to prove the effectiveness of our strategy because we use the HANDOVER message to detect the train arrival in the detection part of our power-saving strategy. Handover is the normal procedure when a vehicle station enters from one cell to another according to the 3GPP Standard. The HANDOVER message is the mandatory message for the handover procedure. We propose the power-saving strategy on the basis of the handover procedure. Therefore, without any modification for PHY layer in the GSM-R railway network, our strategy can be directly carried out on the BSC part as the operators' view. However, for the data and topology (shown in Figure 2), we used the evaluations that are all based on the real scenario. As the data list is too long to show in this article, we directly evoke the database in our evaluation. For the detailed raw data of the topology, real signaling capture and simulation results, please access http://pan.baidu.com/s/1ntoO2gx or http://eaie.bjtu.edu.cn/rawdata_Green_CELL.zip to download, if there are any further questions, please contact ejiying@126.com.

For the same reason, it is also unnecessary to evaluate the robustness of our evaluation. The power-saving strategy will be activated only by the parameter of detecting the HANDOVER message. As mentioned above, the handover procedure is robust and reliable in the GSM-R network because of its high security requirements. Therefore, we can use the message to ensure our solution's robustness.

Competing interests
The authors declare that they have no competing interests.

Acknowledgements
The authors express their thanks for the support from the National Science Foundation of China under Grant 61222105, the Beijing Municipal Natural Science Foundation under Grant 4112048, the Program for New Century Excellent Talents in Universities of China under Grant NCET-09-0206, the NSFC under Grant 60830001, and the Key Project of State Key Lab under Grant no. RCS2011ZZ008. Portions of this article were previously presented at IEEE VTC-2011 Spring, titled, Dynamic green communication strategy for railway cellular network, April 2014.

Author details
[1]School of Electronic and Information Engineering, Beijing Jiaotong University, No. 3 Shangyuancun, Beijing 100044, China. [2]School of Computer and Information Technology, Beijing Jiaotong University, No. 3 Shangyuancun, Beijing 100044, China. [3]China Science and Technology Exchange Center, No. 54 Sanlihe Road, Beijing 100045, China.

References
1. McLaughlin S (2008) Green radio: the key issues programme objectives and overview. Wireless World Research Forum, Stockholm, Sweden, October 13-15, 2008. Available on http://wwrf.ch/files/wwrf/content/files/publications/libraries/Library21.pdf
2. J Hansen, M Sato, P Kharecha, G Russell, DW Lea, M Siddal, Climate change and trace gases. Philos Trans RSoc **365**(1865), 1925–1954 (2007)
3. A Fehske, G Fettweis, J Malmodin, G Biczok, The global footprint of mobile communications: the ecological and economic perspective. IEEE Commun Magazine **49**(8), 55–62 (2011)
4. Z Hasan, H Boostanimehr, VK Bhargava, Green cellular networks: a survey, some research issues and challenges. Commun Surveys Tutorials, IEEE **13**(4), 524–540 (2011)
5. D Antonio De, S Emilio Calvanese, C Antonio, Enabling green cellular networks: a survey and outlook. Comput Commun **37**, 5–24 (2014). ISSN 0140-3664
6. SA Waje, SB Rahane, A survey of Green wireless communications, international journal of electronics communication and instrumentation engineering research and development (IJECIERD), ISSN 2249-684X, Vol.3, Issue2, 25–36, June, 2013
7. X Wang, AV Vasilakos, M Chen, Y Liu, TT Kwon, Taekyoung. A survey of green mobile networks: opportunities and challenges. Networks Appl **17**, 4–20 (2012)
8. S Roy, Energy Logic for Telecommunications, Emerson Network Power Energy Systems, North America, Inc, September, 2008. http://www.emersonnetworkpower.com/documents/en-us/brands/energysystems/documents/netxtend%20pages/energy%20logic%20for%20telecom%20white%20paper.pdf (webpages accessed on February 8, 2014
9. White Paper Ericsson (2007). Sustainable energy use in mobile communications white paper. EAB-07:021801, Ericsson AB, Stockholm, Sweden, 2007
10. A Ani Vincent, O Ani Emmanuel, Assessment of carbon credits for power generation systems at GSM base station site. Int J Renew Energy Res **3**(4), 763–768 (2013)
11. K Son, H Kim, Y Yi, B Krishnamachari, *Toward energy-efficient operation of base stations in cellular wireless networks, Green communications: theoretical fundamentals, algorithms, and applications* (CRC Press, Taylor & Francis, Boca Raton, FL, USA, 2012)
12. E Oh, K Son, B Krishnamachari, Dynamic base station switching-on/off strategies for green cellular networks. IEEE Trans Wirel Commun **12**(5), 2126–2136 (2013)
13. GSM-R. http://en.wikipedia.org/wiki/GSM-R. (webpages accessed on August 31, 2013.)
14. R Md Moklesur, *Overview of energy saving aspects in 2G and 3G mobile communication networks, Master's thesis* (University of Gavle, Sweden, 2009)
15. Ericsson Psi. http://www.ericsson.com/ourportfolio/telecom-operators/psi-coverage
16. Flexenclosure eSite. http://www.flexenclosure.com/esite/
17. L-CWang, S Rangapillai, A survey on green 5G cellular networks, in Proceedings of 2012 International Conference on Signal Processing and Communications (SPCOM), Bangalore, India, pp.1–5, 22-25 July 2012
18. X Zhang, S Zhuowen, Z Yan, W Wang, Energy-efficiency study for two-tier heterogeneous networks (HetNet) under coverage performance constraints. Mobile Networks Appl **18**(4), 567–577 (2013)
19. D Calin, H Claussen, H Uzunalioglu, On femto deployment architectures and macrocell offloading benefits in joint macro-femto deployments. IEEE Commun Mag **48**(1), 26–32 (2010)
20. Y Xiao, van der Schaar M, Energy-efficient nonstationary power control in cognitive radio networks, in Proceedings of IEEE GLOBECOM, IEEE Communications Society, Atlanta, GA, USA, pp. 3029–3034, December 9-13, 2013
21. MA Marsan, L Chiaraviglio, D Ciullo, M Meo, Optimal energy savings in cellular access networks, in Proceedings of 1st International Workshop on Green Communications (IEEE GreenComm 2009), pp. 1–5, IEEE, Dresden, Germany, June 14-18, 2009
22. C Peng, S Lee, S Lu, H Luo, H Li, Traffic-driven power saving in operational 3G cellular networks, in Proceedings of the 17th Annual International Conference on Mobile Computing and Networking, (MOBICOM 2011), pp. 1–12, Las Vegas, Nevada, USA, September 19-23, 2011
23. L Zhang, J Zhu, Z Tian, Research on verification method of mutual authentication in GSM/UMTS inter-system based on finite state machine, in Proceedings of International Conference on IMCCC, 2011, Instrumentation, Measurement, Computer, Communication and Control, (IMCCC 2011), pp. 541–544, Beijing, China, October 21-23, 2011
24. C Hong, A study on intelligent electricity saving technology in GSM system. ZTE Technologies **8**, 24–25 (2008)
25. Divide and conquer algorithms. Available on https://en.wikipedia.org/wiki/Divide_and_conquer_algorithm

26. L Qianhao, Z Jie, Application analysis of light and energy saving GSM BTS. ZTE Technologies **8**, 26–27 (2008)
27. C Lubritto. Telecommunication power system: energy saving, renewable sources and environmental monitoring, Trends in telecommunications technologies, Trends in Telecommunications Technologies, Christos J Bouras (Ed.), Intech, Rijeka, Croatia, March 1, 2010 ISBN: 978-953-307-072-8, 2010. Available from: http://www.intechopen.com/books/trends-in-telecommunications-technologies/telecommunication-power-system-energy-saving-renewable-sources-and-environmental-monitoring

9

On the asymptotic equivalence between stochastic and deterministic energy sources in energy harvesting sensing systems

Jingxian Wu* and Jing Yang

Abstract

In this paper, we seek answer to the question: can a wireless sensing system with energy harvesting power supplies perform as well as the one with conventional power supplies? Conventional sensing systems with deterministic energy sources usually employ uniform sampling. However, due to the stochastic nature of the energy harvested from the ambient environment, uniform sampling is usually infeasible for energy harvesting sensing systems. We thus propose a simple best-effort sensing scheme, which defines a set of equally spaced candidate sensing instants. At a given candidate sensing instant, the sensor will perform sensing if there is sufficient energy available, and it will remain silent otherwise. It is analytically shown that the percentage of silent candidate sensing instants goes to zero as time increases, if and only if the average energy harvesting rate is no less than the average energy consumption rate. Therefore, the difference between the best-effort sensing policy and the uniform sensing policy diminishes as time evolves. The theoretical results are then used to guide the design of a practical sensing system that monitors a time-varying event. Both analysis and simulations show that the energy harvesting system with the best-effort sensing scheme can asymptotically achieve the same mean squared error (MSE) performance as the one with uniform sensing and deterministic energy sources. Therefore, we provide a positive answer to the question by establishing the asymptotic equivalence between stochastic and deterministic energy sources, from both theoretical and practical aspects.

Keywords: Energy harvesting; Stochastic energy sources; Asymptotic equivalence; MSE

1 Introduction

Wireless sensing systems, such as structure health monitoring, biomedical sensing, and environment monitoring, are usually expected to operate uninterruptedly and autonomously over years or decades under extremely stringent energy constraints. Such design objectives necessitate the development of systems powered by energy harvesting devices, such as piezoelectric, thermal, or photovoltaic devices, which can collect energy from the ambient environment. Due to the randomness of the energy harvesting process, the amount of harvested energy can be modeled as a stochastic process. The stochastic nature of the energy sources is fundamentally different from the deterministic energy sources employed by conventional systems, and it imposes formidable challenges on the development of energy harvesting sensing systems.

There have been growing interests in the development of energy harvesting communications and sensing systems. Many of the works are developed to optimize communication-related metrics, such as transmission delay or throughput. In [1], the transmission delay of a point-to-point system is minimized by considering causality constraints on both information and energy, i.e., information cannot be delivered before its arrival and energy cannot be consumed before it is harvested. The formulation in [1] is extended to broadcast channels in [2,3] and to a multiple access channel in [4]. All of the above works use an off-line deterministic scheduling method, which identifies the optimum transmission scheduling based on full knowledge of current and future energy arrivals. The off-line scheduling methods treat future energy arrival

*Correspondence: wuj@uark.edu
Department of Electrical Engineering, University of Arkansas, Fayetteville, AR 72701, USA

as a deterministic process even though the actual energy harvesting process is stochastic.

Online scheduling methods address this problem by using only the statistics of energy arrivals. The off-line and online scheduling for systems with fading channels are discussed in [5], where the online scheduling problem is formulated as a stochastic dynamic programming problem with high complexity. Low complexity sub-optimum online algorithms are presented in [5] for fading channel and in [6] for interference channel. In both works, the performance of all online scheduling policies is strictly worse than that of the off-line scheduling.

Another branch of work focuses on sensing and signal processing-related performance metrics, such as estimation mean squared error (MSE), probability of false alarm and/or missed detection, and detection delay. In [7], the estimation MSE of a sparse signal is minimized by using compressive sensing and a random transmission scheme, where each energy harvesting sensor node transmits with a certain probability under the energy causality constraint. The sensing energy is assumed to be negligible in [7]. Optimum energy allocation schemes are discussed in [8] for the 'quickest detection' of the changing point of an event monitored by energy harvesting sensor networks. The results in [8] do not consider distortions such as sensing noise. Generally speaking, the energy management policies for the optimization of performance metrics related to sensing and inference have been less investigated.

All of the above works show that the performance of energy harvesting systems is in general inferior to systems with conventional deterministic energy sources, even if optimal sensing policies are developed based on the stochastic properties of the energy sources. The performance loss of the energy harvesting systems is mainly due to the stochastic nature of the energy source, such that energy is not always available when needed for sensing or transmission.

Naturally, we would ask the question: can an energy harvesting sensing system with stochastic energy sources perform as well as a conventional sensing system powered by deterministic energy sources? We seek the answer to this question from two aspects: from the theoretical aspect in terms of the stochastic properties of the available energy at a given time, and from the practical aspect in terms of the MSE performance of a sensing system. For a conventional sensing system with deterministic energy sources, uniform sampling is usually employed by the sensors and it is optimum over the duration in which the signal is wide sense stationary (WSS) [9,10]. However, due to the randomness in the energy arrival process, uniform sampling might be infeasible in energy harvesting systems given that there might not be sufficient energy to perform sensing operations at certain time periods. We thus propose a simple best-effort sensing policy that attempts to achieve uniform sampling with its best efforts [11-13]. The best-effort policy defines a set of equally spaced candidate sensing instants. At a given candidate sensing instant, the sensor will perform sensing only if there is sufficient energy to do so, and it will remain silent otherwise.

By analyzing the stochastic properties of the energy sources, we will show that the percentage of silent candidate sensing instants goes to zero as time goes to infinity, if and only if the average energy collection rate is no less than the average energy consumption rate. This means that the difference between the best-effort sensing policy and the uniform sensing policy diminishes as time evolves. The theoretical results indicate that the sensing behavior of systems with stochastic energy sources has the potential to approach that of systems with deterministic energy sources. Guided by the theoretical results, we then develop optimum sensing and detection schemes for a practical energy harvesting sensing system used to monitor a time-varying event. It will be shown through both theoretical analysis and simulations that the MSE performance of the energy harvesting sensing system asymptotically approaches that of a conventional sensing system with uniform sensing as time goes to infinity. Simulation results show that the performance of the two systems converges with as few as 400 candidate sensing instants. Therefore, we demonstrate from both theoretical and practical aspects that there is an asymptotic equivalence between stochastic and deterministic energy sources.

The remainder of this paper is organized as follows. The problem formulation is given in Section 2. Section 3 demonstrates the asymptotic equivalence between stochastic and deterministic energy sources by showing that the percentage of silent sensing instants goes to zero as time goes to infinity. The theoretical results are then applied in Section 4 to guide the development of a practical energy harvesting sensing system that can asymptotically achieve the same MSE performance as a conventional sensing system. Numerical and simulation results are given in Section 5, and Section 6 concludes the paper.

2 Problem formulation

Consider a sensor used to monitor a time-varying event. The sensor is powered by an energy harvesting device, which harvests energy from the ambient environment. Due to the random nature of the energy harvesting process, the amount of harvested energy can be modeled as a random process. The harvested energy is used by the sensor for sensing operations.

The time-varying event being monitored is modeled as a WSS random process $s(t)$, where t is the time variable. It is assumed that $s(t)$ is zero mean with a power-law auto-covariance function

$$\mathbb{E}[\, s(t_1)s(t_2)] = \rho^{|t_1-t_2|} \tag{1}$$

where $\mathbb{E}$ is the mathematical expectation operator, and $0 \leq \rho \leq 1$ is the power-law coefficient. The power-law covariance is an alternative representation of the Ornstein-Uhlenbeck covariance kernel [14].

The sensor attempts to reconstruct the continuous-time time-varying random event by using noise-distorted discrete-time observations of the random process. A sensing policy is defined as a sequence of time instants $\{t_n\}_n$, where t_n is the time instant at which the sensor collects a sample of the random process. The sample collected by the sensor at time t_n is

$$y(t_n) = \sqrt{E_s}s(t_n) + z(t_n) \tag{2}$$

where E_s is the energy allocated for one sensing sample, and $z(t_n)$ is the sensing and/or channel noise with a zero-mean and the auto-covariance function $\mathbb{E}\,[z(t_1)z(t_2)] = \sigma_z^2\delta(t_1 - t_2)$. It should be noted that the noise component is not necessarily Gaussian distributed.

The sensing system attempts to reconstruct the time-varying random field, $s(t)$, by using the sequence of the discrete-time samples, $\{y(t_n)\}_n$. If the sensor is powered by a conventional power supply, then uniform sampling with $t_n = nT_s$ is employed in [9,10] to achieve the optimum sensing performance due to the homogeneous nature of the random field. However, due to the inherent randomness of the energy harvesting process, uniform sampling might be infeasible given that there might not be sufficient energy to perform sensing operations at certain time periods. The performance of uniform sampling with conventional power supply can then serve as a lower bound for systems with energy harvesting devices.

3 Asymptotic equivalence between stochastic and deterministic energy sources

In this section, we first present a best-effort sensing policy for systems with stochastic energy sources. We then study the asymptotic behaviors of the best-effort sensing policy as time goes to infinity. The analytical results demonstrate the asymptotic equivalence between stochastic energy sources and deterministic energy sources.

Due to the random nature of the energy harvesting process, the harvested energy can be modeled as a random process. We model the stochastic energy sources by using the following two assumptions:

Assumption 1. If we divide the time axis into arbitrary small intervals with length $\Delta > 0$, then the energy collected in each interval can be modeled as independently and identically distributed (i.i.d.) random variables, $E_\Delta = P\Delta$, where P is the average energy harvesting rate.

Assumption 2. $\sum_{n=1}^{\infty} \Pr(E_\Delta > n\epsilon) < \infty$ for any $\epsilon > 0$ and $\Delta > 0$.

Assumption 1 indicates that the energy collected at different time intervals are independent, with the average amount of harvested energy proportional to the time duration, and Assumption 2 is used to ensure that the model behaves normally. Such a model is general enough to incorporate many other existing stochastic energy models as special cases. Below are a few examples.

1. *Poisson energy source* [5]. The energy arrives in the form of energy packets with unit energy in each energy packet. The number of energy arrivals in a unit time is a Poisson random variable with mean λ. Thus, the average energy collected in an interval Δ is $E_\Delta = \lambda\Delta$, and the amount of energy collected in different intervals is i.i.d.
2. *Bernoulli energy source* [8]. In each interval Δ, the energy arrival can be modeled as an i.i.d. Bernoulli random variable with parameter $P(E = 1) = p_\Delta$. The average energy collected in an interval Δ is thus $E_\Delta = p_\Delta$.
3. *Gaussian energy source.* The energy source is modeled as $E = |V(t)|^2$, where $V(t)$ is a zero-mean symmetric complex Gaussian random process with covariance function $\mathbb{E}[\, V(t+\tau)V^*(t)] = \sigma_v^2\delta(\tau)$. The average energy collected in an interval Δ is $E_\Delta = \sigma_v^2\Delta$.

The Poisson and Bernoulli energy sources model the energy arrivals with discrete random variables, and they are widely used in the literature. The Gaussian energy source uses continuous random variables to model the energy arrival, and it can be used to model the energy collected from random events that have similar properties as white Gaussian noise. All three models fall in the general stochastic energy model. Other than the two assumptions, the analysis in the rest of this paper does not assume any specific distributions of the energy sources.

The harvested energy is stored in an energy storage device, such as rechargeable batteries or super capacitors. The energy storage device can be modeled as an energy queue, with the input being the random energy arrival from the energy harvesting process, and the output is the energy consumption. Denote the amount of energy available in the energy storage device at time t as $Q(t) \geq 0$. Since the harvested energy is usually very small compared to the capacity of the energy storage device, it is assumed that the energy queue has unlimited capacity. The energy consumption must follow the energy causality constraint, that is, at any time instant, the total amount of harvested energy must be no less than the total amount of consumed energy.

The amount of energy available in the energy queue, $Q(t)$, is a random process due to the randomness in both energy collections and energy consumptions. During some time instants, there might not be enough energy in the energy queue for the sensing operations, that is, $Q(t) < E_s$. Therefore, the uniform sensing policy for systems with a deterministic power supply is not applicable to energy harvesting systems. We propose a best-effort sensing policy defined as follows.

Definition 1 (Best-effort sensing policy). Define a set of candidate sensing instants as $\mathcal{K} = \{k_n | k_n = nT_s, n = 1, 2, \cdots\}$. A sensor performs one sensing operation with energy E_s at time t if and only if: 1) $t \in \mathcal{K}$, and 2) $Q(t) \geq E_s$.

In the best-effort sensing policy, the sensor attempts to mimic the uniform sensing policy with its best efforts. It tries to perform sensing operations at uniform sensing intervals whenever allowed by the energy constraint. However, it will keep silent at a candidate sensing instant nT_s if $Q(nT_s) < E_s$. Denote the information collected at each candidate sensing instant as a sensing symbol, which could be either a silent symbol when $Q(nT_s) < E_s$ or an active symbol when $Q(nT_s) \geq E_s$. With such a sensing mechanism and the stochastic energy source, there might be K silent symbols in the first $N \geq K$ sensing instants $T_s, 2T_s, \cdots, NT_s$. The number of silent symbols is a random variable. The existence of silent symbols might degrade the sensing performance.

Next, we show the first main result of this paper, which states that if certain conditions are satisfied, then there is a diminishing number of silent symbols compared to active symbols as time becomes large enough.

Theorem 1. *Consider an energy harvesting sensing system powered by energy sources satisfying Assumptions 1 and 2. The system employs the best-effort sensing policy described in Definition 1. Define* $K = \sum_{k=1}^{N} \mathbf{1}_{Q(kT_s)<E_s}$ *as the total number of silent symbols in the first N symbol periods, where the indicator function* $\mathbf{1}_{\mathcal{E}} = 1$ *if the event* $\mathcal{E}$ *is true and 0 otherwise.*

If $P \geq \frac{E_s}{T_s}$, *then*

$$\lim_{N\to\infty} \frac{K}{N} = 0, \quad a.s. \tag{3}$$

Specifically, if $P > \frac{E_s}{T_s}$, *then for almost every energy harvesting sample path, there exists* $\bar{K} < \infty$ *such that* $K \leq \bar{K} < \infty$ *as* $N \to \infty$.

Conversely, if $P < \frac{E_s}{T_s}$, *then*

$$\lim_{N\to\infty} \frac{K}{N} \geq 1 - \frac{PT_s}{E_s}, \quad a.s. \tag{4}$$

Proof. Define $E_c = PT_s$ as the average energy harvested in the duration of T_s. Divide the time axis into frames, each of duration LT_s. The m-th frame thus has L candidate sensing instants, and $k_m \leq L$ of them are assumed to be silent. Assume the total amount of energy collected in the m-th frame is E_m, which is random. The amount of energy consumed in the l-th frame can be calculated as $(L-k_m)E_s$. Denote $D_m = E_m - (L-k_m)E_s$ as the difference between the energy harvested and consumed in the m-th frame. It should be noted that D_m could be either positive or negative. The total amount of energy available in the energy queue at the end of the m-th frame is $Q(mLT_s) = \sum_{m=1}^{M} D_m$. It should be noted that $\sum_{m=1}^{M} D_m \geq 0$ due to the energy causality constraint.

With the best-effort sensing policy, the number of silent symbols in the M-th frame must satisfy

$$k_M \leq \max\left\{0, L - \frac{1}{E_s}\sum_{m=1}^{M-1} D_m\right\}, \tag{5}$$

because the energy available at the end of the $(M-1)$-th frame can be used for the sensing of up to $\frac{1}{E_s}\sum_{m=1}^{M-1} D_m$ symbols in the M-th frame.

1) Case 1: $P < \frac{E_s}{T_s}$. With the energy causality constraint, we have $\sum_{m=1}^{M} D_m \geq 0$, or

$$\sum_{m=1}^{M} E_m - MLE_s + E_s\sum_{m=1}^{M} k_m \geq 0 \tag{6}$$

Divide both sides of (6) by MLE_s, and let $M \to \infty$,

$$\lim_{M\to\infty} \frac{\sum_{m=1}^{M} k_m}{ML} \geq 1 - \frac{1}{E_s}\lim_{M\to\infty} \frac{\sum_{m=1}^{M} E_m}{ML} \tag{7}$$

Based on the strong law of large numbers,

$$\lim_{M\to\infty} \frac{\sum_{m=1}^{M} E_m}{ML} = E_c, \quad \text{a.s.} \tag{8}$$

In addition, let $N = LM$, then $K = \sum_{m=1}^{M} k_m$. Thus, (4) can be obtained from (7) and (8).

2) Case 2: $P = \frac{E_s}{T_s}$. Index the frames with at least one silent symbol as $M_1, M_2, \cdots, M_i, \cdots$, i.e.,

$$0 < k_{M_i} \leq L - \frac{\sum_{m=1}^{M_i-1} D_m}{E_s}. \tag{9}$$

If M_i is upper bounded, that is, there exists $\bar{M}$ such that $k_m = 0$ for all $m > \bar{M}$, then K is finite and (3) is true. On the other hand, if M_i is unbounded, then $\lim_{i\to\infty} M_i = \infty$. We have

$$\sum_{m=1}^{M_i} D_m = \sum_{m=1}^{M_i-1} D_m + E_{M_i} - LE_s + k_{M_i}E_s \leq E_{M_i}, \tag{10}$$

where the last inequality is based on (9). Dividing both sides of (10) by LM_i and letting $i \to \infty$ (so as M_i), we have

$$\lim_{M_i\to\infty} \frac{\sum_{m=1}^{M_i} E_m}{LM_i} - E_s + \lim_{M_i\to\infty} \frac{\sum_{m=1}^{M_i} k_m}{LM_i} \le \lim_{M_i\to\infty} \frac{E_{M_i}}{LM_i} \tag{11}$$

Based on the assumption that $\sum_{n=1}^{\infty} P[E > n\epsilon] < \infty$ for any $\epsilon > 0$, and Borel-Cantelli lemma [15], we have

$$\lim_{M_i\to\infty} \frac{E_{M_i}}{LM_i} = 0, \quad \text{a.s.}$$

Combining (11) with (8) yields

$$\lim_{N\to\infty} \frac{K}{N} \le 0, \ \text{a.s.} \tag{12}$$

Since $\frac{K}{N} \ge 0$, (3) is true.

3) Case 3: $P > \frac{E_s}{T_s}$. Proof by contradiction. Assume $\lim_{i\to\infty} M_i = \infty$. When $E_c > E_s$, from (8) and (11), we have

$$\lim_{N\to\infty} \frac{K}{N} \le -(E_c - E_s) < 0, \ \text{a.s.} \tag{13}$$

This contradicts with the fact that $\frac{K}{N} \ge 0$; thus, the assumption $\lim_{i\to\infty} M_i = \infty$ cannot be true when $E_c > E_s$. This means that M_i is finite; thus, K is finite as $N \to \infty$. □

The results in Theorem 1 state that there is an asymptotic equivalence between stochastic energy source and deterministic energy source as time goes to infinity, if and only if the average harvested energy in one sample period is no less than the energy required for one sensing operation. This means that the performance of the best-effort sensing policy with stochastic energy sources has the potential to match that of the uniform sensing policy with deterministic energy sources as time becomes large enough and $P \ge \frac{E_s}{T_s}$. The results hold for a quite general category of energy harvesting processes. It is straightforward to verify that commonly adopted energy harvesting models, such as Poisson energy, Bernoulli energy source, and Gaussian energy source, all fall into this category.

If P is strictly larger than $\frac{E_s}{T_s}$, then the result is stronger because the number of silent symbols is finite even if the total number of symbols goes to infinity. This means that for almost every energy harvesting sample path, there exists $\bar{N} < \infty$ such that $Q(nT_s) \ge E_s$, $\forall n > \bar{N}$, or there exists a time threshold after which the best-effort sensing policy will be exactly the same as the uniform sensing policy.

The converse portion of Theorem 1 states that $P \ge \frac{E_s}{T_s}$ is the necessary and sufficient condition for the asymptotic equivalence between the stochastic and deterministic energy sources, because the percentage of silent symbols is non-diminishing if the condition $P \ge \frac{E_s}{T_s}$ is not satisfied.

Theorem 1 provides the theoretical foundation regarding the asymptotic equivalence between stochastic and deterministic energy sources, with the help of the newly proposed best-effort sensing policy. In the next section, we will show that the asymptotic equivalence can be achieved in a practical sensing system, that is, the proposed best-effort sensing policy with stochastic energy sources can indeed asymptotically achieve the same MSE performance as the uniform sensing policy with constant energy supply.

4 Asymptotically optimum sensing with the best-effort sensing policy

This section studies the optimum design and performance analysis of sensing systems employing the newly proposed best-effort sensing policy.

4.1 Best-effort sensing with stochastic energy sources

Based on the best-effort sensing policy, define the sensing sample at time instant $k_n = nT_s$ as

$$\eta_n = \begin{cases} \sqrt{E_s} s(nT_s) + z(nT_s), & Q(nT_s) \ge E_s, \\ z(nT_s), & Q(nT_s) < E_s. \end{cases} \tag{14}$$

Define $\boldsymbol{\eta} = [\eta_1, \cdots, \eta_N]^T \in \mathcal{R}^N$, and express (14) in a matrix format as

$$\boldsymbol{\eta} = \sqrt{E_s}\mathbf{x} + \mathbf{z} \tag{15}$$

where $\mathbf{x} = [x_1, \cdots, x_n]^T$ with $x_n = s(nT_s)$ if $Q(nT_s) \ge E_s$ and $x_n = 0$ otherwise, and $\mathbf{z} = [z(T_s), \cdots, z(NT_s)]^T$. With the best-effort sensing policy, K out of the N elements in $\boldsymbol{\eta}$ contain only noise components.

The system attempts to reconstruct the time-varying random event $s(t)$ by using the observations, $\boldsymbol{\eta}$. Since we are interested in the reconstruction fidelity of a continuous-time random event, the worst case scenario will be considered by estimating $\{s\left(nT_s + \frac{1}{2}T_s\right)\}_n$, the sequence of points located in the middle between two candidate sensing instants. Define the data vector to be estimated as $\mathbf{d} = \left[s\left(\frac{1}{2}T_s\right), s\left(\frac{3}{2}T_s\right), \cdots, s\left(NT_s - \frac{1}{2}T_s\right)\right]^T$. It should be noted that $s\left(nT_s + \frac{1}{2}T_s\right)$ will be estimated even if $x_n = 0$ and/or $x_{n+1} = 0$.

The linear minimum mean squared error (MMSE) estimation of $\mathbf{d}$ based on $\boldsymbol{\eta}$ is

$$\hat{\mathbf{d}} = \mathbf{W}^T \boldsymbol{\eta}, \tag{16}$$

where $\mathbf{W}$ is the MMSE matrix and it satisfies the orthogonal principle

$$\mathbb{E}\left[(\mathbf{W}^T\mathbf{y}-\mathbf{d})\mathbf{y}^T\right]=\mathbf{0}. \tag{17}$$

The MMSE matrix is $\mathbf{W}^T=\mathbb{E}(\mathbf{d}\boldsymbol{\eta}^T)\left[\mathbb{E}(\boldsymbol{\eta}\boldsymbol{\eta}^T)\right]^{-1}$, which can be expressed as

$$\mathbf{W}^T=\sqrt{E_s}\mathbf{R}_{dx}\left[E_s\mathbf{R}_{xx}+\mathbf{R}_{zz}\right]^{-1}, \tag{18}$$

where $\mathbf{R}_{dx}=\mathbb{E}(\mathbf{d}\mathbf{x}^T)$, $\mathbf{R}_{xx}=\mathbb{E}(\mathbf{x}\mathbf{x}^T)$, and $\mathbf{R}_{zz}=\sigma_z^2\mathbf{I}_N$ with $\mathbf{I}_N$ being a size $N\times N$ identity matrix. If we assume the indices of the sampling instants with $Q(nT_s)<E_s$ being $i_1, i_2, \cdots, i_K$, then the i_k-th column of $\mathbf{R}_{dx}$ is an all-zero column, for $k=1,\cdots,K$. Similarly, the i_k-th row and the i_k-th column of $\mathbf{R}_{xx}$ are all-zero vectors, for $k=1,\cdots,K$.

The covariance matrix of the error vector, $\boldsymbol{\epsilon}=\mathbf{d}-\hat{\mathbf{d}}$, can be written as

$$\mathbf{R}_{\epsilon\epsilon}=\mathbf{R}_{dd}-\mathbf{R}_{dx}\left[\mathbf{R}_{xx}+\frac{\sigma_z^2}{E_s}\mathbf{I}_N\right]^{-1}\mathbf{R}_{xd}. \tag{19}$$

Systems with deterministic energy sources can be considered as a special case of the MMSE receiver described in (15) and (16). With a deterministic energy source, there will be no silent symbols; thus, the system equation can be expressed as

$$\mathbf{y}=\sqrt{E_s}\mathbf{s}+\mathbf{z} \tag{20}$$

where $\mathbf{y}=[y(T_s),\cdots,y(NT_s)]^T$ and $\mathbf{s}=[s(T_s),\cdots,s(NT_s)]^T$ are the received signal vector and data vector, respectively. Denote the MMSE estimation of $\mathbf{d}$ from $\mathbf{y}$ as $\tilde{\mathbf{d}}=\mathbf{W}_y^T\mathbf{y}$, where $\mathbf{W}_y^T$ is the linear MMSE matrix. Similar to (19), the covariance matrix of the error vector, $\mathbf{e}=\mathbf{d}-\tilde{\mathbf{d}}$, can be written as

$$\mathbf{R}_{ee}=\mathbf{R}_{dd}-\mathbf{R}_{ds}\left[\mathbf{R}_{ss}+\frac{\sigma_z^2}{E_s}\mathbf{I}_N\right]^{-1}\mathbf{R}_{sd} \tag{21}$$

where $\mathbf{R}_{ds}=\mathbb{E}(\mathbf{d}\mathbf{s}^T)$ and $\mathbf{R}_{ss}=\mathbb{E}(\mathbf{s}\mathbf{s}^H)$. The cross-covariance matrix $\mathbf{R}_{ds}$ is a Toeplitz matrix with the first row being $\left[r\left(\frac{1}{2}T_s\right), r\left(\frac{3}{2}T_s\right),\cdots,r\left(NT_s-\frac{1}{2}T_s\right)\right]$, and the first column $\left[r\left(\frac{1}{2}T_s\right), r\left(\frac{1}{2}T_s\right), r\left(\frac{3}{2}T_s\right),\cdots, r\left(NT_s-\frac{3}{2}T_s\right)\right]^T$. The covariance matrix $\mathbf{R}_{ss}$ is a symmetric Toeplitz matrix with the first row being $[r(0), r(T_s),\cdots,r(NT_s-T_s)]$.

The error covariance matrices for systems with stochastic and deterministic energy sources are given in (19) and (21), respectively. The average MSE for systems with stochastic and deterministic energy sources can then be calculated, respectively, as

$$\sigma_{\epsilon,N}^2=\frac{1}{N}\text{trace}\left(\mathbf{R}_{\epsilon\epsilon}\right), \tag{22a}$$

$$\sigma_{e,N}^2=\frac{1}{N}\text{trace}\left(\mathbf{R}_{ee}\right). \tag{22b}$$

Intuitively, $\sigma_{\epsilon,N}^2\geq\sigma_{e,N}^2$ since $\mathbf{y}$ contains more information than $\boldsymbol{\eta}$. The MSE of the system with uniform sensing and deterministic energy source can serve as a lower bound for systems with stochastic energy sources. However, due to asymptotic equivalence between the stochastic and deterministic energy sources as presented in Theorem 1, we will show in the next subsection that the proposed best-effort sensing policy can asymptotically achieve the same performance as the uniform sensing policy as $N\to\infty$.

4.2 Asymptotic achievability of the MSE lower bound

The asymptotic equivalence between the two sensing schemes relies on the asymptotic equivalence between sequences of matrices, which is defined as follows.

Definition 2 (Asymptotical equivalence [16]). Two sequences of $N\times N$ matrices, $\{\mathbf{A}_N\}_N$ and $\{\mathbf{B}_N\}_N$ are asymptotically equivalent if

1. $\mathbf{A}_N$ and $\mathbf{B}_N$ are bounded in strong norm:

$$\|\mathbf{A}_N\|, \|\mathbf{B}_N\|<C<\infty, N=1,2,\cdots \tag{23}$$

where $\|\mathbf{A}\|=\max_k\sqrt{\lambda_k}$ is the strong norm of $\mathbf{A}$, λ_k is the eigenvalue of $\mathbf{A}^H\mathbf{A}$, and $\mathbf{A}^H$ is the Hermitian of the matrix $\mathbf{A}$.

2. $\mathbf{A}_N-\mathbf{B}_N$ goes to zero in weak norm as $N\to\infty$

$$\lim_{N\to\infty}|\mathbf{A}_N-\mathbf{B}_N|=0, \tag{24}$$

where $|\mathbf{A}_N|=\sqrt{\frac{1}{N}\sum_{m=1}^N\sum_{n=1}^N|a_{mn}|^2}$ is the weak norm of $\mathbf{A}_N$ with a_{mn} being the (m,n)-th element of the matrix $\mathbf{A}_N$.

The asymptotic equivalence between two sequences of matrices, $\{\mathbf{A}_N\}$ and $\{\mathbf{B}_N\}$, is denoted as $\mathbf{A}_N\sim\mathbf{B}_N$.

Lemma 1. *Consider two sequences of matrices, $\{\mathbf{A}_N\}_N$ and $\{\mathbf{B}_N\}_N$. The squares of the elements on any row or column of the two matrices are absolutely summable as $N\to\infty$. If the two matrices differ in K_1 rows and K_2 columns, and* $\lim_{N\to\infty}\frac{K_1+K_2}{N}=0$, *then $\mathbf{A}_N\sim\mathbf{B}_N$.*

Proof. Denote the (m,n)-th elements of $\mathbf{A}_N$ and $\mathbf{B}_N$ as a_{mn} and b_{mn}, respectively. Since the squares of the elements on each row or column of the two matrices are absolutely summable, the matrices are bounded in strong norm. Assume the two matrices differ in rows $r_1,\cdots,r_{K_1}$ and columns $c_1,\cdots,c_{K_2}$, then

$$|\mathbf{A}_N-\mathbf{B}_N|^2\leq\frac{1}{N}\left[\sum_{k=1}^{K_1}\sum_{n=1}^{N}|a_{r_kn}-b_{r_kn}|^2+\sum_{m=1}^{M}\sum_{k=1}^{K_2}|a_{mc_k}-b_{mc_k}|^2\right] \tag{25}$$

Since the squares of the elements on any row or column are absolutely summable, then there exists $C > 0$ such that

$$\sum_{n=1}^{\infty} |a_{r_k n} - b_{r_k n}|^2 \leq \sum_{n=1}^{\infty} |a_{r_k n}|^2 + \sum_{n=1}^{\infty} |b_{r_k n}|^2 < 2C$$

$$\sum_{m=1}^{\infty} |a_{mc_k} - b_{mc_k}|^2 \leq \sum_{m=1}^{\infty} |a_{mc_k}|^2 + \sum_{m=1}^{\infty} |b_{mc_k}|^2 < 2C$$

Thus,

$$\lim_{N\to\infty} |\mathbf{A}_N - \mathbf{B}_N| < \sqrt{4C} \lim_{N\to\infty} \sqrt{\frac{K}{N}} \tag{26}$$

Since $\lim_{N\to\infty} \frac{K}{N} = 0$, it is straightforward that $\lim_{N\to\infty} |\mathbf{A}_N - \mathbf{B}_N| = 0$. □

From (19) and (21), σ_ϵ^2 and $\mathbf{R}_{\epsilon\epsilon}$ depend on $\mathbf{R}_{dx}$ and $\mathbf{R}_{xx}$, while σ_e^2 and $\mathbf{R}_{ee}$ depend on $\mathbf{R}_{ds}$ and $\mathbf{R}_{ss}$. We have the following lemma regarding the relationships between the two groups of matrices as $N \to \infty$.

Lemma 2. *If $P \geq \frac{E_s}{T_s}$ in the best-effort sensing policy, then $\mathbf{R}_{dx} \sim \mathbf{R}_{ds}$, and $\mathbf{R}_{xx} \sim \mathbf{R}_{ss}$.*

Proof. All matrices are absolutely summable in each row or column. Based on the best-effort sensing policy, $\mathbf{x}$ can be obtained by replacing K elements in $\mathbf{s}$ with zeros. Since $P \geq \frac{E_s}{T_s}$, we have $\lim_{N\to\infty} \frac{K}{N} = 0$ a.s.

If we assume the indices of the sampling instants with $Q(nT_s) < E_s$ being $i_1, i_2, \cdots, i_K$, then $\mathbf{R}_{dx}$ can be obtained by replacing the i_k-th column of $\mathbf{R}_{ds}$ with an all-zero column, for $k = 1, \cdots, K$. From Lemma 1, $\mathbf{R}_{dx} \sim \mathbf{R}_{ds}$.

Similarly, $\mathbf{R}_{xx}$ can be obtained by replacing the i_k-th row and the i_k-th column of $\mathbf{R}_{ss}$ with all-zero vectors, for $k = 1, \cdots, N$. From Lemma 1, $\mathbf{R}_{xx} \sim \mathbf{R}_{ss}$. □

Now we are ready to present the second main result of this paper in the following theorem, which states the asymptotic equivalence between systems with stochastic and deterministic energy sources in terms of MSE performance.

Theorem 2. *Consider two sensing systems, one with the best-effort sensing policy and a stochastic energy source as described in (15), and one with the uniform sensing policy and a deterministic energy source as described in (20). If $P \geq \frac{E_s}{T_s}$, then*

$$\lim_{N\to\infty} \sigma_{\epsilon,N}^2 = \lim_{N\to\infty} \sigma_{e,N}^2 = \sigma_0^2, \tag{27}$$

where $\sigma_{\epsilon,N}$ and $\sigma_{e,N}^2$ are the MSE of the two systems defined in (22a) and (22b), respectively. Specifically, with the power-law time covariance in (1)

$$\sigma_0^2 = \left[\frac{\sigma_z^2}{E_s} + \frac{1-\rho^{T_s}}{1+\rho^{T_s}}\right]^{\frac{1}{2}} \left[\frac{\sigma_z^2}{E_s} + \frac{1+\rho^{T_s}}{1-\rho^{T_s}}\right]^{-\frac{1}{2}} \tag{28}$$

Proof. The Toeplitz matrix, $\mathbf{R}_{ds}$, is uniquely determined by the sequence $\mathbf{r}_{ds} = [r_{-(N-1)}, \cdots, r_0, \cdots, r_{N-1}]^T$, where $r_n = \rho^{\frac{T_s}{2}} \rho^{-(n+1)T_s}$ when $n < 0$, and $r_n = \rho^{\frac{T_s}{2}} \rho^{nT_s}$ otherwise. When $N \to \infty$, the discrete time Fourier transform (DTFT) of the sequence $\mathbf{r}_{ds}$ can be calculated as

$$\Lambda_{ds}(f) = \rho^{\frac{T_s}{2}} \frac{(1-\rho^{T_s})(1+e^{j2\pi f})}{1+\rho^{2T_s} - 2\rho^{T_s}\cos(2\pi f)}. \tag{29}$$

Based on [17] (Lemma 2) , $\mathbf{R}_{ds}$ is asymptotically equivalent to a circulant matrix, $\mathbf{C}_{ds} = \mathbf{U}_N^H \mathbf{D}_{ds} \mathbf{U}_N$, where $\mathbf{U}_N$ is the unitary discrete Fourier transform (DFT) matrix with the (m,n)-th element being $(\mathbf{U}_N)_{m,n} = \frac{1}{\sqrt{N}} \exp\left[-j2\pi \frac{(m-1)(n-1)}{N}\right]$, and $\mathbf{D}_{ds}$ is a diagonal matrix with its k-th diagonal element being $(\mathbf{D}_{ds})_{k,k} = \Lambda_{ds}\left(\frac{k-1}{N}\right)$.

Similarly, the symmetric Toeplitz matrix, $\mathbf{R}_{ss}$, is uniquely determined by the sequence $\boldsymbol{\phi}_{ss} = [\phi_{-(N-1)}, \cdots, \phi_0, \cdots, \phi_{N-1}]^T$, where $\phi_n = \rho^{|n|T_s}$. When $N \to \infty$, the DTFT of the sequence $\boldsymbol{\phi}_{ss}$ can be calculated as

$$\Lambda_{ss}(f) = \frac{1-\rho^{2T_s}}{1+\rho^{2T_s} - 2\rho_s^T\cos(2\pi f)}. \tag{30}$$

The symmetric Toeplitz matrix $\mathbf{R}_{ss}$ is asymptotically equivalent to a circulant matrix, $\mathbf{C}_{ss} = \mathbf{U}_N^H \mathbf{D}_{ss} \mathbf{U}_N$, where $\mathbf{D}_{ss}$ is a diagonal matrix with its k-th diagonal element being $(\mathbf{D}_{ss}) = \Lambda_{ss}\left(\frac{k-1}{N}\right)$.

When $P \geq \frac{E_s}{T_s}$, we have $\lim_{N\to\infty} \frac{K}{N} = 0$ from Theorem 1. Based on Lemma 2 and [16] (Theorem 2.1), we have

$$\mathbf{R}_{dx} \sim \mathbf{R}_{ds} \sim \mathbf{C}_{ss} \tag{31}$$

$$\mathbf{R}_{xx} \sim \mathbf{R}_{ss} \sim \mathbf{C}_{ss} \tag{32}$$

Therefore, based on [16] (Theorem 2.1), we have $\mathbf{R}_{\epsilon\epsilon} \sim \mathbf{R}_{ee} \sim \mathbf{C}_{ee}$, where $\mathbf{C}_{ee}$ is a circulant matrix defined as

$$\mathbf{C}_{ee} = \mathbf{C}_{ss} - \mathbf{C}_{ds}\left[\mathbf{C}_{ss} + \frac{\sigma_z^2}{E_s}\mathbf{I}_N\right]^{-1} \mathbf{C}_{ds}^H \tag{33}$$

The circulant matrix $\mathbf{C}_{ee}$ can be expressed as $\mathbf{C}_{ee} = \mathbf{U}_N^H \mathbf{D}_{ee} \mathbf{U}_N$, where $\mathbf{D}_{ee}$ is a diagonal matrix defined as

$$\mathbf{D}_{ee} = \mathbf{D}_{ss} - \mathbf{D}_{ds}\left[\mathbf{D}_{ss} + \frac{\sigma_z^2}{E_s}\mathbf{I}_N\right]^{-1} \mathbf{D}_{ds}^H. \tag{34}$$

Therefore, both error covariance matrices are asymptotically equivalent to the same circulant matrix. Based on Szego's theorem [17], when $N \rightarrow \infty$, we have

$$\lim_{N\to\infty} \sigma_{\epsilon,N}^2 = \lim_{N\to\infty} \sigma_{e,N}^2 = \int_{-\frac{1}{2}}^{\frac{1}{2}} \left[\Lambda_{ss}(f) - \frac{|\Lambda_{ds}(f)|^2}{\Lambda_{ss}(f) + \frac{\sigma_z^2}{E_s}} \right] df. \tag{35}$$

Substituting (30) and (29) into the above equation and simplifying yields (28). □

Theorem 2 states that systems with the proposed best-effort sensing policy and stochastic energy sources can asymptotically achieve the same MSE performance as systems with the uniform sensing policy and deterministic energy sources, as long as the energy consumption for each sensing operation is not more than the average harvested energy in each sampling period. Theorem 1 provides the theoretical foundation on the asymptotic equivalence between stochastic and deterministic energy sources, and Theorem 2 demonstrates that such equivalence can be achieved in practical systems.

5 Numerical and simulation results

Numerical and simulation results are presented in this section to demonstrate the asymptotic equivalence between systems with stochastic and deterministic energy sources.

Figure 1 shows the value of $\frac{K}{N}$ as a function of N for systems with various stochastic energy sources, where K is the number of silent sensing symbols and N is the total number of candidate sensing symbols. All systems employ the best-effort sensing policy proposed in this paper. The convergence behaviors of $\frac{K}{N}$ are demonstrated for $P = 0.9\frac{E_s}{T_s}$, $P = \frac{E_s}{T_s}$, and $P = 1.1\frac{E_s}{T_s}$, respectively. The simulation results are obtained by averaging over 100 independent runs for each configuration. All three energy sources have similar convergence behaviors as N increases. When $P \geq \frac{E_s}{T_s}$ and $N > 100$, the value of $\log\frac{K}{N}$ decreases almost linearly with respect to $\log N$, and the absolute slope increases as $\frac{PT_s}{E_s}$ increases. The simulation results conform to Theorem 1, which states that $\lim_{N\to\infty}\frac{K}{N} = 0$ when $P \geq \frac{E_s}{T_s}$. On the other hand, when $P = 0.9\frac{E_s}{T_s}$, $\frac{K}{N}$ tends to a constant value 0.1, as $N \rightarrow \infty$. The constant value is the same as the lower bound, $1 - \frac{PT_s}{E_s}$, predicted by Theorem 1. The results in Figure 1 demonstrate through simulations that the percentage of silent sensing symbols diminishes as time goes to infinity, as long as the average energy harvesting rate is no less than the average energy consumption rate. The actual distributions of the energy sources have no impact on the convergence behavior.

Figure 2 compares the MSE performance of systems with both stochastic and deterministic energy sources. The best-effort sensing policy is employed by systems with stochastic energy sources, and uniform sensing is employed by systems with deterministic energy sources. The stochastic energy sources are the Gaussian sources with mean $P = \frac{E_s}{T_s}$. It is assumed that all systems have the same average power $P = \frac{E_s}{T_s}$ with the normalized signal-to-noise ratio (SNR) $\gamma_0 = \frac{P}{\sigma_z^2}$ being 0 dB. The power-law coefficient is $\rho = 0.9$. The simulation results are obtained through Monte Carlo simulations, where the data are generated as a zero-mean Gaussian random process with the covariance function satisfying the power-law relationship. In the simulations, the data are sampled and then estimated, and the estimation results are compared to the true values to calculate the MSE. Each point on the simulation MSE curves is obtained by averaging over 2,000 independent runs for each configuration. It can be seen that

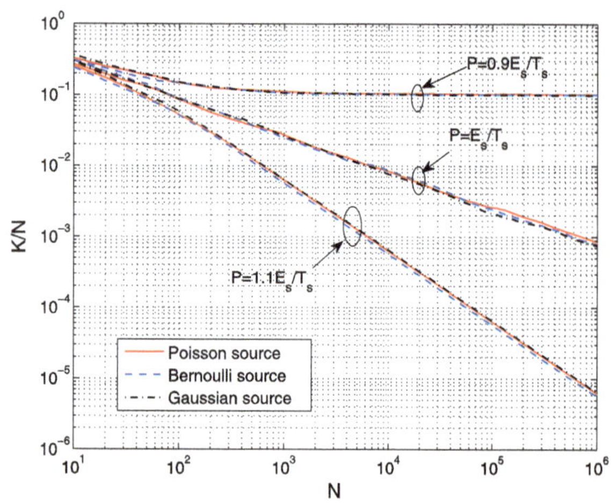

Figure 1 K/N **for various energy sources.**

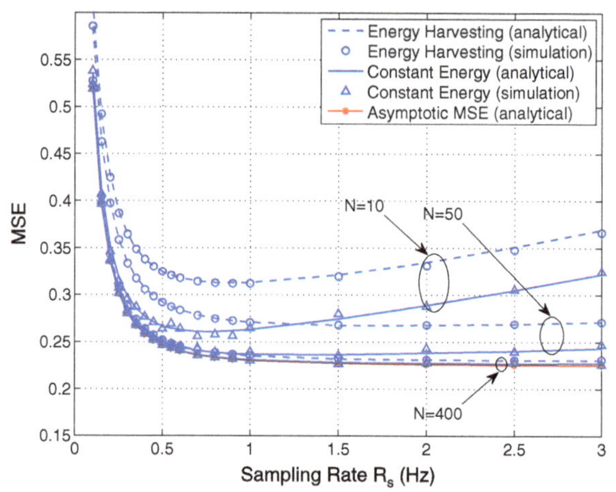

Figure 2 MSE as a function of the sampling frequency ($\rho = 0.9$).

the simulation results match perfectly with their analytical counterparts. As expected, the MSE performance of systems with deterministic energy sources is consistently better than that of systems with stochastic energy sources. However, the performance gap narrows as N increases. When $N = 400$, the MSE performance of the two systems are almost identical, and they coincide with the asymptotic MSE obtained with $N \to \infty$. Therefore, when N is sufficiently large, the best-effort sensing policy can achieve a performance that is almost the same as the uniform sensing policy. Thus, the results in Figure 2 demonstrate the asymptotic equivalence between stochastic and deterministic energy sources in a practical system.

In addition, it can be seen from Figure 2 that the MSE is convex when N is small ($N = 10$ or 50), and it becomes a monotonically decreasing function when N is large ($N = 400$). Under a fixed power, a larger sampling rate means less energy per sample, which might degrade the system performance. On the other hand, a larger sampling rate means a stronger correlation between two adjacent samples, which contributes positively to the MSE performance. Therefore, changing the sampling rate results in different trade-offs between energy per sample and sample correlations. When N is large enough, increasing the sampling rate beyond a certain threshold (e.g., $R_s = 1$ Hz when $N = 400$) has negligible impact on the MSE performance.

Figure 3 shows the simulation and analytical MSE of systems with stochastic energy sources as functions of the SNR. The sampling rate is 1 Hz. The analytical results with finite N are calculated from (22). The simulation results are obtained through Monte Carlo simulations. Each point on the curve is averaged over 2,000 independent simulation runs. The simulation results match the analytical results very well. As expected, the MSE decreases as ρ or N increases. In addition, the MSE with $N = 400$ is almost the same as the asymptotic MSE obtained with $N \to \infty$.

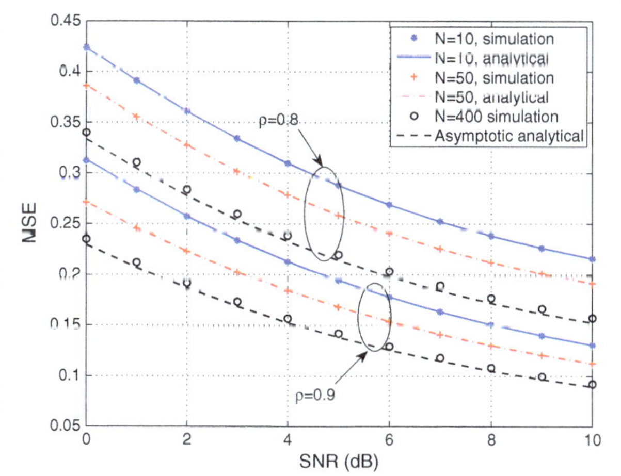

Figure 3 MSE as a function of the SNR for systems with stochastic energy sources ($R_s = 1$ Hz).

6 Conclusions

The asymptotic equivalence between stochastic and deterministic energy sources has been demonstrated through both theoretical analysis and practical examples. To account for the stochastic nature of energy harvested from the ambient environment, a best-effort sensing policy has been proposed for energy harvesting sensing systems. It has been shown that the difference between the best-effort sensing scheme and the ideal uniform sensing scheme diminishes as time goes to infinity, if and only if the average energy collection rate is no less than the average energy consumption rate, regardless of the actual distribution of the stochastic energy source. The asymptotic equivalence has been used for the development of an optimum energy harvesting sensing system. It has been shown through both theoretical analysis and simulation results that systems with the best-effort sensing scheme and stochastic energy sources can achieve almost the same MSE performance as systems with uniform sensing and deterministic energy sources when the number of samples is greater than 400.

Competing interests
The authors declare that they have no competing interests.

Acknowledgements
This work was supported in part by the U.S. National Science Foundation under Grants ECCS-1202075, ECCS-1405403, and ECCS-1454471.

References
1. J Yang, S Ulukus, Optimal packet scheduling in an energy harvesting communication system. IEEE Trans. Comm. **60**(1), 220–230 (2012)
2. MA Antepli, E Uysal-Biyikoglu, H Erkal, Optimal packet scheduling on an energy harvesting broadcast link. IEEE J. Sel. Areas Commun. **29**(8), 1721–1731 (2011)
3. J Yang, O Ozel, S Ulukus, Broadcasting with an energy harvesting rechargeable transmitter. IEEE Trans. Wireless Comm. **11**(2), 571–583 (2012)
4. J Yang, S Ulukus, Optimal packet scheduling in a multiple access channel with energy harvesting transmitters. J. Commun. Networks. **14**, 140–150 (2012)
5. O Ozel, K Tutuncuoglu, J Yang, S Ulukus, A Yener, Transmission with energy harvesting nodes in fading wireless channels: optimal policies. IEEE J Sel. Areas Commun. **29**(8), 1732–1743 (2011)
6. K Tutuncuoglu, A Yener, Sum rate optimal power policies for energy harvesting transmitters in an interference channel. J. Commun. Networks. **14**(2), 151–161 (2012)
7. G Yang, V Tan, CK Ho, SH Ting, YL Guan, Wireless compressive sensing for energy harvesting sensor nodes. Signal Process., IEEE Trans. **61**(18), 4491–4505 (2013)
8. J Geng, L Lai, Non-Bayesian quickest change detection with stochastic sample right constraints. IEEE Trans. Signal Process. **61**(20) (2013)
9. N Sun, J Wu, Optimum sampling in spatial-temporally correlated wireless sensor networks. EURASIP J. Wireless Commun. Networking. **2013**, 5
10. J Wu, N Sun, Optimum sensor density in distortion-tolerant wireless sensor networks. Wireless Commun., IEEE Trans. **11**(6), 2056–2064 (2012)

11. J Wu, J Yang, in *Proc. IEEE Global Telecommun. Conf. Globecom'14*. The asymptotic equivalence between sensing systems with energy harvesting and conventional energy sources (Austin, TX, USA, 8 December 2014)
12. J Yang, J Wu, in *Proc. IEEE Global Telecommun. Conf. Globecom'14*. Optimal sampling of random processes under stochastic energy constraints (Austin, TX, USA, 8 December 2014)
13. J Yang, X Wu, J Wu, in *Proc. IEEE Intern. Conf. Commun. (ICC'15)*. Adaptive sensing scheduling for energy harvesting sensors with finite battery (London, UK, 8 June 2015)
14. K Ritter, GW Wasilkowski, H Wozniakowski, Multivariate integration and approximation of random fields satisfying Sacks-Ylvisaker conditions. Ann. Appl. Probability. **5**, 518–540 (1995)
15. A Klenke, *Probability Theory: A Comprehensive Course (Universitext)*. (Springer, New York, 2007)
16. RM Gray, Toeplitz and circulant matrices: a review. Foundations and Trends in Commun. and Inf. Theory. **2**(3), 155–239 (2006)
17. H Gazzah, P Regalia, J-P Delmas, Asymptotic eigenvalue distribution of block Toeplitz matrices and application to blind SIMO channel identification. Inf. Theory, IEEE Trans. **47**(3), 1243–1251 (2001)

PCAC: Power- and Connectivity-Aware Clustering for Wireless Sensor Networks

Ismail Butun[1], In-ho Ra[2*] and Ravi Sankar[3]

Abstract

In this article, we investigate clustering algorithms that are proposed for wireless sensor networks (WSNs). Then, we propose a clustering algorithm for WSNs called PCAC, that is both power and connectivity aware. Our proposed Power- and Connectivity-Aware Clustering (PCAC) algorithm provides higher energy efficiency and increases the total life time (TLT) of the network. We evaluate our proposed PCAC algorithm in a simulation environment and compare its performance to previously proposed algorithms. According to the simulation results, our proposed PCAC algorithm is energy efficient and also provides longer TLT (in worst-case, up to 85% improvement for a 15-node network with one-hop connectivity) to the network, compared to the previously proposed algorithms.

Keywords: Clustering; Wireless sensor network; Power-aware; Connectivity-aware; Hierarchical; Total life time

1 Introduction

Wireless sensor networks (WSNs) continue to grow as one of the most exciting and challenging research areas of engineering. There are many applications of WSNs, which are intended to monitor physical and environmental phenomena, such as earthquake, pollution, wild fire, water quality and to gather information regarding human activities in health care, manufacturing machinery performance, building safety, military surveillance and reconnaissance, highway traffic, etc. [1].

WSNs are characterized by severely constrained computational and energy resources, and an *ad hoc* network operational environment. They pose unique challenges, due to limited power supplies, low transmission bandwidth, small memory sizes, and limited energy; therefore, networking techniques used in traditional networks cannot be adopted directly [2]. So, new ideas and approaches (algorithms) are needed in order to increase the overall performance of the network, especially in terms of total life time (TLT). Clustering, is one of those techniques that is very useful for WSNs in data aggregation and filtering and is the main focus of this paper.

A clustered WSN is typically as shown in Figure 1. Each cluster is a group of interconnected sensor nodes with a dedicated node called cluster head (CH). CHs are responsible for the management of the cluster such as scheduling of the medium access, dissemination of the control messages, and, the most important, data aggregation. The size of a cluster is defined as the hop distance from the CH to the farthest node in the cluster. For example, in a three-hops cluster, the distance between the CH and the farthest node is three-hops (four nodes are in the path including the end points). The clustered network shown in Figure 1 has a one-hop distance in between CHs and the member sensor nodes.

Clustering is the process of grouping the nodes in a network that are within a specified hop distance or have some shared common properties into clusters and electing CHs for each cluster. This election can be made permanent (static clustering) or repeated in some certain time intervals (dynamic clustering). Clustering is used in many applications of wireless sensor networks in order to reduce the traffic load on the nodes through data aggregation process, to prolong total network life time, to balance the data traffic in the network, and finally to increase the scalability (allows the deployment of hundreds or thousands of nodes). Besides, clustering helps us to increase security of the network by allowing implementation of complex cryptography algorithms. By using a clustered networking approach, power-consuming algorithms (such as data aggregation and encryption) would be run on the CHs,

*Correspondence: ihra@kunsan.ac.kr
[2]Department of Information and Telecommunication Engineering, Kunsan National University, 573-701 Gunsan, South Korea
Full list of author information is available at the end of the article

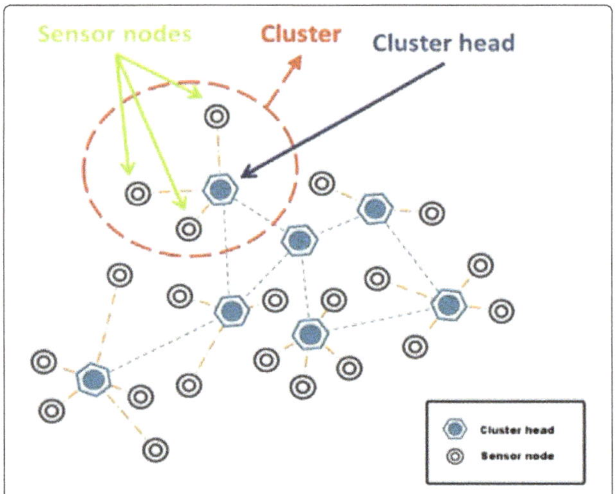

Figure 1 A typical clustered WSN.

and this would help us to improve the TLT of the network significantly.

In this article, we investigate clustering algorithms that are proposed for WSNs and propose a new clustering algorithm that is both power and connectivity aware. The rest of the paper is organized as follows: Section 2 provides a description of the related work available in the literature. Section 3 presents Kachirski *et al.*'s connectivity-aware clustering algorithm, and Section 4 provides the revised and improved version of that algorithm. Our proposed Power- and Connectivity-Aware Clustering (PCAC) algorithm is presented in Section 5. Section 6 provides the comparison of both schemes and also presents the details of our simulation environment. In Section 7, we discuss the observations regarding the effects of the clustering on the performance of the WSNs. Finally, Section 8 concludes the paper and outlines future work.

2 Related work

There are plenty of clustering algorithms available in the literature that are proposed for wireless networks. In this section, we present the most widely used clustering algorithms and mention their advantages and disadvantages:

Low-Energy Adaptive Clustering Hierarchy (LEACH) [3] is a distributed clustering algorithm in which nodes make autonomous decisions without any centralized control. Cluster formation is cyclically performed and history information of the previous CHs are stored. CHs are assigned as a result of a random procedure, where each node can declare itself as a CH with some probability. Energy levels of the nodes are included as a factor in the CH selection while connectivity of the nodes are ignored. Therefore, it is not guaranteed that every node is within K-hops of a CH. This is the main concern of LEACH, which may cause some nodes to be segregated from the rest of the network during the time period in between the two election cycles. Another drawback of LEACH is due to the assumptions that not only the network size and the number of CHs are known in advance but also all nodes are very well synchronized (in order to ensure that CHs can be re-elected periodically to balance the energy consumption). These are very specific assumptions that might not fit well to real-life applications, especially for WSNs.

In [4], Bandyopadhyay *et al.* propose a distributed and randomized clustering algorithm similar to the LEACH. The proposed algorithm also aimed at energy efficiency, and its difference from the LEACH is that it provides hierarchical (multi-level) clustering as well. Other than that, the proposed algorithm holds the same concerns and drawbacks as LEACH does.

In [5], Jia *et al.* present an energy-consumption-balanced clustering algorithm (LEACH-EP) for WSNs that is based upon the LEACH algorithm. It introduces an energy factor in the CH-electing threshold and optimizes the election probability of CH. As in the case of LEACH, LEACH-EP comes with specific assumptions as well. As in the case of LEACH, these specific assumptions might not fit well to real-life applications of WSNs.

Energy-Efficient Clustering Scheme (EECS) [6] is also based upon the LEACH algorithm and aims at energy efficiency. Its difference from the LEACH is the setup phase of the clusters (cluster formation). LEACH-C [7] and SECA [8] are also based on the LEACH algorithm. Both algorithms consider node-positions into account while calculating the CHs and try to cumulate the clusters. All of these LEACH-based proposed algorithms hold the same concerns and drawbacks as LEACH does.

In Hybrid, Energy-Efficient, Distributed Clustering (HEED) [9] approach, CHs are periodically selected according to a hybrid of their residual energy and a secondary parameter, such as a node's proximity to its neighbors or node degree. HEED does not make any assumptions about the distribution or the density of the nodes, nor their connectivities.

Evenly Distributed Clustering (EDC) algorithm [10] distributes clusters uniformly and minimizes the number of clusters. It considers the connectivity of the nodes with the K-hops parameter. It is a heuristic approach, in which each node only exchanges its head selection with its neighbors. Based on neighbors' selection results, each node chooses the nearest head as its CH. The drawback of this algorithm is that it does not consider the density of the nodes in a network. In order to increase the TLT of the network, it is important to elect more CHs in the dense areas of the network. However, the algorithm is aimed at distributing the cluster heads evenly to the network deployment field.

In [11], Brust *et al.* present algorithms for cluster head candidate selection that are based on topology (location) of the nodes. The algorithms aim to avoid selecting nodes located close to the network partition border because those nodes are more likely to move out of the partition, thus, causing a CH re-election. By using the connectivity information, they propose three algorithms to find the strong, weak, bridge, and board nodes in the network. Authors do not provide any information on how to select the CHs among their selection of nodes (strong, weak, bridge, and board nodes). Overall, this classification of nodes for CH selection would be useful for the mobile *ad hoc* networks (MANETs) where mobility is the prime factor that changes the network topology. However, the network topology in WSNs is quite stable compared to MANETs, and therefore, this kind of node classification is unnecessary and also expensive (power consuming) for CH selection.

Energy-Efficient Unequal Clustering (EEUC) [12] is proposed for periodically data-gathering WSNs. It partitions the nodes into clusters of unequal size, and clusters closer to the base station have smaller sizes than those farther away from the base station. This way, CHs closer to the base station can preserve some energy for inter-cluster data forwarding.

Hierarchical clustering proposed in [13] is a framework based on two-level clustering, multi-hops clusters for data aggregation (the first-level clustering), and one-hop clusters for intrusion detection (the second-level clustering). Although the idea sounds promising for some applications of WSNs (especially for the industrial applications), the details of the formation algorithms for the multi-level clustering were missing (we assume that this was left as a future work).

Kachirski *et al.*'s [14] clustering algorithm is based on the connectivity of the nodes in the network. The higher the connectivity (neighbors) a node has, the higher the probability of it to be elected as the CH of a certain neighborhood (cluster). This algorithm is one of the best choices for us to work on for several reasons: first of all, it did not require probabilistic approach on clustering, and therefore, the result of the clustering would cover the whole network. Secondly, the connectivity of the nodes are the main concern on the election of CHs, which is reasonable. In general, the nodes that have more connections would be rather elected as CHs. Finally, the algorithm is easily implementable, which allows the proof of the theoretical work on both hardware and simulation environment.

There are flaws in Kachirski *et al.*'s [14] clustering algorithm: 1) nodes were not able to vote for themselves through the election phase of the CHs. Thus, we fixed this problem and called it as 'revised version of Kachirski *et al.*'s algorithm'. We showed geographically how this change would result in a more logical cluster formation. 2) There was no power awareness in their proposed algorithm. Therefore, in this article, we propose our clustering algorithm that is built upon the revised version of Kachirski *et al.*'s algorithm. Our proposed PCAC algorithm is both power and connectivity aware, that is why it provides maximum throughput while saving the energies of the nodes, therefore, significantly increasing the TLT of the network.

3 Kachirski *et al.*'s connectivity-based approach for clustering

In order to demonstrate the principles of the Kachirski *et al.*'s [14] clustering algorithm, consider the network shown in Figure 2. Here, we assume that each node has one-hop[a] connectivity, meaning that each node can communicate with its direct neighbors that are in one-hop communication distance (in terms of radio range). In order to elect the CHs, these are the steps to be followed:

1. Let C_i denote the number of established connections (nodes that are one-hop away in our case) for node i, with total number of N nodes in the network. Each node calculates its own C_i value (as shown in Figure 3; note that the number written in each node represents the total number of neighbors for each node) and sends it to all its neighbors.
2. After receiving C_k values from its neighbors k (where $k \neq i$ for all $i = 1 \dots N$), a node i calculates the connectivity index (S_i) as shown in Equation 1:

$$S_i = C_i + \sum_k C_k \quad (1)$$

   Each node calculates its own connectivity index according to Equation 1. For the network shown in Figure 3, the connectivity indices would be as shown in Figure 4.
3. Each node broadcasts its connectivity index (S_i) to all other nodes with a time to live (TTL) value equivalent to time spent through one-hop communication.

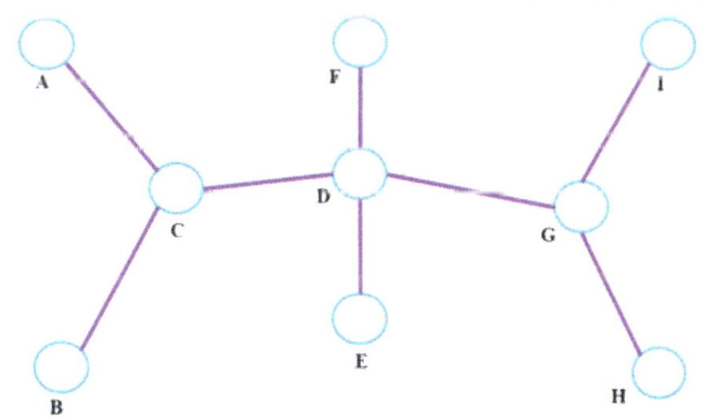

Figure 2 A typical 9-node WSN.

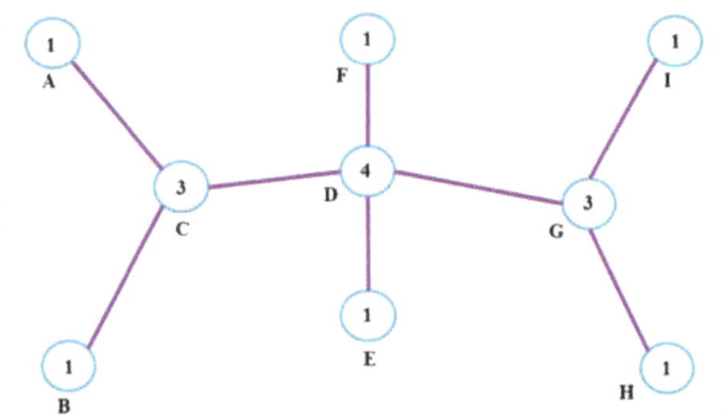

Figure 3 Established connections graph, indicating total number of one-hop neighbors for the WSN shown in Figure 2.

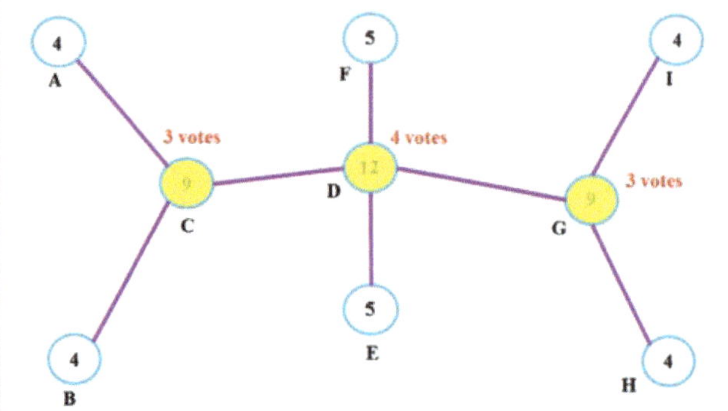

Figure 5 Elected CHs (one-hop) and their votes, after the voting session for the WSN shown in Figure 2.

4. Each node then has to participate in a voting session in which the CH will be determined. Each node votes for the node that has the highest S_i value, as a result of the broadcast operation in step 3.
5. After the voting procedure, if a node receives at least one vote, it is assigned as the CH. After the voting session, the network members in Figure 4 select their CHs as shown in the Figure 5[b].

4 Revised version of Kachirski *et al.*'s connectivity-based approach for clustering

In the specific case of the network shown in Figure 2, there are nine members of the network, and three members (out of nine) are elected as cluster heads, as a result of the voting procedure (see Figure 5). As the network connectivity increases, we expect to have more connected members in the network resulting in less number of selected CHs. As an example, for the same configuration of the network in Figure 2, if we use two-hops connectivity for the node communications, we obtain the neighborhood graph as shown in Figure 6. By applying Equation 1 and then performing the voting session, the connectivity index graph (denoted on the nodes) and the CH selections would be as shown in Figure 7[b].

This is quite an interesting result, since we were expecting to have less CHs by increasing the connectivity (number of maximum hops). This happens because of a fault in the voting procedure of Kachirski *et al.*'s [14] clustering algorithm. We realized that throughout the voting procedure, nodes are not voting for themselves even though they may have the highest connectivity. This may result in more CHs to be elected than needed. In order to fix this problem, we revised Kachirski *et al.*'s clustering algorithm by letting the nodes vote for themselves (if they have the highest connectivity index).

We applied the revised scheme to our example network (see Figure 2), and the result of the voting scheme is as shown in Figure 8[b]. Here, the total number of CHs is one, resulting in less CHs (instead of three) compared to the original Kachirski et al.'s clustering algorithm, as expected (the fault is fixed).

5 PCAC: Power- and Connectivity-Aware Clustering

In WSNs, energy is one of the scarce resources that needs to be conserved. As a result of the clustering algorithms, elected cluster heads become the highest energy-consuming nodes of the network, since they perform

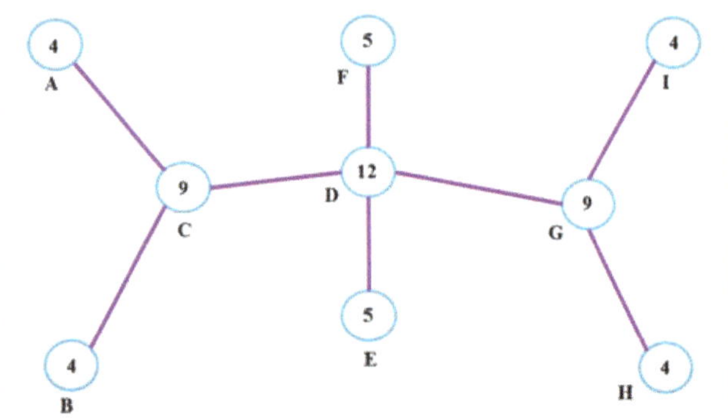

Figure 4 Connectivity index graph (one-hop) of the WSN shown in Figure 2.

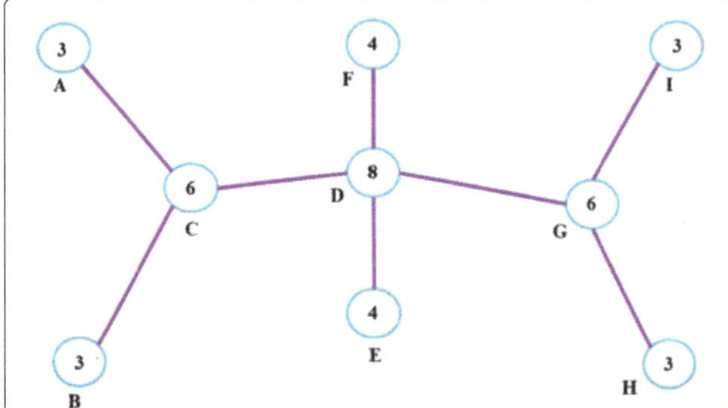

Figure 6 Established connections graph, indicating total number of two-hops neighbors for the WSN shown in Figure 2.

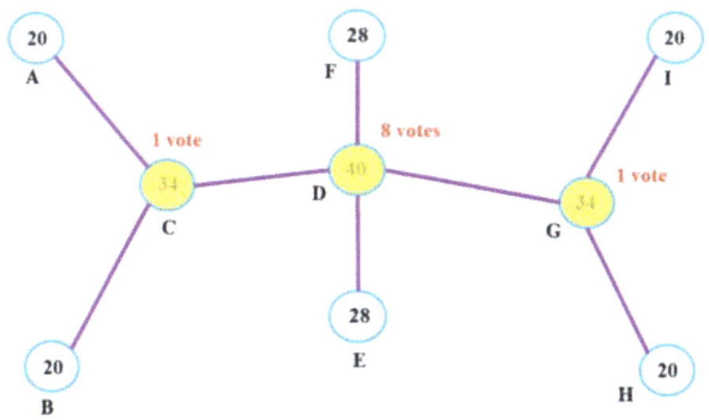

Figure 7 Connectivity index graph and elected CHs (two-hops) of the WSN shown in Figure 2.

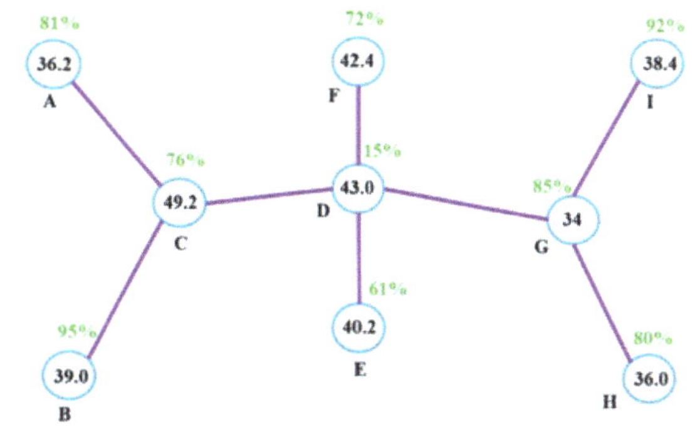

Figure 9 Connectivity index graph (two-hops) of WSN shown in Figure 2, as a result of our PCAC algorithm.

operations related to data aggregation, security, routing, etc., on behalf of other nodes.

Kachirski *et al.*'s [14] clustering algorithm (see Section 3), and it's revised version (see Section 4) only consider a node's connectivity with its neighbors while determining a CH. But they do not consider any parameter regarding the energies (nor the powers) of the nodes.

In order to increase the TLT of a WSN, energy (power) levels of the nodes also should be considered while determining the CHs. Therefore, in this section of our paper, we propose the PCAC algorithm built upon the revised version of (see Section 4) Kachirski *et al.*'s [14] clustering algorithm. We achieved this by introducing the power-level readings through connectivity index calculations (see Section 3, step 2). Our scheme determines the CHs according to these calculations. Besides, the voting scheme follows the revised version of Kachirski *et al.*'s clustering algorithm (nodes may vote for themselves).

The description of our proposed scheme is as follows:

1. Let C_i denote the number of established connections for node i, with total number of N nodes in the network. Each node calculates its own C_i value and sends it to all its neighbors.
2. After receiving C_k values from its neighbors k (where $k \neq i$, for all $i = 1 \ldots N$), a node i calculates the connectivity index (S_i) as shown in Equation 2:

$$S_i = C_i + \sum_k C_k + \beta \times P_i \quad (2)$$

In this equation, β is called power factor, and the P_i value represents the battery power level of each node i. β represents the correlation of the power level of each node (P_i) to the connectivity index (S_i) of that particular node. By this way, whenever a centrally located node starts to deplete its battery power, it is guaranteed that it will not be elected as a CH in the next CH-election phase. Equation 2 describes how, in our PCAC algorithm, each node's connectivity index (S_i) not only carries information regarding its connectivity (C) with its neighbors, but also the battery power level (P_i) of that particular node.
In our calculations, maximum value for P_i is chosen as 1.00, meaning that the battery power level of the

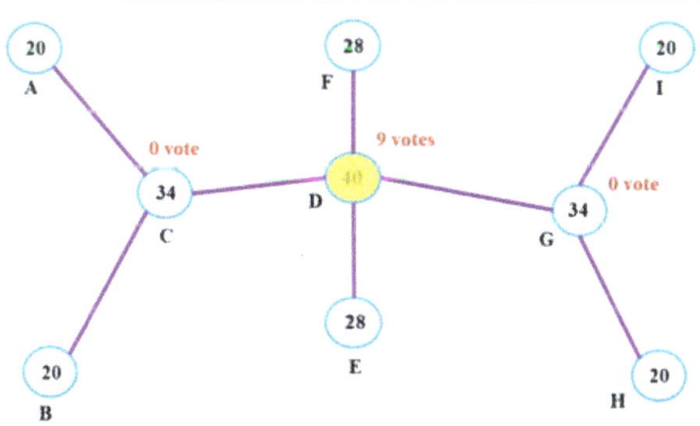

Figure 8 Elected CHs (two-hops) of the WSN shown in Figure 2 by using the revised clustering scheme.

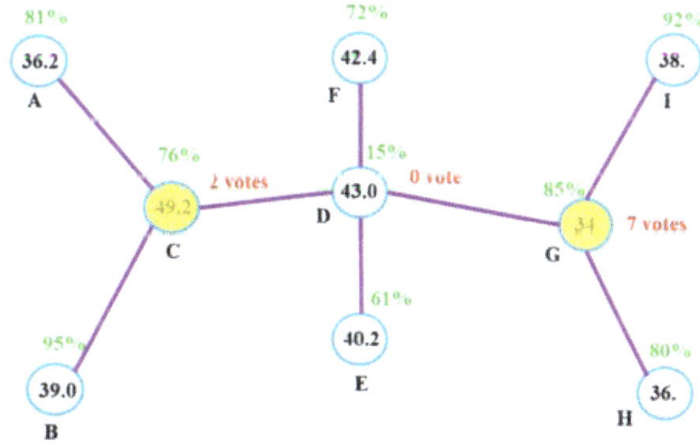

Figure 10 Elected CHs (two-hops) of the WSN shown in Figure 2 by using our PCAC algorithm.

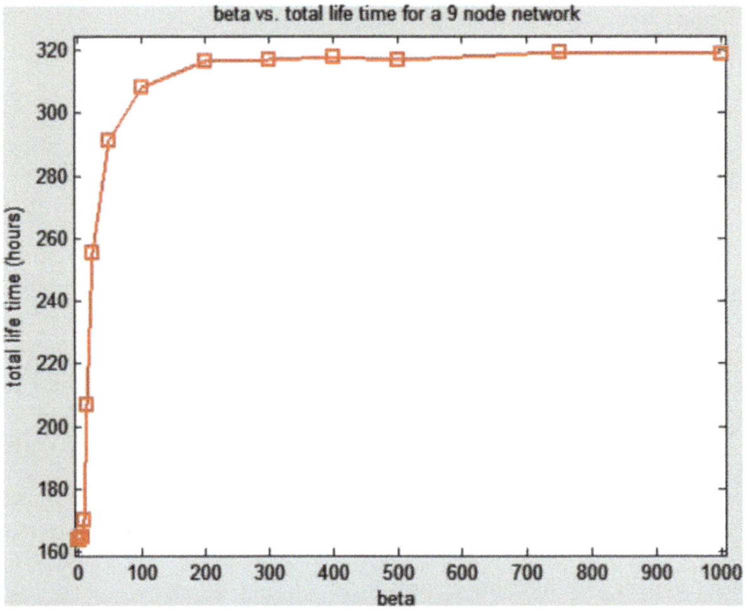

Figure 11 Total life time vs. beta, for the WSN shown in Figure 2 by using our PCAC algorithm.

node is 100% of its maximum level. For the given WSN shown in Figure 2, the connectivity indices and the voting results are shown in Figure 8. We recalculate the connectivity indices for that network according to Equation 2, and the new results are shown in Figure 9. Here, each green writing over the nodes represents the battery power level (percentage) of that node at the time that the clustering calculation is done.

3. Each node broadcasts its connectivity index (S_i) to all other nodes with a TTL value equivalent to time spent through one-hop communication. TTL helps the network to prevent packet-replay attacks. Besides, TTL is used to discard late-arriving votes (because of network congestion) and lets the voting procedure be completed in a timely manner.
4. Each node then has to participate in a voting session in which the CH will be determined. Each node votes

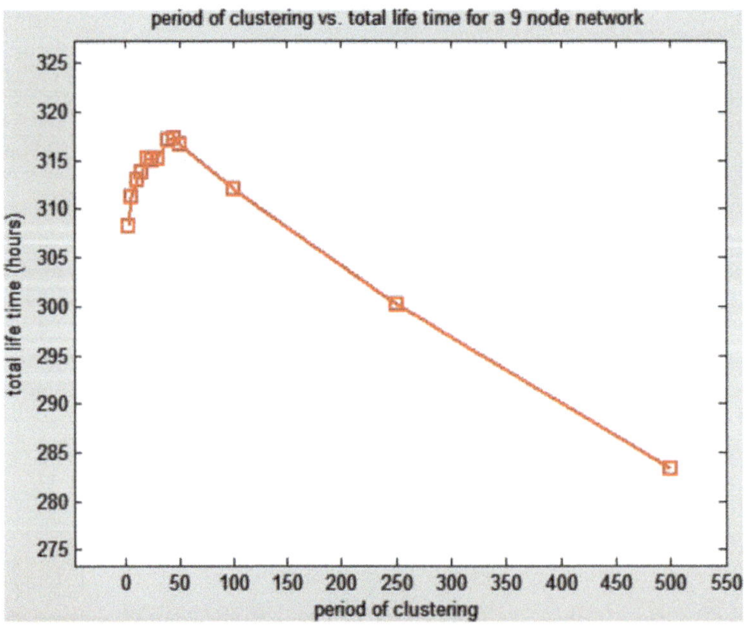

Figure 12 Total life time vs. period of clustering, for WSN shown in Figure 2 by using our PCAC algorithm.

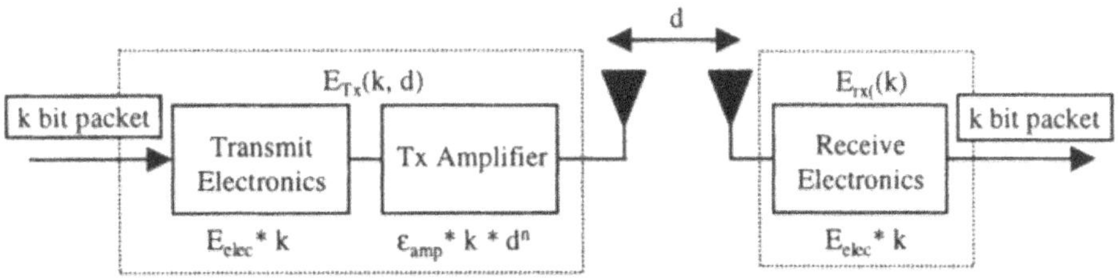

Figure 13 Radio-energy-dissipation model used in our simulations [7].

for the node that has the highest S_i value (nodes are allowed to vote for themselves), as a result of the broadcast operation in step 3.

5. After the voting procedure, if a node receives at least one vote, it is assigned as the CH. After the voting session, the network members in Figure 9 select their CHs as shown in the Figure 10[b].

5.1 Observations on TLT, β, and τ

TLT is the most important parameter in WSNs. It determines the life time of the network. In our calculations and throughout this paper, TLT is determined as follows: whenever a node's battery power depletes and disconnects from the network, its life time is considered as TLT of the network. In order to distinguish dead nodes from disconnected nodes, we use battery-low signal; whenever a node's battery goes into weak stage, it broadcasts a 'low battery' message to all neighbors. If a neighbor disconnects after this message, then we consider this as the end of the network operation. Otherwise, the network resumes operation.

Energy consumption and TLT are inversely proportional: the more energy consumption a node has, the less TLT it will have. Both parameters are correlated to connectivity as follows: the higher the connectivity index (S_i) of a node, the more messages (or packets) will pass through it. This means more transmissions consuming more node energy, thereby depleting its battery power (P_i) faster. Therefore, TLT of the network is also reduced.

When a WSN uses our proposed PCAC algorithm, we expect two parameters to effect the TLT of thenetwork:

- *Power factor* (β): an optimum value of β can be determined by keeping every parameter in the network fixed and then by observing the TLT of the network with the change of β.
- *Period of clustering* (τ): it is the time period that determines the renewal of the cluster heads by re-applying the clustering algorithm. An optimum value of τ can be determined by keeping every parameter in the network fixed and then by observing the TLT of the network with the change of τ.

As an example, we simulated our PCAC algorithm on the network shown in Figure 9 with the simulator discussed in the next section (Section 6). Figure 11 shows the behavior of the TLT with the change of β, whereas Figure 12 shows the behavior of the TLT with the change of τ. According to the result of the simulations, we may conclude that, for the network configuration of Figure 9 and the parameter selection shown in Section 6.2, the optimum value of β is 200 (the TLT curve in Figure 11 saturates for $\beta \geq 200$) and the optimum value of τ is 45 (the TLT curve in Figure 12 gets the maximum value at $\tau = 45$ and starts decreasing as τ becomes bigger or smaller than this value).

5.2 Applicability of our proposed PCAC algorithm to nowadays WSNs

Our proposed PCAC algorithm is very applicable to nowadays WSNs. Because, current Commercial Off-The-Shelf (COTS) nodes, such as Wasp motes [15], provide the power reading of it's batteries (as a percentage) as an available information which could be sent to other nodes. This information would be used directly by our proposed PCAC algorithm in order to determine the cluster heads.

6 Comparison of both schemes in terms of total life time of the WSN

In order to evaluate and compare the effect of both Kachirski *et al.*'s (revised) clustering algorithm and our proposed PCAC algorithm on the total life time of the WSNs, we created a simulation environment in *MATLAB*^{TM}[16]. The details of the simulation environment and the theoretical background are as follows:

Table 1 Values for the energy-consumption-related parameters used through our simulations

Parameter	Value
E_{elec}	50 nJ/bit
ϵ_{fs}	10 pJ/bit/m^2
ϵ_{mp}	0.0013 pJ/bit/m^4
d_0	87.7m
E_{DA}	5 nJ/bit/data

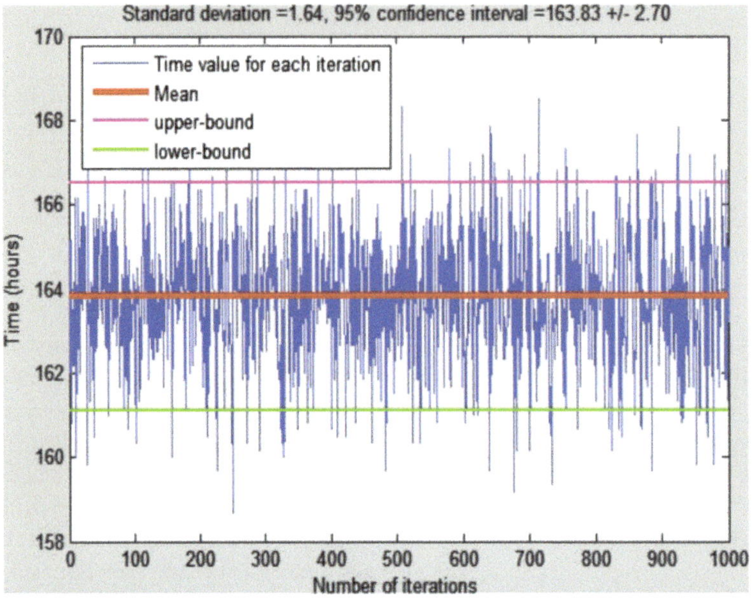

Figure 14 Distribution of the simulation time.

6.1 Energy-consumption calculations

For energy-consumption calculations, we followed Heizelman *et al.*'s work [7]. We assume a simplified model (since radio wave propagation is mostly non-stable and difficult to model) for the radio hardware energy dissipation, where the transmitter dissipates energy by running the radio electronics and the power amplifier, whereas the receiver dissipates energy by running the radio electronics only, as shown in Figure 13.

We consider two different channel models depending on the distance between the transmitter and the receiver:

1. *Near-field (free space - fs) channel model*: if the distance between the transmitter and the receiver is less than a threshold (d_0), then this model is used (also called d^2 power-loss model).
2. *Far-field (multi-path - mp) channel model*: if the distance between the transmitter and the receiver is greater than a threshold (d_0), then this model is used (also called d^4 power-loss model).

According to [5], the threshold value for the distance is calculated as follows:

$$d_0 = \frac{\sqrt{\epsilon_{\text{fs}}}}{\sqrt{\epsilon_{\text{mp}}}} \tag{3}$$

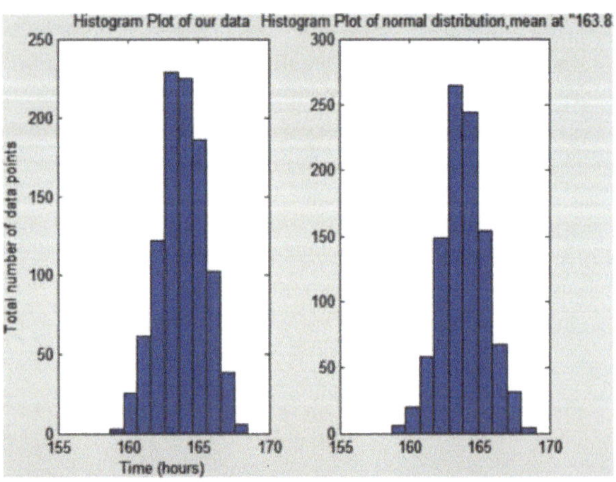

Figure 15 Histogram plots of the simulation time compared to the normal distribution.

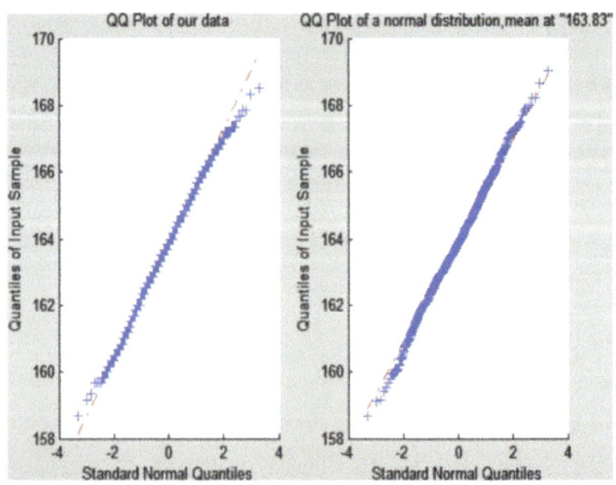

Figure 16 Quantile-quantile plots of the simulation time compared to the normal distribution.

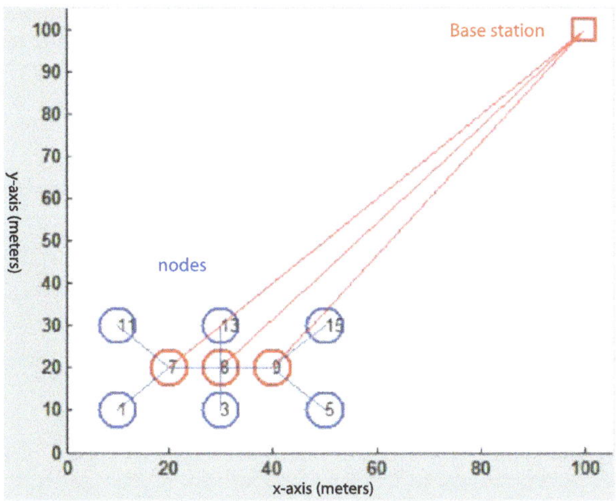

Figure 17 Location of the nodes and the *BS* throughout the simulations.

where ϵ_{fs} and ϵ_{mp} are constants related to free-space loss and multi-path loss, respectively.

In order to transmit m-bit data to a distance of d, the radio spends:

$$E_{Tx}(m,d) = E_{Tx-\text{elec}}(m) + E_{Tx-\text{amp}}(m,d) = \begin{cases} mE_{\text{elec}} + m\epsilon_{\text{fs}}d^2, & d < d_0 \\ mE_{\text{elec}} + m\epsilon_{\text{mp}}d^4, & d \geq d_0 \end{cases} \quad (4)$$

In order to receive the same m-bit data, the radio spends:

$$E_{Rx}(m) = E_{Rx-\text{elec}}(m) = mE_{\text{elec}} \quad (5)$$

The energy spent on the radio electronics circuitry, E_{elec}, is due to the digital modulation (transmitter side), digital demodulation (receiver side), error correction codes, and filtering, whereas the amplifier energy, $E_{Tx-\text{amp}}$, is due to the electromagnetic spreading of the signal into the air and depends on the distance as mentioned above (see Equation 3).

Assume that each CH has N member nodes. CH dissipates energy by receiving the data from member nodes, aggregating those data, and finally transmitting the aggregate data to the base station (BS). We assume that BS is located far away from the nodes, and therefore, transmission between the CH and the BS follows the far-field channel model (d^4 power-loss model). During a single data frame, we calculate the energy dissipated in the CH as follows:

$$\begin{aligned} E_{CH} &= \{E_{\text{aggregating\_data\_from\_member\_nodes}}\} \\ &\quad + \{E_{\text{transmit\_aggregate\_data\_to\_BS}}\} \\ &= \{N(mE_{\text{elec}} + mE_{\text{DA}})\} + \left\{mE_{\text{elec}} + m\epsilon_{\text{mp}}d^4_{\text{toBS}}\right\} \end{aligned} \quad (6)$$

where m represents the total number bits in a data frame, mE_{DA} represents the energy dissipated during aggregating m-bit data, and finally d_{toBS} represents the distance between the CH and the BS.

Assume that each member node is located in the near field of the CH so that near-field channel model (d^2 power-loss model) will be used for calculating the energy dissipated during data transmission from the member node towards the CH. Therefore, we calculate the energy dissipated in each member node as follows:

$$E_{\text{member\_node}} = mE_{\text{elec}} + m\epsilon_{\text{fs}}d^2_{\text{toCH}} \quad (7)$$

where d_{toCH} represents the distance between the member node and the CH, and therefore, it takes different values for each node.

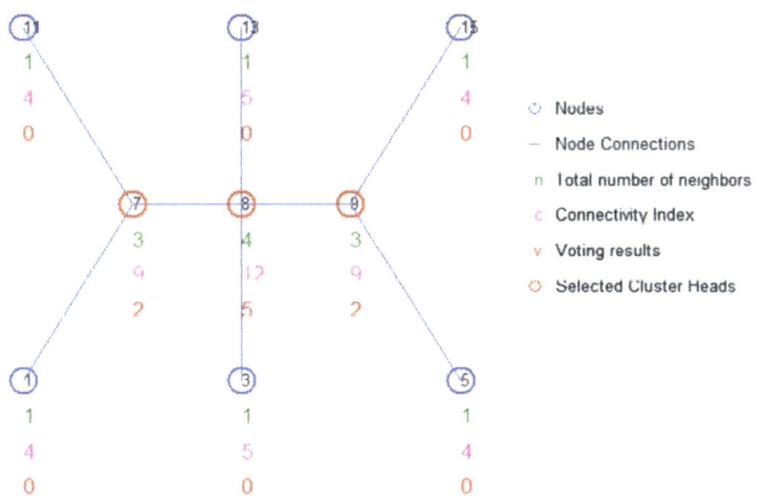

Figure 18 CH selection of a nine-node WSN with Kachirski *et al.*'s algorithm (revised version).

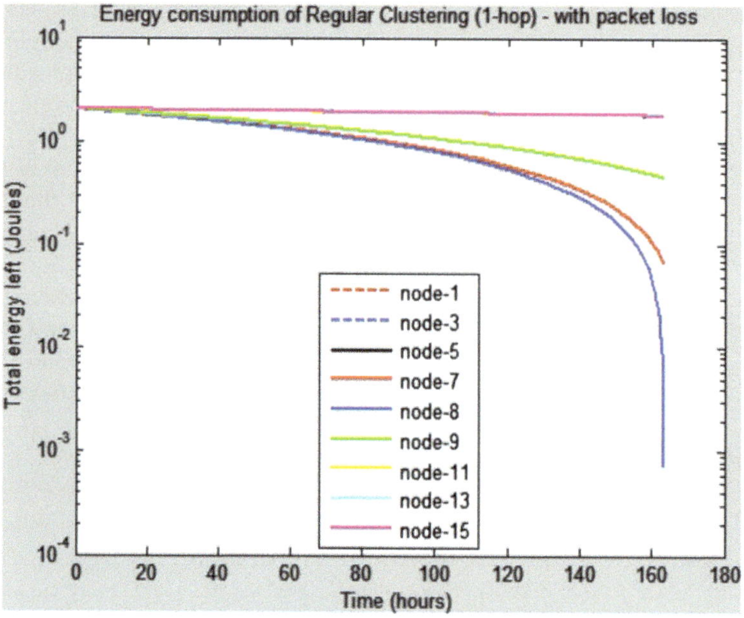

Figure 19 Energy consumption graph of Kachirski *et al.*'s clustering algorithm (revised version).

For all our simulations referred in our work, we followed the energy-consumption models of [7] and used the values for the energy-consumption-related parameters as shown in Table 1.

6.2 Simulation parameters

The parameters that we used during the simulation are as follows:

- Each node in the network is identical to each other and has a starting energy of 2 *J*.
- There is a *BS* located outside of the network to collect the data from CHs.
- The deployment area is 100 *m* × 100 *m*.
- Data flow from nodes to CHs. CHs aggregate the data and then forward to the *BS*.
- The header size for each frame is 200 bits.
- The data size for each frame is 4,000 bits.
- Data rate is 1 frame per 10 min (0.1 frames/min).
- We consider a packet drop rate of 5% for the transmission of each data frame due to the collisions and multi-path fading.
- We consider a stationary network meaning that both *BS* and the nodes are not moving.
- Each simulation is run 1,000 times, and an average value of the total life time (that falls into 95% confidence interval) is calculated. For example, for the simulation result presented in Section 6.4 (energy

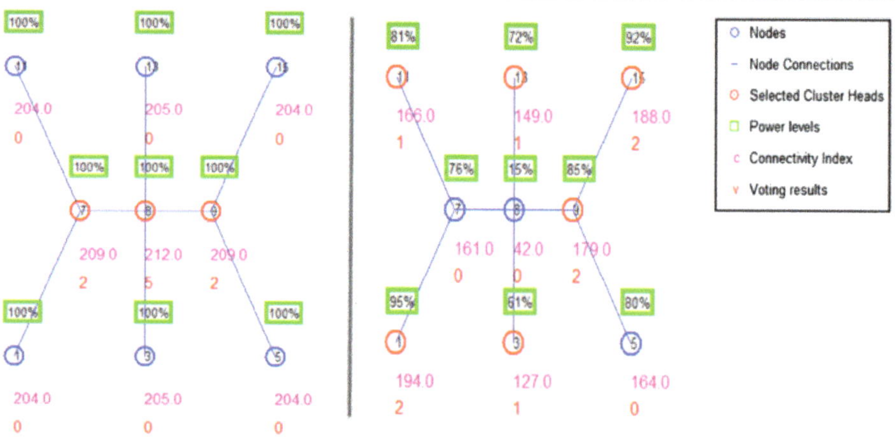

Figure 20 CH selection of a nine-node WSN with our algorithm; (left) at time $t = 0$, (right) at time $t = t_1$ ($t_1 > 0$).

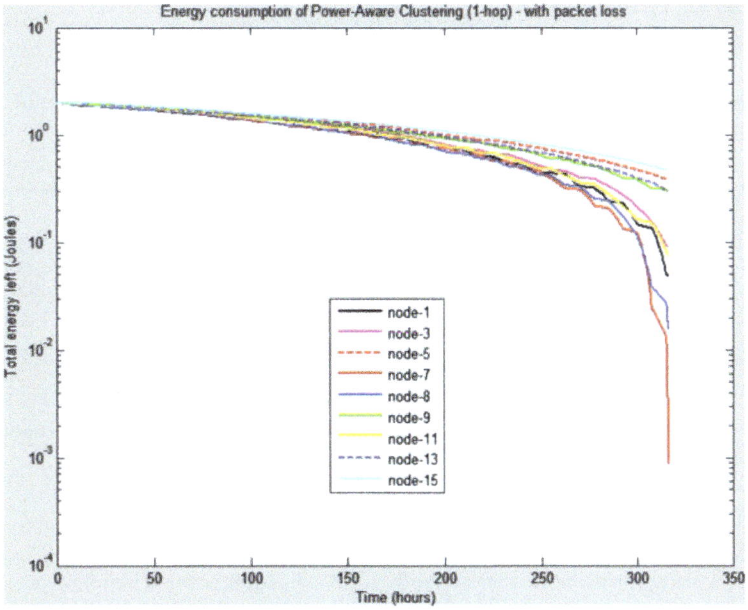

Figure 21 Energy-consumption graph of our PCAC algorithm.

consumption graph of Kachirski *et al.*'s clustering algorithm), the distribution of the total life time is shown in Figure 14. Here, the lower bound and upper bounds of the confidence interval (95%), as well as the mean value, are shown on the graph. Figures 15 and 16 show the corresponding histogram plot and quantile-quantile plot, respectively. In these figures, the first plot represents our simulation data, and the second plots represent the normal distribution (Gaussian) that has the same mean value as our data. From these figures, we observe that our simulation result shows a normal distribution (in the quantile-quantile plot, our data cumulates on the $x = y$ line). Therefore, we calculated the mean value for each simulation in this text to represent all the values resulted in 1,000 iterations.

6.3 Location of the nodes

Throughout our simulations, we assumed that both *BS* and the nodes are stationary; therefore, their positions are fixed. For the following sections, positions of the nodes and the *BS* will be as shown in Figure 17. Here, circular shapes represent the nodes (blue ones are the member nodes and the red ones are the CHs) CHs)

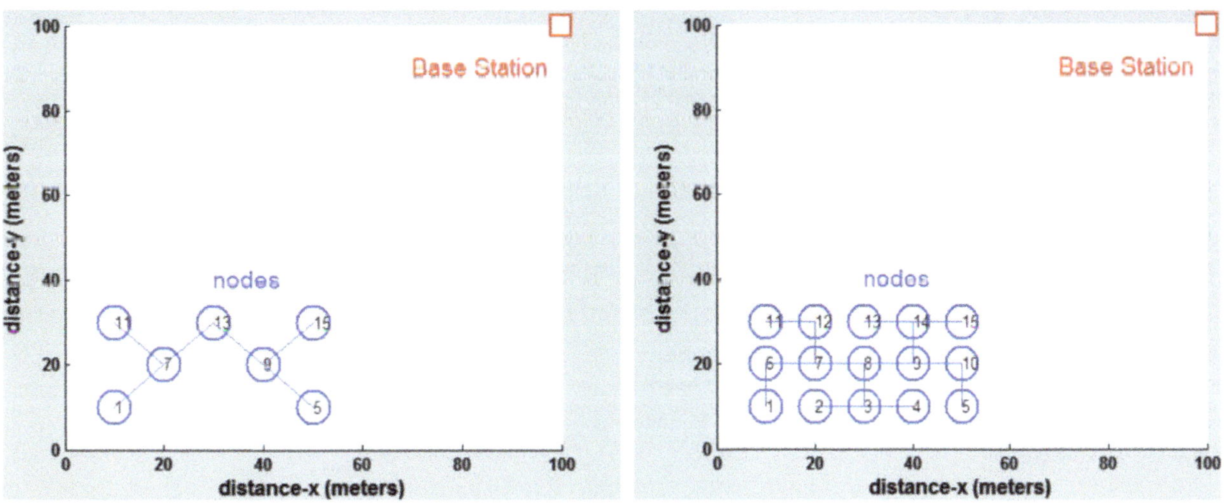

Figure 22 Different network topologies with 7 and 15 nodes.

whereas the square shape represents the *BS*. The red lines represent the connection between the CHs and the *BS*, whereas the blue lines represent the connections between the CHs and their member nodes. The whole deployment area is 100 $m \times 100$ m, and the location of the *BS* is $[100\ m,\ 100\ m]$. The nodes are deployed to the area with the following boundaries: $[10\ m,\ 10\ m], [10\ m,\ 30\ m], [50\ m,\ 10\ m]$, and $[50\ m,\ 30\ m]$.

6.4 Energy consumption of Kachirski *et al.*'s clustering algorithm (revised version)

We ran the revised version of Kachirski *et al.*'s clustering algorithm on our simulator with the parameters shown in Section 6.2 and the positions shown in Section 6.3. We considered one-hop connectivity for all nodes in the network. Figure 18 shows the total number of neighbors for each node (including the connection paths), connectivity indices, and results of the voting along with the elected CHs.

Figure 19 shows the energy-consumption performance of the mentioned algorithm with respect to time. We stopped the simulation whenever a single node dies (runs out of battery power), and we call this time as the 'total life time of the network', since at this point the network starts to disintegrate (segregation starts).

In Figure 19, we can see that node 8 depleted its energy faster than other nodes and therefore determined the network's total life time as 163.8 h.

6.5 Energy consumption of our proposed PCAC algorithm

We ran our proposed PCAC algorithm on our simulator with the parameters shown in Section 6.2 and the positions shown in Section 6.3. We consider one-hop connectivity for all nodes in the network. As mentioned in Section 5, we selected β as 200 and τ is 45, in order to achieve the maximum TLT. Figure 20 shows the result of the clustering algorithm at times $t = 0$ and $t = t_1 (t_1 > 0)$. Figure 21 shows the energy-consumption performance of our proposed algorithm with respect to time. In Figure 21, we can see that node 7 depleted its energy faster than other nodes and therefore determined the network's TLT as 316.66 h. When compared to the revised version of Kachirski *et al.*'s clustering algorithm (see previous subsection), relative performance improvement in TLT of the network is 93%.

Table 2 Relative performance improvements (%) on the life time of the network when our PCAC algorithm is used

Maximum hops	For 7-node network	9-node	15-node
One-hop	86	93	85
Two-hops	234	313	256
Three-hops	366	463	438

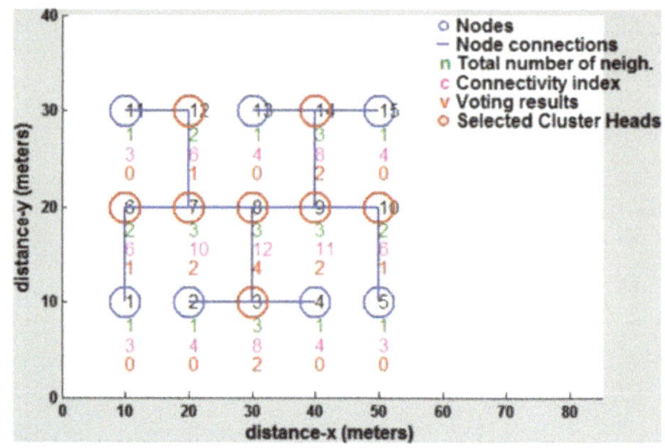

Figure 23 Clustering of 15-node network, one-hop communication case.

6.6 Further comparisons

In order to provide further proof of performance improvement on life time, we repeated the same simulation setup with different network topologies with 7 nodes, 9 nodes, and finally 15 nodes (see Figure 22). We ran both clustering algorithms on these networks in three different maximum number of hops: one, two, and three. The resulting relative performance improvements on the total life time of the network are as shown in Table 2. Accordingly, we conclude that, as the maximum number of hops increases, our clustering algorithm becomes more beneficial. This is because more nodes become eligible to be elected as CHs as the maximum number of hops increases. According to our simulations, the relative performance improvement of clustering algorithm varies between 85% and 93% for one-hop neighborhood, 234% and 313% for two-hops neighborhood, and finally 366% and 463% for three-hops neighborhood, respectively.

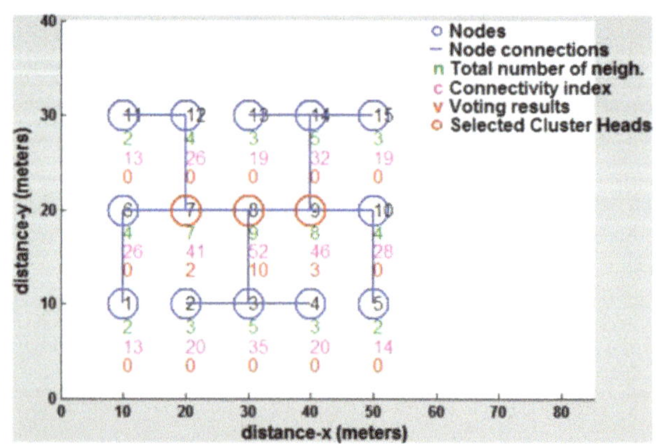

Figure 24 Clustering of 15-node network, two-hops communication case.

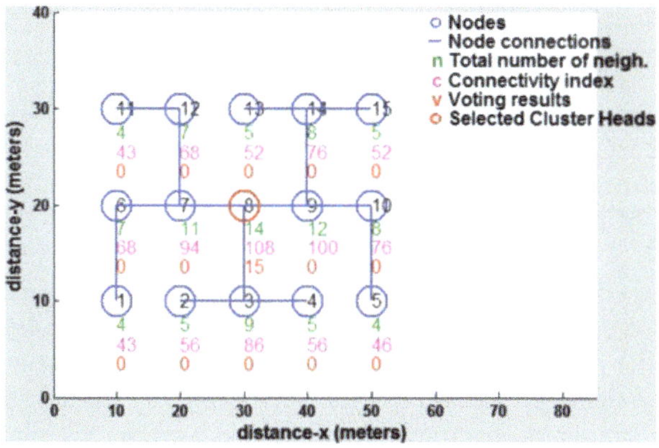

Figure 25 Clustering of 15-node network, three-hops communication case.

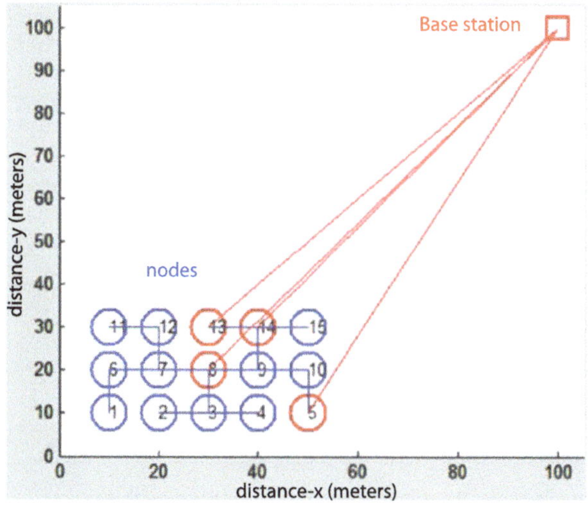

Figure 27 Location of the nodes and the BS.

7 Some observations on the effect of clustering to the network performance

7.1 Effect of maximum number of hops on total number of cluster heads

As the maximum number of hops increases, the nodes in the network achieve more communications with the other member nodes, and as a result, the network requires less number of CHs. To support this hypothesis, we ran the CH selection algorithm on a 15-node network for three different maximum number of hops: one, two, and three. The resulting total number of neighbors, connectivity indices, voting results, and the elected CHs are shown in Figures 23, 24, and 25, respectively.

Accordingly, Figure 26 shows the plot of maximum number of hops versus the total number of elected CHs, for a 15-node network.

7.2 Effect of total number of cluster heads (maximum hops) on the total life time of the network

The positions of the BS and the WSN nodes are as shown in Figure 27. BS is located in the far field of the WSN, meaning that the distance between CHs and the BS is greater than 87.7 m. The simulation parameters are the same as shown in Section 6.2. β is chosen as 200 and the τ as 40 frames. By using these parameters and positions, we ran the simulation for four cases of the maximum hops: zero, one, two, and three. Figure 28 shows the behavior of TLT of the network with respect to maximum hops. Accordingly, we conclude that, as the maximum hops increases, the TLT of the network increases. From the slope of the curve, we can deduce that this increase saturated at a certain number of hops. This is the point where

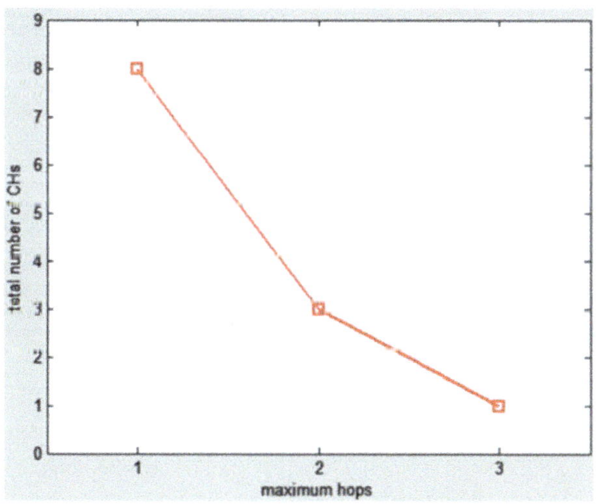

Figure 26 Maximum number of hops vs. total number of CHs for a 15-node network.

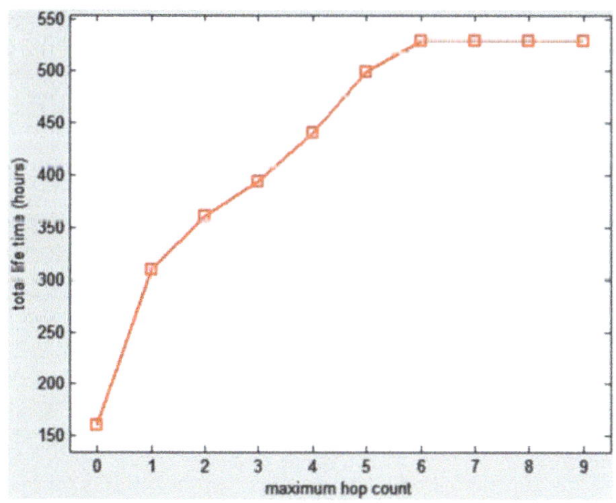

Figure 28 Maximum hops vs. total life time of the network.

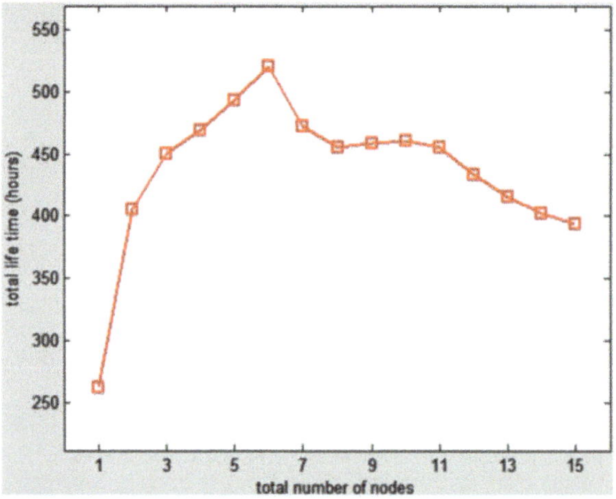

Figure 29 Total number of nodes vs. total life time of the network.

each node can reach any node in the network (six-hops in this case).

7.3 Effect of total number of nodes in a cluster on total life time of the network

We wondered about the effect of total number of the nodes in the network on the TLT of the network. To investigate this, we considered the same network shown in Figure 27 along with the simulation parameters the same as Section 7.2. The only parameter that is different here is the maximum hops. We kept it constant and equal to three. Then, we started the simulation with 15 nodes, and then, each time, we removed one of the end nodes and repeated the simulation till we are left with one node in the network. As a result, Figure 29 shows the behavior of the TLT of the network with respect to the total number of nodes in the network. Accordingly, we conclude that there is a certain number of nodes (six nodes in our case) in the network that allow the network to achieve maximum TLT (519.15 h in our case).

7.4 Effect of data rate on the total life time of the network

Here, we investigated the effect of frame rate of the nodes in the network on the TLT of the network. We considered the same network shown in Figure 27 along with the simulation parameters the same as Section 7.2. The only parameter that is different here is the maximum hops. We kept it constant and equal to three. Then, we started the simulation with 15 nodes, and then, each time, we changed the frame rate and repeated the simulation for various frame rates. As a result, Figure 30 shows the behavior of the TLT of the network with respect to the frame rate of the nodes in the network. Accordingly, we conclude that, as the frame rate increases, TLT decreases. This is an expected result, because, as the frame rate increases, more packets are sent between the nodes thus more energy is consumed.

7.5 Conclusions from the observations

There are three major trade-off situations that need to be balanced when implementing solutions (i.e., security, etc.) to a clustered WSN:

1. There is a trade-off between 'maximum hop count' and 'total number of CHs'. As the maximum hop

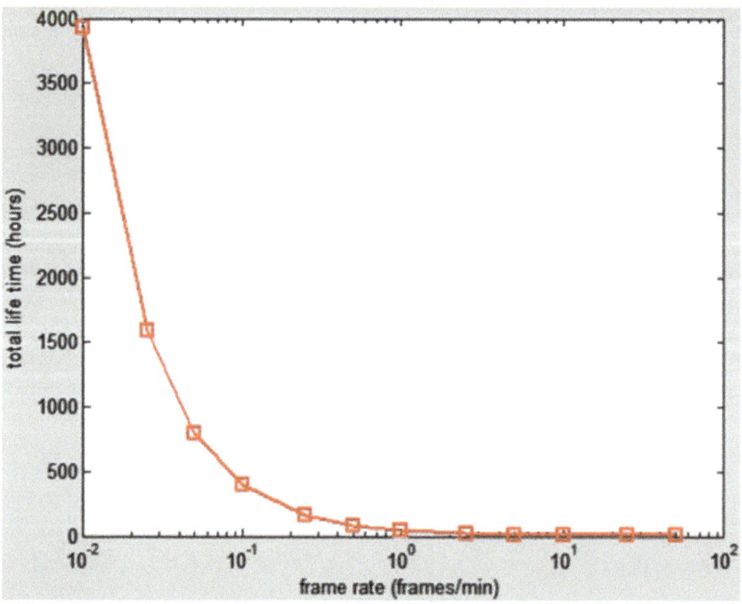

Figure 30 Frame rate vs. total life time of the network.

count increases, total number of CHs decreases and vice versa.

2. There is a trade-off between 'total number of CHs' and 'TLT of the network'. There is an optimum number of CHs which lead the network to survive the most TLT possible (without having any partioning/segregation).
3. There is a trade-off between 'data rate (frames/minute)' and 'TLT of the network'. As the data rate increases, more data need to be processed and more packets need to be transmitted causing more power to be spent; therefore, the TLT of the network decreases.

8 Conclusions

In this work, we provided the energy-consumption simulation results of the revised version of Kachirski *et al.*'s clustering algorithm and our proposed PCAC algorithm. According to these results, our proposed PCAC algorithm out-performed the revised version of Kachirski *et al.*'s clustering algorithm in terms of energy efficiency and also total life time of the network.

According to our simulation results, with our proposed PCAC algorithm, relative performance improvement (compared to the revised version of Kachirski *et al.*'s clustering algorithm) in total life time of the network varies between 85% and 93% for one-hop neighborhood, 234% and 313% for two-hops neighborhood, and finally 366% and 463% for three-hops neighborhood, respectively.

Here, we note that mobility can also be included as an another parameter in CH calculations (in Equation 2) for MANETs. For example, highly mobile nodes (Wasp motes [17] provide three-axis accelerometer reading which would be used to measure mobility) may be elected as CHs, because they might be in contact with most of the nodes in a certain amount of time. Since WSNs are mostly stationary, we did not consider any mobility in our calculations and left this part as a future work to be considered.

Endnotes

[a]The same method would be applied in the case of multiple-hops (2,3,..., etc.) connections if needed.

[b]CHs are highlighted with yellow color and also the votes they received are noted on top of them in red color writing.

Competing interests

The authors declare that they have no competing interests.

Authors' information

Ismail Butun, ismail.butun@btu.edu.tr; In-ho Ra, ihra@kunsan.ac.kr; Ravi Sankar, sankar@usf.edu.

Acknowledgements

This research was supported by Basic Science Research Program through the National Research Foundation of Korea (NRF) funded by the Ministry of Education, Science and Technology (2013054460).

Author details

[1]Department of Mechatronics Engineering, Bursa Technical University, Gaziakdemir M. Mudanya C. No:4/10, 16190 Osmangazi, Bursa, Turkey. [2]Department of Information and Telecommunication Engineering, Kunsan National University, 573-701 Gunsan, South Korea. [3]Department of Electrical Engineering, University of South Florida, 4202 E. Fowler Avenue, Tampa, FL 33620, USA.

References

1. I Butun, Y Wang, Y Lee, R Sankar, Intrusion prevention with two-level user authentication in heterogeneous wireless sensor networks. Int. J. Secur. Netw. **7**(2), 107–121 (2012)
2. I Butun, S Morgera, R Sankar, A survey of intrusion detection systems in wireless sensor networks. Commun. Surv. Tutorials. **16**(1), 266–282 (2013)
3. WR Heinzelman, A Chandrakasan, H Balakrishnan, in *System Sciences, 2000. Proceedings of the 33rd Annual Hawaii International Conference On*. Energy-efficient communication protocol for wireless microsensor networks (IEEE Piscataway, New Jersey, 2000), p. 10
4. S Bandyopadhyay, EJ Coyle, in *INFOCOM 2003. Twenty-Second Annual Joint Conference of the IEEE Computer and Communications. IEEE Societies*. An energy efficient hierarchical clustering algorithm for wireless sensor networks, vol. 3 (IEEE Piscataways, New Jersey, 2003), pp. 1713–1723
5. J Jia, Z He, J Kuang, Y Mu, in *Wireless Communications Networking and Mobile Computing (WiCOM), 2010 6th International Conference On*. An energy consumption balanced clustering algorithm for wireless sensor network (IEEE Piscataway, New Jersey, 2010), pp. 1–4
6. M Ye, C Li, G Chen, J Wu, in *Performance, Computing, and Communications Conference, 2005. IPCCC 2005. 24th IEEE International*. EECS: an energy efficient clustering scheme in wireless sensor networks (IEEE Piscataway, New Jersey, 2005), pp. 535–540
7. WB Heinzelman, AP Chandrakasan, H Balakrishnan, An application-specific protocol architecture for wireless microsensor networks. IEEE Trans. Wireless Commun. **1**(4) (2002)
8. J-Y Chang, P-H Ju, An efficient cluster-based power saving scheme for wireless sensor networks. J. Wireless Commun. Netw. **2012**(1), 1–10 (2012)
9. O Younis, S Fahmy, HEED: a hybrid, energy-efficient, distributed clustering approach for ad hoc sensor networks. Mobile Comput. IEEE Trans. **3**(4), 366–379 (2004)
10. Q Chen, J Ma, Y Zhu, D Zhang, L Ni, An energy-efficient k-hop clustering framework for wireless sensor networks. Wireless Sens. Networks. **4373**, 17–33 (2007)
11. MR Brust, A Andronache, S Rothkugel, Z Benenson, in *Communication Systems Software and Middleware, 2007. COMSWARE 2007. 2nd International Conference On*. Topology-based clusterhead candidate selection in wireless ad-hoc and sensor networks (IEEE Piscataway, New Jersey, 2007), pp. 1–8
12. C Li, M Ye, G Chen, J Wu, in *Mobile Adhoc and Sensor Systems Conference, 2005. IEEE International Conference On*. An energy-efficient unequal clustering mechanism for wireless sensor networks (IEEE Piscataway, New Jersey, 2005)
13. S Shin, T Kwon, GY Jo, Y Park, H Rhy, An experimental study of hierarchical intrusion detection for wireless industrial sensor networks. Ind. Inform. IEEE Trans. **6**(4), 744–757 (2010)
14. O Kachirski, R Guha, in *System Sciences, 2003. Proceedings of the 36th Annual Hawaii International Conference On*. Effective intrusion detection using multiple sensors in wireless ad hoc networks (IEEE Piscataway, New Jersey, 2003), p. 8
15. Libelium, Wasp mote - wireless sensor networks 802.15.4 ZigBee Mote. (Libelium Inc.) Available At: http://www.libelium.com/products/waspmote
16. Matlab, MATLAB simulation program by Mathworks Inc. (Mathworks Inc.) Available At: http://www.mathworks.com/
17. Libelium, Libelium Wasp mote with 3D accelerometer. (Libelium Inc.) Available At: http://www.libelium.com/video-accelerometer/

Distributed back-pressure scheduling with opportunistic routing in cognitive radio networks

Yuan Le[1*], Xiuzhen Cheng[1], Dechang Chen[2], Nan Zhang[1], Taieb Znati[3], Mznah A Al-Rodhaan[4] and Abdullah Al-Dhelaan[4]

Abstract

A challenging problem in cognitive radio networks (CRNs) is to design a throughput efficient routing and scheduling algorithm for end-to-end communications between secondary users (SUs) in a distributed manner. Due to the opportunistic nature of CRN routing, we consider a framework with randomized path selection in this paper. Motivated by the back-pressure scheduling policy, we adopt the differential queue backlog as the routing metric to achieve throughput efficiency. Moreover, we demonstrate the sufficient condition for our framework to achieve throughput optimality by analyzing the Lyapunov drift. We further propose a distributed medium access control algorithm that can approximately satisfy the required condition when the transmission attempt probability is low. The performance of our proposed scheme is evaluated in a simulation platform, and the evaluation results verify the effectiveness of our distributed scheme.

Keywords: Cognitive radio networks; Opportunistic routing; Back-pressure scheduling; Throughput optimal routing

1 Introduction

Cognitive radio networking has been extensively studied in recent years [1-6]. The cognitive radio (CR) technique allows unlicensed users (secondary users (SUs)) to opportunistically access the licensed spectrum without interfering with licensed users (primary users (PUs)) to exploit the under-utilized portion of the spectrum [7-9]. The coexistence of PUs and SUs makes it infeasible to directly adopt the routing algorithms designed for general multi-hop wireless networks, as the links can be highly dynamic due to the existence of PUs. In the worst case, it may not be possible to deliver a packet along a certain path without interfering with any PU. With such a dynamic environment, designing a reliable routing protocol for a CR network (CRN) becomes a challenging problem.

There exist many routing algorithms that simultaneously consider spectrum sensing and management. For instance, a spectrum-aware routing algorithm was developed in [10] to integrate spectrum discovery with route establishment and then assign channels to links to minimize interference. But because the network topology in a CRN varies over time due to the dynamic nature of PU activities, many existing routing algorithms are rendered inapplicable as their computed paths usually do not adapt to the network dynamism. Thus whenever the network topology changes, routes may need to be recalculated, leading to a high computational cost. To address this challenge, opportunistic routing (OR) is introduced to CRN, in which SUs select only the next hop relay for their transmissions based on the routing metric computed from the current link states. For instance, a delay efficient opportunistic routing algorithm is proposed in [11].

On the other hand, the back-pressure (BP) framework proposed by Tassiulas et al. [12] introduces throughput optimal routing and scheduling to support multiple flows in wireless *ad hoc* networks. Observing that OR provides an opportunity to overcome the challenges caused by network dynamics (OR selects the next relay for routing based on the current link quality) and that BP scheduling has the potential to guarantee throughput

*Correspondence: ppzt@gwmail.gwu.edu
[1] Department of Computer Science, The George Washington University, Washington, DC, USA
Full list of author information is available at the end of the article

optimality (BP selects the flow with the highest back pressure), we decide to combine the concepts of back pressure and opportunistic routing, aiming at achieving throughput optimal routing in dynamic multi-hop cognitive radio networks. Furthermore, to avoid the centralized link controller in BP, we adopt CSMA/CA for collision avoidance among all selected flows. To the best of our knowledge, we are the first to combine BP and OR to take advantage of the benefits brought by OR and BP in a distributed fashion. Because the general back-pressure framework consists of a max-weight optimization problem, which is NP complete, we have to seek a greedy algorithm to implement our framework of distributed BP scheduling with OR routing in CRNs for a suboptimal solution.

In this paper, we adopt the queue model employed by the back-pressure algorithm in [12] to combine OR with BP and employ the differential queue backlog as our routing metric. We further analyze the Lyapunov drift of our framework and derive the sufficient condition to achieve throughput optimality. Leveraging this condition, we propose a distributed medium access control (MAC) algorithm that stabilizes the network when the traffic load is light. The core idea of our algorithm is to tune the contention window size (CW) corresponding to the maximum queue backlog to achieve the required transmission probability. We evaluate our protocol in a simulation platform, and our results indicate that the proposed protocol can achieve both throughput optimality and reliability under a light traffic load.

The rest of the paper is organized as follows. Section 2 discusses the most related work. The distributed back-pressure scheduling with opportunistic routing framework, together with its throughput optimality analysis, are presented in Section 3. We propose a distributed MAC algorithm that can stabilize the network when the transmission probability is low in Section 4. The evaluation settings and simulation results are reported in Section 5. We conclude this paper with a discussion on our future research in Section 6.

2 Related work

There exist many studies on the routing protocol design for cognitive radio networks [10,13-17]. A high-throughput spectrum-aware routing protocol was investigated in [10]. In [13], the authors developed a tree-based routing algorithm that can achieve low end-to-end delay. These protocols all use a predefined single path for all transmissions. Since the connectivity of a link may be highly dynamic in a cognitive radio network, the performance can be significantly decreased if a crucial link is frequently unavailable. To address this problem, a multi-path-based routing algorithm was proposed in [14]. To establish a more reliable end-to-end path, an opportunistic routing framework was adopted in [15]. To enhance throughput and obtain a global-path stable routing path, a network formation game was introduced to model the process of multi-hop relay selection in a cooperative CRN [16]. Mobility-assisted routing to deal with the intermittently connected characteristic of mobile *ad hoc* CRNs was investigated in [17].

Opportunistic routing has also been investigated in wireless mesh networks (WMNs) [11,18-23]. The widely studied ExOR routing protocol was proposed by Biswas et al. in [18]. ExOR selects the next hop relay node after the transmission is done. It adopts sequential acknowledgements to exchange priority information among all the receivers when a transmission occurs and selects the receiver with the highest priority as the next hop relay to forward the packet towards its destination. In contrast, Chachulski et al. developed a MAC-independent opportunistic routing scheme named MORE in [19], which combines routing with network coding to avoid node coordination. A contention-aware delay analysis of opportunistic routing was presented in [20], which provides a mathematical model about the contention to compute the expected delay. The inherent properties of information dissemination in opportunistic networks was studied in [21], and [22] proposes two algorithms for intermittently connected mobile P2P networks, which exploit the spatial locality, spatial regularity, and activity heterogeneity of human mobility to select relays.

The original back-pressure algorithm proposed by Tassiulas et al. in [12] has become an appealing research area due to its guarantee of throughput optimality. However, it involves a max-weight optimization problem, making the algorithm computationally inefficient. To address this challenge, many suboptimal solutions have been developed, which can be generally categorized into two classes. The first class [24,25] requires a centralized controller to make the scheduling decisions and focuses on developing scheduling algorithms whose stability can be guaranteed in a fraction of the capacity region. The impact of an imperfect scheduling policy on the capacity region was demonstrated by Lin et al. in [24]. Dimakis et al. [25] indicated that the full capacity region can be achieved by a greedy scheduling policy for certain networks. The second class algorithms [26-28] study distributed scheduling policies. In [26], the authors considered a datagram network and developed a distributed scheduling policy using only local information. Neely et al. [27] proposed a distributed approach using randomized transmission selection. In [28], a distributed greedy protocol was presented to achieve stability in a reduced capacity region. A learning-based distributed back-pressure scheduling algorithm requiring no information exchange was proposed in [29]; but it requires a relatively long convergence time.

In addition to finding suboptimal solutions for the original max-weight optimization problem, utility optimization based protocol design using the back-pressure framework has also being extensively studied. In [30], Georgiadis et al. claimed that the back-pressure algorithm performs within $O(1/V)$ of optimal throughput utility, with $O(V)$ average delay, where V is a control parameter that can be chosen as large as desired. This gives an explicit $[O(1/V), O(V)]$ utility-delay tradeoff. With different definitions of the utility functions, [31-33] proposed different back-pressure policies that may jointly consider power control, fairness, end-to-end delay, and flow control.

In this paper, we propose a distributed framework that combines opportunistic routing with back pressure scheduling and derive the sufficient condition for our framework to achieve throughput optimality. Based on this condition, we design a MAC protocol that stabilizes the network when the transmission probabilities are low. To our best knowledge, this paper is the first to take advantage of the benefits provided by BP and OR to tackle the challenges faced by cognitive radio networks with dynamic link statuses.

3 A framework of distributed back-pressure scheduling with opportunistic routing

3.1 Our distributed scheduling framework

In this section, we detail our framework of distributed back-pressure scheduling with opportunistic routing for a CRN with multiple simultaneous flows. Back-pressure scheduling is introduced to select the flow with maximum back pressure at each node while opportunistic routing is introduced to address the high dynamism of CRNs, such that the next hop relay for a flow can be selected opportunistically from all the feasible paths. Since the original back-pressure algorithm assumes the existence of a global controller that prevents the simultaneous activation of potentially conflicting links/flows for collision avoidance, we adopt CSMA/CA in our distributed framework for medium access control (MAC) as it is not feasible to implement a centralized link controller in a CRN. In summary, our distributed framework takes back-pressure scheduling for flow selection, opportunistic routing for path determination, and CSMA/CA for MAC regulation.

We consider a multi-hop CR network with N nodes and L links. Assume that there exist F flows in the network. For simplicity, we further assume that there are F flow queues at each node, with the cth queue serving the cth flow. The exogenous packet arrival rate at node i for an arbitrary flow c is denoted by a_i^c, and the corresponding arrival rate vector is denoted by A^c. Denote by $X_i^c(t)$ the queue length of flow c in node i. Let $s^c(l)$ and $d^c(l)$ be the sending and receiving nodes of link l for flow c, respectively. The connectivity and routing matrix $R^c = \{r_{il}^c\}$ for flow c is defined by a $N \times L$ matrix, with r_{il}^c representing the lth element in the ith row of R^c:

$$r_{il}^c = \begin{cases} 1, & \text{if } i = d^c(l), \\ -1, & \text{if } i = s^c(l), \\ 0, & \text{otherwise.} \end{cases} \tag{1}$$

Note that the link connectivity in a CR network may vary over time and thus cause the R^c matrix to change.

Let p_l^c be the probability that link l is used to transmit packets for flow c. Thus, the queue length in time slot $(t+1)$ can be represented as follows:

$$X_i^c(t+1) = X_i^c(t) + \sum_{l=1}^{L} r_{il}^c \cdot b_l \cdot p_l^c + a_i^c, \tag{2}$$

where a_i^c is the arrival rate of flow c at node i and b_l is the transmission rate of link l.

Let P_i be the probability that node i can successfully transmit a packet. By adopting opportunistic routing, a packet can be successfully transmitted if at least one of its feasible relays can receive the packet correctly. Thus, P_i can be represented as

$$P_i = \sum_{c} \sum_{l:s^c(l)=i} p_l^c. \tag{3}$$

Let B be a diagonal matrix in which the lth diagonal element is b_l. Then, the equivalent matrix expression of $X^c(t+1)$ becomes:

$$X^c(t+1) = X^c(t) + R^c B P^c + A^c. \tag{4}$$

A link backlog D_l for each link is defined as follows. The original back-pressure routing policy [12] adopts the same definition.

$$D_l^c = b_l \cdot \left(X_{s^c(l)} - X_{d^c(l)} \right), \tag{5}$$

$$D_l = \max_c D_l^c \tag{6}$$

The best candidate flow and the corresponding backlog for node i, denoted by c_i^* and D_i^*, respectively, are selected according to the following policy:

$$c_i^* = \arg_c \max_{l:s^c(l)=i} D_l^c, \tag{7}$$

$$D_i^* = \max_{l:s^c(l)=i} D_l^c, \tag{8}$$

That is, the maximum D_l is selected from all the outbound links of node i, and its corresponding flow is selected as the best candidate flow.

In the next subsection, we derive the sufficient condition for our framework to achieve throughput optimality.

3.2 Throughput optimality of the proposed framework

We first define the stability of a network as follows.

Definition A network is stable under a scheduling policy π, if the queue length $X_i^c(t)$ satisfies:

$$\limsup_{t\to\infty} \sum_{c=1}^{F}\sum_{i=1}^{N} X_i^c(t) < D, \tag{9}$$

where D is a constant that is dependent on the arrival rate vectors A^c and transmission rate b_l.

Let C_π be the set of arrival rates for which the network is stable under policy π. The stability region C of a network is defined as $C = \bigcup_\pi C_\pi$.

Definition A scheduling policy is *throughput optimal* if the network is stable for any arrival rate $a \in C$ under the policy.

In the following, we present a lemma to show that our distributed scheduling framework proposed in Section 3.1 is throughput optimal under a certain success transmission probability P_i.

Lemma 1. *In a CSMA/CA based CRN, a back-pressure scheduling with distributed routing framework is throughput optimal if the following equation holds for all the nodes:*

$$P_i = \left(1 - O\left(\frac{1}{D_i^*}\right)\right). \tag{10}$$

where $\mathbf{O}(\cdot)$ stands for the big-O notation.

The detailed proof of Lemma 1 can be found in the Appendix.

Lemma 1 provides the sufficient condition for a CSMA/CA-based scheduling framework to achieve throughput optimality. In the next section, we propose a distributed approximately throughput optimal MAC algorithm based on Equation 10.

4 A distributed approximate MAC algorithm

In this section, we first derive the contention window size that can satisfy Equation 10. Then, a detailed protocol of our scheduling framework is proposed.

4.1 Contention window size derivation

In our previous section, we show that a CSMA/CA-based CRN is stable when its successful transmission probabilities $\{P_i\}$ satisfy a certain constraint. In a network with a centralized transmission controller, it is not hard to achieve the required successful transmission probabilities. However, in a wireless network where stations compete for medium access in a distributed manner, it is difficult to guarantee the transmission probabilities. In this subsection, we propose a distributed medium access control algorithm that approximately achieves the required transmission probabilities for throughput optimality.

In a CSMA/CA network, the transmission probability P_i of station i is determined by the attempt probabilities $\{P_a^i, i = 1 \dots N\}$. We adopt the model in [34], in which

$$P_i = P_a^i * \prod_{j\neq i}\left(1 - P_a^j\right), \tag{11}$$

and the attempt probability can be expressed as the following function of the contention window size CW_i

$$P_a^i = \frac{2}{CW_i + 1}. \tag{12}$$

Notice that when the attempt probabilities approach to 0, the product $\prod_{j\neq i}\left(1 - P_a^j\right)$ converges to 1, leading to $P_i = P_a^i$. In a typical CRN, transmissions occur sparsely and opportunistically due to the existence of PUs. Therefore, we can approximately represent P_i by P_a^i. On the other hand, according to Lemma 1, the following equation should hold for every P_a^i and D_i^*:

$$P_a^i = 1 - \mathbf{O}\left(\frac{1}{D_i^*}\right). \tag{13}$$

This implies that when the maximum backlog D_i^* becomes larger, the attempt probability P_a^i is enlarged. Recall that P_i can be represented by P_a^i only when the attempt probabilities approach to 0. Hence, the maximum attempt probability should be bounded when the maximum backlog is increasing, so that the sparse transmission constraint can be guaranteed. Otherwise, high attempt probabilities could decrease the successful transmission probabilities and consequently cause congestions. And the congestions may significantly increase the queue, causing even higher attempt probabilities due to the retransmissions.

To address this problem, we limit the queue length to a finite number. When the queue length reaches its upper bound, all newly arriving packets are dropped. Let $Q_{\max}$ be the maximum queue length. Then, the maximum backlog $D_{\max}$ is equal to $Q_{\max}$. The maximum attempt probability is thus bounded by

$$P_a^{\max} = 1 - \mathbf{O}\left(\frac{1}{Q_{\max}}\right). \tag{14}$$

Given a $Q_{\max}$, we can have different values for $P_a^{\max}$ depending on $\mathbf{O}\left(\frac{1}{Q_{\max}}\right)$. To find a better $P_a^{\max}$, we consider the scenario where all queues in all stations reach $Q_{\max}$. In such a situation, an optimal $P_a^{\max}$ should maximize the total throughput among the network. In [34,35], it has been shown that in a saturated network, the aggregated throughput can be maximized by finding the optimal attempt probabilities. By adopting the analysis in [35], we can calculate the optimal attempt probability P_a^{opt} given the maximum collision duration T_c, the slot duration T_{slot}, and the number of stations N. The relationship between N and P_a^{opt} can be expressed as

$$P_a^{\text{opt}} = \frac{\zeta_0}{N}, \tag{15}$$

where ζ_0 is the root of the equation

$$1 - \zeta = \left(1 - \frac{T_{\text{slot}}}{T_c}\right) e^{-\zeta}.$$

For a given CRN, T_c and T_{slot} are predefined constants; thus, P_a^{opt} can be calculated by each station. By setting $P_a^{\max} = P_a^{\text{opt}}$, we have

$$P_a^{\max} = 1 - \mathbf{O}\left(\frac{1}{D_{\max}}\right) = \frac{\zeta_0}{N}. \tag{16}$$

Let $\mathbf{O}\left(\frac{1}{D_i^*}\right)$ be a function of the form $\frac{c_1}{D_i^*+c_2}$. Notice that when the maximum backlog D_i^* is equal to 0, the station should not attempt to transmit. This can be expressed as:

$$\begin{cases} 1 - \dfrac{c_1}{0 + c_2} = 0 & (17) \\ 1 - \dfrac{c_1}{D_{\max} + c_2} = \dfrac{\zeta_0}{N}. & (18) \end{cases}$$

By solving the above equations, we have

$$c_1 = c_2 = \frac{D_{\max}\left(N - \zeta_0\right)}{\zeta_0}, \tag{19}$$

and thus the corresponding contention window size $CW\left(D_i^*\right)$ can be expressed as

$$CW\left(D_i^*\right) = 1 + 2 \cdot \frac{D_{max}\left(N - \zeta_0\right)}{\zeta_0 \cdot D_i^*}. \tag{20}$$

4.2 Approximation analysis

In this subsection, we show the approximation analysis of our algorithm, comparing to the optimal solution. As shown in Equation 11, the difference between the attempt probability P_a^i and the transmission probability P_i, denoted by the collision probability P_f^i can be expressed as:

$$P_f^i = P_a^i - P_i = P_a^i \cdot \left(1 - \prod_{j \neq i}\left(1 - P_a^j\right)\right). \tag{21}$$

Recall that the attempt probability is upper-bounded by the maximum attempt probability $P_a^{\max}$, thus we have

$$\begin{aligned} P_f^i &= P_a^i \cdot \left(1 - \prod_{j \neq i}\left(1 - P_a^j\right)\right) \\ &\leq P_a^{\max} \cdot \left(1 - \prod_{j \neq i}\left(1 - P_a^{\max}\right)\right) \\ &= P_a^{\max} \cdot \left(1 - \left(1 - P_a^{\max}\right)^{N-1}\right) \\ &= \frac{\zeta_0}{N}\left(1 - \left(1 - \frac{\zeta_0}{N}\right)^{N-1}\right). \end{aligned} \tag{22}$$

Considering the impact of a collision, it is trivial that a collision can lead to a retransmission. If we consider such additional transmissions as newly arrived packets, the equivalent arrival rate for these retransmissions can be expressed as $\sum_{i=1}^N P_f^i$. Thus, comparing to the ideal solution with zero collision probability, our proposed algorithm decreases the feasible arrival rate for an amount of $\sum_{i=1}^N P_f^i$. Hence, the difference between the maximum capacity region and the feasible capacity region of our proposed algorithm is no larger than

$$\sum_{i=1}^{N} P_f^i \leq N \cdot \frac{\zeta_0}{N}\left(1 - \left(1 - \frac{\zeta_0}{N}\right)^{N-1}\right). \tag{23}$$

4.3 Protocol description

Our protocol consists of four major steps: i) flow selection, ii) random backoff period, iii) transmission and acknowledgement exchange, and iv) forwarding decision. Similar to the original opportunistic routing protocol [36], we assume that a distance matrix is maintained by each station, which is used to calculate the distance to other stations.

When a station is ready to transmit, it first selects the flow with the maximum backlog. After the flow is selected, the CW value is calculated based on the maximum backlog, as described in Section 4.1, and a random backoff period is generated. The packet is then broadcast when the backoff period is passed. The distance between the station and the destination of the packet, as well as the current queue length of the flow, is appended to the packet header. Each station that receives the packet should compare the distance and queue length from the header with its own values. If the receiver is closer to the destination and the queue length of the corresponding flow is smaller than that of the sender, the receiver sends an acknowledgement back and considers itself as a candidate relay. Candidate relays decide whether or not to forward the received packet after all the acknowledgements are received. This procedure is repeated until the packet reaches the destination station.

4.3.1 Flow selection

In our protocol, each station selects the flow with the maximum backlog to serve. To calculate the backlog, each station maintains a matrix that includes the approximate queue length of its neighbors. When the station receives a packet or an acknowledgement message, the corresponding queue length value in the matrix is synchronized to the value included in the header.

4.3.2 Random backoff period

After a flow is selected, the maximum backlog D_i^* is used to calculate the CW size as defined in Equation 20. During the backoff period, the station continues monitoring the spectrum. If a packet transmission is sensed or a PU occupies the spectrum, the station should pause the backoff timer and wait until the channel becomes available again.

4.3.3 Transmission and acknowledgement

When the backoff period is passed, the station pops a packet from the flow queue and immediately starts transmitting. The distance between the current station and the destination station and the current queue length of the flow are both stored in the header. Then, the station waits for acknowledgement messages from its neighbors. Different from the default 802.11 MAC algorithm, we reserve multiple time slots for ACK exchanges in our protocol. In the *i*th slot, the recipient with the *i*th smallest queue length sends the ACK. This order can be calculated from the queue length matrix stored at each station. In the case where multiple stations have the same queue length, the stations send their ACKs in the order of their station IDs.

To indicate that a packet has been received successfully, the ACK also includes the station ID and the queue length of the best candidate relay known to the ACK sender. For instance, assume the queue length values of the flow *c* in stations *A*, *B*, and *C* are 2, 4, and 5, respectively. In this case, station *A* sends the ACK message first with its own ID and queue length value. In the second time slot, station *B* sends its ACK, reporting that station *A*, with queue length value 2, is the best candidate relay. In the last time slot, station *C* sends its ACK carrying the same best candidate relay information. This acknowledgement mechanism can prevent duplicate forwarding when an ACK message from a higher priority recipient is not received by a lower priority one. As shown in Figure 1, assume that the ACK message from station *A* is not received by station *B* due to the hidden terminal problem. At the second time slot, station *B* considers itself as the best candidate and sends an ACK message reporting *B* as the best candidate with queue length 4 as the minimum queue length.

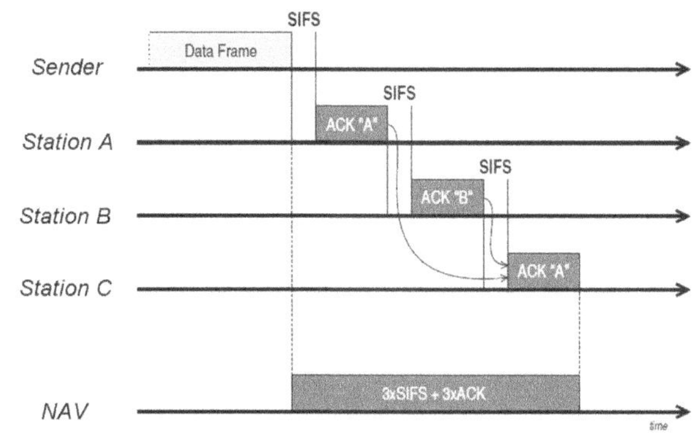

Figure 1 Illustration of sequential ACKs.

Station *C* receives the two ACKs and compares the two queue lengths. The smaller one, which is from station *A* in this example, is chosen as the best candidate and is broadcasted with the ACK message. Finally, station *B* receives the ACK from station *C* and updates its best candidate information.

4.3.4 Forwarding decision

After exchanging the ACK messages, the stations make the decision regarding whether or not to forward the received packet. Specifically, after all ACK messages are received, if a station does not receive any ACK message with a smaller queue length value than itself, it pushes the received packet into the corresponding queue and later forwards it toward its destination. Otherwise, the station discards the received packet.

5 Performance evaluation

In this section, we evaluate the performance of our proposed transmission protocol using the *OMNET++* platform [37]. The average queue length, packet loss rate, throughput, and end-to-end delay are all evaluated in our simulation tests.

5.1 Simulation settings

We adopt the *INETMANET* framework in OMNET++ [37] as our cornerstone. The MAC and IP modules are modified to implement our transmission protocol. In our simulation, the parameters of the MAC and PHY layer adopt the default values same as those in the 802.11b standard. Additionally, the transmit power and signal-to-interference-plus-noise ratio (SINR) thresholds are defined so that the transmission range is equal to 250 m. Table 1 lists the basic parameters employed in our simulation.

In our network, multiple stations are placed in the simulation area and set to be static. We assume the stations always transmit packets using the maximum transmission rate, which implies that rate adaptation is disabled.

Table 1 The 802.11*b* parameters in Omnet++

Slot time (μs)	SIFS (μs)	DIFS (μs)
20	10	50
ζ_0	MTU (bytes)	ACK length (bits)
0.1622	1,500	112
IP header (bits)	MAC header (bits)	PHY header (bits)
160	272	192
Bitrate	Bitrate of ACK	Bitrate of PHY
11 Mbps	2 Mbps	1 Mbps
Transmit power (mW)	SINR threshold (dB)	Sensitivity (dBm)
2	4	−85

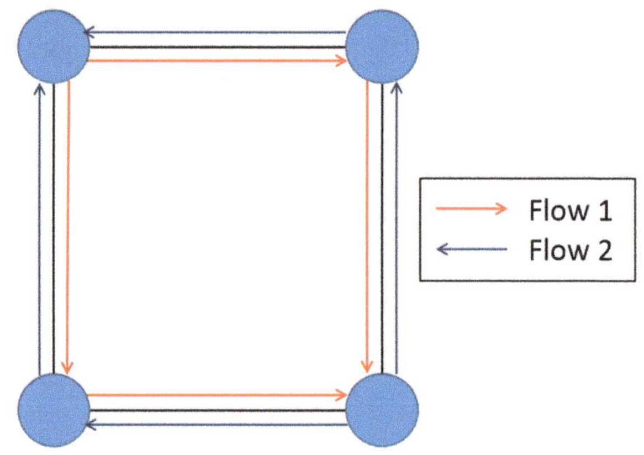

Figure 2 Network topology and the two flows for scenario I.

The flows are defined at the beginning of the simulation, and their arrival rates remain the same throughout the entire simulation period. Additionally, we disable RTS/CTS exchanges. The maximum queue length of each queue is set to 15 in our simulation.

5.2 Methodology

Our evaluation is conducted under two different network scenarios. We first consider a simple multi-hop network with four stations located at the four vertices of a square. The edge length of the square is set to 200 m; thus, the two diagonal vertices cannot communicate directly. We define two flows in this network: the first flow transmits packets from the upper-left station to the lower-right station; and the other flow goes from the lower-right station to the upper-left one. Each flow can have two paths from their source nodes to their destinations. To evaluate the impact of link availability on network performance, we consider a simple PU deploy model, in which each station can be interrupted by one and only one PU. In addition, each PU individually occupies the channel with a certain occupancy rate. Please notice that the random PU deploy model is not adopted in this scenario because all the four stations are close to each other, and thus a randomly deployed PU could simultaneously interfere with all the stations at a high probability. The impact of the packet arrival rate and that of the primary user occupancy rate[a] on the network performance are evaluated for this scenario. The network topology and the flows are illustrated in Figure 2.

In our second scenario, we randomly deploy N stations and five primary users in a 1,500 by 1,500 m simulation area. The interference range of the primary users is set to 300 m; thus when a primary user occupies the channel, the stations within its interference range cannot transmit or receive any packets. In addition, L flows are defined in the network by randomly selecting L pairs of sources and destinations from the N stations. In this scenario, we study the impact of the packet arrival rate A, the total number of stations N, and the total number of flows L, on the performance of our protocol.

5.3 Evaluation results

The evaluation results are detailed in terms of the average queue length, the average end-to-end delay, the average packet loss rate, and the average throughput. All of our simulation results are the averaged value of ten runs.

5.3.1 *Average queue length*

Figure 3 shows the impact of the packet arrival rate on the average queue length in the simple network. The queue length calculated from Equations 10 and 19 is shown in the figure as well (labeled by *ideal queue length*). From this figure, we observe that the average queue length curve is comprised of three distinct phases. In the first phase, the time gap between the arrivals of two packets is longer than the length of the maximum backoff period; and thus,

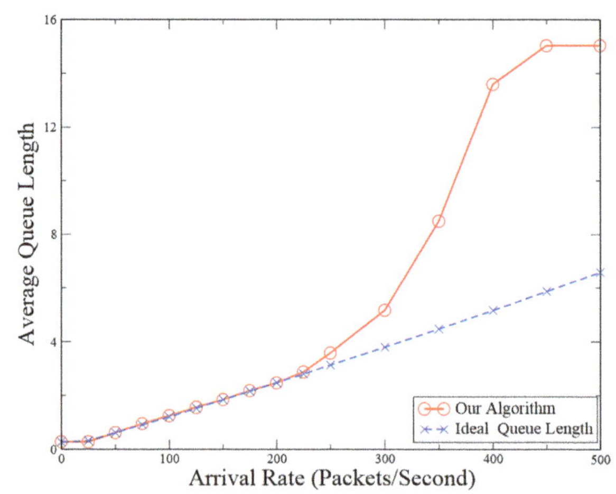

Figure 3 Average queue length vs. arrival rate.

the first packet is transmitted before the second packet arrives, although the backoff period of each packet is long. As a result, the average queue length remains 0 or 1 for the majority of the time. In the second phase, the average queue length grows linearly while the arrival rate increases. In this phase, packets reach their destinations at approximately the same rate as the arrival rate; thus, the transmission probability is proportional to the arrival rate. In the last phase, the low transmission attempt probability condition does not hold after the arrival rate exceeds some threshold. Accordingly, packets collide with others frequently, and the average queue length grows quickly to the upper bound. We also evaluate the variation of the average queue length for the second phase in Figure 4. From the figure, we observe that at the beginning, the queue length of the source station rapidly increases. The increasing queue length then reduces the CW size and thus increases the transmission probability. As a result, the queue length becomes stable after the transmission probability reaches the arrival rate.

In Figure 5, we evaluate the impact of the PU occupancy rate on the average queue length. In this evaluation, the arrival rate is set to 150 packets per second. We observe that the curve is similar to that of Figure 3. We can infer this result by considering the impact of PU occupancy as a reduction of the arrival interval. In other words, increasing the PU occupancy rate is approximately equivalent to increasing the packet arrival rate, although the increasing ratio may not necessarily be equal. Similar results are observed in the simulation of the other three metrics as well; thus, the impact of the PU occupancy rate will not be reported in the following subsections. Furthermore, the PU occupancy rate is set to 0.4 without further clarification.

The performance of our protocol in the large network is shown in Figures 6 and 7. From the figures, we can observe that the three phases exist in this scenario as well. In addition, when the arrival rate is low, the increased number of stations does not obviously impact on the performance, as the attempt probabilities of all the stations are very low, and thus the increasing number of stations does not obviously enlarge the collision probability. When the arrival rate grows, the impact of the amount of stations on the average queue length becomes more apparently. This is because the increasing number of stations lead to a higher collision probability and thus increase the queueing time for each packet. On the other hand, by increasing the number of flows, the threshold to transfer to phase three can be reached more quickly, and the average queue length grows faster. This result can be easily inferred as the aggregated arrival rate is proportional to the number of flows.

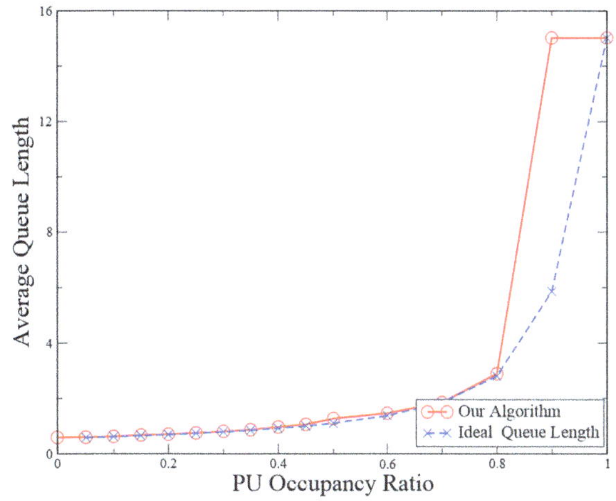

Figure 5 Average queue length vs. PU occupancy rate.

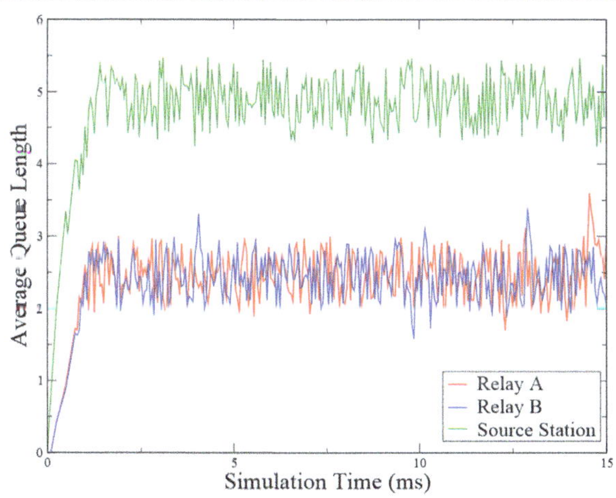

Figure 4 Average queue length vs. time.

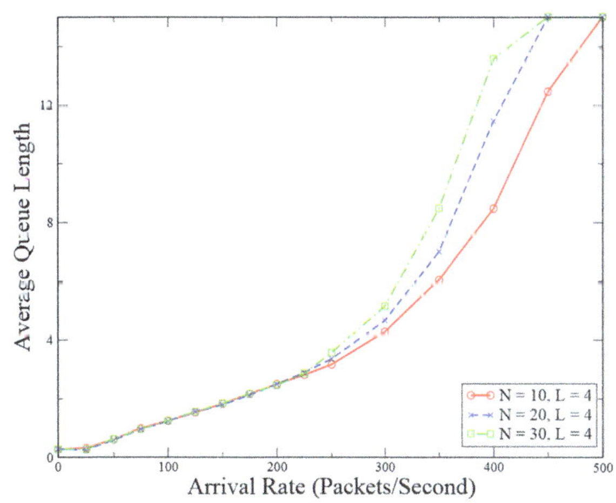

Figure 6 Average queue length vs. arrival rate with different number of stations.

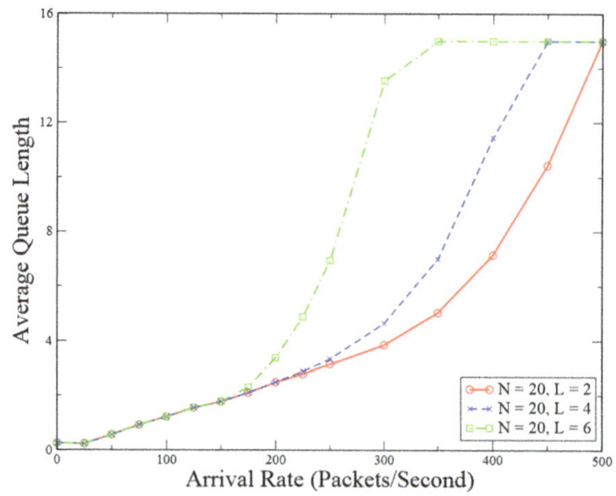

Figure 7 Average queue length vs. arrival rate with different number of flows.

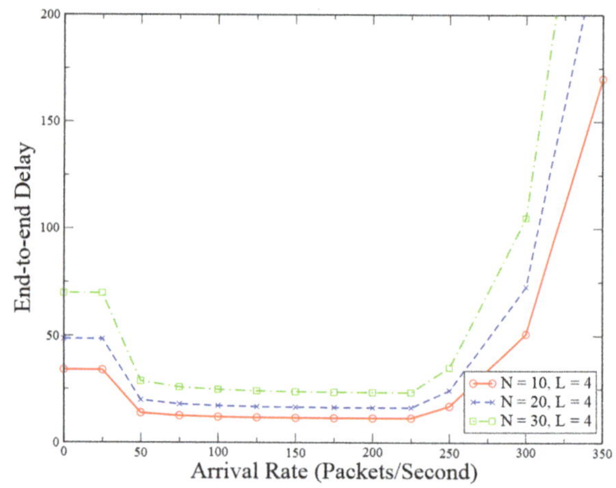

Figure 9 End-to-end delay vs. arrival rate with different number of stations.

5.3.2 End-to-end delay

We first illustrate the end-to-end delay in the simple scenario in Figure 8. From the figure, one can see that when the packet arrival rate is lower than 50 packets per second, the end-to-end delay is approximately equal to 20 ms. The major time cost in this case comes from the maximal backoff period. As the arrival rate increases, the end-to-end delay decreases rapidly, which is caused by the shortened backoff period corresponding to the increased queue backlog. Following this, the delay remains unchanged until the arrival rate exceeds the threshold for phase three. Once this threshold is reached, the end-to-end delay then converges to infinity. As the queue length exceeds the upper bound, packets start being dropped.

Figures 9 and 10 illustrate the end-to-end delay of our protocol in the second scenario. We notice that the end-to-end delay increases when N grows since the average number of hops for each flow is related to the network size, and the backoff period for each transmission is related to the number of stations within its transmission range. Also, the end-to-end delay increases when L becomes larger, as the queueing time in each hop is increased. Similar to the average queue length, the threshold to reach phase three becomes smaller while L grows.

5.3.3 Packet loss ratio

The packet loss ratio is measured by the ratio of the amount of lost packets to the total number of transmitted packets. Similar to the average queue length, we can observe the existence of a threshold and divide the curve into three phases. As shown in Figure 11, the packet loss ratio remains low until the threshold is reached. Then, the ratio increases rapidly to 1 as the entire network becomes congested; and consequently, most packets

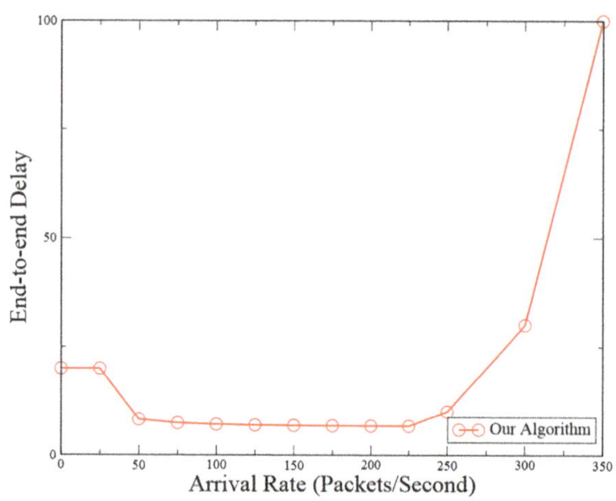

Figure 8 End-to-end delay vs. arrival rate.

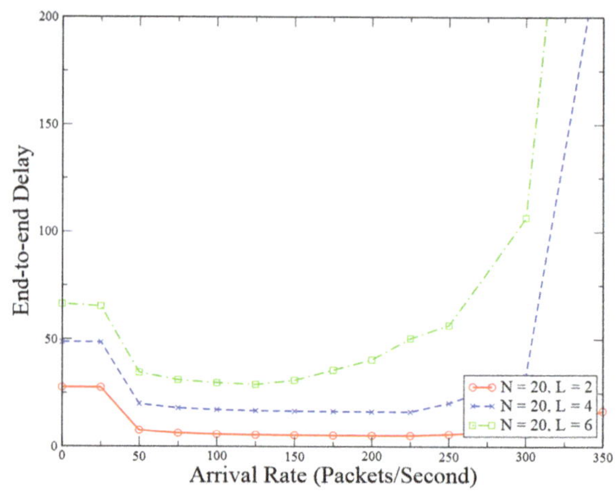

Figure 10 End-to-end delay vs. arrival rate with different number of flows.

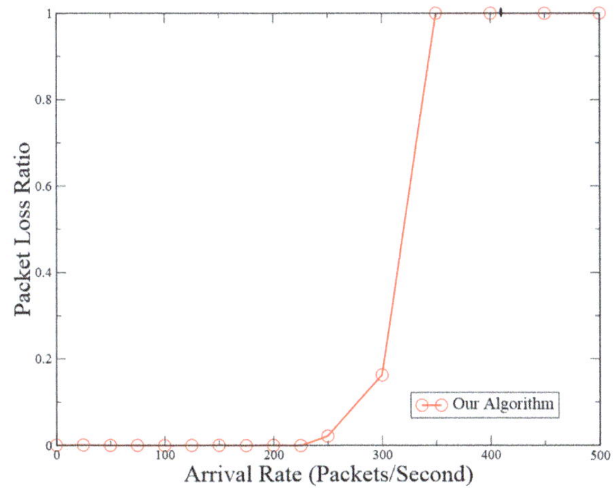

Figure 11 Packet loss ratio vs. arrival rate.

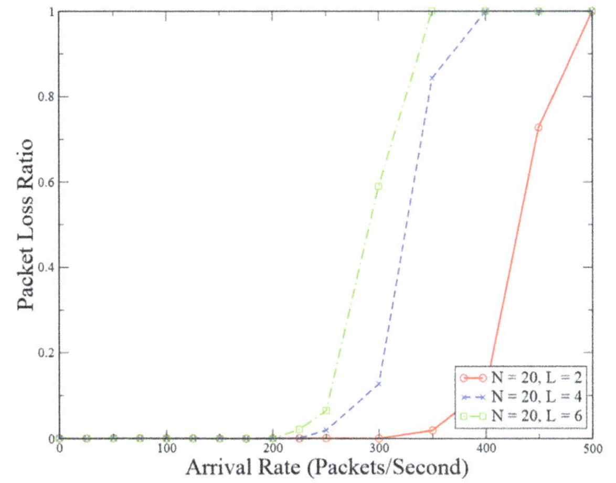

Figure 13 Packet loss ratio vs. arrival rate with different number of flows.

either are dropped by the stations or have collided with other packets.

The impacts of N and L on the packet loss ratio are reported in Figures 12 and 13, respectively. It can be observed that the impact of N and L on the packet loss ratio is in accordance to that on the average queue length.

5.3.4 Throughput

The throughput of the network can be measured as the total number of packets arrived at their destinations within a unit of time. The results are shown in Figures 14, 15, and 16. For the simple scenario, we notice that the throughput is equal to the aggregated arrival rate before the threshold is exceeded. After that, the throughput drops quickly to 0, as no packet can be delivered to the destination.

For the second scenario, from Figure 15, we can see that the thresholds are not affected by the total number of stations. When the arrival rate approaches to the threshold, the collision probability becomes higher, so some of the packets are dropped after multiple retransmission failure. A higher N can lead to a higher collision probability and thus can cause more packets being dropped. On the other hand, the increasing stations provide more queueing space for storing packets when the network is congested, as well as more feasible paths to the destination. So when the network is not fully congested, the performance of our algorithm does not have much difference when N increases. However, when the arrival rate becomes large enough, a network with a greater N is easier to be fully congested, as the collision probability grows faster than a network with a smaller N. In contrast, Figure 16 shows the

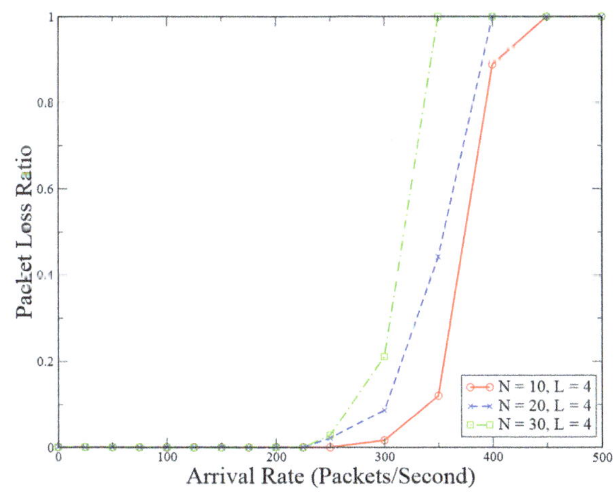

Figure 12 Packet loss ratio vs. arrival rate with different number of stations.

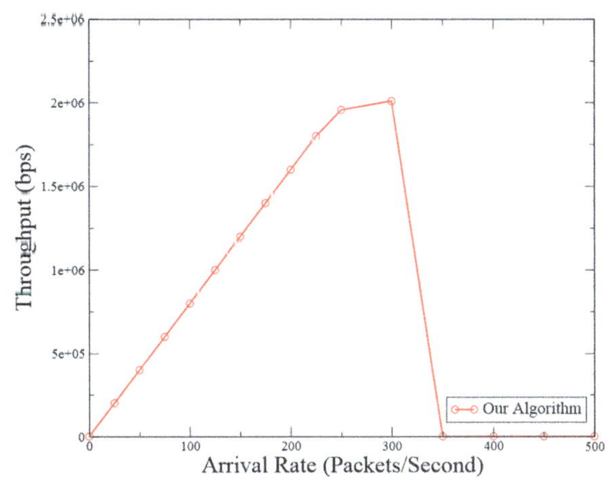

Figure 14 Throughput vs. arrival rate.

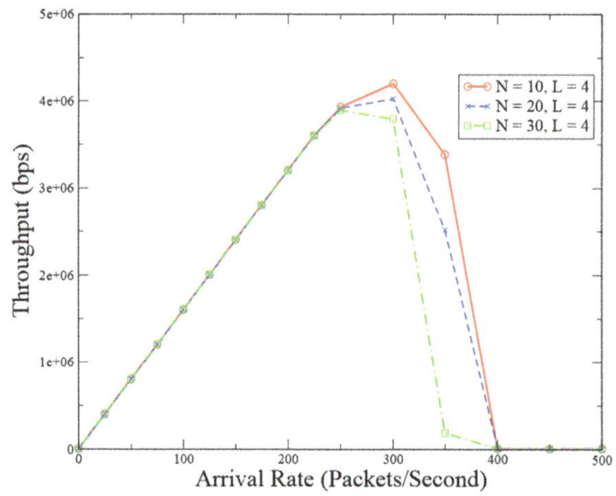

Figure 15 Throughput vs. arrival rate with different number of stations.

impact of L on the performance of our algorithm. It is clear that by increasing the number of flows, a higher aggregated throughput can be achieved when the arrival rate is low, because the aggregated arrival rate is proportional to the number of flows. At the same time, the network becomes much more congested, and thus the threshold is reached earlier.

5.3.5 *Summary*

The above evaluation results demonstrate that our proposed algorithm performs well when the arrival rate is smaller than a certain threshold. This threshold depends on the number of flows in the network but is insensitive to the network size.

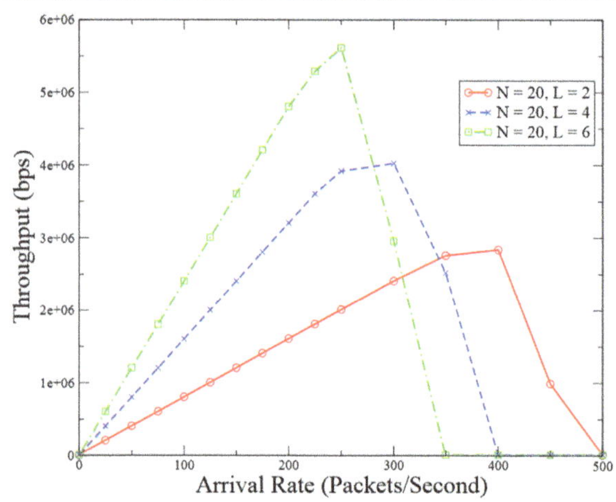

Figure 16 Throughput vs. arrival rate with different number of flows.

6 Conclusions

In this paper, we have proposed a framework of distributed back-pressure scheduling with opportunistic routing for CRNs and studied the sufficient condition for the proposed framework to achieve throughput optimality. To the best of our knowledge, we are the first to combine opportunistic routing with back-pressure scheduling in a distributed fashion. We also have proposed a novel distributed routing and medium access control protocol for end-to-end communications in a CRN. We have evaluated the performance of our protocol in a simulation-based study, and our evaluation results demonstrate that the proposed protocol is able to achieve both throughput optimality and reliability with the existence of PUs. As part of our future work, we plan to improve the end-to-end delay of our proposed protocol.

Endnote

[a]The PU occupancy rate refers to the probability of a PU occupying its channel at each time slot.

Appendix

[Proof of Lemma 1]

Proof Similar to the proof in [12], we consider a Lyapunov function

$$V(X(t)) = \sum_{c=1}^{F}\sum_{i=1}^{N}(X_i^c(t))^2.$$

To prove the throughput optimality, we need to show that the following relationships hold:

$$E\left[V(X(t+1)) - V(X(t))|X(t)\right] < \infty, \forall X(t), \quad (24)$$

and for any given $\epsilon > 0$ and arrival rate $a \in C$, there exists a positive number b, such that

$$E\left[V(X(t+1)) - V(X(t))|X(t)\right] < -\epsilon, \text{for } \forall V(X(t)) > b \quad (25)$$

In the following, we first prove that the inequality (Equation 24) holds.

$$\begin{aligned}
&E\left[V(X(t+1)) - V(X(t))|X(t)\right] \\
&= E\left[\sum_{c=1}^{F}\sum_{i=1}^{N}\left(\left(X_i^c(t+1)\right)^2 - \left(X_i^c(t)\right)^2\right)|X(t)\right] \\
&= E\left[\sum_{c=1}^{F}\sum_{i=1}^{N}\left(X_i^c(t+1) - X_i^c(t)\right)^2 |X(t)\right] \quad (*) \\
&+ 2\cdot E\left[\sum_{c=1}^{F}\sum_{i=1}^{N}\left(X_i^c(t+1) - X_i^c(t)\right)\cdot X_i^c(t)|X(t)\right] \quad (**)
\end{aligned}$$

For part $(**)$, it is trivial to show that the amount of data directed in or out of a queue is bounded by the total capacity of the links connected to the queue. Thus, part $**$ is bounded by a constant, say, b_1.

Considering part $(*)$, we have

$$\begin{aligned}
&E\left[\sum_{c=1}^{F}\sum_{i=1}^{N}\left(X_i^c(t+1)-X_i^c(t)\right)\cdot X_i^c(t)|X(t)\right]\\
&=E\left[\sum_{c=1}^{F}\sum_{i=1}^{N}\left(\sum_{l=1}^{L}r_{il}^c\cdot b_l\cdot p_l^c+a_i^c\right)\cdot X_i^c(t)|X(t)\right]\\
&=E\left[\sum_{c=1}^{F}\sum_{i=1}^{N}\sum_{l=1}^{L}X_i^c(t)\cdot r_{il}^c\cdot b_l\cdot p_l^c|X(t)\right]\\
&\quad+E\left[\sum_{c=1}^{F}\sum_{i=1}^{N}X_i^c(t)\cdot a_i^c|X(t)\right]\\
&=-\sum_{c=1}^{F}\sum_{l=1}^{L}D_l^c(t+1)p_l^c+\sum_{c=1}^{F}(X^c)^TA^c
\end{aligned}\tag{26}$$

Clearly, the term $-\sum_{c=1}^{F}\sum_{l=1}^{L}D_l^c(t+1)P_l^c$ is bounded by 0. The proof in [12] indicates that $\sum_{c=1}^{F}(X^c)^TA^c$ is bounded by $D(t+1)^T\sum_{i=1}^{|S|}\lambda_i s_i$. Here, $S=\{s_i, i=1,2,\ldots,|S|\}$ is the set of all feasible activation vectors, and $\{\lambda_i\}$ is a set of positive factors that satisfy $\sum_{i=1}^{|S|}\lambda_i<1$. Thus, we conclude that inequality (24) holds for every $X(t)$.

In the following, we show that there exists a positive value b such that the inequality (Equation 25) holds when $V(X(t))>b$. In the previous proof, we have shown that the following inequalities hold:

$$\sum_{c=1}^{F}(X^c)^TA^c\le D(t+1)^T\sum_{i=1}^{|S|}\lambda_i s_i,\tag{27}$$

$$E\left[\sum_{c=1}^{F}\sum_{i=1}^{N}\left((X_i^c(t+1))^2-(X_i^c(t))^2\right)|X(t)\right]<b_1\tag{28}$$

Assume that there exists an activation vector $\hat{s}$ that can maximize $D(t+1)^Ts$. Let $\hat{s_l^c}$ be the indicator for flow c on link l. Then, equation $\hat{s_l^c}=1$ holds if c is the corresponding flow to D_l and $\hat{s}_l$ is equal to 1. Considering the difference between $\sum_{c=1}^{F}\sum_{l=1}^{L}D_l^c(t+1)p_l^c$ and $D(t+1)^T\hat{s}$, we have:

$$\begin{aligned}
&D(t+1)^T\hat{s}-\sum_{c=1}^{F}\sum_{l=1}^{L}D_l^cp_l^c\\
&=\sum_c\sum_l\hat{s_l^c}D_l^c-\sum_l\sum_cD_l^cp_l^c\\
&\le\sum_c\sum_lD_l\left(\hat{s_l^c}-p_l^c\right)\\
&=\sum_lD_l\sum_c\left(\hat{s_l^c}-p_l^c\right)
\end{aligned}\tag{29}$$

If we represent a link l by its source and destination node pair, we can rewrite $\sum_lD_l\sum_c\left(\hat{s_l^c}-p_l^c\right)$ as:

$$\sum_lD_l\sum_c\left(\hat{s_l^c}-p_l^c\right)=\sum_{i=1}^{N}\sum_{j\in N(i)}D_{ij}\sum_c\left(\hat{s_{ij}^c}-p_{ij}^c\right),\tag{30}$$

where $N(i)$ is the set of neighboring nodes of i. Then, we have:

$$\begin{aligned}
&D(t+1)^T\hat{s}-\sum_{c=1}^{F}\sum_{l=1}^{L}D_l^c(t+1)p_l^c\\
&\le\sum_{i=1}^{N}\sum_{j\in N(i)}D_{ij}\sum_c\left(\hat{s_{ij}^c}-p_{ij}^c\right)\\
&\le\sum_i\sum_jD_i^*\sum_c\left(\hat{s_{ij}^c}-p_{ij}^c\right)\\
&=\sum_iD_i^*\sum_j\sum_c\left(\hat{s_{ij}^c}-p_{ij}^c\right).
\end{aligned}\tag{31}$$

Notice that for any node i, the total number of links that can transmit simultaneously is no more than 1. Thus, we have

$$\sum_j\sum_c\hat{s_{ij}^c}\le1.\tag{32}$$

From Equations 31 and 32, we obtain:

$$\begin{aligned}
&D(t+1)^T\hat{s}-\sum_{c=1}^{F}\sum_{l=1}^{L}D_l^c(t+1)p_l^c\\
&\le\sum_iD_i^*\left(1-\sum_j\sum_cp_{ij}^c\right)\\
&=\sum_iD_i^*\left(1-P_i\right).
\end{aligned}$$

By setting P_i to $\left(1-\mathrm{O}\left(\frac{1}{D_i^*}\right)\right)$, we have

$$-\sum_{c=1}^{F}\sum_{l=1}^{L}D_l^c(t+1)p_l^c\le-D(t+1)^T\hat{s}+C_0\cdot N,\tag{33}$$

where C_0 is a constant that is related to the function $\mathbf{O}\left(\frac{1}{D_i^*}\right)$. Then, with Equations 26, 27, 28, and 33, we have

$$\begin{aligned}
&E\left[V(X(t+1)) - V(X(t))|X(t)\right] \\
&\le b_1 + 2\left(-\sum_{c=1}^{F}\sum_{l=1}^{L} D_l^c(t+1)p_l^c + D(t+1)^T\sum_{i=1}^{|S|}\lambda_i s_i\right) \\
&\le b_1 + 2\left(-D(t+1)^T\hat{s} + C_0\cdot N\right) + 2D(t+1)^T\sum_{i=1}^{|S|}\lambda_i s_i \\
&\le b_1 + 2C_0\cdot N - 2D(t+1)^T\hat{s} + 2D(t+1)^T\hat{s}\sum_{i=1}^{|S|}\lambda_i \\
&\le b_1 + 2C_0\cdot N - 2D(t+1)^T\hat{s}\left(1-\sum_{i=1}^{|S|}\lambda_i\right).
\end{aligned}$$

As shown in [12], we have:

$$(D(t+1)^T\hat{s}) \ge \frac{1}{N}\sqrt{\frac{b}{FN}}\cdot\min_l\{b_l\}, \quad \text{when } V(X(t)) > b \tag{34}$$

Thus, when $V(X(t)) > b$, we have

$$\begin{aligned}
&E\left[V(X(t+1)) - V(X(t))|X(t)\right] \\
&\le b_1 + 2C_0\cdot N - 2\left(\left(1-\sum_{i=1}^{|S|}\lambda_i\right)\frac{1}{N}\sqrt{\frac{b}{FN}}\cdot\min_l\{b_l\}\right).
\end{aligned} \tag{35}$$

By setting $b = NF\cdot\left[\frac{N(\epsilon+b_1+2C_0\cdot N)}{2\left(1-\sum_{i=1}^{|S|}\lambda_i\right)\min_l\{b_l\}}\right]^2$, we prove that the inequality (Equation 25) holds. Thus, the throughput optimal is achieved when Equation 10 holds for all the nodes.

Competing interests
The authors declare that they have no competing interests.

Acknowledgements
This work is supported in part by the National Science Foundation of the US under grants CNS-0831939, CNS-1162057, CNS-1162159, and CNS-1265311. The authors would also extend their appreciation to the Deanship of Scientific Research at King Saud University for funding this work through research group no. RGP-VPP-264.

Author details
[1]Department of Computer Science, The George Washington University, Washington, DC, USA. [2]Uniformed Services University of the Health Sciences, Bethesda, MD, USA. [3]Department of Computer Science, University of Pittsburg, Pittsburgh, PA, USA. [4]College of Computer and Information Sciences, King Saud University, Riyadh, Saudi Arabia.

References
1. M Song, C Xin, Y Zhao, X Cheng, Dynamic spectrum access: from cognitive radio to network radio. IEEE Wireless Commun. **19**(1), 23–29 (2012)
2. T Jing, X Chen, Y Huo, X Cheng, in *IEEE INFOCOM*. Achievable transmission capacity of cognitive mesh networks with different media access control (IEEE, New York, 2012), pp. 1764–1772
3. X Xing, T Jing, Y Huo, H Li, X Cheng, in *IEEE INFOCOM*. Channel quality prediction based on Bayesian inference in cognitive radio networks (IEEE, New York, 2013), pp. 1465–1473
4. X Xing, T Jing, W Cheng, Y Huo, X Cheng, Spectrum prediction in cognitive radio networks. IEEE Wireless Commun. **20**(2), 90–96 (2013)
5. Z Cai, S Ji, J He, AG Bourgeois, in *Distributed Computing Systems (ICDCS), 2012 IEEE 32nd International Conference On*. Optimal distributed data collection for asynchronous cognitive radio networks (IEEE, New York, 2012), pp. 245–254
6. Z Cai, S Ji, J He, L Wei, AG Bourgeois, Distributed and asynchronous data collection in cognitive radio networks with fairness consideration. Parallel Distributed Syst. IEEE Trans. **25**(8), 2020–2029 (2014)
7. W Li, X Cheng, T Jing, Y Cui, K Xing, W Wang, Spectrum assignment and sharing for delay minimization in multi-hop multi-flow crns. IEEE J. Selected Areas Commun. (JSAC). **31**(11), 2483–2493 (2013)
8. W Li, X Cheng, RZhao Bie F, *An extensible and flexible truthful auction framework for heterogeneous spectrum markets* (ACM, New York, 2014), pp. 175–184
9. W Li, S Wang, X Cheng, R Bie, Truthful multi-attribute auction with discriminatory pricing in cognitive radio networks. ACM SIGMOBILE Mobile Comput. Commun. Rev. **18**(1), 3–13 (2014)
10. A Sampath, L Yang, L Cao, H Zheng, BY Zhao, in *Proc. of international conference on cognitive radio oriented wireless networks and communications (CROWNCOM)*. High throughput spectrum-aware routing for cognitive radio networks (IEEE, New York, 2007)
11. Z Cai, Y Duan, A Bourgeois, Delay efficient opportunistic routing in asynchronous multi-channel cognitive radio networks. J. Comb. Optimization, 1–21 (2013). doi:10.1007/s10878-013-9623-y
12. L Tassiulas, A Ephremides, Stability properties of constrained queueing systems and scheduling policies for maximum throughput in multihop radio networks. Automatic Control IEEE Trans. **37**(12), 1936–1948 (1992). doi:10.1109/9.182479
13. B Zhang, Y Takizawa, A Hasagawa, A Yamaguchi, S Obana, in *Wireless Communications and Networking Conference, 2007. WCNC 2007, IEEE*. Tree-based routing protocol for cognitive wireless access networks, (2007), pp. 4204–4208. doi:10.1109/WCNC.2007.768
14. X Wang, TT Kwon, Y Choi, in *Proceedings of the 2009 ACM Workshop on Cognitive Radio Networks*. CoRoNet '09. A multipath routing and spectrum access (mrsa) framework for cognitive radio systems in multi-radio mesh networks (ACM, New York, NY, USA, 2009), pp. 55–60. doi:10.1145/1614235.1614249. http://doi.acm.org/10.1145/1614235.1614249
15. S-C Lin, K-C Chen, in *Global Telecommunications Conference (GLOBECOM 2010), 2010 IEEE*. Spectrum aware opportunistic routing in cognitive radio networks, (2010), pp. 1–6. doi:10.1109/GLOCOM.2010.5683924
16. W Li, X Cheng, T Jing, X Xing, in *INFOCOM, 2013 Proceedings IEEE*. Cooperative multi-hop relaying via network formation games in cognitive radio networks, (2013), pp. 971–979. doi:10.1109/INFCOM.2013.6566886
17. J Huang, S Wang, X Cheng, M Liu, Z Li, B Chen, Mobility-assisted routing in intermittently connected mobile cognitive radio networks. IEEE Trans. Parallel Distributed Syst. **25**(11), 2956–2968 (2014). doi: 10.1109/TPDS.2013.291
18. S Biswas, R Morris, in *Proceedings of the 2005 Conference on Applications, Technologies, Architectures, and Protocols for Computer Communications*. SIGCOMM '05. Exor: opportunistic multi-hop routing for wireless networks (ACM New York, NY, USA, 2005), pp. 133–144. doi:10.1145/1080091.1080108. http://doi.acm.org/10.1145/1080091.1080108
19. S Chachulski, M Jennings, S Katti, D Katabi, in *Proceedings of the 2007 Conference on, Applications, Technologies, Architectures, and Protocols for Computer Communications*. SIGCOMM '07. Trading structure for randomness in wireless opportunistic routing (ACM, New York, NY, USA,

2007), pp. 169–180. doi:10.1145/1282380.1282400. http://doi.acm.org/10.1145/1282380.1282400

20. A Jindal, K Psounis, Contention-aware performance analysis of mobility-assisted routing. IEEE Trans. Mobile Comput. **8**(2), 145–161 (2009). doi:10.1109/TMC.2008.98
21. S Wang, X Wang, X Cheng, J Huang, R Bie, in *IEEE ICDCS*. The tempo-spatial information dissemination properties of mobile opportunistic networks with levy mobility (IEEE, New York, 2014)
22. S Wang, M Liu, X Cheng, Z Li, J Huang, B Chen, Opportunistic routing in intermittently connected mobile p2p networks. IEEE J. Selected Areas Commun. (JSAC). **31**(9), 369–378 (2013)
23. Y Duan, G Liu, Z Cai, in *Wireless Algorithms, Systems, and, Applications*. Opportunistic channel-hopping based effective rendezvous establishment in cognitive radio networks (Springer, New York, 2012), pp. 324–336
24. X Lin, N. B Shroff, in *INFOCOM 2005. 24th Annual, Joint Conference of the IEEE Computer and Communications Societies. Proceedings IEEE*, vol. 3. The impact of imperfect scheduling on cross-layer rate control in wireless networks, (2005), pp. 1804–18143. doi:10.1109/INFCOM.2005.1498460
25. A Dimakis, J Walrand, Sufficient conditions for stability of longest-queue-first scheduling: Second-order properties using fluid limits. Adv. Appl. Probability. **38**(2), 505–521 (2006)
26. L Tassiulas, Adaptive back-pressure congestion control based on local information. Automatic Control IEEE Trans. **40**(2), 236–250 (1995). doi:10.1109/9.341781
27. MJ Neely, E Modiano, CE Rohrs, Dynamic power allocation and routing for time-varying wireless networks. Selected Areas Commun. IEEE J. **23**(1), 89–103 (2005). doi:10.1109/JSAC.2004.837349(410)23
28. X Wu, R Srikant, JR Perkins, Scheduling efficiency of distributed greedy scheduling algorithms in wireless networks. Mobile Comput. IEEE Trans. **6**(6), 595–605 (2007). doi:10.1109/TMC.2007.1061
29. L Jiang, J Walrand, A distributed csma algorithm for throughput and utility maximization in wireless networks. Netw. IEEE/ACM Trans. **18**(3), 960–972 (2010). doi:10.1109/TNET.2009.2035046
30. L Georgiadis, MJ Neely, L Tassiulas, Resource allocation and cross-layer control in wireless networks. Found. Trends Netw. **1**(1), 1–144 (2006). doi:10.1561/1300000001
31. MJ Neely, E Modiano, CE Rohrs, in *IEEE Journal on Selected Areas in Communications*. Dynamic power allocation and routing for time varying wireless networks (IEEE, New York, 2003), pp. 89–103
32. MJ Neely, E Modiano, C-P Li, in *Proceedings of IEEE INFOCOM*. Fairness and optimal stochastic control for heterogeneous networks (New York, 2005), pp. 1723–1734
33. L Huang, S Moeller, MJ Neely, B Krishnamachari, in *Modeling and Optimization in Mobile, Ad Hoc and Wireless Networks (WiOpt) 2011 International Symposium On*. Lifo-backpressure achieves near optimal utility-delay tradeoff, (2011), pp. 70–77. doi:10.1109/WIOPT.2011.5930067
34. Y Le, L Ma, W Cheng, X Cheng, B Chen, in *INFOCOM, 2012 Proceedings IEEE*. Maximizing throughput when achieving time fairness in multi-rate wireless lans, (2012), pp. 2911–2915. doi:10.1109/INFCOM.2012.6195728
35. M Heusse, F Rousseau, R Guillier, A Duda, in *SIGCOMM '05. Proceedings of the 2005 Conference on Applications, Technologies, Architectures, and Protocols for Computer Communications*. Idle Sense: an optimal access method for high throughput and fairness in rate diverse wireless lans (ACM, New York, NY, USA, 2005), pp. 121–132
36. S Biswas, R Morris, Opportunistic routing in multi-hop wireless networks. SIGCOMM Comput. Commun. Rev. **34**(1), 69–74 (2004). doi:10.1145/972374.972387
37. OMNeT++ Home Page. http://www.omnetpp.org

12

User scheduling for MU-MIMO transmission with active CSI feedback

Kyu-haeng Lee and Chong-kwon Kim*

Abstract

User scheduling boosts the multi-user multi-input multi-output (MU-MIMO) gain by selecting an optimal set of users to increase the 802.11 Wi-Fi system capacities. Many kinds of user scheduling algorithms, however, fail to fully realize the advantages of MU-MIMO due to considerable channel state information (CSI) overhead. In this paper, we propose a new MU-MIMO MAC protocol, called 802.11ac+, including a novel user scheduling algorithm. Unlike most proposals, where user scheduling is performed after an access point (AP) receives CSI from all users, 802.11ac+ determines the best user set during the CSI feedback phase. In particular, the AP broadcasts a channel hint about previously scheduled users, and the remaining users actively send CSI reports according to their effective channel gains (ECGs) calculated from the hint. Based on the proposed scheme, we develop two fair scheduling protocols, Round-Robin 802.11ac+ (RR-11ac+) and Proportional-Fair 802.11ac+ (PF-11ac+). Through trace-driven MATLAB simulations, we prove that the proposed schemes not only improve the throughput gain but also enhance the fairness among users.

1 Introduction

Most significant throughput gains of recent wireless communication systems come from multi-user multi-input multi-output (MU-MIMO) transmission using beamforming, which focuses energy toward a user so that the SNR and data rates increase. Wi-Fi standards as well as LTE-based cellular systems [1,2] have been adopting the MU-MIMO as a key technology for the next-generation wireless communication, and recently, several commercial products have been successfully deployed on the market.

The performance of MU-MIMO can be significantly improved when proper user scheduling is performed by multi-antenna nodes such as access points (APs) or base stations (BSs). Users with high received power or with an orthogonal channel can be chosen for increasing diversity gain or multiplexing gain, respectively. However, determining an optimal user set is nontrivial since it is generally integrated with finding optimal beamforming weights and powers. Even though we employ a simple beamforming strategy (e.g., zero-forcing beam forming (ZFBF)), a brute force search over all possible user sets is required and thus still creates a heavy burden. This motivates greedy user selection algorithms aimed at providing sub-optimal performance with a practical low complexity [3-9].

To reap the benefit of the MU-MIMO user selection in the aforementioned algorithms, the *channel state information* (CSI) feedback overhead issue should be addressed. The downlink CSI of the candidate users must be efficiently fed back to the AP. In contrast to cellular systems [1] where separate control channels are used to report the CSI, current 802.11ac Wi-Fi systems use a series of poll-based CSI feedbacks for each user [2], as shown in Figure 1. Furthermore, the CSI feedback is transmitted at the low basic rate (e.g., 6.5 Mbps) and also grows as the number of transmitter/receiver antennas, quantization level, and subcarrier group size increases. The CSI feedback overhead can reach up to 25× compared to the data transmission time in the case of 160 MHz of bandwidth and 4×1 MIMO [10]. Such excessive overhead could easily overwhelm the multi-user diversity gain even under optimal user selection. Figure 2 shows the throughput gains of a user selection scheme over 802.11ac [2] as a function of the number of polls. As shown in the figure, throughput gain in the downlink decreases with the number of polls. In particular, when the AP accesses the CSI of all users, the loss increases to about 70%. Significant CSI overhead in the downlink also brings a slight throughput loss in the uplink, thereby degrading the whole system performance.

*Correspondence: ckim@snu.ac.kr
Department of CSE, Seoul National University, 151-744 Seoul, Korea

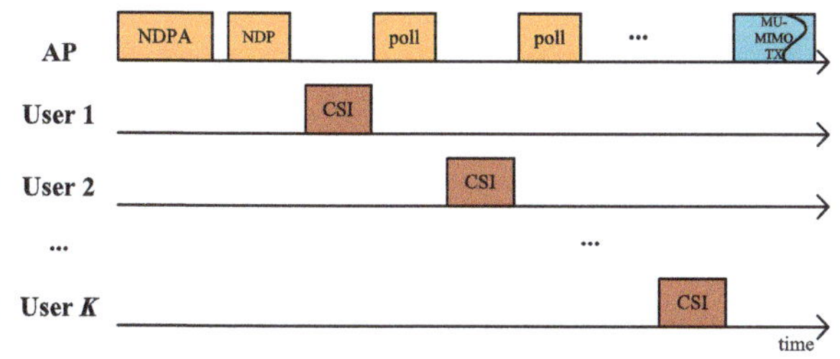

Figure 1 Channel sounding exchange for MU-MIMO transmission in 802.11ac. NDP and NDPA stand for 'Null Data Packet frame' and 'NDP Announcement frame', respectively.

In this paper, we present an MU-MIMO MAC protocol for Wi-Fi systems, called 802.11ac+. In particular, 802.11ac+ provides a novel and practical user scheduling solution. Our idea is very intuitive: perform the user scheduling during the CSI feedback phase. To do this, the AP broadcasts channel information about previously scheduled users by appending it to a poll frame. We refer to this channel information as *a channel hint*. Then, users calculate their effective channel gains (ECGs) from the channel hint, and the user with the largest gain *actively* sends a CSI report back to the AP. Upon receiving the CSI report, the AP includes the user in the multi-user transmission schedule and repeats this process while its degrees of freedom (DoF) constraint is satisfied. As a result, the performance of 802.11ac+ is comparable to other user scheduling heuristics, with the key difference that it requires a much smaller amount of CSI feedback.

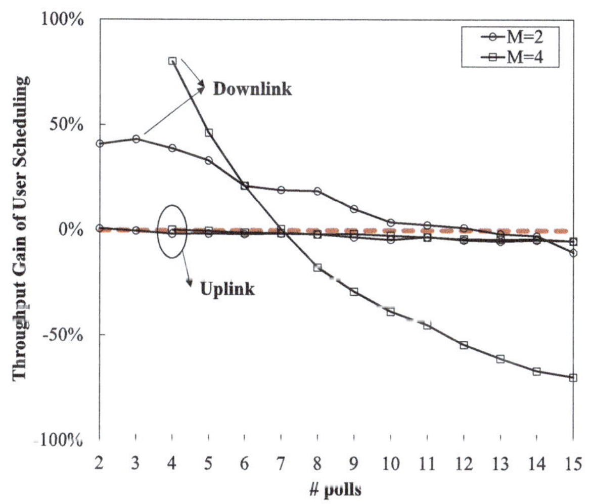

Figure 2 Throughput gain of user scheduling over 802.11ac. *M* and *K* stand for the number of AP antennas and the number of users, respectively. In this simulation, we set $K = 15$, and all users are assumed to have the same average SNR of 16.7 dB. We set 5 and 500 Mbps of traffic sending rates for downlink and uplink, respectively. As a user scheduling algorithm, we choose semiorthogonal user selection (SUS) [6], where a multi-antenna node selects a user if it has the highest ECG (Effective Channel Gain) among users in each selecting round. The red horizontal dashed line indicates the zero gain.

The design of 802.11ac+ is challenging for the following reasons. First, the channel hint should be well designed to balance between overhead and efficiency. In order to reduce the overhead, the channel hint should be succinct, but at the same time, it should include all channel information of already scheduled users such that the remaining users can estimate their ECGs. To deal with such a trade-off, we propose an efficient channel hint broadcasting mechanism. More specifically, an AP extracts the effective channel vector of the last scheduled user from the received CSI reports and uses it as a channel hint. The second challenging issue comes from the fact that users hardly know how much better their ECGs are in a fully distributed manner, which leads us to adopt a contention for users to get a feedback opportunity. Due to the nature of contention, there may be a frame collision if more than one user sends a CSI report at the same time. To address the collision problem, we use a delayed feedback approach, where users delay their CSI feedback according to their ECGs: the bigger the ECG a user has, the faster the CSI report will be. To improve the performance, we focus on maximizing the probability of feedback success. Additionally, we extend our user selection algorithm to well-known fair scheduling protocols, Round-Robin (RR) and Proportional-Fair (PF), to achieve fairness among users.

To evaluate the performance of our approach, we implement the 802.11ac+ and its fair scheduling protocols on a MATLAB simulator. The extensive trace-driven simulation results show that 802.11ac+ obtains much higher throughput gain than 802.11ac and a MAC protocol employing a well-known user scheduling algorithm. Also, we demonstrate that two fair scheduling protocols of 802.11ac+ give a much better throughput fairness than 802.11ac, especially when users experience different channel qualities.

We summarize our main contributions as follows. First, we propose the design of an MU-MIMO MAC protocol, called 802.11ac+, which accomplishes user scheduling gain with a far smaller amount of CSI feedback. Second, in order to realize the protocol, we provide a channel hint broadcasting mechanism and an active CSI feedback scheme. Third, we build two fair scheduling protocols based on the user scheduling of 802.11ac+ to achieve fairness among users. Lastly, the performance of 802.11ac+ has been evaluated through trace-driven simulations in MATLAB.

The remainder of the paper is organized as follows. Section 2 provides the related work, and we next describe the 802.11ac+ mechanism in greater detail in Section 3. In Section 4, we discuss the fair scheduling protocols of

802.11ac+, and Section 5 shows the performance evaluation. We finally conclude our paper in Section 6.

2 Related work

We survey the research results on the user scheduling schemes and channel feedback overhead reduction techniques related to 802.11ac+.

2.1 User scheduling schemes

User selection has been highlighted since it can improve the MU-MIMO performance significantly. In user scheduling, it is challenging and often impractical to determine the optimal user set due to the large search space. Therefore, some protocols based on heuristics have been proposed. Many publications study the problem of maximizing sum capacity [3,7-9]. SUS [6] selects one user in each round by exploiting channel orthogonality between users. However, the selected users cannot guarantee that they always increase the sum rate. GUSS [7] considers 'delete' and 'swap' operations to guarantee a positive increment of channel capacity in each selecting round. Jin et al. propose a volume metric as the product of diagonal elements of an upper-triangular matrix by performing QR factorization to the selected user channels [8]. Shen et al. [9] propose and compare two algorithms for both approaches with block diagonalization [11] which is a generalized concept of channel inversion.

In addition to the sum rate maximization, the rate balancing problem aimed at maximizing the throughput subject to the constraint that the rates of the different users need to have certain fixed ratios has also been considered in many literatures [12,13]. Hellings et al. propose a gradient projection-based solution [12], and Guthy et al. propose a user classification scheme based on perturbation analysis [13], in order to handle the rate balancing problem. Along this line, Lima et al. consider a frequency allocation problem for MIMO-OFDMA networks [14]. Unfortunately, these schemes come with a nonnegligible performance loss due to the excessive feedback overhead.

2.2 Channel feedback overhead reduction techniques

One way of overcoming the CSI overhead is to use the compression techniques for reducing CSI feedback bits. Codebook and quantization are already adopted in LTE and MIMO-based Wi-Fi systems [1,2,15-17]; however, selecting the optimal quantization level is still an open problem. In Wi-Fi systems, it is proposed to compress the CSI report along three dimensions: time, frequency, and quantization level. Even though compression is used, it is reported that the feedback can take about 25 times longer than the data transmission time when sending a small packet with a high data rate [10]. A vast literature of work has aimed to design an efficient CSI quantization mechanism [18-22]. Recently, compressive sensing has also been used for feedback reduction in MIMO communications [23,24]. The downside of compressions is, as expected, a throughput loss of MU-MIMO transmission: fewer bits may offer diminishing returns.

In addition to the compression, the reduction of feedback loads by adaptively sending feedback can be used. One possible solution is to allow only users whose signal quality (e.g., SNR, carrier-to-noise ratio (CNR)) is higher than a pre-defined threshold to report their CSI [10,25-28]. However, these approaches may fail to obtain higher effective channel gain because their CSI reports are sent without considering the relationship between user channels.

3 802.11ac+

3.1 System model

In this paper, we consider a single basic service set (BSS) Wi-Fi network, where an M-antenna AP and K single-antenna user stations communicate with each other, as shown in Figure 3. We assume that MIMO channels satisfy the i.i.d. Rayleigh fading condition: the real and imaginary components of $\mathbf{h_{km}}$ are i.i.d. Gaussian with unitary variance and zero mean, where $\mathbf{h_{km}}$ is a channel gain from the mth antenna of the AP to the kth user. Then, we can characterize the channel of user k from the AP, i.e., $\mathbf{h_k}$, as a zero mean complex Gaussian channel vector.

In MU-MIMO, the multi-antenna transmitter, e.g., AP, uses precoding to send multiple data streams among several users at a given instant. Similar to most recent schemes, we incorporate zero-forcing beamforming (ZFBF) as the precoding strategy, since it effectively removes the mutual interference among concurrent transmissions by using a low-complexity precoding matrix computation. In ZFBF, the precoding vector of one user is selected to be orthogonal to the channel vector of the other users. Let S be the group of selected users. Then, the precoding matrix for S, denoted by $W(S)$, is obtained as:

$$W(S) = H(S)^{\dagger} = H(S)^{*}(H(S)H(S)^{*})^{-1} \tag{1}$$

where $(\cdot)^{\dagger}$, $H(S)$, and H^{*} stand for a pseudo-inverse, the channel matrix of S, and the conjugate transpose of H, respectively.

3.2 Overview

As explained earlier, we do not separate CSI feedback from the user scheduling procedure. This is the key difference between the user scheduling in 802.11ac+ and other approaches. Figure 4 illustrates the main idea. At the beginning, an AP polls the first user (i.e., user 1 in this example) and it replies with its CSI. Then, the AP immediately joins user 1 to the scheduled user set S. We note that the first user is selected by the AP's queuing policy. By using the first CSI, the AP makes a channel hint, and it broadcasts a polling frame with the channel hint to all

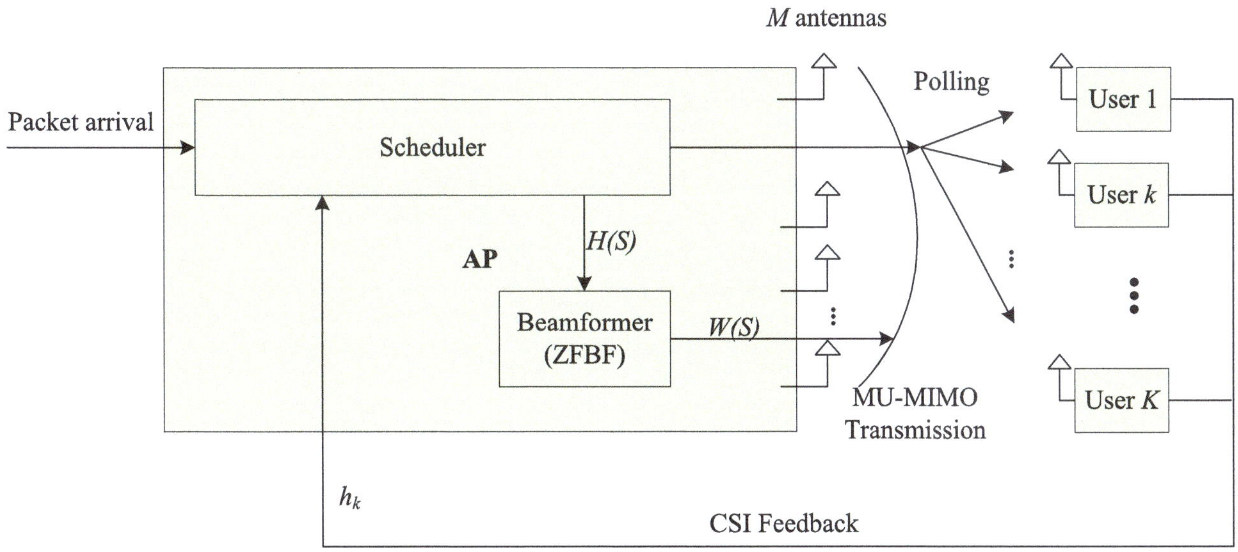

Figure 3 MU-MIMO downlink system with the *M*-antenna AP and *K* single-antenna user stations.

users. Upon receiving the poll, users compute their ECGs from the channel hint. Then, the user with the highest gain sends its CSI report to the AP through a contention. In this example, user 2 has the largest channel gain so it can feed back CSI for the second time. This step repeats until the AP successfully receives *M* CSI reports or the feedback timeout is triggered.

The intuition behind our user scheduling method is that users actively participate in scheduling decisions, unlike in most user scheduling algorithms, which are executed at only the multi-antenna transmitters. As a result, user scheduling in 802.11ac+ can limit the number of sounding exchanges to the number of AP antennas, while still keeping the scheduling gain. In the next section, we introduce two main mechanisms to realize 802.11ac+: channel hint broadcasting and active CSI feedback.

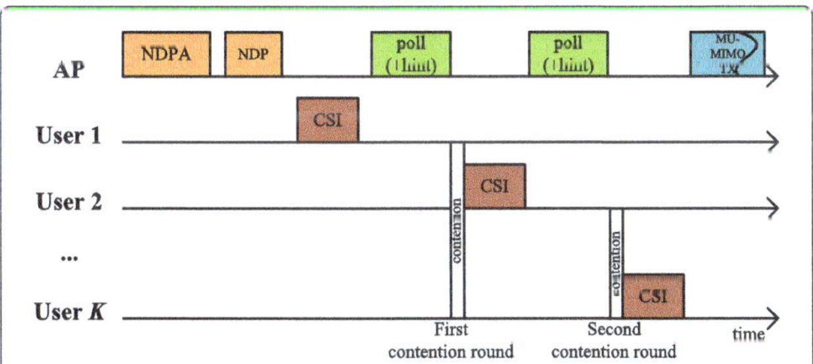

Figure 4 Operation example of 802.11ac+. Both AP and users participate in scheduling decisions. Except for the first scheduled user, all users should send a feedback report through a contention. If the AP fails in reception of any CSI reports, then it does not go to the next contention round and finishes the feedback procedure. Therefore, there could be maximum ($M - 1$) contention rounds in the case of *M* AP antennas.

3.3 Channel hint broadcasting

3.3.1 *Effective channel gain*

Since the user scheduling in 802.11ac+ is based on harnessing the ECG, we first look into the concept of it. For a user channel $\mathbf{h_k}$, its effective channel vector[a] is defined as $\mathbf{e_k}$ and can be calculated by projecting $\mathbf{h_k}$ onto the orthogonal complement of the subspace spanned by set $\{\mathbf{e}_{(1)}, ..., \mathbf{e}_{(|S|)}\}$, being interpreted as the effective channel vectors of previously selected users [6,7]:

$$\mathbf{e_k} = \mathbf{h_k} - \sum_{j=1}^{|S|} \frac{\mathbf{h_k}\mathbf{e}^*_{(j)}}{\|\mathbf{e}_{(j)}\|^2}\mathbf{e}_{(j)} \quad (2)$$

$$= \mathbf{h_k}\left(I - \sum_{j=1}^{|S|} \frac{\mathbf{e}^*_{(j)}\mathbf{e}_{(j)}}{\|\mathbf{e}_{(j)}\|^2}\right) \quad (3)$$

$$= \mathbf{h_k}Q(S) \quad (4)$$

where I is the identity matrix and $Q(S)$ is a projection matrix. Then, the ECG of user k is denoted as $\|\mathbf{e_k}\|^2$.

Specifically, for OFDM systems, which divide the bandwidth into orthogonal subcarriers and treat each of the subcarriers as an independent narrowband channel, the ECG should be averaged over all subcarriers [b]. Let $\mathbf{e_k}[\mathbf{c}]$ be the ECG on subcarrier c, then we have:

$$\|\mathbf{e_k}\|^2 = \frac{1}{N_c}\sum_{c=1}^{N_c} \|\mathbf{e_k}[\mathbf{c}]\|^2 \quad (5)$$

where N_c is the total number of subcarriers.

3.3.2 *Channel hint design*

There are many ways to implement a channel hint. In a simple way, the AP can use a projection matrix as a channel hint. Although this simplifies the computation burden at the receiver, it always consumes significant bits for representing an $M \times M$ matrix per each subcarrier (the maximum required number of appended channel vectors is $(M-1)MN_c$.). By contrast, if we use the received CSI as a channel hint, then this will increase the receiver side computing complexity.

Our proposed design exploits a compromise between the above two schemes. Every time the AP receives a CSI report, it computes the effective channel vector from that CSI and uses it as a channel hint. Hence, in every poll frame, the effective channel vector of the last scheduled user will be included. On the receiver side, users should maintain the value of summation in Equation 3 during the feedback phase so that they can update a projection matrix and obtain their ECGs much faster.

We illustrate the poll frame structure used in 802.11ac+ in Figure 5. As shown in the figure, the poll frame requires only one additional channel vector, and thus, the maximum number of appended vectors is $(M-1)N_c$. Since in each selection round, each user and the AP require one $(1 \times M) \times (M \times M)$ vector-matrix multiplication per subcarrier, we conclude that the computational complexity is $C(M-1)N_c$, where C is a computational cost corresponding to one vector-matrix multiplication.

One may claim that polling additional $(M-1)$ users and taking CSI from them for user selection would be more effective than using $(M-1)$ channel hints. However, this may have a large feedback overhead, which brings MU-MIMO performance degradation. To clarify this, we compare the overhead of two schemes. Let τ_{ECV}, τ_{POLL}, and τ_{CSI} be the transmission time of sending one effective channel vector in channel hint, one polling frame, and one CSI feedback, respectively. Note that $\tau_{\mathrm{ECV}} + \tau_{\mathrm{POLL}}$ is the transmission time for one channel hint, and $\tau_{\mathrm{ECV}} < \tau_{\mathrm{CSI}}$. First, in 802.11ac+, the AP will send maximum $(M-1)$ channel hints, and maximum M users will report their CSI reports. Then, we have the following:

$$\mathrm{overhead}_{802.11ac+} = \tau_{\mathrm{POLL}} + (M-1)(\tau_{\mathrm{ECV}} + \tau_{\mathrm{POLL}}) + M\tau_{\mathrm{CSI}} + 2M\tau_{\mathrm{SIFS}} + (M-1)\tau_{\mathrm{contention}} \quad (6)$$

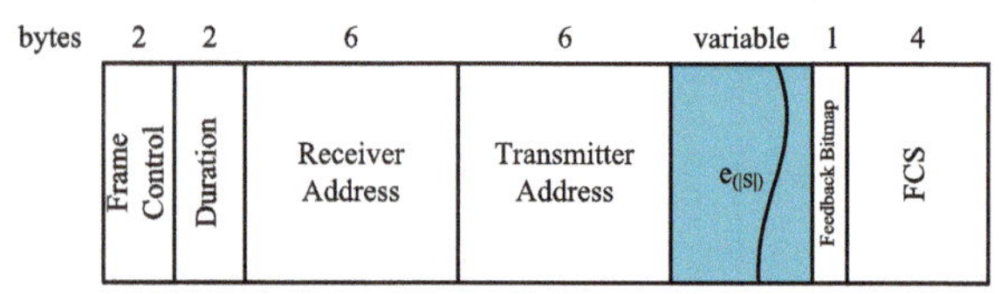

Figure 5 The change in use of the polling frame. When using a channel hint, polling is no longer destined for only one target user, so the receiver address should be the broadcast address. The length of the channel hint varies depending on the quantization level.

$$= M\tau_{\mathrm{POLL}} + (M-1)\tau_{\mathrm{ECV}} + M\tau_{\mathrm{CSI}} + 2M\tau_{\mathrm{SIFS}} + (M-1)\tau_{\mathrm{contention}} \quad (7)$$

where τ_{SIFS} stands for 802.11 Short InterFrame Space, which is 16 μs in 802.11ac [2]. Here, $\tau_{\mathrm{contention}}$ is an overhead from the feedback contention of 802.11ac+, which will be discussed in the next section. As will be discussed later, the maximum value of a $\tau_{\mathrm{contention}}$ is about $2\tau_{\mathrm{SIFS}}$ (four slots per each round).

Now, assume that the AP polls $(M-1)$ users after taking CSI from M users. In other words, the AP takes CSI from total $(2M-1)$ users for user selection. Then, we have the following:

$$\mathrm{overhead}_{\mathrm{add}} = (2M-1)(\tau_{\mathrm{POLL}} + \tau_{\mathrm{CSI}}) + 2(2M-1)\tau_{\mathrm{SIFS}} \quad (8)$$

$$= (2M-1)\tau_{\mathrm{POLL}} + (2M-1)\tau_{\mathrm{CSI}} + 2(2M-1)\tau_{\mathrm{SIFS}}. \quad (9)$$

Even if taking $\tau_{\mathrm{ECV}} \simeq \tau_{\mathrm{CSI}}$, $\mathrm{overhead}_{802.11ac+} < \mathrm{overhead}_{\mathrm{add}}$. Also, since the AP has no choice but to poll users randomly in this scheme, the user selection gain could be small, as shown in the case of $M = 4$ in Figure 2.

When sending a channel hint, since the AP may not intend all users to participate in the feedback procedure, it requires a method to notify only some dedicated users. For example, an AP should prevent users from sending CSI reports when it has no frames for them. For target user notification, we utilize the legacy Null Data Packet Announcement (NDPA) frame, by using fewer bits for the Association ID in the STA Info field and mapping bit positions to each user ID. This method maintains the same frame format and thus is efficient. A detailed description of the notification is out of the scope of this paper.

3.4 Active CSI feedback

3.4.1 *Feedback contention*

In 802.11ac+, only the user with the highest ECG should respond to the poll. However, this is a challenging task since users cannot know the ECG of the others in a distributed manner. If two users, who might think their gains are the best, feed back their CSI reports simultaneously, a frame collision will occur. Without any coordination between users, they should get opportunities to send feedback through a contention like the 802.11 distributed coordinate function (DCF). We call this a *feedback contention* in this paper.

In order to resolve the feedback contention, we adopt a delayed transmission approach. In this approach, users delay their feedback transmission regarding their ECGs, such that a user with higher value can access the feedback opportunity faster. To realize this, we first apply slotted time in our scheme, where the slot duration is fixed and each slot has a pre-defined threshold. We refer to this threshold as a *slot threshold*. A user is assigned to the specific slot according to its ECG and is allowed to transmit feedback only at the slot. If a user senses a feedback transmission earlier than its slot, it gives up and waits until the next poll.

Since users should listen to the medium until they obtain a transmission opportunity, the minimum duration of the slot requires at least one slot time of 802.11 (e.g., 4 μs for CCA + 5 μs for RX/TX turnaround). To limit the expense of delayed feedback, an AP uses the timeout threshold (γ). If the AP does not receive any feedback reports before γ, then, it finishes the scheduling. When feedback collision or timeout occurs in the first contention round, the scheduling may be stopped with only one user (i.e., the first user) remaining in S. In this case, the AP just sends a frame to that user in diversity mode, which guarantees the minimal level of system performance.

Figure 6 shows an example of an active feedback scheme with three users and five feedback slots for the first contention round. The associated slot thresholds are given in Table 1, which will be explained in the following section. Assume that user 2 and user 1 can transmit CSI feedback in the second and fourth slots, respectively, while user 3 cannot get an opportunity due to the low gain. User 2 sends the CSI feedback in the second slot. The others listen to the feedback transmission and wait for the next contention round.

3.4.2 Slot threshold optimization

The performance of the active feedback mainly relies on how to select the thresholds: γ and slot thresholds. As described before, active feedback fails when a collision or timeout occurs; however, their impacts are different from each other. The loss from collision is much bigger than that from timeout, due to the relatively long time wasted in the collision. Thus, we first fix γ and find optimal values of the slot thresholds for each contention round. Now, we describe how to determine the thresholds as below.

Provided that a random channel vector on subcarrier c, $\mathbf{h_k}[\mathbf{c}]$, is a zero-mean circularly symmetric complex Gaussian random variable, from Equation 5, $\|\mathbf{e_k}\|^2$ is a Gamma-distributed random variable with N_cL (shape

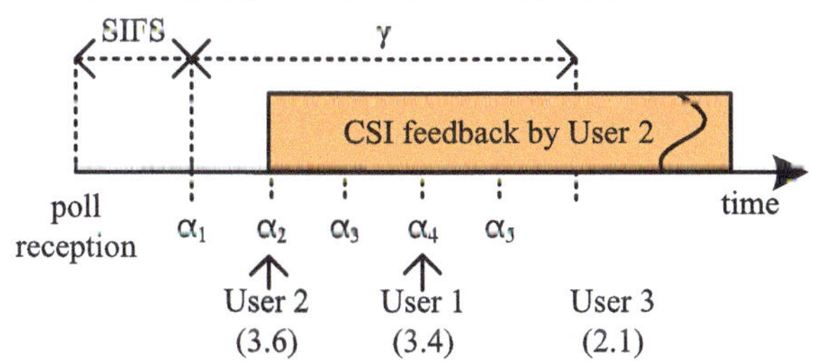

Figure 6 Example of the active CSI feedback with γ of five slots. Slot thresholds are given in Table 1. User 2 wins the contention because its ECG is higher than that of the others. Other users are notified of User 2's transmission and wait for the next round.

Table 1 An example of the thresholds

CR	(w_s, w_c, w_t)	α_1	α_2	α_3	α_4	α_5
1	(1,1,1)	3.667	3.543	3.453	3.371	3.280
	(0.4,0.4,0.2)	3.680	3.558	3.472	3.396	3.318
2	(1,1,1)	2.541	2.436	2.360	2.291	2.215
	(0.4,0.4,0.2)	2.552	2.449	2.376	2.312	2.246
3	(1,1,1)	1.384	1.304	1.247	1.196	1.140
	(0.4,0.4,0.2)	1.392	1.314	1.259	1.212	1.163

parameter) and $1/N_c$ (scale parameter), where $L = \mathrm{rank}(Q(S))$. Formally, for a constant α, $\|\mathbf{e_k}\|^2$ satisfies:

$$P\{\|\mathbf{e_k}\|^2 \leq \alpha\} = F(\alpha; N_cL, 1/N_c) \tag{10}$$

where $F(\alpha; N_cL, 1/N_c)$ is the cumulative distribution function (CDF) of the Gamma distribution for α, N_c and L. For fixed N_c and L, we use $F(\alpha)$ and $F(\alpha; N_cL, 1/N_c)$ in the same sense. Note that L varies every contention round.

For a timeout γ (= G slots) in a certain contention round, let $\alpha_1, \alpha_2, ..., \alpha_G$ be the threshold corresponding to each slot, where $\alpha_1 \geq \alpha_2 \geq, ..., \geq \alpha_G \geq 0$. We assume that γ is fixed over all contention rounds. We define the probability of successful feedback as the probability that only one user will send a CSI report in γ. Then, it can be calculated by summing up all success probabilities over all time slots:

$$p^{\mathrm{suc}} = \sum_{g=1}^{G} p_g^{\mathrm{suc}} \tag{11}$$

$$\begin{aligned} &= K'\{1 - F(\alpha_1)\}F(\alpha_1)^{(K'-1)} \\ &+K'\{F(\alpha_1) - F(\alpha_2)\}F(\alpha_2)^{(K'-1)} \\ &+... \\ &+K'\{F(\alpha_{G-1}) - F(\alpha_G)\}F(\alpha_G)^{(K'-1)} \end{aligned} \tag{12}$$

where K' is the number of contending users in the contention round. We note that $K' \leq K - 1$ because the first user is already selected by AP, and K' can be estimated by using the target user notification.

The timeout and collision probabilities are defined as follows:

$$p^{\mathrm{tout}} = F(\alpha_G)^{K'} \tag{13}$$

$$p^{\mathrm{col}} = 1 - p^{\mathrm{suc}} - p^{\mathrm{tout}}. \tag{14}$$

Then, our objective is to find a set of thresholds as follows:

$$\underset{\alpha_1,...,\alpha_G}{\mathrm{argmax}} w_s p^{\mathrm{suc}} - w_c p^{\mathrm{col}} - w_t p^{\mathrm{tot}} \tag{15}$$

where w_s, w_c, and w_t are weights for the probabilities of success, collision, and timeout, respectively.

The optimization problem described in Equation 15 is nonconvex and is solved using nonlinear optimization technique. Note that it only needs to be solved whenever the network configuration (N_c, M, K) is changed or we can construct a set of solutions offline. Once a set of thresholds has been constructed, users can refer to them at any time.

Table 1 shows an example of the thresholds in the case of $\gamma = 5, M = 4, K' = 14$, and $N_c = 30$, where CR stands for contention round. Recall that there are maximum $(M-1)$ contention rounds, and each round consists of γ time slots. From the table, we can see that the thresholds decrease during the course of a contention round because successful feedback reduces L one by one and hence decreases the DoF of the chi-square distribution. In addition, the reduced number of contending users also results in lower thresholds in further contention rounds. The thresholds of nonequal weights are slightly bigger than those of equal weights. In this case, the probability of success as well as the probability of collision decreases.

We compare the downlink throughput and success feedback probability of 802.11ac+ as a function of γ in Figure 7. In particular, to observe the impact of the threshold optimization according to channel distributions, we use two different channel distributions. Figure 7a shows the ECG distributions of two traces used in this simulation: 'real trace' and 'synthetic trace (Rayleigh)'. Note that the slot thresholds are computed from the real trace. As shown in Figure 7b, a throughput loss (maximum 4 Mbps when $\gamma = 4$) happens in the synthetic trace scenario. This result is predictable: the thresholds are optimized only for the real trace. Even though the performance loss might seem small in this result, using a more exact channel model will improve the robustness of the proposed scheme. Note that in this section, we provide an analysis of the Rayleigh fading case only, but we can apply the same technique for different distributions, e.g., Rician, by changing Equation 10 in the optimization.

The graph also demonstrates that nonequal weights reduce both the feedback success probability and the collision probability (omitted here), as expected, thereby decreasing the scheduling gain. As a result, the throughput of nonequal weights is lower than that of the equal weights over all cases. Additionally, we can see that too large a γ decreases the effective channel gain while too small a γ decreases the user diversity gain. For the real trace scenario, both downlink throughputs increase as γ reaches four; after that, they decrease (for the synthetic trace, maximum downlink throughput is obtained when γ is three).

4 Fair scheduling

In this section, we develop two fair scheduling approaches of the 802.11ac+ based on the Round-Robin and Proportional-Fair algorithms. Two main changes are adopted in the fair scheduling protocols to maintain the property of the 802.11ac+ user scheduling. The first change is to make use of a simple fair queue in the AP of both RR-11ac+ and PF-11ac+. In particular, the AP selects the first user by using the following:

$$\underset{k \in A}{\operatorname{argmax}} \frac{1}{\mathcal{R}_k} \tag{16}$$

where $\mathcal{R}_k$ is the average data rate of user k and A is a target user set. For the remaining users, each fair scheduling protocol uses a different scheduling strategy, but the main idea is the same as that of 802.11ac+. We describe the details of RR-11ac+ and PF-11ac+ in the next subsections.

4.1 RR-11ac+

This is the simplest way to obtain the fairness that gives users equal transmission opportunities. Most Round-Robin schedulers use the average data rates of all users, and they can be monitored at the scheduler, which is very easy to implement.

In RR-11ac+, we use the same metric to schedule users, but a simple change is made for multi-user transmission: before the active CSI feedback, an AP limits the contending users according to their average data rates. Specifically, among remaining users in A, the AP selects top $(M - 1 + o)$ users in ascending order of average data rates such that users who have been served less have more feedback opportunities. If the nonnegative parameter o is set to 0, then RR-11ac+ is reduced to the 802.11ac with the legacy Round-Robin scheduling; however, it still considers the ECG of users. For example, if one's ECG is too low so joining it into S is no longer helpful for multi-user transmission, then it will be discarded.

4.2 PF-11ac+

The proportional fairness maintains a balance between maximizing the sum rate and allowing all users at least a minimal level of service. Basically, both the original Proportional-Fair algorithms and PF-11ac+ use the following criterion to select users:

$$\frac{\mathcal{T}_k}{\mathcal{R}_k} \tag{17}$$

where $\mathcal{T}_k$ is the current available data rate of user k.

The difficulty of legacy Proportional-Fair scheduling is that the scheduler cannot know the exact data rates of users $(\mathcal{T}_k)$ when the scheduling decision has to be made. However, the active feedback of 802.11ac+ provides a simple implementation of the Proportional-Fair scheduling since users can actively join the scheduling decision. In particular, they can estimate $\mathcal{T}_k$ by using their ECGs (i.e., $\mathcal{T}_k \sim \log(\|\mathbf{e_k}\|^2)$). Note that in PF-11ac+, users should maintain their average data rates, and the slot thresholds

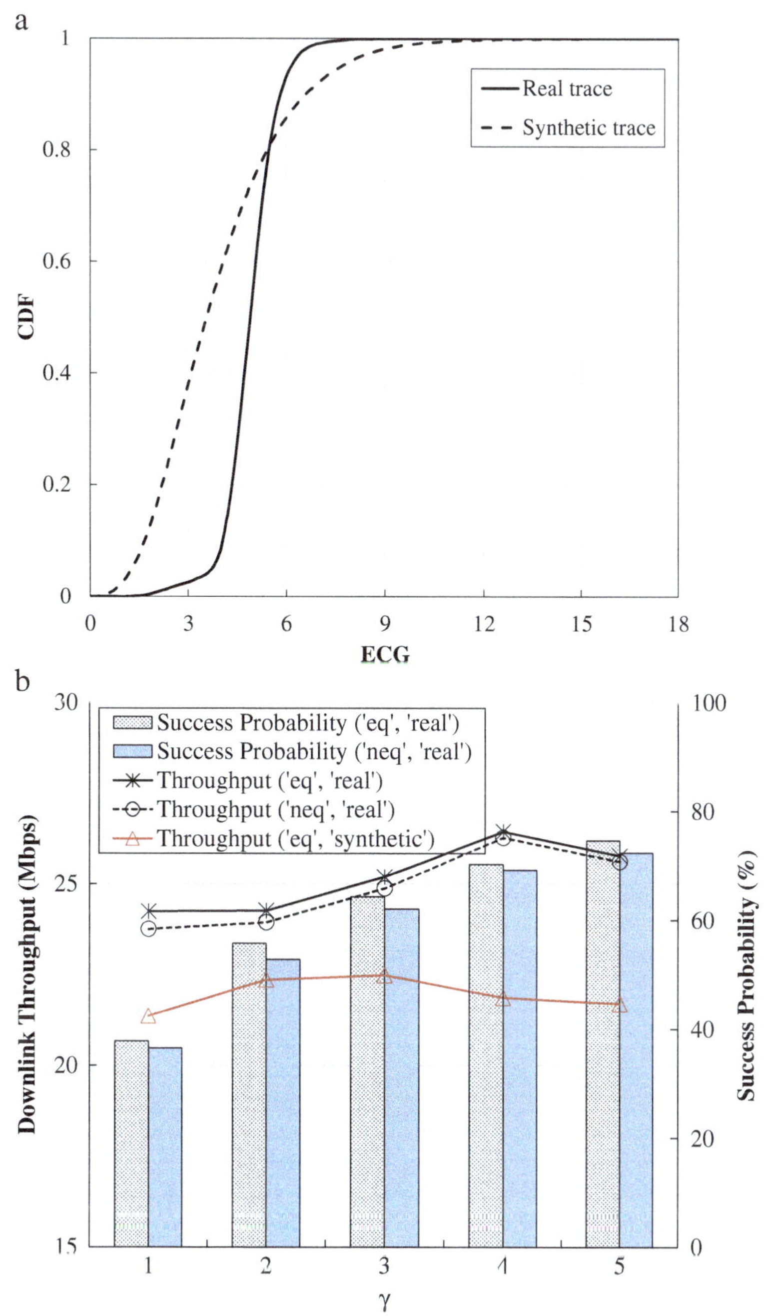

Figure 7 Performance of the active CSI feedback. **(a)** We use two different CSI traces (real and synthetic traces). The channel coefficients of the synthetic trace follow the Rayleigh distribution explained in Section 3.1. **(b)** We use weights of (0.4,0.4,0.2) as the nonequal weights cases. The results for ('neq', 'synthetic') are omitted because they show a similar pattern as those for 'real'. **(a)** ECG distributions. **(b)** Performance comparison with respect to γ.

should be redefined according to Equation 17 by adopting the technique in Section 3.

4.3 Summary

Table 2 shows the comparison between 802.11ac+, RR-11ac+, and PF-11ac+. As shown in the table, there are notable differences in three aspects: target user set, first user selection, and *n*th user selection. First, in 802.11ac+ and PF-11ac+, all users are considered for scheduling, while in RR-11ac+, the number of target users is bounded to $M - 1 + o$, as explained earlier. Second, in 802.11ac+, basically, the first user is selected randomly according to AP's queuing policy, while in RR-11ac+ and PF-11ac+, the user with maximum utility Equation 16 will be chosen.

Table 2 Comparison between three schemes

	Size of target user set	First user selection	Metric for $n(>1)$th user selection
802.11ac+	K	random or $\text{argmax}\|\mathbf{e_k}\|^2$	$\|\mathbf{e_k}\|^2$
RR-11ac+	$M-1+o$	$\text{argmax}\frac{1}{\mathcal{R}_k}$	$\frac{\mathcal{T}_k}{\mathcal{R}_k}$
PF-11ac+	K		

To maximize the sum capacity, the user with the largest ECG will be selected as a first user, similar to SUS [6]. In using this method, one possible concern is that optimally selecting the first user is difficult since no CSI is available at the time of user selection. To handle this issue, we propose referring to the statistics of previous channel status of the users, for the first user selection. This method should work well when transmissions of the AP happen in channel coherent time, which typically ranges from 15 to 100 ms [10,29]. Lastly, for nth user selection, 802.11ac+ uses ECG as a selection metric, while the fair scheduling protocols use Equation 17 as a selection metric.

5 Performance evaluation

In this section, simulation results for the throughput and fairness performance of the proposed schemes are presented.

5.1 Setting

We implement 802.11ac, an SUS [6] based MAC protocol, 802.11ac+, RR-11ac+, and PF-11ac+ on the MATLAB simulator. To conduct high-fidelity emulation of real-world settings, we use the 802.11n data traces provided by the authors of [30]. The traces contain per-subcarrier (30 subcarriers for 20 MHz) CSI and SNR readings (ranging from 4 to 43 dB) for 18 users. By using the traces, we reconfigure 25 users and a maximum of four AP antennas in our simulations.

Except for the set of fair scheduling protocol simulations, we assume that all users have the same average SNR of 16.7 dB (according to the trace). We set the simulation parameters to the default values in Table 3 and 802.11ac [2]. The AP and users generate traffic for each other according to their average sending rates (traffic intensity). We measure the average aggregate throughput of the downlink, uplink[c], and system. Note that the system throughput is defined as the sum of downlink and uplink throughput. We also assume that all protocols used in the simulation enable a packet aggregation scheme.

Table 3 Default simulation parameters

Parameter	Value
Downlink traffic intensity	5 (Mbps)
Uplink traffic intensity	0.5 (Mbps)
K	15
The number of polls used in SUS	10
M	4
SNR	16.7 (dB)
γ	4
o	4

5.2 802.11ac+ performance

5.2.1 *Impact of K*

The user diversity will be more effective as the number of users to be considered for the scheduling increases; however, the significant CSI feedback overhead may limit the gain. In order to examine the performance of 802.11ac+, we illustrate the downlink and system throughput of three protocols according to the number of users in Figure 8. Both throughputs increase until the number of users reaches a certain point, after which they start to decrease. 802.11ac+ shows significant throughput gain over the other protocols from the high channel utilization via its user selection scheme. As a result, it achieves the maximum downlink (system) throughput gain of 100% (69%) and 168% (97%) over 802.11ac and SUS, respectively. On

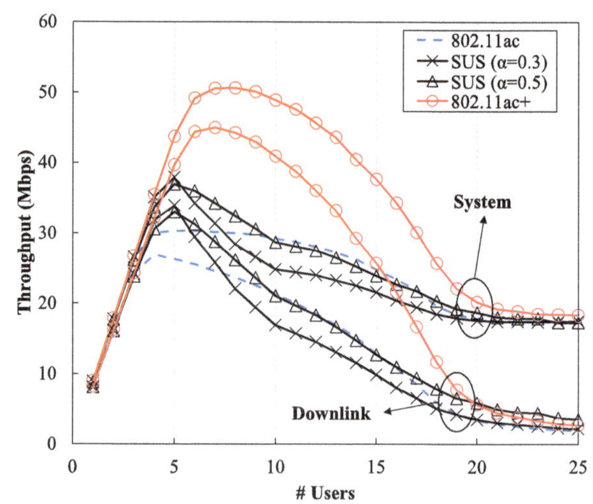

Figure 8 Throughput according to the number of users. Due to dominant uplink traffic and heavy contention, the network throughput decreases with the number of users. We note that α used in SUS controls the trade-off between diversity gain and effective channel gain [6].

the other hand, SUS shows a similar or worse performance compared to 802.11ac except for a moderate number of users. Note that these maximum throughput gains are computed as follows:

$$\max \frac{\text{throughput}_k^{802.11ac+} - \text{throughput}_k^{802.11ac(\text{SUS})}}{\text{throughput}_k^{802.11ac(\text{SUS})}} \quad (18)$$

where $\text{throughput}_k^{\text{scheme}}$ denotes the target scheme throughput when $K = k$.

5.2.2 *Impact of M*

More AP antennas provide better spatial multiplexing gain while increasing the CSI overhead because a large number of bits are required for representing CSI. In this simulation, we investigate and compare the performance according to the number of AP antennas, as shown in Figure 9. As expected, the throughput of protocols increases with the number of AP antennas. However, the CSI overhead limits the performance of SUS while 802.11ac+ can obtain much higher throughput gain due to small CSI overhead. As a result, it achieves the maximum downlink (system) throughput gain of 98% (51%) and 161% (75%) over 802.11ac and SUS, respectively.

5.2.3 *Impact of the number of polls*

To investigate the user scheduling gain affected by the CSI overhead, we compare the three system throughput results according to the number of polls in Figure 10. As expected, SUS suffers from the CSI overhead as the number of polls increases. As a result, SUS shows poor throughput performance compared to 802.11ac after the number of polling frames exceeds seven in the case of $M = 4$. In contrast to SUS, 802.11ac+ outperforms other protocols by using far less CSI feedback while fully harnessing the user scheduling gain. As a result, in the case of $M = 4$ and four polling frames, its throughput gains over 802.11ac and SUS are 51% and 20%, respectively.

5.3 Fair scheduling protocol performance

5.3.1 *System throughput*

We compare the system throughput of RR-11ac+ and PF-11ac+ with 802.11ac and 802.11ac+ in Figure 11. As shown in the graph, 802.11ac+ and its fair approaches show similar throughputs, although they use different user selection criteria from each other. In particular, the system throughputs of RR-11ac+ and PF-11ac+ are much bigger than that of 802.11ac, since the user scheduling adopted in them still considers the effective channel gain, unlike 802.11ac. Recall that all users experience similar average SNRs in this simulation.

5.3.2 *Fairness*

To evaluate the fairness performance of RR-11ac+ and PF-11ac+, users are subjected to different average SNRs from 4 to 43 dB, and a user with a higher ID has a bigger SNR than one with a lower ID, unlike in the previous set of simulations.

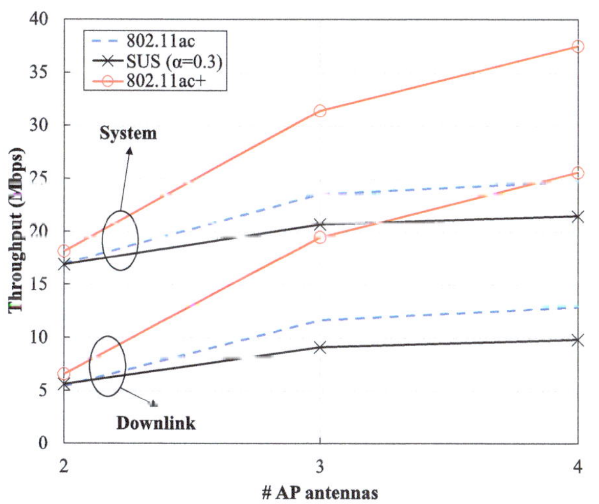

Figure 9 Throughput according to the number of AP antennas. The large number of AP antennas increases the size of the CSI feedback frame. Thus, requesting many users' CSI for scheduling may limit the scheduling gain significantly, as shown in the result of SUS. However, 802.11ac+ can obtain higher throughput gain over both protocols due to the limited feedback.

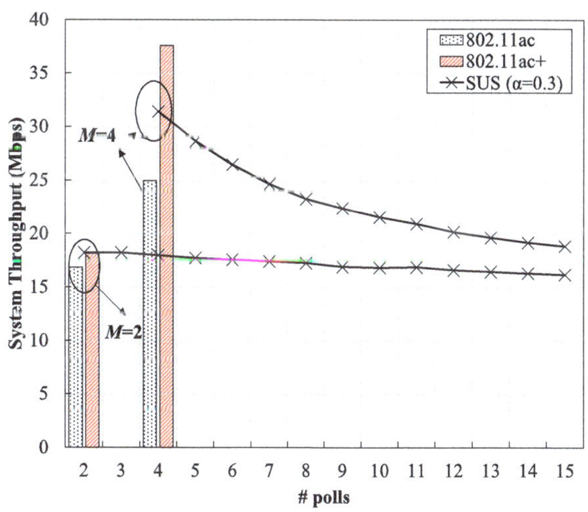

Figure 10 System throughput according to the number of polls. 802.11ac+ obtains higher throughput gain from the user scheduling with a fixed number of polls, while in the case of SUS, the performance degrades as the number of polls increases.

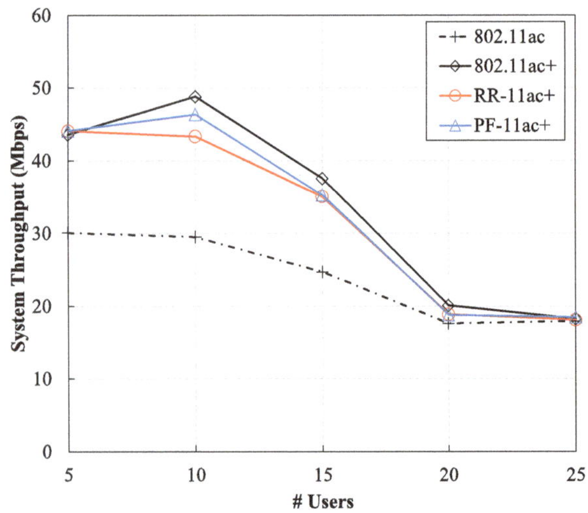

Figure 11 System throughput comparison of 802.11ac, 802.11ac+, and two fair scheduling protocols. 802.11ac+, RR-11ac+, and PF-11ac+ show a similar performance under the same average user SNR environment due to the fact that they consider the effective channel gain.

Figure 12 illustrates Jain's Fairness Index of downlink throughput of four protocols. From the result, we can see that the MU-MIMO transmission with more antennas gives better fairness because it can serve more users at once. Over all cases, RR-11ac+ shows the best fairness performance, followed by PF-11ac+. In particular, RR-11ac+ with $M = 4$ achieves close-to-perfect fairness among users. In the meantime, 802.11ac and 802.11ac+ bring low fairness levels compared with the two fair scheduling protocols. The performance of 802.11ac+ is similar to or higher than that of 802.11ac, and yet the sum rate of 802.11ac+ is much higher than that of 802.11ac.

Finally, we plot the downlink throughput that each user attains under each protocol in Figure 13. From the figure, we can clearly see that RR-11ac+ and PF-11ac+ achieve fairly similar throughputs over all users except some users with low SNRs. In addition, in this result, PF-11ac+ shows better aggregate downlink throughput than 802.11ac+, since users in 802.11ac+ have different chances of being scheduled regarding only their ECGs: the user scheduling of 802.11ac+ only favors a user to enhance the sum rate. As a result, there exist some users that have low throughput in 802.11ac+. Surprisingly, in the case of 802.11ac, this phenomenon is more obvious; more users suffer from starvation and a few users with high SNRs enjoy the high throughput. The main reason is that since the user scheduling in 802.11ac+ cares about the channel orthogonality between users, the probability of creating a low-benefit scheduling group is lowered.

6 Conclusions

In this work, we propose a new MU-MIMO MAC protocol, 802.11ac+, which obtains significant user scheduling gain with a far smaller amount of CSI feedback by exploiting channel hint-based polling and active CSI feedback. Trace-driven MATLAB simulation results show that 802.11ac+ achieves downlink throughput gains of up to 100% and 168% over 802.11ac and SUS-based MAC protocols. Additionally, two fair scheduling protocols of 802.11ac+ give close-to-perfect user fairness even when the users experience different channel qualities.

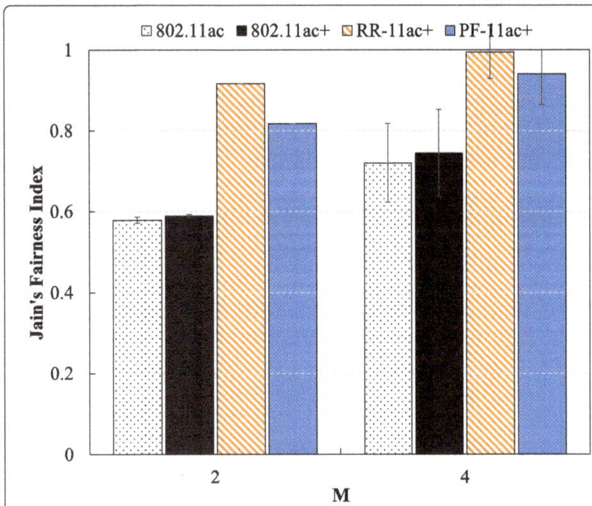

Figure 12 Jain's fairness index according to *M*. RR-11ac+ and PF-11ac+ can achieve high throughput fairness compared to 802.11ac and 802.11ac+.

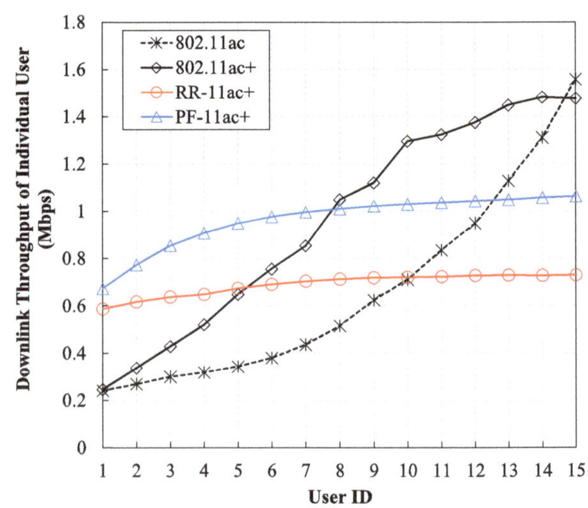

Figure 13 Downlink throughput comparison of 802.11ac, 802.11ac+, RR-11ac+, and PF-11ac+. RR-11ac+ and PF-ac+ achieve fairly similar throughputs over all users though they experience different channel qualities.

Endnotes

[a]The effective channel vector of the first user is the same as its channel vector, $\mathbf{e}_{(1)} = \mathbf{h}_{(1)}$.

[b]In this paper, we do not consider a per-subcarrier decision. We leave this issue to our future work.

[c]For better graph representation, we omit the graphs for the uplink.

Competing interests
The authors declare that they have no competing interests.

Acknowledgements
This work was supported in part by the IT R&D program of MSIP/KEIT. [10041861, Development of Wired/Wireless i-AVB System Technology for Concurrent Transmission of HD Media and Control Data].

References
1. Evolved Universal Terrestrial Radio Access (E-UTRA), Downlink Multiple Input Multiple Output (MIMO) enhancement for LTE-Advanced (Release 11). 3GPP TR V11.0.0. http://www.3gpp.org/ (2015)
2. IEEE Approved Draft Standard for IT - Telecommunications and Information Exchange Between Systems - LAN/MAN - Specific Requirements - Part 11: Wireless LAN Medium Access Control and Physical Layer Specifications. IEEE P802.11ac/D7.0, 1–456 (2013)
3. G Dimic, ND Sidiropoulos, On downlink beamforming with greedy user selection: performance analysis and a simple new algorithm. Signal Process. IEEE Trans. **53**(10), 3857–3868 (2005)
4. J Wang, DJ Love, MD Zoltowski, User selection with zero-forcing beamforming achieves the asymptotically optimal sum rate. Signal Process. IEEE Trans. **56**(8), 3713–3726 (2008)
5. M Kobayashi, G Caire, Joint beamforming and scheduling for a multi-antenna downlink with imperfect transmitter channel knowledge. IEEE J Sel A Commun. **25**(7), 1468–1477 (2007)
6. T Yoo, A Goldsmith, On the optimality of multiantenna broadcast scheduling using zero-forcing beamforming. Selected Areas Commun. IEEE J. **24**(3), 528–541 (2006)
7. S Huang, H Yin, J Wu, VCM Leung, User selection for multi-user MIMO downlink with zero-forcing beamforming. Vehicular Techn. IEEE Trans. **62**(7), 3084–3097 (2012)
8. L Jin, X Gu, Z Hu, Low-complexity scheduling strategy for wireless multiuser multiple-input multiple-output downlink system. Commun. IET. **5**(7), 990–995 (2011)
9. Z Shen, R Chen, JG Andrews, RW Heath, BL Evans, in *Signals, Systems and Computers, 2005. Conference Record of the Thirty-Ninth Asilomar Conference On*. Low complexity user selection algorithms for multiuser MIMO systems with block diagonalization (Pacific Grove, California, USA, 2006)
10. X Zhang, X Xie, K Sundaresan, in *ACM MOBICOM*. Adaptive feedback compression for MIMO networks (Miami, Florida, USA, 2013)
11. QH Spencer, AL Swindlehurst, M Haardt, Zero-forcing methods for downlink spatial multiplexing in multiuser MIMO channels. Signal Process. IEEE Trans. **52**(2), 461–471 (2004)
12. C Hellings, M Joham, W Utschick, in *Smart Antennas (WSA), 2011 International ITG Workshop On*. Gradient-based rate balancing for mimo broadcast channels with linear precoding (IEEE Aachen, Germany, 2011), pp. 1–8
13. C Guthy, W Utschick, in *Proceedings of ITG/IEEE Workshop on Smart Antennas (WSA)*. Sum throughput maximization in quality of service constraint multiuser mimo systems based on perturbation analysis (Vienna, Austria, 2007)
14. FRM Lima, T Ferreira Maciel, WC Freitas, FR Porto Cavalcanti, Improved spectral efficiency with acceptable service provision in multiuser MIMO scenarios. Vehicular Tech. IEEE Trans. **63**(6), 2697–2711 (2014)
15. DJ Love, RW Heath, W Santipach, ML Honig, What is the value of limited feedback for MIMO channels? Commun. Mag. IEEE. **42**(10), 54–59 (2004)
16. P Frank, A Muller, J Speidel, in *Wireless Conference (EW), 2010 European*. Fair performance comparison between CQI- and CSI-based MU-MIMO, for the LTE downlink (Lucca, Italy, 2010)
17. IEEE Std 802.11n. IEEE Standard for Information technology - Local and metropolitan area networks - Specific requirements - Part 11: Wireless LAN Medium Access Control (MAC) and Physical Layer (PHY) Specifications. IEEE P802.11 (2009)
18. K Huang, RW Heath, JG Andrews, Limited feedback beamforming over temporally-correlated channels. Signal Process. IEEE Trans. **57**(5), 1959–1975 (2009)
19. V Pohl, PH Nguyen, V Jungnickel, C Von Helmolt, in *IEEE Global Telecommunications Conference*. How often channel estimation is needed in MIMO systems (San Francisco, USA, 2003)
20. W Wang, A Harada, Enhanced limited feedback schemes for DL MU-MIMO ZF precoding. Wireless Commun. IEEE Trans. **12**(4), 1554–1561 (2013)
21. JH Lee, W Choi, Optimal feedback rate sharing strategy in zero-forcing MIMO broadcast channels. Wireless Commun. IEEE Trans. **12**(6), 3000–3011 (2013)
22. DJ Love, RW Heath, VKN Lau, D Gesbert, BD Rao, M Andrews, An overview of limited feedback in wireless communication systems. Selected Areas Commun. IEEE J. **26**(8), 1341–1365 (2008)
23. EJ Candes, MB Wakin, An introduction to compressive sampling. Signal Process. Mag. IEEE. **25**(2), 21–30 (2008)
24. ST Qaseem, TY Al-Naffouri, S Alghadhban, in *Telecommunications (ICT), IEEE International Conference On*. Compressive sensing for feedback reduction in MIMO broadcast channels (Doha, Qatar, 2010)
25. D Gesbert, M-S Alouini, in *Signal Processing and Information Technology, IEEE International Symposium On*. Selective multi-user diversity (Darmstadt, Germany, 2003)
26. D Gesbert, M-S Alouini, in *Communications, IEEE International Conference On*. How much feedback is multi-user diversity really worth? (Paris, France, 2004)
27. V Hassel, D Gesbert, M-S Alouini, GE Oien, A threshold-based channel state feedback algorithm for modern cellular systems. Wireless Commun. IEEE Trans. **6**(7), 2422–2426 (2007)
28. W Zhang, KB Letaief, MIMO broadcast scheduling with limited feedback. Selected Areas Commun. IEEE J. **25**(7), 1457–1467 (2007)
29. Q Yang, X Li, H Yao, J Fang, K Tan, W Hu, J Zhang, Y Zhang, in *Proc. of ACM SIGCOMM*. Bigstation: enabling scalable real-time signal processing in large MU-MIMO systems (Hong Kong, China, 2013)
30. A Bhartia, Y-C Chen, S Rallapalli, L Qiu, in *ACM MOBICOM*. Harnessing frequency diversity in Wi-Fi networks (Las Vegas, Nevada, USA, 2011)

Joint power and feedback bit allocation for energy- efficient design in limited-feedback coordinated beamforming systems

Fang Yuan* and Chenyang Yang

Abstract

Energy efficiency (EE) has become an important design goal for future cellular systems. In this paper, we study the EE problem for limited-feedback coordinated beamforming systems providing real-time services with constant date rate. To maximize the EE, we jointly optimize the transmit power among users and the feedback bits of each user to quantize the desired and interfering channel directions, subject to the constraints of outage probability and number of feedback bits. An iterative optimization algorithm is proposed for alternating power and bit allocation by employing Yates' framework. To further reduce the computational complexity, a suboptimal algorithm is proposed in closed form by decoupling the problem into subproblems under asymptotical analysis. Simulation results show that the suboptimal algorithm performs closely to the iterative optimization algorithm under high-quantization resolution. Moreover, the proposed two algorithms provide substantial EE gain over existing schemes with equal power and bit allocation, and the EE depends on the target data rate and outage probability of each user as well as the circuit power consumption at the base stations.

1 Introduction

Explosive growth of high-quality wireless services prefigures that energy saving has become a critical issue [1,2]. As a result, it is an urgent task to design energy-efficient communication systems.

Coordinated multi-point transmission with coordinated beamforming (CoMP-CB) can improve spectral efficiency (SE) of cellular networks by avoiding inter-cell interference (ICI), whose performance largely depends on the quality of channel direction information (CDI) available at the base station (BS) [3]. When the CDI cannot be estimated at the BS, e.g., in frequency division duplex systems [4], limited-feedback techniques are often applied, where the CDI is quantized with finite number of bits by each users and fed back to the BS [5,6]. Limited-feedback techniques have been extensively studied for various multi-input multi-output (MIMO) systems [7-9]. In multi-cell systems, equally allocating the bits to quantize the CDIs from multiple BSs is not efficient. Therefore, the bit allocation for quantizing multiple CDIs becomes an important issue in multi-cell systems [5,6,10], which differs from the single-cell systems. Various bit allocation schemes were proposed for CoMP-CB systems to minimize the rate loss led by the quantized CDI [5,6] or to maximize the lower bound of per-user rate [10]. These schemes are designed toward maximizing SE, which do not necessarily provide high energy efficiency (EE) [1].

The EE of downlink system is defined as the ratio of system's rate to the power consumption, where the power consumption includes the circuit power and the transmit power [1]. An EE-oriented design needs to consider the feature of the specific traffic in system. As revealed in [11-13], three major capacity measures for system's rate have been defined in the literature to differentiate real-time services from non-real-time services in the physical layer: ergodic capacity, delay-limited capacity [13,14], and outage capacity [15]. Ergodic capacity is suitable for the best-effort services with no delay constraint, and the EE problems for different systems with ergodic capacity were considered in [1,16]. The delay-limited capacity characterizes the impact of coding length [15] or the

*Correspondence: yuanfang@ee.buaa.edu.cn
School of Electronics and Information Engineering, Beihang University, XueYuan Road No.37, 100191, Beijing, China

queueing length [13], and the EE problems for different systems with delay-limited capacity were considered in [17,18]. In this paper, we study the energy-efficient limited-feedback CoMP-CB systems with real-time services measured by the outage capacity, where each user has a required quality-of-service (QoS) on the data rate and outage probability. As revealed in [12], the deep fades in wireless channels can be detrimental to real-time services in wireless systems. Compared to the other two capacity measures, the outage capacity is more suitable for constant-rate real-time services under a block-fading channel model [15]. In multi-user limited-feedback systems, each user may need different number of bits for feeding back the CDI, since the feedback channel conditions differ from user to user [19]. These motivate to maximize the EE for limited-feedback CoMP-CB under the constraints on the number of feedback bits for each user in addition to the QoS.

For the considered constant-rate systems, by the definition of EE, maximizing the EE amounts to minimizing the total power consumption at the coordinated BSs [20]. When the circuit power is fixed, minimizing the total power consumption is further equivalent to minimizing the total transmit power. Transmit power minimization problems have been extensively studied in literature under perfect CDI [21]. For example, the works in [22,23] studied the power minimization problems under individual signal-to-interference-and-noise ratio (SINR) constraint. Moreover, the work in [24-26] considered the power minimization problems with outage probability constraints. However, all these works are developed assuming perfect CDI, which do not perform well in practical system with quantized CDI. In limited-feedback systems, the power minimization problems become more challenging, because the impact of imperfect CDI on the system performance should be properly evaluated. The imperfect CDI not only results in many difficulties to analyze the system performances, e.g., outage probability [10,27], but also may make the optimization problems nonlinear even in single-cell systems [28]. In [28], a power allocation algorithm is proposed to minimize the total transmit power for limited-feedback multi-user MIMO systems under individual outage probability constraints. Nevertheless, only the inter-user interference (MUI) is considered in single-cell systems and the impact of ICI is ignored. In addition, the proposed algorithm is given by semi-definite programming (SDP), which is of high complexity [29].

As far as the authors know, no previous work has considered the joint power and feedback bit allocation to improve the system's EE for limited-feedback CoMP-CB under constant date rate services. The EE performance of CoMP-CB is severely limited by two types of residual interferences, i.e., MUI and ICI, which results from the inaccurate coordinated beamforming led by imperfect CDI. The residual interferences can be reduced by allocating more feedback bits for quanting the corresponding CDI or less transmit power to the corresponding interferer. Therefore, allocating the feedback bits and transmit powers in an efficient way eventually improves the system's EE. Motivated by this, we jointly optimize the allocation of transmit powers among multiple users and the feedback bits of each user between quantizing the desired and interfering channels under the constraints on individual feedback overhead and outage probability. To find the optimal solution for the problem, we derive an iterative algorithm using Yates' framework [22]. Since high computational complexity will cause high circuit power consumption, a suboptimal algorithm is developed for the system with high-quantization resolution, where both the power and bit allocation are with closed form expressions of low complexity.

The rest of this paper is organized as follows. We first present the system model in Section 2. In Section 3, we formulate the EE maximization problem and find the optimal solution. Then, in Section 4, a low complexity algorithm is derived. Simulation results are provided in Section 5, and the paper is concluded in the last section.

Notations: $(\ )^T$ and $(\ )^H$ are respectively the transpose and transpose-conjugate operation. $|\ |$ denotes the magnitude, $\|\ \|$ is the two-norm, $[x]^+ = \max\{0, x\}$, $\boldsymbol{x} \succ \boldsymbol{y}$ ($\boldsymbol{x} \succeq \boldsymbol{y}$) means the vector $\boldsymbol{x}$ is element-wise larger (no less) than the vector $\boldsymbol{y}$, and $\text{Exp}(c)$ means the exponential distribution with mean of $1/c$.

2 System model

2.1 Signal model

Consider a cluster of N_b BSs each equipped with N_t antennas to cooperatively serve K single antenna users in each cell with CoMP-CB [6].

For the kth user served by BS b, user b_k, the received signal is expressed as

$$y_{b_k} = \alpha_{b,b_k}\boldsymbol{h}_{b,b_k}^H\sqrt{p_{b_k}}\boldsymbol{v}_{b_k}s_{b_k} + \sum_{m=1,m\neq k}^{K}\alpha_{b,b_k}\boldsymbol{h}_{b,b_k}^H\sqrt{p_{b_m}}\boldsymbol{v}_{b_m}s_{b_m} + \sum_{c=1,c\neq b}^{N_b}\sum_{n=1}^{K}\alpha_{c,b_k}\boldsymbol{h}_{c,b_k}^H\sqrt{p_{c_n}}\boldsymbol{v}_{c_n}s_{c_n} + n_{b_k}$$

where p_{b_k}, $\boldsymbol{v}_{b_k}$, s_{b_k}, and n_{b_k} are respectively the transmit power, unit-norm precoder, transmit data, and receiver noise, α_{b,b_k} is the large-scale fading channel gain between BS b and user b_k, and $\boldsymbol{h}_{b,b_k}$ is the small-scale fading channel vector. Without loss of generalities, we assume s_{b_k} and n_{b_k} are with unit variance, and $\boldsymbol{h}_{b,b_k}$ is subject to independent and identically distributed (i.i.d.) Rayleigh fading.

The SINR at user b_k is

$$\mathrm{SINR}_{b_k} = \frac{p_{b_k}\alpha^2_{b,b_k}|\boldsymbol{h}^H_{b,b_k}\boldsymbol{v}_{b_k}|^2}{1+\sum\limits_{m=1,m\neq k}^{K} I_{b_m,b_k} + \sum\limits_{c=1,c\neq b}^{N_b}\sum\limits_{n=1}^{K} I_{c_n,b_k}}$$

where $I_{b_m,b_k} = p_{b_m}\alpha^2_{b,b_k}|\boldsymbol{h}^H_{b,b_k}\boldsymbol{v}_{b_m}|^2$ is the MUI, and $I_{c_n,b_k} = p_{c_n}\alpha^2_{c,b_k}|\boldsymbol{h}^H_{c,b_k}\boldsymbol{v}_{c_n}|^2$ is the ICI. We consider $K > 1$ and $N_b > 1$ such that both MUI and ICI exist.

In the following, we call $\boldsymbol{h}_{b,b_k}$ and $\boldsymbol{h}_{c,b_k}$ respectively as the desired and interfering channels for user b_k ($c \neq b$).

2.2 Limited-feedback transmission model

We assume that the users have perfect CDI after channel estimation. The CDI is quantized through a given codebook with proper size known by both the BSs and users. Then the codebook indices representing the quantized CDIs of desired and interfering channels of each user are fed back to its serving BS in uplink, after which the CDIs are shared among the BSs for coordinated beamforming.

Take the desired channel $\boldsymbol{h}_{b,b_k}$ as an example. Denote the perfect CDI as $\bar{\boldsymbol{h}}_{b,b_k} = \boldsymbol{h}_{b,b_k}/\|\boldsymbol{h}_{b,b_k}\|$ and $\hat{\boldsymbol{h}}_{b,b_k}$ as the quantized CDI with unit-norm. The relation between the perfect CDI and quantized CDI can be modeled as [9]

$$\bar{\boldsymbol{h}}_{b,b_k} = \cos\theta_{b,b_k}\hat{\boldsymbol{h}}_{b,b_k} + \sin\theta_{b,b_k}\boldsymbol{q}_{b,b_k} \tag{1}$$

where $\boldsymbol{q}_{b,b_k}$ is the error vector of unit-norm, and $\cos\theta_{b,b_k} = |\bar{\boldsymbol{h}}^H_{b,b_k}\hat{\boldsymbol{h}}_{b,b_k}|$ reflects the quantization accuracy.

The average CDI distortion led by quantization is defined as $\mathrm{E}\{\sin^2\theta_{b,b_k}\} = 1 - \mathrm{E}\{|\bar{\boldsymbol{h}}^H_{b,b_k}\hat{\boldsymbol{h}}_{b,b_k}|^2\}$ in [8], whose value depends on the employed codebooks, which is intractable for analysis. We therefore consider a quantization cell approximation method used in [9], whose average CDI quantization distortion is

$$\mathrm{E}\left\{\sin^2\theta_{b,b_k}\right\} = \frac{N_t-1}{N_t}2^{-\frac{B_{b,b_k}}{N_t-1}} \tag{2}$$

where B_{b,b_k} is the number of bits for quantizing the CDI of the desired channel $\boldsymbol{h}_{b,b_k}$. It was shown in [9] via simulation that Eq. 2 provides an accurate approximation for other codebooks designed for i.i.d. Rayleigh fading channels.

Similarly, we can obtain the average CDI quantization distortion for the interfering channel $\boldsymbol{h}_{c,b_k}$ as $\mathrm{E}\{\sin^2\theta_{c,b_k}\} = \frac{N_t-1}{N_t}2^{-\frac{B_{c,b_k}}{N_t-1}}$ for $c \neq b$, where B_{c,b_k} is the number of bits for quantizing the CDI of $\boldsymbol{h}_{c,b_k}$.

As in most limited-feedback literature [6,8,19], we consider the zero-forcing beamforming (ZFBF) at the BSs which is low complexity but suboptimal and consider full multiplexing, i.e., $N_t = N_bK$, due to its analytic tractability. Extra transmit antennas at the BSs, i.e., $N_t > N_bK$, do not necessarily provide higher EE for limited-feedback systems. On one hand, the case $N_t > N_bK$ introduces extra circuit powers because of extra radio-frequency (RF) chains at the BSs [30]. On the other hand, the quantization distortion in (2) grows exponentially as the number of antennas increases, which can greatly degrade the user SINR [31] and eventually bring down the system EE. Therefore, in this work, we consider only the case of full multiplexing. pause

Denote $\boldsymbol{H}_{b,\ell} = [\hat{\boldsymbol{h}}_{b,\ell_1}, \cdots, \hat{\boldsymbol{h}}_{b,\ell_K}]$ as the channel matrix from BS b to the users served by BS ℓ ($b, \ell = 1, \ldots, N_b$). Then the ZFBF precoder at BS b for user b_k in the CoMP-CB system, $\boldsymbol{v}_{b_k}$, is obtained from normalizing the kth column of $(\mathcal{H}^H_b\mathcal{H}_b)^{-1}\mathcal{H}_b$ where $\mathcal{H}_b = [\boldsymbol{H}_{b,b}, \boldsymbol{H}_{b,1}, \cdots, \boldsymbol{H}_{b,b-1}, \boldsymbol{H}_{b,b+1} \cdots, \boldsymbol{H}_{b,N_b}]$.

3 EE optimization for limited-feedback CoMP-CB system

In this section, we study the bit and power allocation problem to maximize the downlink EE of the CoMP-CB system, where the total number of feedback bits of each user is given. We consider the system providing real-time services under a block fading channel, where each user requires a constant data rate. Since the data rate requirement of each user may not be supported under fading channel with imperfect CDI, to guarantee the QoS, a constraint on outage probability is considered, which is assumed less than a target value.

3.1 Problem formulation

Denote γ_{b_k} as the targeted SINR of user b_k for the constant data rate in downlink. Then, the outage probability of user b_k is defined as

$$\mathrm{P}_{\mathrm{out}}(\gamma_{b_k}) = \Pr\{\mathrm{SINR}_{b_k} < \gamma_{b_k}\}.$$

Considering the outage probability is required less than a target value O_{b_k}, i.e., $\mathrm{P}_{\mathrm{out}}(\gamma_{b_k}) \leq O_{b_k}$, the required data rate by user b_k, i.e., outage capacity, is given as

$$R_{b_k} = \log(1+\gamma_{b_k})(1-O_{b_k}).$$

The EE of downlink CoMP-CB system is defined as the ratio of the system's sum rate to the total power consumption in downlink [32,33], which is,

$$EE = \frac{\sum_{b=1}^{N_b}\sum_{k=1}^{K} R_{b_k}}{N_bp_c + \sum_{b=1}^{N_b}\sum_{k=1}^{K} p_{b_k}} \tag{3}$$

where p_c is the circuit power at each BS. The modeling of circuit power can be found in [34], which includes the power consumed by RF chain, signal processing, cooling, etc. at each BS as well as that by the backhaul links for channel sharing.

With zero-forcing beamforming (ZFBF), it was shown in [35] that the interference powers $|\boldsymbol{h}^H_{b,b_k}\boldsymbol{v}_{b_m}|^2$ and $|\boldsymbol{h}^H_{c,b_k}\boldsymbol{v}_{c_n}|^2$ are exponentially distributed respectively with probability density distribution (PDF) of $\mathrm{Exp}(1/\delta_{b,b_k})$ and $\mathrm{Exp}(1/\delta_{c,b_k})$, where their expectations are

$$\delta_{b,b_k} = 2^{\frac{-B_{b,b_k}}{N_t-1}} \text{ and } \delta_{c,b_k} = 2^{\frac{-B_{c,b_k}}{N_t-1}}. \tag{4}$$

Moreover, when $N_b K = N_t$, the precoder $\boldsymbol{v}_{b_k}$ is independent with $\boldsymbol{h}_{b,b_k}$ and the variable $|\boldsymbol{h}_{b,b_k}^H \boldsymbol{v}_{b_k}|^2$ is exponentially distributed with PDF of Exp(1) [9]. Then, the outage probability of user b_k under given transmit powers is [10]

$$\mathrm{P_{out}}(\gamma_{b_k}) = 1 - \frac{e^{-\frac{\gamma_{b_k}}{\alpha_{b,b_k}^2 p_{b_k}}}}{\prod_{m=1,m\neq k}^{K}\left(1+\frac{\gamma_{b_k}\delta_{b,b_k}p_{b_m}}{p_{b_k}}\right)\prod_{c=1,c\neq b}^{N_b}\prod_{n=1}^{K}\left(1+\frac{\gamma_{b_k}\delta_{c,b_k}\alpha_{c,b_k}^2 p_{c_n}}{\alpha_{b,b_k}^2 p_{b_k}}\right)} \tag{5}$$

which depends on the transmit power, average CDI quantization distortion and target SINR of the user.

By observing (5), we can find that limited feedback has a large impact on the outage probability. The impact of MUI and ICI is conveyed by the denominator of (5). Given the transmit powers, the larger average CDI distortion δ_{b,b_k} or δ_{c,b_k} is, the larger MUI or ICI the user b_k experiences, and the larger the outage probability becomes. In contrast, under perfect CDI, the MUI and ICI become zero and the denominator of (5) is unit, which means the bit allocation is not necessary.

To maximize the EE of downlink CoMP-CB, the problem of jointly allocating the transmit powers among multiple users and the total number of bits of each user between quantizing its desired and interfering CDIs can be formulated as

$$\max_{p_{b_k}, \boldsymbol{B}_{b_k}} \quad EE \tag{6}$$

$$\text{s.t.} \quad \mathrm{P_{out}}(\gamma_{b_k}) \le O_{b_k}, \quad \forall b_k \tag{6a}$$

$$\boldsymbol{1}^T \boldsymbol{B}_{b_k} \le B_{b_k}, \quad \forall b_k \tag{6b}$$

$$p_{b_k} > 0, \quad \forall b_k \tag{6c}$$

$$\boldsymbol{B}_{b_k} \succeq \boldsymbol{0}, \quad \forall\, b_k \tag{6d}$$

where $\boldsymbol{B}_{b_k} \triangleq [B_{1,b_k}, B_{2,b_k}, \cdots, B_{N_b,b_k}]^T$ is the bit allocation profile of user b_k for quantizing the CDIs from desired and interfering BSs, B_{b_k} is the total number of feedback bits of user b_k, $\boldsymbol{1}$ and $\boldsymbol{0}$ are respectively the all-one and all-zero vectors. The QoS requirement parameters γ_{b_k} and O_{b_k} are constant and pre-specified by the system. In the constraint (6b), as shown in (4), allocating more bits for B_{b,b_k} in $\boldsymbol{B}_{b_k}$ can lower down the average MUI, while allocating more bits for B_{c,b_k} $(c \neq b)$ in $\boldsymbol{B}_{b_k}$ can lower down the cth average ICI. Therefore, the bit allocation in (6) is to improve the system EE by considering both the impact of MUI and ICI.

The optimization problem in (6) does not associate with the instantaneous channel information. As a result, the power and bit allocation does not need a frequent design, which will not causing too much signaling overhead. Nonetheless, if we consider the maximum transmit power constraint at each BS, the problem will become infeasible when the target date rate is high, even with perfect CDI. To demonstrate the EE potential of the CoMP-CB system and differentiate the infeasibility led by the transmit power and imperfect CDI, we do not take into account the maximum transmit power constraint.

3.2 Optimal solution of joint power and bit allocation

Since each user's date rate R_{b_k} is constant in the numerator of EE and the circuit power does not depend on the variables to be optimized, problem (6) is equivalent to the following joint power and bit allocation problem that minimizes the total transmit power

$$\min_{p_{b_k}, \boldsymbol{B}_{b_k}} \quad \sum_{b=1}^{N_b} \sum_{k=1}^{K} p_{b_k} \tag{7}$$

$$\text{s.t.} \quad (6a),(6b),(6c),(6d).$$

It should be noted that although the equivalence between the problems (6) and (7) is applied here to reduce the complexity of optimization problem, the work in this paper is motivated by improving the system EE as defined in (3). Moreover, to show the impact of proposed algorithms for joint power and bit allocation, and the other impacts such as circuit power and outage probability, the system EE in (3) rather than total transmit power in (7) will be mainly evaluated in the simulation section.

The optimization problem in (7) is nonlinear. To see this, from (5), we can rewrite the constraint (6a) as

$$p_{b_k} \ge \left(\frac{\gamma_{b_k}}{\alpha_{b,b_k}^2} + \sum_{m=1,m\neq k}^{K} p_{b_k} \log\left(1+\frac{\gamma_{b_k}\delta_{b,b_k}p_{b_m}}{p_{b_k}}\right) + \sum_{c=1,c\neq b}^{N_b}\sum_{n=1}^{K} p_{b_k}\log\left(1+\frac{\gamma_{b_k}\delta_{c,b_k}\alpha_{c,b_k}^2 p_{c_n}}{\alpha_{b,b_k}^2 p_{b_k}}\right)\right)\frac{1}{\log\frac{1}{1-O_{b_k}}} \tag{8}$$

$$\triangleq \boldsymbol{T}_{b_k}(\boldsymbol{p}, \boldsymbol{B}_{b_k}) \tag{9}$$

where $\boldsymbol{p} \triangleq [p_{1_1}, \cdots, p_{1_K}, \cdots, p_{N_{b1}}, \cdots, p_{N_{bK}}]^T$ denotes the power profile for all KN_b users. All the variables in (8), including the transmit powers and the bits conveyed in average CDI distortions, are nonlinear in the constraint.

Solving the problem (7) is quite challenging due to its nonlinearity. Instead of solving it directly, here, we study an equivalent problem for (7) and reveal that the nonlinear optimization problem can be solved by using a standard iteration algorithm.

The nonlinear optimization problem (7) is equivalent to the following problem

$$\min_{p_{b_k}} \quad \sum_{b=1}^{N_b} \sum_{k=1}^{K} p_{b_k} \tag{10}$$

$$\text{s.t.} \quad p_{b_k} \geq \boldsymbol{T}_{b_k}(\boldsymbol{p}), \quad \forall b_k \tag{10a}$$

$$p_{b_k} > 0, \quad \forall b_k$$

where the function $\boldsymbol{T}_{b_k}(\boldsymbol{p})$ is defined through the optimization of bit allocation

$$\boldsymbol{T}_{b_k}(\boldsymbol{p}) \triangleq \min_{\boldsymbol{1}^T\boldsymbol{B}_{b_k} \leq B_{b_k}, \boldsymbol{B}_{b_k} \succeq 0} \boldsymbol{T}_{b_k}(\boldsymbol{p}, \boldsymbol{B}_{b_k}). \tag{11}$$

The equivalence comes from the fact that the bit allocation profile $\boldsymbol{B}_{b_k}$ of user b_k only affects $\boldsymbol{T}_{b_k}(\boldsymbol{p}, \boldsymbol{B}_{b_k})$. Considering the inequality (8), the minimum value of p_{b_k} is the one satisfying the constraint in (10a) among all the possible values of $\boldsymbol{T}_{b_k}(\boldsymbol{p}, \boldsymbol{B}_{b_k})$ with different bit allocation profile $\boldsymbol{B}_{b_k}$ for user b_k. Therefore, the two problems are equivalent in the sense that they have the same globally optimal solution.

The optimal solution to problem (10) can be found by alternatively optimizing the bit and power allocation.

3.2.1 Bit allocation with given power allocation

In each iteration with given transmit powers, the bit allocation for quantizing the desired and interfering CDIs of user b_k can be found from the following problem (i.e., (11)),

$$\min_{\boldsymbol{B}_{b_k}} \quad \boldsymbol{T}_{b_k}(\boldsymbol{p}, \boldsymbol{B}_{b_k})$$

$$\text{s.t.} \quad \boldsymbol{1}^T\boldsymbol{B}_{b_k} \leq B_{b_k}, \boldsymbol{B}_{b_k} \succeq 0 \tag{12}$$

It is easy to verify that this problem is convex on the $\boldsymbol{B}_{b_k}$. Therefore, the optimal solution can be obtained by using standard convex optimization tools [29].

3.2.2 Power allocation with given bit allocation

When the bit allocation for each user is obtained, the power allocation for multiple users can be optimized by using Yates' framework [22], which is a fixed-point iteration problem. To show the convergence of the proposed iterative algorithm, we provide the following proposition.

Proposition 1. *$\boldsymbol{T}_{b_k}(\boldsymbol{p})$ is a standard interference function of the vector $\boldsymbol{p}$, i.e., it satisfies the following three properties:*

1. *Positivity: $\boldsymbol{T}_{b_k}(\boldsymbol{p}) > 0$.*
2. *Monotonicity: $\boldsymbol{T}_{b_k}(\boldsymbol{p})$ is monotonic with each element in $\boldsymbol{p}$.*
3. *Scalability: $\boldsymbol{T}_{b_k}(\tau\boldsymbol{p}) < \tau\boldsymbol{T}_{b_k}(\boldsymbol{p})$ for any constant $\tau > 1$.*

The proposition can be proved using a similar method as that in [25].

Yates found that for any function used for iteration, if the function satisfies all the properties of a standard interference function [22], the iteration will converge to the globally optimal solution. Using Yates' framework, we can update the transmit powers with the solution of bit allocation obtained from problem (12) as

$$p_{b_k} = \boldsymbol{T}_{b_k}(\boldsymbol{p}) \tag{13}$$

The iterative algorithm to solve problem (10) is summarized in Algorithm 1, which converges to optimal solution and provides a performance baseline. However, the iteration in the algorithm increases the computational complexity for signal processing and thus may increase the circuit power consumption. Therefore, we further propose a non-iterative algorithm with lower complexity in the sequel.

Algorithm 1 Iterative algorithm for finding optimal solution

Initialize: $i = 0, \boldsymbol{p}^0_{b_k}, \boldsymbol{B}^0_{b_k} \succeq 0$

Iterate for each user b_k

1. update $i = i + 1$;
2. update bit allocation $\boldsymbol{B}^i_{b_k}$ for user b_k by solving problem (12) with given $\boldsymbol{p}^{i-1}_{b_k}$;
3. update power allocation $\boldsymbol{p}^i_{b_k}$ by (13) with given $\boldsymbol{B}^i_{b_k}$ and $\boldsymbol{p}^{i-1}_{b_k}$;

Terminate until convergence or reaching maximum number of iterations.

4 Low complexity algorithm

In this section, we propose a suboptimal but low complexity algorithm to solve problem (10). By using asymptotical analysis and introducing upper bound, we decouple the joint bit and power allocation problem into two subproblems: 1) optimizing power allocation with given bit allocation, and 2) optimizing bit allocation with optimized power allocation. The two subproblems are solved without the need for iteration.

4.1 Power allocation with given bit allocation

We first optimize the power allocation with given bit allocation for each user.

Using the inequality (14) in [36], we can obtain a upper bound for outage probability in (5) as

$$\mathrm{P}_{\mathrm{out}}(\gamma_{b_k}) \leq 1 - e^{-\frac{\gamma_{b_k}}{p_{b_k}} \boldsymbol{T}^r_{b_k}(\boldsymbol{p}, \boldsymbol{B}_{b_k})} \tag{14}$$

where the function $\boldsymbol{T}^r_{b_k}(\boldsymbol{p}, \boldsymbol{B}_{b_k})$ is written as

$$\boldsymbol{T}^r_{b_k}(\boldsymbol{p}, \boldsymbol{B}_{b_k}) = \frac{1}{\alpha^2_{b,b_k}} + \delta_{b,b_k} \sum_{m=1, m\neq k}^{K} p_{b_m} + \sum_{c=1, c\neq b}^{N_b} \sum_{n=1}^{K} \frac{\alpha^2_{c,b_k}\delta_{c,b_k}}{\alpha^2_{b,b_k}} p_{c_n}. \tag{15}$$

With this upper bound, we formulate a power allocation problem with given bit allocation as

$$\min_{p_{b_k}} \quad \sum_{b=1}^{N_b} \sum_{k=1}^{K} p_{b_k} \tag{16}$$

$$\text{s.t.} \quad 1 - e^{-\frac{\gamma_{b_k}}{p_{b_k}} T^r_{b_k}(\boldsymbol{p}, \boldsymbol{B}_{b_k})} \le O_{b_k}, \quad \forall b_k \tag{16a}$$

$$p_{b_k} > 0, \quad \forall b_k \tag{16b}$$

whose solution is suboptimal for the original problem (10). Specially, we can rewrite the constraint (16a) as

$$p_{b_k} \ge \gamma^r_{b_k} T^r_{b_k}(\boldsymbol{p}, \boldsymbol{B}_{b_k}) \tag{17}$$

where $\gamma^r_{b_k} \triangleq \frac{-\gamma_{b_k}}{\log(1-O_{b_k})}$.

Now both the objective function and constraints in (16) are expressed as linear functions of the optimization variables, i.e., the elements in $\boldsymbol{p}$. Define a constant matrix $\boldsymbol{A}$ of size $KN_b \times KN_b$ with the (i,j)th element as

$$\boldsymbol{A}_{i,j} = \begin{cases} 0, & \text{if } j = i \\ \gamma^r_{b_k} \delta_{b,b_k}, & \text{if } K(b-1) < j \le Kb, j \ne i \\ \frac{\gamma^r_{h_k} \alpha^2_{c,h_k} \delta_{c,b_k}}{\alpha^2_{b,b_k}}, & \text{if } j \le K(b-1), \text{or } j > Kb \end{cases} \tag{18}$$

and a constant vector $\boldsymbol{w}$ of size $KN_b \times 1$ with the ith element as

$$\boldsymbol{w}_i = \frac{\gamma^r_{b_k}}{\alpha^2_{b,b_k}} \tag{19}$$

where $c \ne b$ and the index $i = (b-1)K+k, b = 1, \cdots, N_b$, $k = 1, \cdots, K$. The matrix $\boldsymbol{A}$ describes the coefficients of the variables (e.g., transmit powers) in the constraint (17). The vector $\boldsymbol{w}$ describes the constants irrespective of variables in the constraint (17). Therefore, with the upper bound in (14), problem (16) becomes a linear programming problem for the transmit powers, which has a much lower complexity than the problem (7).

With the definition of $\boldsymbol{A}$ and $\boldsymbol{w}$, we can immediately write the problem (16) into a standard linear programming problem, which is

$$\min_{\boldsymbol{p}} \quad \boldsymbol{1}^T \boldsymbol{p}$$
$$\text{s.t.} \quad \boldsymbol{p} \succeq \boldsymbol{A}\boldsymbol{p} + \boldsymbol{w}, \text{ and } \boldsymbol{p} \succ \boldsymbol{0}. \tag{20}$$

From [29], the optimal solution of problem (20) can be obtained with closed form as

$$\boldsymbol{p} = (\boldsymbol{I} - \boldsymbol{A})^{-1} \boldsymbol{w} \tag{21}$$

and the corresponding minimum sum transmit power of all the BSs becomes

$$p_S = \boldsymbol{1}^T (\boldsymbol{I} - \boldsymbol{A})^{-1} \boldsymbol{w}. \tag{22}$$

There are two benefits of the solution in (21). One is that linear programming is desirable for low complexity in computation. The other is that the solution is of explicit form, which is desirable for the performance analysis in the following.

4.2 Bit allocation with optimized power allocation

We then optimize the bit allocation between the desired and interfering CDIs to further reduce the minimum total transmit power p_S. Unfortunately, the optimal bit allocation that minimizes p_S is hard to find because the expression of p_S in (22) is complicated and nonlinear with the CDI quantization distortion. In the sequel, we study the bit allocation problem resorting to asymptotic analysis.

Proposition 2. *The minimum total transmit power in (22) satisfies*

$$\lim_{\delta \to 0} \frac{p_S}{p_{\mathrm{CDI}} + \sum_{b=1}^{N_b} \sum_{k=1}^{K} \frac{\gamma^r_{b_k}}{\alpha^2_{b,b_k}} \Delta_{b_k}} = 1$$

where $\delta = \max_{\ell,b_k} \delta_{\ell,b_k}$, $p_{\mathrm{CDI}} = \sum_{b=1}^{N_b} p_{s,b}$ *with* $p_{s,b} = \sum_{m=1}^{K} \frac{\gamma^r_{bm}}{\alpha^2_{b,bm}}$, $\Delta_{b_k} = \sum_{\ell=1}^{N_b} g_{s,\ell} \alpha^2_{\ell,b_k} \delta_{\ell,b_k}$ *with*

$$g_{s,\ell} = \begin{cases} p_{s,b} - \frac{\gamma^r_{b_k}}{\alpha^2_{b,b_k}}, & \ell = b \\ p_{s,\ell}, & \ell \ne b \end{cases} \tag{23}$$

Proof: When $\delta \to 0$, it can be verified that $\|\boldsymbol{A}\| \to 0$, using the Taylor series, we obtain the equivalent infinitesimal of $(\boldsymbol{I} - \boldsymbol{A})^{-1}$ from the first order expansion as [37]

$$(\boldsymbol{I} - \boldsymbol{A})^{-1} \approx \boldsymbol{I} + \boldsymbol{A}. \tag{24}$$

Using the equivalent infinitesimal replacement in the limit operation and the definition of $\boldsymbol{A}$ and $\boldsymbol{w}$, we have

$$\lim_{\delta \to 0} \frac{p_S}{\boldsymbol{1}^T (\boldsymbol{I} - \boldsymbol{A})^{-1} \boldsymbol{w}} = \lim_{\delta \to 0} \frac{p_S}{\boldsymbol{1}^T (\boldsymbol{I} + \boldsymbol{A}) \boldsymbol{w}}$$
$$= \lim_{\delta \to 0} \frac{p_S}{p_{\mathrm{CDI}} + \sum_{b=1}^{N_b} \sum_{k=1}^{K} \frac{\gamma^r_{b_k}}{\alpha^2_{b,b_k}} \Delta_{b_k}}. \tag{25}$$

Now, we complete the proof for the proposition. ■

The condition $\delta \to 0$ requires that the system is with high-quantization resolution for each CDI. With perfect CDI, it is easy to show that the minimum total transmit power is $p_{\mathrm{CDI}} = \sum_{b=1}^{N_b} p_{s,b}$. When the CDI is imperfect, the transmit power will increase in order to achieve the same target SINR and outage probability under the residual MUI and ICI, which leads to the term $\sum_{b=1}^{N_b} \sum_{k=1}^{K} \frac{\gamma^r_{b_k}}{\alpha^2_{b,b_k}} \Delta_{b_k}$ in the dominator in (25). This fraction of transmit power reduces with the improved CDI quantization accuracy, and thereby can be minimized by the bit allocation.

According to Proposition 2, to improve the EE for the system with high-quantization resolution, we only need to minimize $\sum_{b=1}^{N_b}\sum_{k=1}^{K}\frac{\gamma_{b_k}^r}{\alpha_{b,b_k}^2}\Delta_{b_k}$. Considering the individual constraint imposed on the number of bits for each user, and considering that only Δ_{b_k} is related with the bit allocation of user b_k, the power allocation problem can be formulated as

$$\min_{\boldsymbol{B}_{b_k}} \Delta_{b_k} = \sum_{\ell=1}^{N_b} g_{s,\ell}\alpha_{\ell,b_k}^2\delta_{\ell,b_k} \quad (26)$$
$$\text{s.t.} \quad \mathbf{1}^T\boldsymbol{B}_{b_k} \le B_{b_k}, \text{ and } \boldsymbol{B}_{b_k} \succeq 0.$$

This is a convex optimization problem. First, the constraints are linear and thus convex. Second, the objective function is a weighted sum of exponential functions with positive weights, which is also convex. The solution can be found from the Karush-Kuhn-Tucker (KKT) conditions, which turns into a water-filling algorithm with the unique solution as follows:

$$B_{\ell,b_k} = (N_t-1)[\log_2 g_{s,\ell} + \log_2 \alpha_{\ell,b_k}^2 - \log_2 \nu(N_t-1)]^+ \quad (27)$$

where $\nu > 0$ should satisfy $\sum_{\ell=1}^{N_b} B_{\ell,b_k} = B_{b_k}$.

The result in (27) shows that the allocated number of bits for quantizing the CDI from the ℓth BS to user b_k not only increases with the large-scale fading gain α_{ℓ,b_k}^2 but also grows with the parameter $g_{s,\ell}$ defined in (23), which is associated with the QoS requirements (i.e., γ_{ℓ_k} and O_{ℓ_k}) for the users served in the ℓth cell.

To illustrate this, we consider the case that the QoS requirements for users in the same cell are identical, i.e., $\gamma_{\ell_k}^r = \gamma_\ell^r$ for all k, then (23) reduces to

$$g_{s,\ell} = \begin{cases} \gamma_b^r\left(\sum_{m=1}^{K}\frac{1}{\alpha_{b,bm}^2} - \frac{1}{\alpha_{b,b_k}^2}\right), & \ell = b \\ \gamma_\ell^r \sum_{m=1}^{K}\frac{1}{\alpha_{\ell,\ell m}^2}, & \ell \ne b \end{cases}$$

It means that the parameters $g_{s,\ell}$ are linear with its own QoS requirement γ_ℓ^r in the ℓth cell. Consequently, we can find that (27) tends to allocate more bits to the cell with larger QoS requirement. By contrast, the bit allocation algorithm for improving SE proposed in [5,6] only exploits the receive powers in the desired and interfering channels but does not consider the different QoS requirements, and thus cannot be applied in this work.

For general setting of QoS requirement $\gamma_{\ell_k}^r$, the parameter $g_{s,\ell}$ measures the interference condition from the ℓth BS to the user b_k quantitatively. Observe $g_{s,\ell}$ defined in (23): when $\ell = b$, it equals to $g_{s,b} = \sum_{m=1,m\ne k}^{K}\frac{\gamma_{bm}^r}{\alpha_{b,bm}^2}$, which is a sum of $K-1$ items coming from the $K-1$ MUIs experienced by user b_k from the bth BS; when $\ell \ne b$, it equals to $g_{s,\ell} = \sum_{m=1}^{K}\frac{\gamma_{\ell m}^r}{\alpha_{\ell,\ell m}^2}$, which is a sum of K items coming from the ICI of K users served by the ℓth BS. Therefore, both impacts of MUI and ICI are reflected in the bit allocation (27).

The low complexity algorithm is summarized in Algorithm 2, where the power allocation (21) depends on the bit allocation algorithm (27). It will be shown in the simulation section that under high-quantization resolution, the performance of suboptimal algorithm is close to that of optimal algorithm.

Algorithm 2 Low complexity algorithm

For each user b_k,

1. bit allocation: obtain $\boldsymbol{B}_{b_k}$ with (27);
2. power allocation with the bit allocation in (27): obtain p_{b_k} with (21).

4.3 Feasibility of the proposed algorithms

The bit and power allocation algorithms are derived under the implicit assumption that the optimal problems in (10) and (20) are feasible, i.e., the allocated powers are neither negative nor infinity. However, as indicated by [22], in some cases with strong interference, the optimal solutions are not feasible.

A sufficient and necessary condition for the feasibility of problem (20) is that the maximum eigenvalue of $\boldsymbol{A}$ should be less than one [22]. Therefore, the solution in (21) is applicable only when the maximum eigenvalue of $\boldsymbol{A}$ is less than one.

This is also a sufficient condition for the feasibility of problem (10) (equivalently, original problem (6)), since the solution of problem (20) is an upper bound of the solution of problem (10). The exact sufficient and necessary condition for problem (10) is generally unknown. As a result, in some cases the suboptimal power allocation in (21) may not be feasible when the optimal power allocation in (13) is feasible.

5 Simulation results

In this section, we evaluate the performance of the proposed algorithms through simulations.

5.1 Simulation setup

We consider a CoMP-CB system with a cooperative cluster size of three, i.e., $N_b = 3$. The inter-cluster interference is modeled as noise, which is reflected in the noise power set as -104 dBm. The cell radius is 250 m, the path loss factor is 3.76, and the downlink transmission bandwidth is 10 Mhz. The small-scaling fading channels are subject to i.i.d. Rayleigh flat fading. Random vector quantization (RVQ) codebooks are applied for quantizing each CDI.

We assume there are two users in each cell, i.e., $K = 2$, and the number of antennas at each BS is six, i.e.,

$N_t = 6$. It means each user experiences one MUI interference from the other user in the same cell and four ICI interferences from the users in two neighboring cells.

The system topology is shown in Figure 1, where two scenarios about user locations are studied. The first scenario is with fixed user location, which is applied in Figures 2, 3, 4, 5, 6, 7. In this scenario, two users in each cell are located on the line between their serving BS and the point O and with a distance d to their serving BS and a distance d_I to the other two interfering BSs. Moreover, the users in three cells are symmetrically distributed. The major benefit of this scenario is that we only need to employ one single parameter d to reflect user location, which determines the large-scale channel gains of desired and interfering channels. When changing the distance d to the serving BS, the large-scale channel gains of desired and interfering channels are changed simultaneously which are unequal except for the user location $d = 250$ m. The different impacts of desired and interfering channels are reflected by the heterogeneous large-scale channel gains from desired and interfering BSs, which are exploited in the power and bit allocation by the two proposed algorithms. The second scenario is with random user locations, which is applied in Figure 8 to show the impact of general user locations. In the second scenario, each user is randomly dropped in the CoMP region shown in Figure 1. The CoMP region is defined as the cell-edge region, where the inter-cell interference is strong and coordinated beamforming is necessary.

To reflect different data rate requirements of each user, we set the data rate constraint in the three cells as $R_{1_1} = R_{1_2} = \rho_1 R(1-O), R_{2_1} = R_{2_2} = \rho_2 R(1-O)$, and $R_{3_1} = R_{3_2} = \rho_3 R(1-O)$ bps/hz, where $\sum_{i=1}^{3} \rho_i = 1$. Then, the total date rate of all the six users is $2R(1-O)$. Unless otherwise specified, the target outage probability for each user is $O = 0.1$.

Similar simulation results can be observed for more complex cases with arbitrary size of cluster, random located users and arbitrary data rate requirement, which are not shown for conciseness.

For simplicity, in the following, we refer to the optimal power and bit allocation obtained from the iterative algorithm in Algorithm 1 as 'OPA+OBA', and the suboptimal power and bit allocation obtained from the low complexity algorithm in Algorithm 2 as 'SPA+SBA', both in the main text and in the legend. The obtained bit allocation results in the optimization problems are rounded into nearest integers as the number of allocated bits.

We assume that the total number of feedback bits for each user is conveyed via N_s slots in uplink. The number of feedback bits in each slot is adaptive to the average uplink signal-to-noise ratio (SNR), which is determined by the constraint $\mathrm{P}_r \left\{ \log_2(1 + \mathrm{SNR}^u_{b_k}) \le \frac{B_{b_k}}{N_s} \right\} \le 0.1$, where $\mathrm{SNR}^u_{b_k}$ is the instantaneous receive SNR at the BS b from user b_k. We set $N_s = 3(N_t - 1)$ in Subsection 5.2 and the impact of the value of N_s on the uplink and downlink energy consumptions will be investigated in Subsection 5.3. It is worthy to note that when the value of N_s changes, the total number of feedback bits for each user changes.

Denote uplink transmit power as p^u, which is set as 26 dBm (approximately 400 mw). Let $\boldsymbol{h}^u_{b,b_k}$ be the instantaneous uplink small-scale fading channel vector. Then, the uplink receiving SNR is $\mathrm{SNR}^u_{b_k} = p^u \alpha^2_{b,b_k} \|\boldsymbol{h}^u_{b,b_k}\|^2$. The large-scale fading gains of the uplink and downlink channels are equal and the small-scaling fading uplink channels are subject to i.i.d. Rayleigh flat fading and independent with the downlink channels.

5.2 Validation of upper bound and asymptotic analysis

To validate the upper bound and asymptotic analysis, we compare the total transmit power at all coordinated BSs obtained from the iterative algorithm in Algorithm 1, the low complexity algorithm in Algorithm 2, and the asymptotical analysis in Proposition 2 using the bit allocation in (27).

As shown in Figure 2, the low complexity algorithm performs close to the iterative optimization algorithm as the number of feedback bits for each user increases, which indicates that using the upper bound of outage probability in (14) leads to minor performance loss. Moreover, as the number of bits increases, the performance of both algorithms converge to the limit result of asymptotic analysis in Proposition 2.

5.3 Performance evaluation of the proposed algorithms

5.3.1 Convergence of the optimal algorithm

Figure 3 shows the total transmit power obtained from the iterative algorithm in Algorithm 1 and the low complexity algorithm in Algorithm 2 versus the number of iterations. It is shown that the total transmit power obtained by the iterative algorithm converges to the optimality rapidly in the iteration, and the low complexity algorithm performs close to the iterative algorithm. Moreover, the total transmit power increases with the data rate requirement.

5.3.2 Comparison with other schemes

The EE achieved by the proposed iterative algorithm is shown in Figure 4 versus the system's rate, where the system's rate is given as the numerator of the EE defined in (3). Considering that there are no comparable methods available in the literature, we compare with equal bit allocation (EBA) between the desirable and interfering CDIs for each user and equal power allocation (EPA) among multiple users, which are widely applied in prevalent limited feedback systems. To implement the scheme of 'EBA+EPA', we solve optimization problem (6) under two additional constraints: $p_{b_k} = p$ and $B_{\ell,b_k} = B_{b_k}/N_b$

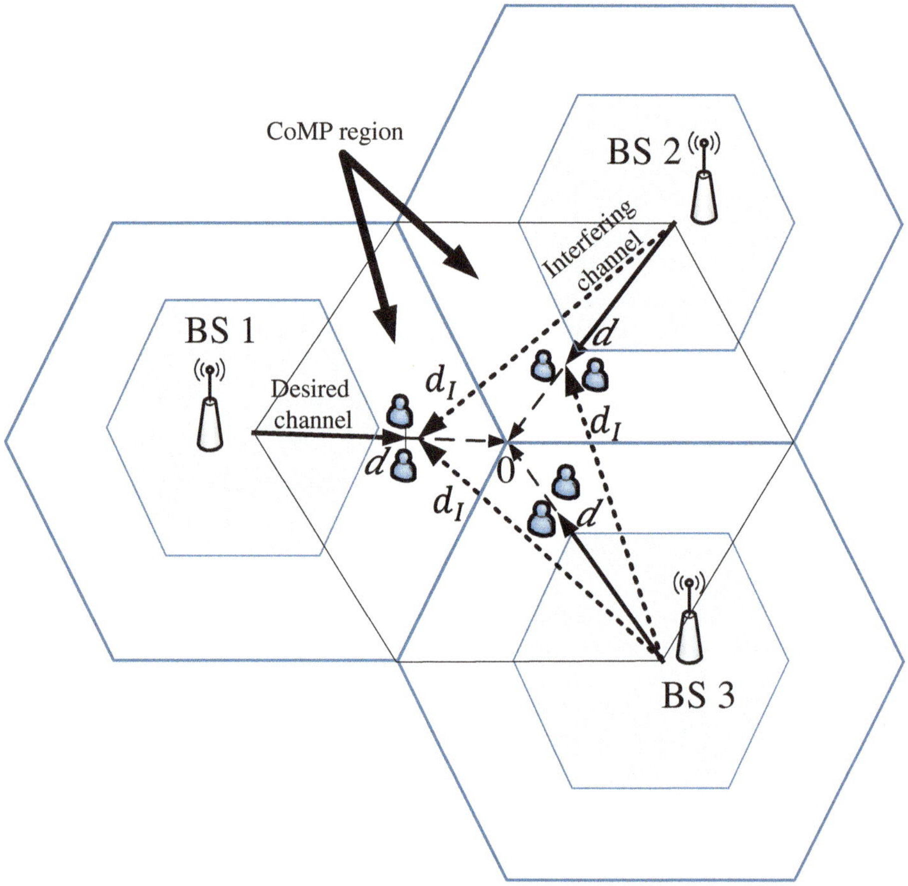

Figure 1 Topology of the simulated CoMP-CB systems.

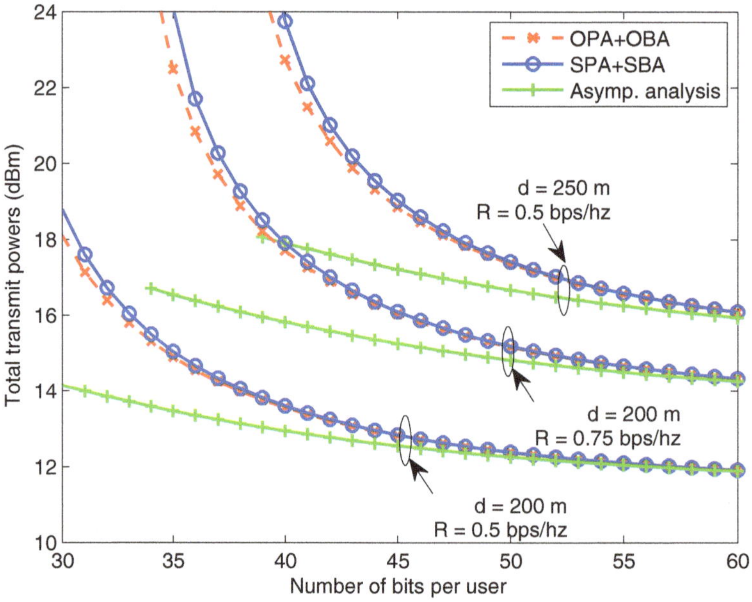

Figure 2 Total transmit power versus the number of bits for the CDI feedback of each user.

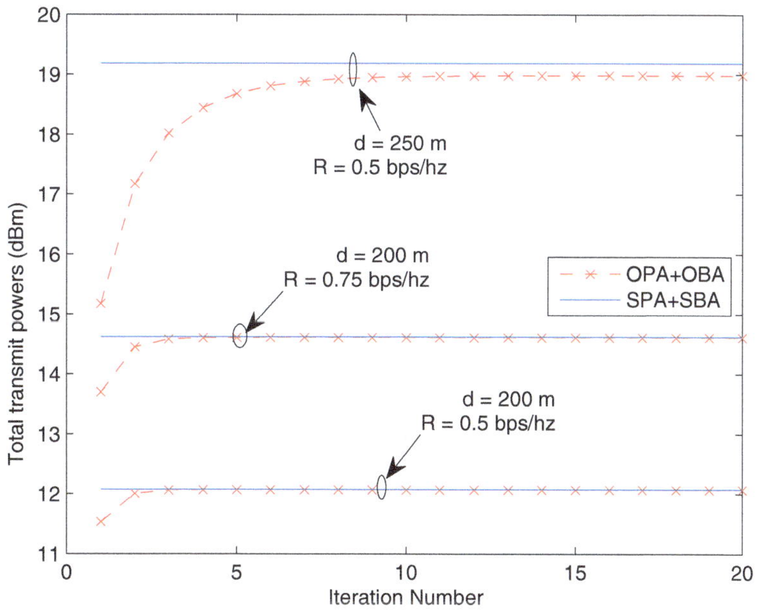

Figure 3 Convergence of the iterative algorithm.

for user b_k ($\ell = 1, 2, 3$). To implement the scheme of 'OPA+EBA,' we fix the equal bit allocation in the algorithm of Algorithm 1, and iterate the transmit powers only. The scheme of 'OPA+EBA' provides the optimal solution for power allocation when the bit allocation is fixed, which can serve as the performance upper bound for any other power allocation algorithms [21,25].

The simulation results show that with equal bit allocation, the optimal power allocation can remarkably improve EE in spite of user locations. Moreover, the EE can be further improved by using the optimal bit allocation. From $d = 250$ to $d = 200$ m, the EE gain of the optimal bit allocation over equal bit allocation increases significantly, since it is more beneficial to use

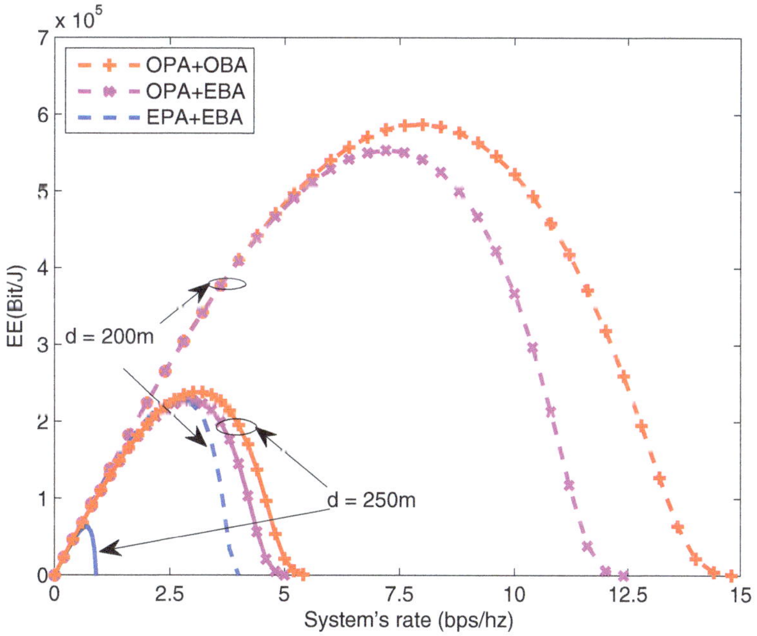

Figure 4 EE under different user locations. $p_c = 50$ W [38], $\rho_1 = 1/2$, $\rho_2 = \rho_3 = \frac{1}{4}$.

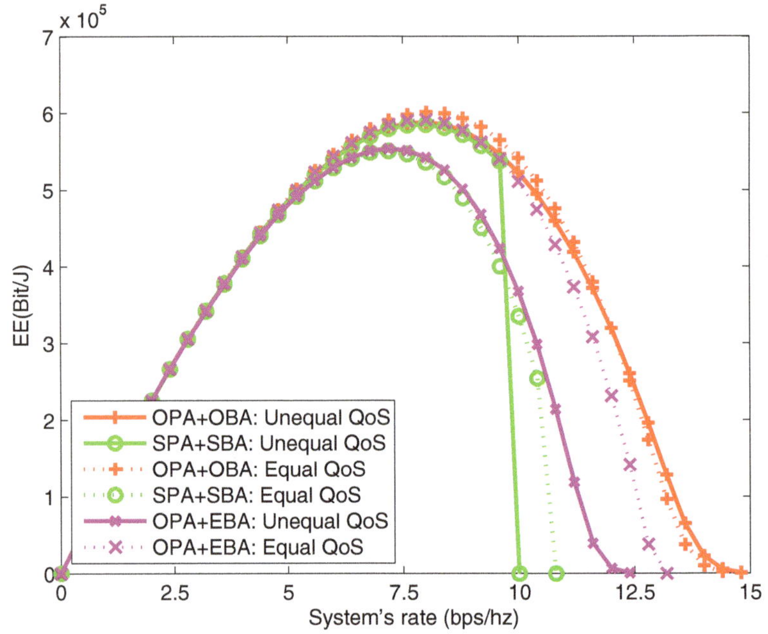

Figure 5 EE of optimal and suboptimal algorithms, $p_c = 50$ W, $d = 200$ m. Equal QoS: $\rho_1 = \rho_2 = \rho_3 = 1/3$; unequal QoS: $\rho_1 = 1/2$, $\rho_2 = \rho_3 = \frac{1}{4}$.

the bit allocation for exploiting the heterogeneity of large-scale channel gains from desired and interfering BSs. The results highlight the importance of joint power and bit allocation to improve the EE in limited feedback CoMP system.

The EE achieved by the proposed iterative algorithm (OPA+OBA) and low complexity algorithm (SPA+SBA) versus the system's rate is shown in Figure 5, where $p_c = 50$ W and $d = 200$ m. We consider two QoS requirements for the users: the equal QoS, where $\rho_1 = \rho_2 = \rho_3 = 1/3$; the unequal QoS, where $\rho_1 = 1/2$ and $\rho_2 = \rho_3 = \frac{1}{4}$.

The results show that the EE gap between the iterative and low complexity algorithms is small for low system's rate but large for high system's rate. For the case of high

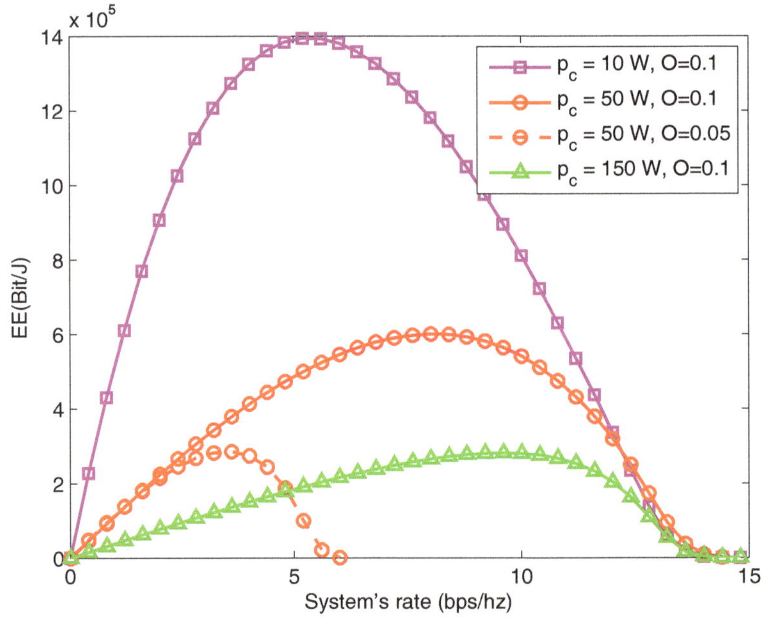

Figure 6 EE under different circuit powers and outage probability constraints. Equal QoS: $\rho_1 = \rho_2 = \rho_3 = \frac{1}{3}$.

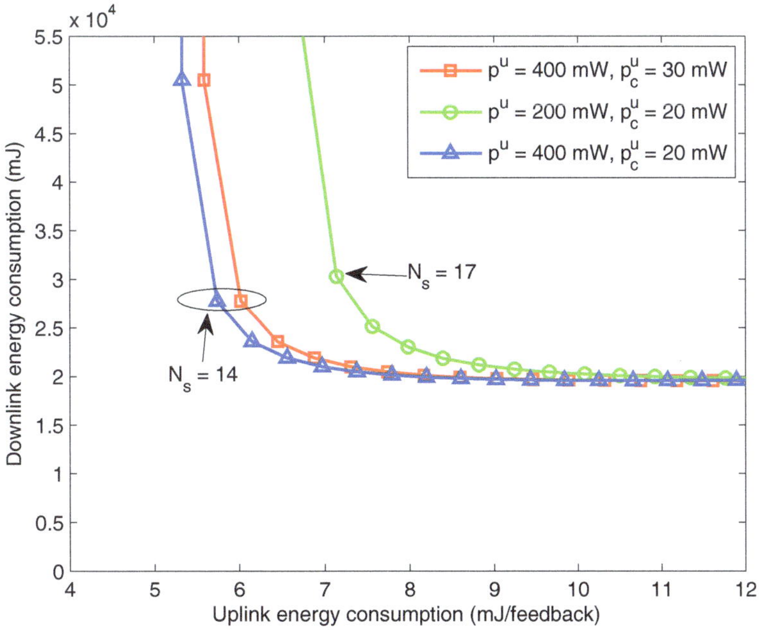

Figure 7 Trade-off between the downlink and uplink energy consumptions. Equal QoS: $\rho_1 = \rho_2 = \rho_3 = 1/3$.

system's rate under equal QoS, the users with high data rate requirement are with large elements in $\boldsymbol{A}$ as defined in (23). This makes the maximum eigenvalue of $\boldsymbol{A}$ close to 1 and results in the infeasibility for power allocation in (21) as analyzed in Section 4.3. The infeasibility events occur more frequently for the case with unequal QoS, since the users in the first cell with $\rho_1 R$ are with even larger elements in $\boldsymbol{A}$. As shown in the figure, the performance gain of the optimal bit allocation over the equal bit allocation is larger under unequal QoS requirement than under equal QoS requirement. Moreover, under equal bit allocation, we can find that the optimal bit allocation under unequal QoS is inferior to that under equal QoS. This is because unequal QoS makes the interference among the users more heterogeneous, such that equal bit allocation becomes more improper.

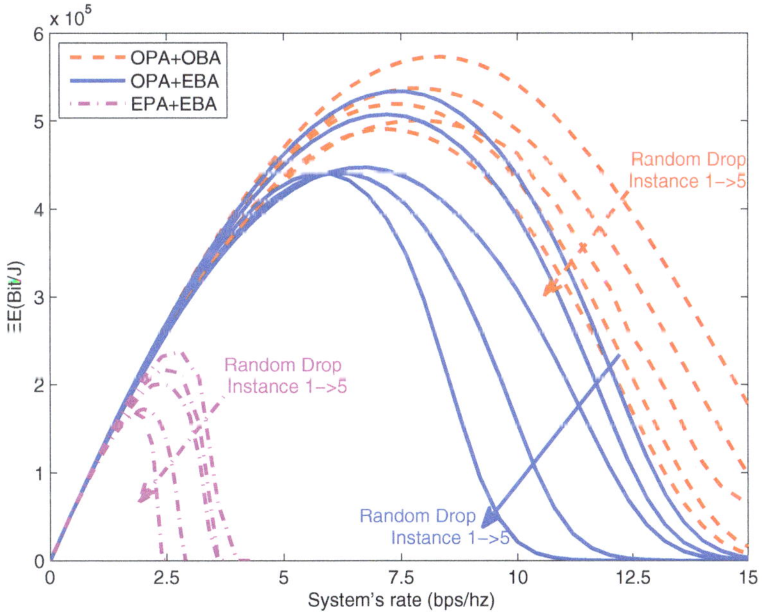

Figure 8 Random drops of users in CoMP-CB system. Equal QoS: $\rho_1 = \rho_2 = \rho_3 = 1/3$.

5.3.3 *Impact of circuit power consumption and outage probability*

In Figure 6, we present the EE achieved by the proposed optimal algorithm under different circuit power and outage probability for the case with equal QoS requirement where $d = 200$ m. In addition to $p_c = 50$ W, two other circuit power consumptions are considered: low circuit power case where $p_c = 10$ W and high circuit power case where $p_c = 150$ W [2]. Two target outage probabilities are considered: 0.1 and 0.05. The simulation results show that the EE decreases with the outage probability. When the outage probability is the same, the EE decreases as the circuit power consumption increases. Since the circuit power consumption does not affect the feasibility of the proposed algorithm, the EE with the same outage probability under different circuit power consumptions drops to zero at the same value of system's rate.

5.4 Trade-off between downlink and uplink energy consumptions

In Figure 7, we show the downlink and uplink energy consumptions, where the iterative algorithm in Algorithm 1 is used. We assume that there are $T = 100$ slots in downlink transmission per uplink feedback and consider $d = 250$ m and individual data rate requirement of 0.5 bps/hz. Since in the uplink the BSs receive the CDI feedback from the users, we also consider the circuit power consumed at the BS for reception. The downlink circuit power for transmission and reception are respectively $p_c = 50$ W and $p_c = 42$ W. The energy consumption in uplink is the energy consumed by the circuit power p_c^u and transmit power p^u in the N_s feedback slots at each user. The circuit power consumption for users are from [39]. We set the duration of each slot as 1 ms for both downlink transmission and uplink feedback.

As shown in the figure, for the same values of p_c^u and p^u, the more energy are consumed at each user, the less energy are consumed at the BSs. The energy consumption in downlink when $p_c^u = 20$ mW, $p^u = 200$ mW and $N_s = 17$ is approximately equal to that when $p_c^u = 20$ mW, $p^u = 400$ mW, and $N_s = 14$. This is because with higher uplink transmit power more bits can be fed back in each time slot and thereby the number of the time slots reduces.

In all three cases, the downlink energy consumption first drops dramatically and then remains constant as the uplink energy consumption increases. This is because when the uplink energy consumption is low, the total number of bits of each user is small such that the MUI and ICI in the system severely degrades the EE performance. The nonlinear impact of the number of bits for feedback on the downlink performance leads to a dramatic change on the downlink energy consumption. However, as the uplink energy consumption increases, the total number of bits of each user becomes large such that the MUI and ICI in the system approaches zero, then the downlink energy consumption becomes unchanged.

In practice, we can judiciously select the number of feedback slots N_s, i.e., the total number of feedback bits for each user, to trade-off the downlink and uplink energy consumptions.

5.5 Performance evaluation for general user locations

We investigate the impact of general user locations on the performance of the proposed algorithm in Figure 8. The second scenario for random user locations in Figure 1 is considered. We consider equal QoS requirement and assume that each user is randomly dropped in the CoMP region with a minimum distance of 200 m to their serving BSs in the cell-edge.

Figure 8 provides five instances of EE performance under different date rates for CoMP system with randomly distributed users, where the schemes of 'EPA+EBA,' 'OPA+EBA,' and 'OPA+OBA' are evaluated. The five instances are selected randomly, and in each instance six users are dropped independently and uniformly in the CoMP region. As shown in Figure 8, similar to Figure 4 where users are symmetrically distributed at fixed locations, it is also of significance to improve the EE performance for limited feedback CoMP-CB system with users at general user locations by using the proposed joint power and bit allocation algorithm.

6 Conclusions

We studied energy-efficient power and bit allocation for downlink limited-feedback CoMP-CB systems under the individual date rate requirement and feedback overhead constraint for each user. An optimal solution and a low complexity suboptimal algorithm were respectively proposed to allocate the transmit powers among multiple users and allocate the total number of bits of each user for quantizing the desired and interference channel directions. Simulation results have shown that the proposed algorithms improves the EE of the CoMP-CB systems significantly with respect to the equal power and bit allocation. The target data rate and outage probability of each user, the location of the users as well as circuit power consumption have a large impact on the EE of the system. We found that the energy consumed at the coordinated BSs first reduces rapidly with the increase of energy consumed at each user but remains almost constant when the uplink energy consumption further increases. This suggests that the energy consumption at the BSs can be reduced significantly with a minor increase of the energy consumption at each user if the total number of feedback bits for the user is carefully assigned.

Competing interests

The authors declare that they have no competing interests.

Acknowledgements

This work is supported by the National Natural Science Foundation of China (NSFC) under Grant 61120106002, the National Basic Research Program of China, 973 Program 2012CB316003 and the National High Technology Research and Development Program of China (No. 2014AA01A703).

References

1. GY Li, Z Xu, C Xiong, C Yang, S Zhang, Y Chen, S Xu, Energy-efficient wireless communications: tutorial, survey, and open issues. IEEE Wireless Commun. Mag. **18**(6), 28–35 (2011)
2. ICT-EARTH, Energy Efficiency Analysis of the reference systems, areas of improvements and target breakdown (2011). https://www.ict-earth.eu
3. C Yang, S Han, X Hou, AF Molisch, How to design CoMP to achieve its promised potential?. IEEE, Wireless Commun. Mag. (invited paper). **20**(1), 67–74 (2013)
4. DJ Love, RW Heath Jr, Equal gain transmission in multiple-input multiple-output wireless systems. IEEE, Trans. Commun. **51**(7), 1102–1110 (2003)
5. R Bhagavatula, RW Heath, Adaptive limited feedback for sum-rate maximizing beamforming in cooperative multicell systems. IEEE, Trans. Signal Process. **59**(2) (2011)
6. N Lee, W Shin, Adaptive feedback scheme on K-cell MISO interfering broadcast channel with limited feedback. IEEE Trans. Wireless Commun. **10**(2), 401–406 (2011)
7. DJ Love, RW Heath Jr, T Strohmer, in *Proc. IEEE Int. Communi. Conf. (ICC)*. Grassmannian beamforming for multiple-input multiple-output wireless systems (Alaska, US, 2003)
8. N Jindal, MIMO broadcast channels with finite-rate feedback. IEEE, Trans. Inform. Theory. **52**(11), 5045–5060 (2006)
9. T Yoo, N Jindal, A Goldsmith, Multi-antenna downlink channels with limited feedback and user selection. IEEE J. Select. Areas Commun. **25**(7), 1478–1491 (2007)
10. F Yuan, C Yang, G Wang, M Lei, Adaptive channel feedback for coordinated beamforming in heterogeneous networks. IEEE Trans. Wireless Commun. **12**(8), 3980–3994 (2013)
11. E Biglieri, J Proakis, S Shamai, Fading channels: information-theoretic and communications aspects. IEEE, Trans. Inform. Theory. **44**(6), 1998
12. J Luo, R Yates, P Spasojevic, Service outage based power and rate allocation for parallel fading channels. IEEE Trans. Inform. Theory. **51**(7), 2594–2611 (2005)
13. D Wu, R Negi, Effective capacity: a wireless link model for support of quality of service. IEEE Trans. Wireless Commun. **2**(4), 630–643 (2003)
14. SV Hanly, DNC Tse, Multiaccess fading channels part II: delay-limited capacities. IEEE Trans. Inform. Theory. **44**(7), 2816–2831 (1998)
15. G Caire, G Taricco, E Biglieri, Optimum power control over fading channels. IEEE Trans. Inform. Theory. **45**(5), 1468–1489 (1999)
16. C Jiang, LJ Cimini, in *Proc. IEEE Wireless Communi. and Netw. Conf. (WCNC)*. Energy-efficient transmission for MIMO interference channels (Paris, FR, 2012)
17. MC Gursoy, D Qiao, S Velipasalar, Analysis of energy efficiency in fading channels under QoS constraints. IEEE Trans. Wireless Commun. **8**(8), 4252–4263 (2009)
18. A Helmy, L Musavian, T Le-Ngoc, in *IEEE Int. Conf. Communi. (ICC)*. Energy-efficient power allocation for multicarrier systems with delay-outage probability constraints (Budapest, HU, 2013)
19. G Caire, N Jindal, M Kobayashi, N Ravindran, Multiuser MIMO downlink achievable rates with downlink training and channel state feedback. IEEE Trans. Inform. Theory. **56**(6), 1478–1491 (2010)
20. Z Xu, C Yang, GY Li, Y Liu, S Xu, Energy-efficient CoMP precoding in heterogeneous networks. IEEE Trans. Signal Process. **62**(4), 1005–1017 (2014)
21. M Chiang, P Hande, T Lan, CW Tan, *Power Control in Wireless Cellular Networks*, vol. 2. (Now publishers, US, 2008)
22. Yates RD, A framework for uplink power control in cellular radio systems. IEEE J. Sel. Areas Commun. **13**(7), 1341–1347 (1995)
23. H Dahrouj, W Yu, Coordinated beamforming for the multicell multi-antenna wireless system. IEEE, Trans. Wireless Commun. **9**(5), 1748–1759 (2010)
24. S Kandukuri, S Boyd, Optimal power control in interference limited fading wireless channels with outage-probability specifications. IEEE Trans. Wireless Commun. **1**(1), 46–55 (2002)
25. J Papandriopoulos, J Evans, S Dey, Optimal power control for Rayleigh-faded multiuser systems with outage constraints. IEEE Trans. Wireless Commun. **4**(6), 2705–2715 (2005)
26. CW Tan, in *Proc. IEEE INFOCOM*. Optimal power control in Rayleigh-fading heterogeneous networks (Shanghai, CN, 2011), pp. 2552–2560
27. D Jaramillo-Ramirez, M Kountouris, E Hardouin, in *Proc. IEEE Int. Symp. on Pers., Indoor and Mobile Radio Communi.(PIMRC)*. Coordinated multi-point transmission with imperfect channel knowledge and other-cell interference (Sydney, AU, 2012)
28. B Khoshnevis, W Yu, Bit allocation laws for multi-antenna channel quantization: multi-user case. IEEE Trans. Signal Process. **59**(5), 2270–2283 (2011)
29. S Boyd, L Vandenberghe, *Convex Optimization*. (Cambridge University Press, New York, NY, US, 2004)
30. Z Xu, C Yang, GY Li, S Zhang, Y Chen, S Xu, Energy-efficient configuration of spatial and frequency resources in MIMO-OFDMA systems. IEEE Trans. Commun. **61**(2), 564–575 (2013)
31. Y Zhang, RS Cheng, in *Proc. IEEE Wireless Communi. and Netw. Conf. (WCNC)*. High SNR performance of antenna selection in MISO interference channel with zero-forcing and limited feedback (Paris, FR, 2012)
32. C Li, J Zhang, KB Letaief, in *Proc. IEEE Int. Conf. Commun. (ICC)*. Energy efficiency analysis of small cell networks (Budapest, HU, 2013)
33. JB Rao, AO Fapojuwo, On the tradeoff between spectral efficiency and energy efficiency of homogeneous cellular networks with outage constraint. IEEE Trans. Vehi. Technol. **62**(4), 1801–1814 (2013)
34. S Han, C Yang, AF Molisch, Spectrum and energy efficient cooperative base station doze. IEEE J. Select. Areas Commun. **32**(2), 285–296 (2013)
35. M Kountouris, JG Andrews, Downlink SDMA with limited feedback in interference-limited wireless networks. IEEE Trans. Wireless Commun. **11**(8), 2730–2741 (2012)
36. Y Huang, CW Tan, BD Rao, in *Proc. IEEE Globe Telcommun. Conf.(GLOBECOM)*. Outage balancing in multiuser MISO networks: network duality and algorithms (California, US
37. J-C Kuang, *Applied Inequalities (Chinese)*, 3rd edn. (Hunan Education Press, Hunan, CN, 1993)
38. G Auer, V Giannini, C Desset, I Godor, P Skillermark, M Olsson, MA Imran, D Sabella, MJ Gonzalez, O Blume, A Fehske, How much energy is needed to run a wireless network?. IEEE Wireless Commun. Mag. **18**(5), 40–49 (2011)
39. G Miao, Energy-efficient uplink multi-user MIMO. IEEE, Trans. Wireless Commun. **12**(5), 2302–2313 (2013)

14

Experimental evaluation of the WiMAX downlink physical layer in high-mobility scenarios

Pedro Suárez-Casal*, José Rodríguez-Piñeiro, José A García-Naya and Luis Castedo

Abstract

The experimental evaluation of the WiMAX downlink physical layer in high-mobility scenarios is extremely difficult to carry out because it requires the realization of measurement campaigns with expensive fast-moving vehicles. In this work, however, we succeeded in doing such an experimental evaluation with a low-mobility vehicle. The key idea is the enlargement of the symbol period prior to its transmission over the air. Such enlargement reduces the frequency spacing between the orthogonal frequency division multiplexing (OFDM) subcarriers in WiMAX transmissions and hence induces significant inter-carrier interference (ICI) on the received signals. The performance impact of such ICI in terms of error vector magnitude (EVM) and throughput is analyzed under different conditions like the use of multiple antennas, the placement of the receive antennas, or the accuracy of the channel information to adapt the transmission rate.

Keywords: ICI; High-speed; WiMAX; MIMO

Introduction

Existing mobile communication networks are primarily designed for low user speeds below 15 km/h. Nowadays, however, there is an increased number of wireless terminals mounted on high-speed vehicles such as cars, trains, buses, subways, or airplanes. High-mobility communication networks (HMCN) aim at interconnecting such terminals to convey information not only for human users but also for machines to send all sorts of command, control, and safety information.

WiMAX is a communication standard suitable for the provision of wireless broadband connectivity. WiMAX is a term coined by the WiMAX Forum to promote the interoperability between the IEEE 802.16 family of wireless communication standards. WiMAX is the first commercially available and deployed technology for delivering mobile fourth generation (4G) services. IEEE 802.16 standardization activities started in 1999 and were the first ones to address broadband for wireless metropolitan area networks. Although long-term evolution (LTE) is currently being more widely used by 4G mobile network operators, specially in Europe and the United States, there is still a significant amount of network operators based on WiMAX, specially in Asia [1]. More specifically, WiMAX is being used in high-mobility scenarios like the metropolitan transportation system of Japan [1]. In addition, the WiMAX Forum has recently instituted an Aviation Working Group within its organizational structure to collaborate on the adaptation of WiMAX to the specific needs of the aviation community [2].

The physical layer (PHY) of the WiMAX radio interface uses orthogonal frequency division multiplexing (OFDM) as the modulation scheme. OFDM is particularly suitable to carry data over broadband frequency-selective channels because it allows for a low-cost channel equalization. OFDM waveforms, on the contrary, are rather sensitive when transmitting over time-selective channels such as those encountered in high-mobility scenarios. Time-selectivity causes the OFDM subcarriers to loose their orthogonality property and produces inter-carrier interference (ICI). Time-selectivity is particularly harmful when considering complex scenarios with multiple antennas and/or multiple users.

*Correspondence: pedro.scasal@udc.es
Department of Electronics and Systems, University of A Coruña, Facultade de Informática, Campus de Elviña, A Coruña 15071, Spain

Different techniques have been proposed in the literature to mitigate the effects of ICI. More specifically, issues like pilot pattern design [3], window design [4], channel estimation [5], and channel equalization have been thoroughly studied.

Basis expansion models (BEMs) have been extensively used to capture the double selectivity of the wireless channel and to implement cost-effective channel estimation and equalization schemes [6]. The rationale under BEM is the efficient representation of either the impulse or the frequency response of a channel by means of a linear combination of some basis vectors. Different BEMs have been proposed to be used under doubly-selective channels, such as the complex exponential BEM (CE-BEM) [7], polynomial BEM (P-BEM) [8,9], discrete prolate spheroidal BEM (DPS-BEM) [10], and Karhunen-Loève BEM (KL-BEM) [11,12]. Regarding pilot patterns for OFDM under ICI conditions, the Kronecker delta model, where guard subcarriers are allocated around pilot subcarriers, has been proposed as an optimal design to minimize channel estimation errors [13,14]. A detailed review of the algorithms and techniques to combat double-selectivity of the wireless channel can be found in [15].

Communication standards such as WiMAX or LTE define the pilot structure to be used by the elements in the network. Such pilot structures typically do not correspond to those assumed in many ICI estimation and equalization techniques proposed in the literature. In this work, we focus on specific algorithms which do not depend on such assumptions [12,16,17]. These methods typically estimate the ICI from a set of consecutive frequency response estimations and exploit time variations per subcarrier to estimate its spreading.

Experimental performance evaluations of 4G technologies in high-mobility situations are scarce in the literature due to the huge difficulties of carrying out experiments in such scenarios. In this work, we follow an approach where the time-selective wireless channels of high-mobility situations are recreated from experiments carried out at low speeds and hence are more cost-efficient to implement [18]. The approach consists of time-interpolating OFDM symbols prior to its transmission over-the-air (OTA), which leads to a reduction of the bandwidth of the whole OFDM signal. This produces OFDM waveforms which convey exactly the same information as the original one but with a reduced subcarrier spacing, hence artificially increasing the waveform sensitivity to ICI. At reception, and prior to its demodulation, the interpolation operation is inverted via a simple decimation operation. The resulting OFDM symbols are affected by ICI similarly as if they were transmitted over a high-mobility wireless channel. In fact, interpolating the original signal by a factor I will affect the transmitted signal similar to what would happen if it were transmitted at I times the original speed.

In this paper, we extend preliminary results obtained using this technique and already reported in [18,19]. The main contributions of this paper are as follows:

- Evaluation of ICI estimation and cancellation algorithms with experimental measurements of multiple-input multiple-output (MIMO) transmissions under time-selective conditions. These results are compared to simulations with comparable model parameters.
- Experimental study of the impact of ICI and feedback delay on performance metrics such as throughput.
- Improved measurement methodology with respect to that used in previous works ([18,19]). Instead of carrying out a dedicated measurement for each emulated speed value, the signals corresponding to all emulated speeds (time interpolation factors) are transmitted sequentially right after each other, thus making possible their latter comparison under similar conditions. Additionally, two receivers are used simultaneously, and therefore outdoor-to-outdoor as well as outdoor-to-indoor measurements are recorded under similar conditions.

This work focuses on OFDM waveforms based on the WiMAX physical layer. However, note that the experimental methodology also applies to other OFDM-based transmissions, such as those used in LTE. Although the results are obtained for a particular standard, this work provide hints on the behavior of other OFDM-based transmissions on high-mobility conditions and on the performance of the ICI cancellation techniques in different scenarios.

Signal model

We consider the transmission of MIMO-OFDM symbols synthesized according to the PHY layer specifications of the IEEE 802.16e (Mobile WiMAX) standard. We assume OFDM symbols with N subcarriers and a cyclic prefix of N_g samples are transmitted. In total, each OFDM symbol occupies $N_t = N + N_g$ samples. OFDM symbols are grouped in frames of K consecutive symbols, including a symbol reserved as a preamble. We also assume multiple antennas at transmission and reception, i.e., MIMO-OFDM transmissions. The number of transmit and receive antennas is M_T and M_R, respectively. OFDM symbols are spatially multiplexed over the M_T transmit antennas except the preamble which is transmitted only through the first antenna.

Let $\mathbf{s}_k^{(m)} \in \mathbb{C}^{N\times 1}$, $k = 1,\dots,K$, $m = 1,\dots,M_T$ be the column vector that represents the N complex-valued information symbols transmitted in the kth OFDM symbol over the mth antenna. Such vectors contain data, pilot, and guard symbols. Similarly, $\mathbf{x}_k^{(m)} \in \mathbb{C}^{N_t\times 1}$ contains the N_t samples corresponding to the kth OFDM symbol transmitted over the mth antenna.

Elaborating the signal model of a MIMO-OFDM system, the discrete-time representation of the transmitted MIMO-OFDM symbols is:

$$\mathbf{x}_k = \left(\mathbf{I}_{M_T} \otimes \mathbf{G}_1\mathbf{F}^H\right)\mathbf{s}_k, \quad k = 1, \ldots, K, \tag{1}$$

where $\mathbf{s}_k = \left[\mathbf{s}_k^{(1)T}, \mathbf{s}_k^{(2)T}, \cdots, \mathbf{s}_k^{(M_T)T}\right]^T$ is the $NM_T \times 1$ column vector containing the information, pilot, and guard symbols transmitted in the kth MIMO-OFDM symbol; $\mathbf{F}$ is the standard $N \times N$ discrete Fourier transform (DFT) matrix; $\mathbf{G}_1$ is a $N_t \times N$ matrix which appends the N_g samples of the cyclic prefix; $\otimes$ denotes the Kronecker product; $\mathbf{I}_{M_T}$ is the $M_T \times M_T$ identity matrix; and $\mathbf{x}_k = \left[\mathbf{x}_k^{(1)T}, \mathbf{x}_k^{(2)T}, \cdots, \mathbf{x}_k^{(M_T)T}\right]^T$ is the $N_tM_T \times 1$ column vector with the samples of the kth MIMO-OFDM symbol. We assume the samples in $\mathbf{x}_k^{(m)}$ are serially transmitted over the mth antenna at a sampling rate $F_s = 1/T_s$ where T_s is the sampling period.

The information symbol vectors are constructed as $\mathbf{s}_k^{(m)} = \mathbf{P}_k^{(m)}\mathbf{p}_k^{(m)} + \mathbf{D}_k^{(m)}\mathbf{d}_k^{(m)}$ where $\mathbf{p}_k^{(m)}$ is a $P \times 1$ vector containing the pilot symbols in the kth OFDM symbol transmitted over the mth transmit antenna, whereas $\mathbf{P}_k^{(m)}$ is the $N \times P$ matrix that defines the positions of the pilots in such a symbol. Similarly, $\mathbf{d}_k^{(m)}$ is the $D \times 1$ vector containing the data symbols, and $\mathbf{D}_k^{(m)}$ is the $N \times D$ matrix that defines their positions. Data symbols are the output of a quadrature amplitude modulation (QAM) constellation mapper whose inputs are the channel encoded source bits. Note that $P+D < N$, with $N-(P+D)$ being the number of guard subcarriers. Matrices $\mathbf{P}_k^{(m)}$ and $\mathbf{D}_k^{(m)}$ consist of ones and zeros only and are designed so that data and pilots are assigned to different subcarriers. According to the Mobile WiMAX standard, pilot subcarriers allocated in the mth antenna are set to 0 in all other antennas.

We next assume that the previous OFDM waveforms are transmitted over a time-varying MIMO channel. Such a channel is represented by:

$$h^{i,j}(t,\tau) = \sum_{l=1}^{L} h_l^{i,j}(t)\delta\left(\tau - \tau_l^{i,j}\right), i = 1, \ldots, M_R \text{ and } j = 1, \ldots, M_T, \tag{2}$$

where $h^{i,j}(t,\tau)$ is the channel impulse response between the jth transmit antenna and ith receive antenna, while $h_l^{i,j}(t)$ and $\tau_l^{i,j}$ are the lth path gain and delay of $h^{i,j}(t,\tau)$, respectively. Elaborating on the discrete-time signal model of a MIMO-OFDM system, the MIMO channel during the transmission of the kth OFDM symbol is represented by the block matrix:

$$\mathbf{H}_k = \begin{bmatrix} \mathbf{H}_k^{1,1} & \mathbf{H}_k^{1,2} & \cdots & \mathbf{H}_k^{1,M_T} \\ \mathbf{H}_k^{2,1} & \mathbf{H}_k^{2,2} & \cdots & \mathbf{H}_k^{2,M_T} \\ \cdots & \cdots & \cdots & \cdots \\ \mathbf{H}_k^{M_R,1} & \mathbf{H}_k^{M_R,2} & \cdots & \mathbf{H}_k^{M_R,M_T} \end{bmatrix}, \tag{3}$$

where $\mathbf{H}_k^{i,j} \in \mathbb{C}^{N_t \times N_t}$ are matrices representing the channel impulse response between the (i,j) antenna pair whose entries are:

$$\mathbf{H}_k^{i,j}(r,s) = h^{i,j}(((k-1)N_t + N_g + r - 1)T_s, \text{mod}(r-s, N)T_s). \tag{4}$$

Note that $\mathbf{H}_k^{i,m}\mathbf{x}_k^{(m)}$ represents the time-convolution between the signal transmitted over the mth antenna and $h^{i,m}(t,\tau)$.

Having in mind the previous channel representation, the discrete-time received signal is given by:

$$\mathbf{r}_k = \left(\mathbf{I}_{M_R} \otimes \mathbf{F}\mathbf{G}_2\right)\mathbf{H}_k\mathbf{x}_k + \mathbf{w}_k = \mathbf{G}_k\mathbf{s}_k + \mathbf{w}_k, \tag{5}$$

where $\mathbf{G}_2$ is a $N \times N_t$ matrix which represents the removal of the cyclic prefix, $\mathbf{G}_k$ is a block matrix containing the channel frequency response between each antenna pair, and $\mathbf{w}_k \in \mathbf{C}^{N_t \times 1}$ is a vector of independent complex-valued Gaussian-distributed random variables with variance σ_w^2. When the channel is time invariant, all submatrices in $\mathbf{G}_k$ are diagonal. However, when the channel is time-variant, entries different than zero appear outside their main diagonals, hence introducing ICI between the transmitted subcarriers. In such case, Equation 5 is rewritten as:

$$\mathbf{r}_k = \bar{\mathbf{G}}_k\mathbf{s}_k + \mathbf{z}_k + \mathbf{w}_k, \tag{6}$$

where $\bar{\mathbf{G}}_k$ is a block matrix with the main diagonals of the submatrices of $\mathbf{G}_k$ and $\mathbf{z}_k = (\mathbf{G}_k - \bar{\mathbf{G}}_k)\mathbf{s}_k$ represents the ICI in the received signal. Note that in multi-antenna systems, ICI occurs not only among subcarriers but also among different transmit antennas.

When transmitting MIMO-OFDM symbols over time-varying channels, the amount of ICI relates to the normalized Doppler spread given by $D_n = f_dT$, where f_d is the maximum Doppler frequency and $T = NT_s$ is the duration of an OFDM symbol excluding the cyclic prefix. According to our previous proposal [18], it is possible to adjust the parameter T by time interpolation by a factor I, yielding an OFDM symbol duration $T^I = IT_sN$. Therefore, the normalized Doppler spread impacting the time-interpolated OFDM signal is:

$$D_n^I = f_dT^I = f_dIT_sN = \frac{T_sNIf_cv}{c} = \frac{T_sNf_c}{c}v^I, \tag{7}$$

with f_c the carrier frequency, c the speed of light, and $v^I = Iv$ the emulated speed as a result of an actual measurement speed v and a time-interpolation factor I. Consequently, enlarging the symbol length T^I allows for the emulation of a velocity v^I, which is I times higher than the actual speed of the receiver, namely v.

Figure 1 shows the graphical representation of a measurement setup designed according to the previous premises. Interpolating the MIMO-OFDM waveforms prior to its transmission allows for emulating high-velocity conditions while conducting low-velocity experiments. Due to the interpolation step, signals over the air will suffer from severe ICI degradation, although the maximum Doppler frequency is low.

It should be noticed that interpolation does not allow for a perfect recreation of high-mobility channels because the signals over-the-air have a reduced bandwidth and are less sensitive to the channel frequency selectivity.

Nevertheless, note that in this work, we are mostly interested in conducting experiments to test the performance of ICI cancellation methods in WiMAX receivers over real-world channels rather than channel equalization methods which can be tested in static experiments. Time interpolation does not reproduce the exact conditions of high mobility scenarios but provide a cost-efficient approximation to them.

Receiver structure

Figure 2 shows the block diagram of the receiver structure utilized along this work. The samples captured with the hardware equipment are first decimated to undo the interpolation applied at the transmitter to induce the ICI. The decimated samples are input to an OFDM receiver that performs frame detection, and time and fractional carrier frequency offset (CFO) estimation before the DFT; and channel estimation, ICI cancellation, data subcarriers equalization, and channel decoding after the DFT. Figure 2 shows the general structure of the receiver. The same OFDM receiver is used to obtain the experimental results and the simulation results.

The ensuing subsections present a more detailed description of each processing block represented in Figure 2.

Frame detection and synchronization

Frame detection is carried out using the correlation properties of the preamble symbol. WiMAX defines a preamble symbol with pilot subcarriers generated from a pseudo-noise sequence modulated as binary phase-shift keying (BPSK), with a spacing of two guard subcarriers between them. This structure leads to a threefold repetition in the time domain, which can be exploited for frame detection and time and fractional carrier frequency offset (CFO) estimation. Integer frequency shifts are corrected after the DFT by performing a cross-correlation between the differential sequences of the transmitted and received preamble pilot sequences [20,21].

ICI estimation

The frequency response matrix $\mathbf{G}_k$ in the received MIMO-OFDM signal model given by Equation 5 needs to be estimated. We start determining a noisy frequency response estimation as a least squares (LS) estimation on the pilot subcarriers, affected by the interferences $\mathbf{z}_k$ and $\mathbf{w}_k$, followed by a linear minimum mean squared error (LMMSE) interpolation of the channel coefficients on the data subcarriers. Recall that pilot subcarriers are transmitted over disjoint sets of subcarriers at different antennas.

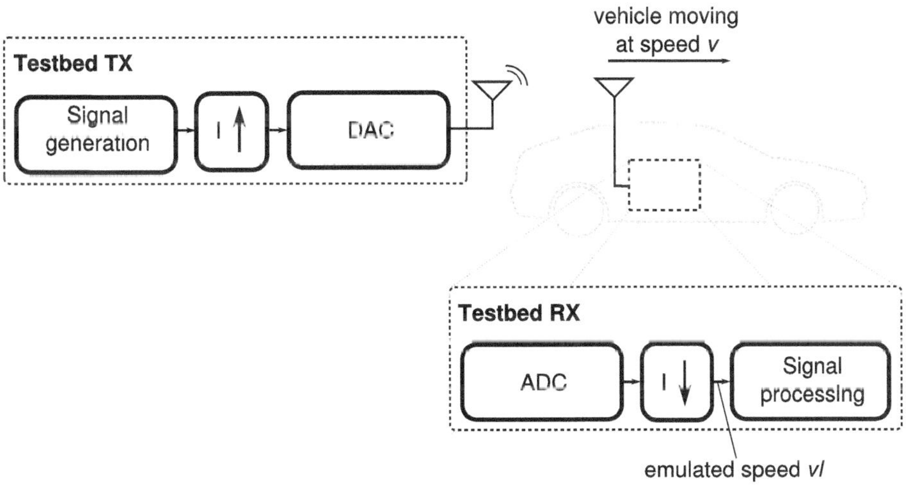

Figure 1 Block diagram of the outdoor-to-outdoor measurement setup. The transmit antennas are placed outdoors, and the receive antennas are installed outside the car on its roof. The interpolator at the transmitter side and the decimator at the receiver side enable the recreation of high-speed conditions during low-speed experiments. The corresponding outdoor-to-indoor setup is basically the same. The only difference is that receiving antennas are placed inside the car instead of outside.

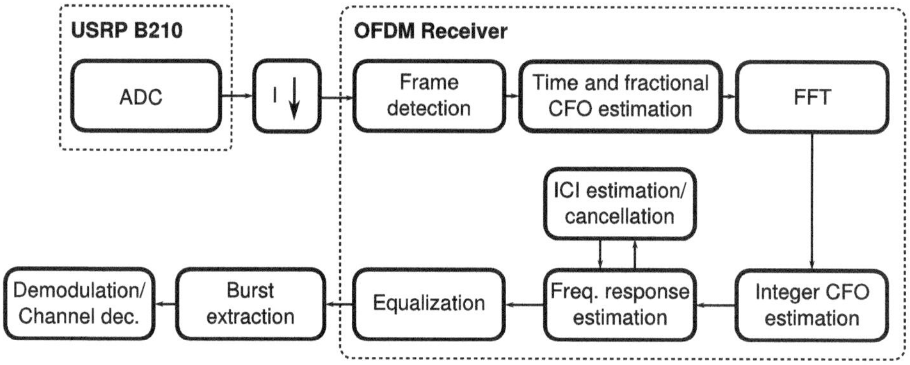

Figure 2 Block diagram of the receiver structure with decimation and the OFDM receiver with Inter-Carrier Interference (ICI) cancellation. The OFDM receiver is the same for the experimental and the simulation results.

Assuming spatially uncorrelated channels, this estimation can be done independently for each transmit-receive antenna pair. For simplicity reasons, antenna indices will be dropped in the following expressions.

The previous frequency response estimations can be expressed as:

$$\hat{\mathbf{g}}_k = \mathbf{A}\mathrm{diag}\left\{\mathbf{p}_k\right\}^{-1}\mathbf{P}_k^T\mathbf{r}_k, \tag{8}$$

where $\mathbf{A} = \mathbf{C}_{hh_p}(\mathbf{C}_{h_ph_p} + \sigma_I^2\mathbf{I}_P)^{-1}$ is a $N \times P$ matrix with the LMMSE interpolation coefficients, $\mathbf{C}_{hh_p} = \mathbf{P}_k^T\mathrm{E}\{\mathbf{g}_k\mathbf{g}_k^T\}$ and $\mathbf{C}_{h_ph_p} = \mathbf{P}_k^T\mathrm{E}\{\mathbf{g}_k\mathbf{g}_k^T\}\mathbf{P}_k$ are submatrices of the covariance matrix of the channel frequency response, and $\hat{\mathbf{g}}_k$ is a $N \times 1$ vector with the estimated channel coefficients per subcarrier. The second-order statistics of the channel are estimated by using all the received pilot subcarriers in a frame. Also, the power σ_z^2 of the ICI arising from $\mathbf{z}_k$ in Equation 6 is estimated according to [22], and it is considered as part of $\sigma_I^2 = \sigma_w^2 + \sigma_z^2$. As explained below, these frequency response estimations play a fundamental role in the estimation of the ICI.

Indeed, the ICI frequency response matrix $\mathbf{G}_k$ can be estimated by expressing the subcarrier spreading in terms of a BEM represented by a $RN_t \times Q$ matrix $\mathbf{B} = [\mathbf{B}_1^T, \mathbf{B}_2^T, \cdots, \mathbf{B}_R^T]^T$, where the dimensions of $\mathbf{B}_r$ are $N_t \times Q$, with $r = 1, \ldots, R$. Matrix $\mathbf{B}$ represents the basis vector for the time evolution of the channel during R consecutive OFDM symbols, with Q the order of the BEM. The BEM allows for expressing the frequency response matrix $\mathbf{G}_k$ in terms of a matrix $\mathbf{C}$ of size $Q \times N$. Using BEM, the estimate of the rth symbol in the group can be written as:

$$\hat{\mathbf{G}}_r = \sum_{q=1}^{Q} \mathbf{F}\mathbf{G}_2\mathrm{diag}\left\{\mathbf{b}_{r,q}\right\}\mathbf{G}_1\mathbf{F}^H\mathrm{diag}\left\{\hat{\mathbf{c}}_q\right\}, \tag{9}$$

where $\mathbf{b}_{r,q}$ is the qth column of $\mathbf{B}_r$, $\mathbf{G}_1$ and $\mathbf{G}_2$ are the matrices which append and remove the cyclic prefix in Equation 5, and $\hat{\mathbf{c}}_q$ is a $N \times 1$ vector with the columns of the estimate $\hat{\mathbf{C}}^T$.

We next define the $N \times R$ matrix $\hat{\mathbf{J}}_r = \left[\hat{\mathbf{g}}_r, \hat{\mathbf{g}}_{r+1}, \cdots, \hat{\mathbf{g}}_{r+R-1}\right]$, with $r \in [1, K - R + 1]$. Such a matrix enables us to estimate the matrix $\mathbf{C}$ of BEM coefficients by means of the following LS regression [17]:

$$\hat{\mathbf{C}}_r = \left(\mathbf{K}^H\mathbf{K}\right)^{-1}\mathbf{K}^H\hat{\mathbf{J}}_r^T, \tag{10}$$

where $\mathbf{K}$ is a $R \times Q$ matrix whose columns are obtained from $\mathbf{B}$ as:

$$\mathbf{k}_q = \left[b_q\left(N_g + N/2\right), b_q\left(N_t + N/2\right), \cdots, b_q\left((N/2)+(R-1)N_t\right)\right]. \tag{11}$$

Finally, the estimates $\hat{\mathbf{c}}_q$ in Equation 9 are obtained from the columns of $\hat{\mathbf{C}}_r^T$, which provide the coefficients for the channel matrices $\mathbf{G}_r$ of the group.

The results in this work are all obtained by using the discrete prolate spheroidal BEM (DPS-BEM) [10]. Such a BEM is built on the Slepian sequences arising from time sequences whose energy is localized on a given frequency interval. For the purpose of ICI estimation, the frequency interval is the one corresponding to the Doppler spectrum, whose domain is bounded by the maximum Doppler frequency. Recall that the use of DPS-BEMs is equivalent to assuming a time-selective channel with a flat Doppler spectrum. Consequently, to determine the specific DPS-BEM appropriate for a given scenario, it is necessary to know the mobile speed, and more specifically, the emulated speed after interpolation [10].

ICI cancellation and equalization

Once the full frequency response matrices for all antenna pairs are obtained, both ICI cancellation and equalization of the received signal are done to obtain estimates of the information symbols $\mathbf{s}_k$. In the literature, both block interference cancellation (BIC) and sequential interference cancellation (SIC) schemes have been proposed [15,23]. In this work, a LMMSE SIC receiver of seven taps is used to remove the ICI and equalize the channel. Such a method has been chosen due to its good trade-off between computational cost and performance.

The ICI estimation algorithm described in the previous subsection works on groups of OFDM symbols. Nevertheless, it should be noticed that the estimation error inside the group is not uniform along the OFDM symbols. Lower errors are achieved in the central symbols, and this is taken into account by the equalizer implemented in our receiver. ICI cancellation and channel equalization is carried out according to the following steps:

1. Obtain the matrices $\hat{\mathbf{C}}_r$ for the $K - R + 1$ groups in a frame, as explained in the previous subsection.
2. Remove ICI from each OFDM symbol as follows:
   (a) For symbols $k \in [1, R/2]$, ICI is suppressed with the coefficients obtained from $\hat{\mathbf{C}}_1$.
   (b) For symbols $k \in [R/2 + 1, K - R/2]$, ICI is suppressed with the coefficients obtained from $\hat{\mathbf{C}}_{k-R/2+1}$.
   (c) For symbols $k \in [K - R/2 + 1, K]$, ICI is suppressed with the coefficients obtained from $\hat{\mathbf{C}}_{K-R+1}$.
3. Obtain a new frequency response estimate from the ICI-reduced received signal and return to step 1.

As can be seen, except for the first and last symbols of the frame, ICI is estimated and cancelled for the central OFDM symbol of each set. The final equalization of the ICI-reduced signal is done by zero forcing. Detected information symbols are demapped and sent to a Viterbi decoder to obtain the information bits.

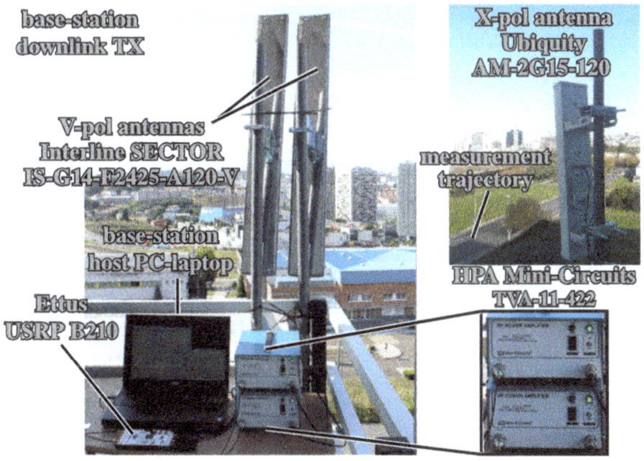

Figure 3 Base station downlink transmitter is placed outdoors. Located on the second floor of the CITIC building located in the Campus de Elviña at the University of A Coruña. Note that only one of the two vertically polarized antennas is used for 1 × 1 transmissions, while both cross polarized antennas are used for the 2 × 2 MIMO transmissions.

Experimental setup

The testbed employed for the experimental evaluation in this work is an upgrade of that employed in the measurement campaigns described in [18,19] which, at the same time, has evolved from the one described in [24]. The testbed consists of three USRP B210 boards [25] (see Figures 3 and 4) built from the AD9361 chip [26] by Analog Devices, which supports a continuous frequency coverage from 70 MHz to 6 GHz; full-duplex MIMO operation with up to 56 MHz of bandwidth; USB 3.0 connectivity; on-chip 12 bit analog-to-digital converters (ADC) and digital-to-analog converters (DAC) up to 61.44 M sample/s; automatic gain control; and configurable transmit and receive gain values.

A single board is used in continuous transmit-only mode for implementing the base station transmitter for the downlink. The base station is equipped with two Mini-Circuits TVA-11-422 high-power amplifiers [27], two Interline SECTOR IS-G14-F2425-A120-V vertically polarized transmit antennas [28], and an Ubiquity AM-2G15-120 cross polarized transmit antenna [29]. Notice that a single vertically polarized antenna was used for the single-input single-output (SISO) transmissions, while the cross polarized antenna was employed for measuring the 2 × 2 MIMO ones.

The remaining two USRP B210 boards are used for implementing two different mobile receivers, both mounted on a car. The first one is connected to a couple of eRize ERZA24O-09MBR omnidirectional 9 dBi-gain antennas placed outdoors, on the roof the car. The second mobile receiver is connected to another two omnidirectional antennas placed inside the car, between the two front seats. Using both mobile receivers allows us to capture, at the same time, the signals transmitted by the base station to both outdoor and indoor receivers.

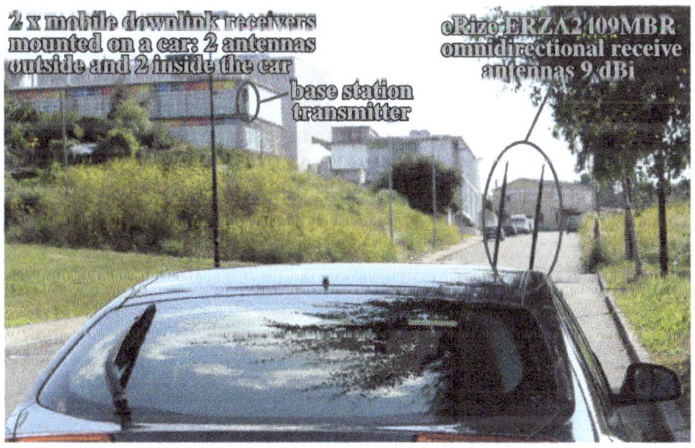

Figure 4 Mobile receivers mounted on a car. Two antennas corresponding to the outdoor receiver are placed on the roof of the car, while another two antennas for the second receiver are inside the car, between the two front seats.

With respect to the software, we use a multi-thread receiver implemented in C++ with Boost and using the Ettus USRP Hardware Driver (UHD) software. The main thread of the receiver is responsible for retrieving the samples coming from the USRP through the USB 3.0 bus and store them in a set of buffers in the main memory of the host laptop. The second thread reads the samples from such buffers and saves them persistently in a dedicated solid-state drive. Finally, there is a low-priority thread for logging important information for documenting the measurement campaign. On the other hand, the transmitter is a single-thread process since the same signals are cyclically transmitted over-the-air. Therefore, the signals to be radiated are first stored in a temporary buffer and next transmitted in a loop to the USRP through the USB 3.0. The rest of the software was implemented in MATLAB.

For the measurements, we use the WiMAX profile corresponding to 7 MHz of bandwidth at a sampling rate of $F_b = 8$ M sample/s, $N = 1,024$ subcarriers, and a cyclic prefix of 1/8 ($N_g = 128$). Table 1 summarizes the parameter values of this WiMAX profile. At the base station, once the OFDM signals were generated, they are time interpolated with the interpolation factors $I \in \{4, 12, 20\}$, hence producing signals with bandwidth ranging from 1.75 MHz for $I = 4$ to 350 kHz for $I = 20$. Next, the signals corresponding to each interpolation factor I are scaled by a factor $\sqrt{I}$ in order that signals corresponding to all interpolation factors be transmitted with the same energy. The three time-interpolated frames are then frequency-shifted 1.2 MHz to avoid the DC leakage at both the transmitter and the receiver. Hence, the DC subcarrier is frequency shifted 1.2 MHz to avoid the distortion caused by the DC leakage to the subcarriers around the DC, since the subcarrier spacing is reduced and we are dealing with direct-conversion transceivers. In order to facilitate the signal-to-noise ratio (SNR) estimation at the receiver, after transmitting the three time-interpolated frames, the transmitter is kept silent during a small time lapse which allows for estimating the noise variance at the receiver.

Table 1 Parameter values of the WiMAX profile used in the experiments

Parameter	Value
Sampling frequency	8 MHz
Useful bandwidth	7 MHz
Number of subcarriers	1,024
Number of data subcarriers	720
Number of pilot subcarriers	120
Subcarrier spacing	7.81 kHz
Cyclic prefix length	128 samples
Permutation zone	PUSC

Finally, all generated signals are stored in a file in the disk of the base station transmitter.

Once the transmit signals have been generated, the base station is notified and the over-the-air transmission process starts. First, the signals are read from the corresponding source file and transferred to the USRP, where they are again interpolated in the FPGA before reaching the (DAC). Note that this interpolation stage is needed for adapting the signal sampling rate to the sampling frequency of the DAC, thus not affecting the signal bandwidth. Next, the signals are up-converted to the central frequency $f_c = 2.6$ GHz, pre-amplified inside the USRP (configured with a gain value of 60 dB out of 89.5 dB), amplified by the two Mini-Circuits TVA-11-422 high-power amplifiers (one per transmit antenna) at their maximum gain of 40 dB, and finally radiated by the antennas (we use the vertically polarized transmit antennas for SISO transmissions, while the cross polarized ones are employed for 2×2 MIMO ones). The mean transmit power value measured at each antenna input is +17.5 dBm when a single transmit antenna is used. In case both transmit antennas are employed, the transmit gain is reduced to 3 dB per antenna to keep the total transmit power equal despite of the number of transmit antennas.

Figure 3 shows the base station placed outdoors, on the second floor of the CITIC building located in the Campus de Elviña at the University of A Coruña. It also shows the power amplifiers, the antennas (vertically or cross polarized), the USRP B210, and the laptop running the software for the base station. Looking at the picture of the cross-polarized antennas in Figure 3, one can also see part of the road traversed by the car during the measurements.

The two receivers are also built around the USRP B210 and the UHD with the software installed on two laptops, one per receiver. Notice that during the measurements, the acquired signals are persistently stored in dedicated solid state drives attached to laptop receivers, but they are not processed using the WiMAX receiver at that moment. As shown in Figure 4, the two outdoor receive antennas are sticked on the roof of the car used for measuring and they are connected directly to the USRP B210, which is powered by its corresponding laptop. The receive gain of the USRP is set to 35 dB (out of 73 dB), ensuring a linear operation. The second receiver is completely installed inside the car used for the measurements, with the antennas between the two front seats of the car. We use another laptop for running the receive software and for persistently storing the signals acquired by the indoor receiver. Unlike the outdoor receiver, the receive gain of the indoor receiver is set to 45 dB to better accommodate the amplitude of the acquired signals to the range of the ADC, while ensuring that the receive signal is not distorted by the amplifiers.

Figure 5 shows the measurement scenario, including the position of the base station and the path followed by the mobile receivers in the car. From the starting to the end points, there is a distance of 210 m, which is traversed by the car in 25.2 s at a constant speed of 30 km/h.

Physical layer configuration

OFDM symbol structure of the transmitted signal follows the downlink of mobile WiMAX standard regarding subcarrier allocation, symbol mapping, and channel encoding. The frames consists of $K = 25$ symbols, with the first one reserved for the preamble signal transmitted only by the first antenna. The other symbols carry the information of six bursts with the modulation and channel coding profiles: 4-QAM 1/2, 4-QAM 3/4, 16-QAM 1/2, 16-QAM 3/4, 64-QAM 1/2, and 64-QAM 3/4. Each burst spans during four consecutive OFDM symbols along all the available data subcarriers. The permutation matrices $\mathbf{P}_k^{(m)}$ and $\mathbf{D}_k^{(m)}$ are generated according to the partial usage of subcarriers (PUSC) zone with the corresponding modifications for MIMO as specified in the standard and pilot subcarriers on $\mathbf{p}_k^{(m)}$ generated as a boosted BPSK sequence obtained by mapping the output of a linear feedback shift register. Channel coding is performed by the convolutional encoder defined in the standard, which features tail-biting to terminate coding blocks.

Regarding the receiver, a DPS-BEM with $Q = 5$ is used for ICI estimation, given its good properties to estimate ICI assuming that only the maximum Doppler frequency is known.

In order to ensure similar channel characteristics for all the interpolation factors in each measurement, a super-frame consisting of four different parts is built. The first part is a small time period during which the base station is kept silent to allow for noise variance estimation at the receiver as it was already explained. The remaining three parts are built from the same mobile WiMAX frame consisting of the six burst plus the preamble that was described above. Consequently, the second part of the super-frame corresponds to the aforementioned mobile WiMAX frame interpolated in time by the factor $I = 4$. Next, the third part is the same as the second one but using $I = 12$, while the fourth part employs $I = 20$. Finally, the super-frame is cyclically transmitted from the base station to both mobile receivers, one placed outside the car and the other installed inside.

Results

This section presents the results obtained from the measurements described above. We also carried out simulations to support the experimental results. Simulations were designed to create scenarios comparable to those obtained from the measurements. Parameters estimated from the real-data captured are the mean SNR of each frame, and the K-factor of the wireless channel. The mean SNR is estimated taking as a reference the noise variance estimated during the silence preceding each frame; and the K-factor is estimated and averaged for all frames. Regarding simulations, the estimated mean SNRs for each frame in the real scenario are applied to the simulated transmissions, all with the same average K-factor. Finally, full characterization of the channel frequency responses or spatial covariance matrices have not been obtained, so simplified assumptions have been made while conducting the simulations. Basically, we assumed a frequency-flat spatially uncorrelated MIMO channel. The frequency-flat

Figure 5 Measurement scenario at the Campus de Elviña, A Coruña. The measurement trajectory as well as the location of the base station are specified.

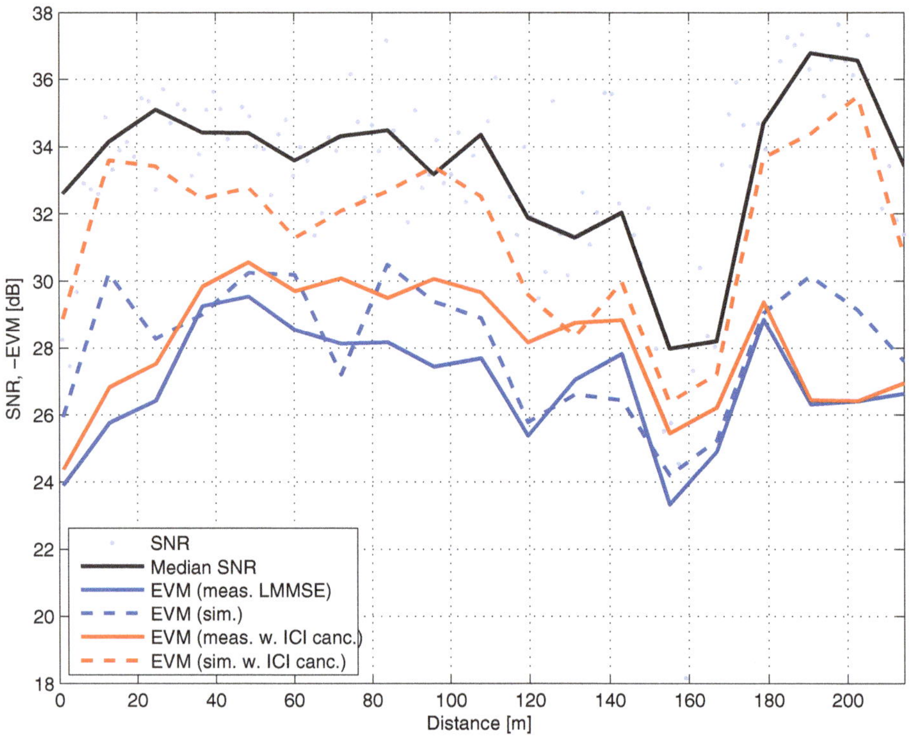

Figure 6 EVM and SNR for 1 × 1 and $v' = 120$ km/h. The receive antennas are placed outside the car.

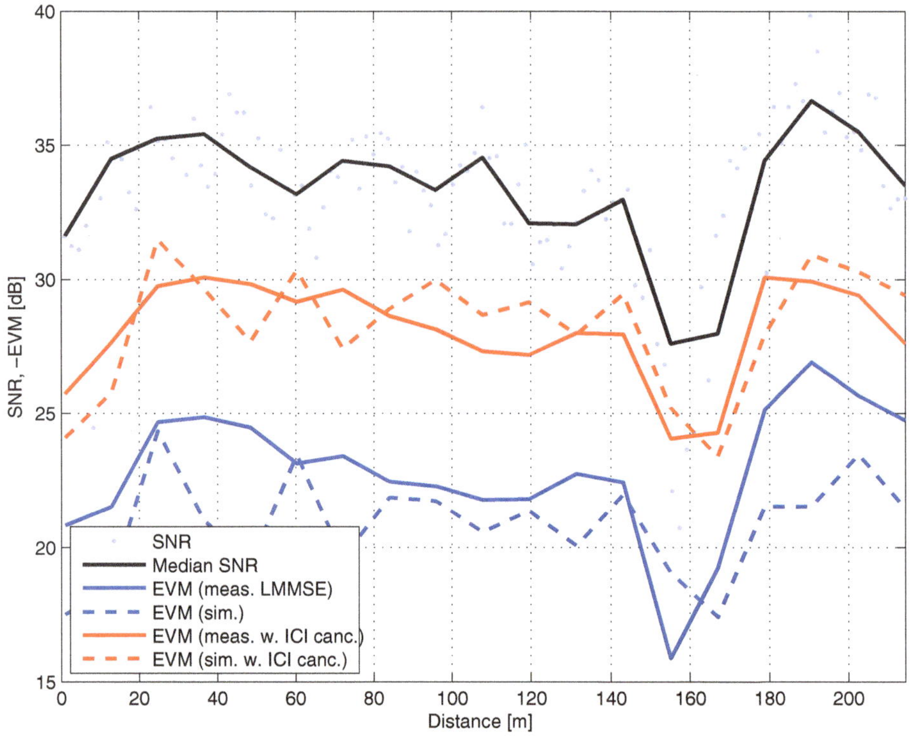

Figure 7 EVM and SNR for 1 × 1 and $v' = 360$ km/h. The receive antennas are placed outside the car.

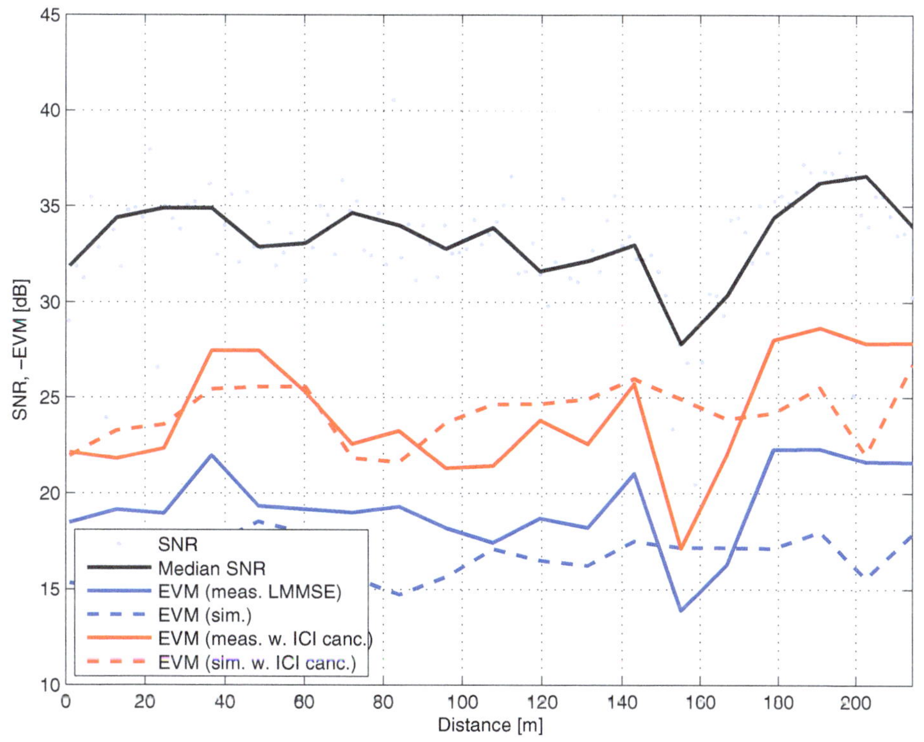

Figure 8 EVM and SNR for 1 × 1 and $v' = 600$ km/h. The receive antennas are placed outside the car.

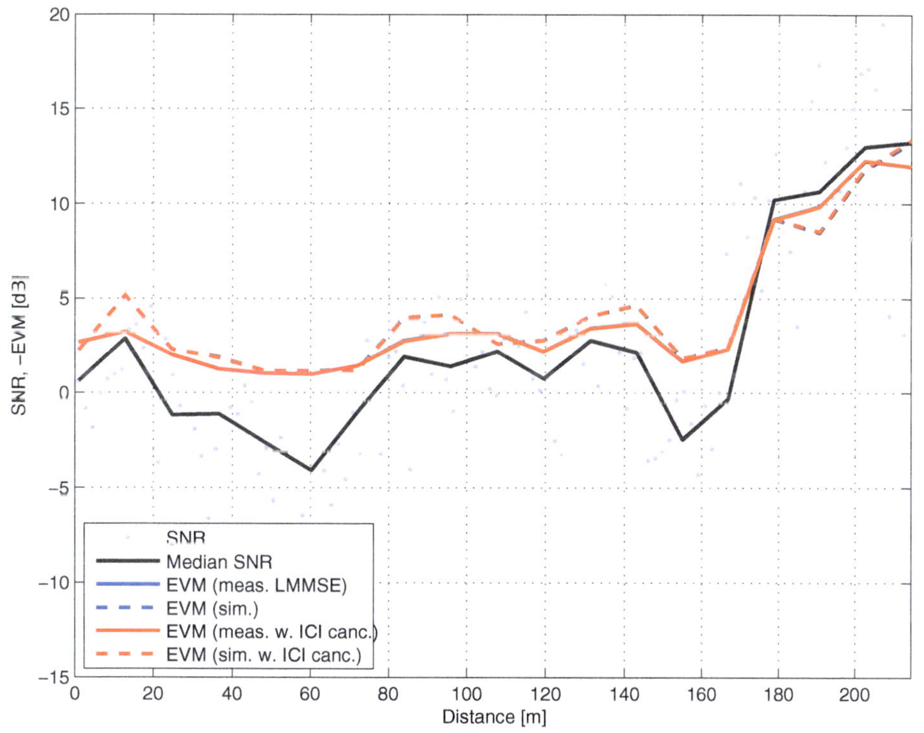

Figure 9 EVM and SNR for 1 × 1 and $v' = 120$ km/h. The receive antennas are placed inside the car.

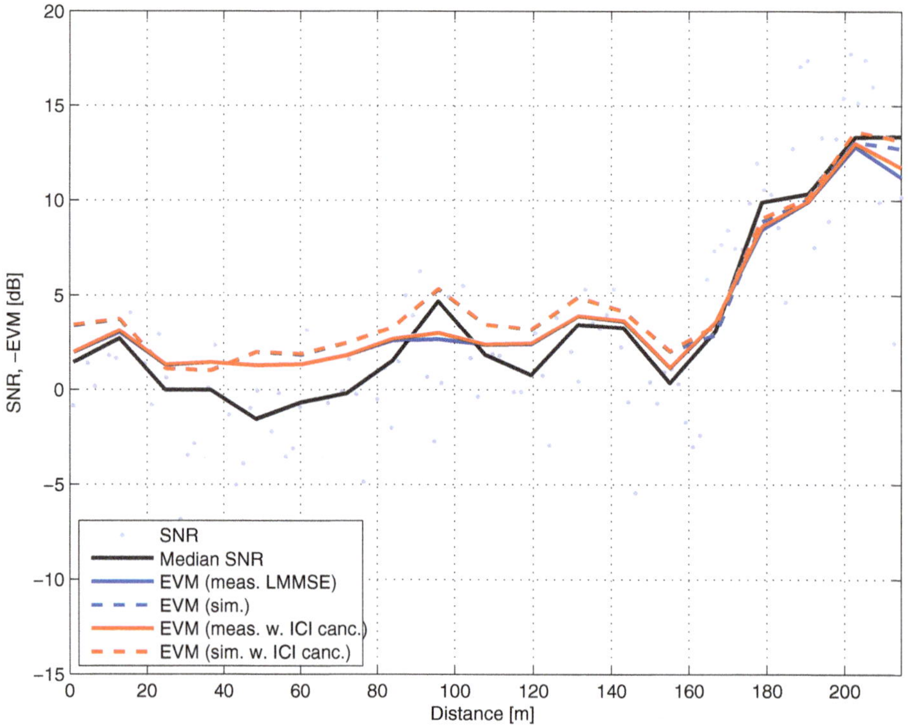

Figure 10 EVM and SNR for 1 × 1 and $v' = 360$ km/h. The receive antennas are placed inside the car.

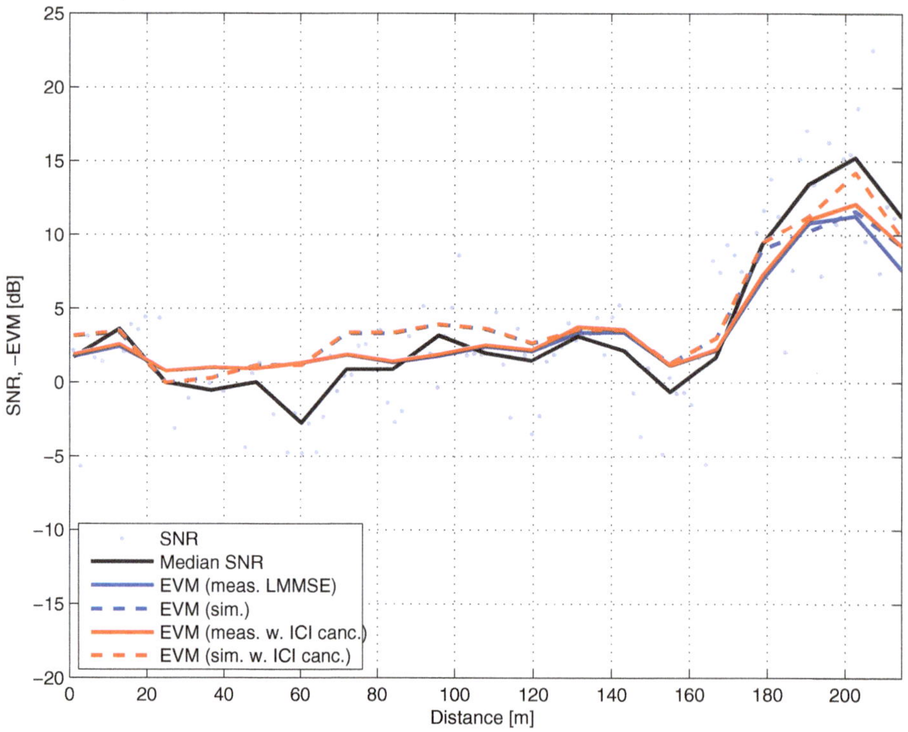

Figure 11 EVM and SNR for 1 × 1 and $v' = 600$ km/h. The receive antennas are placed inside the car.

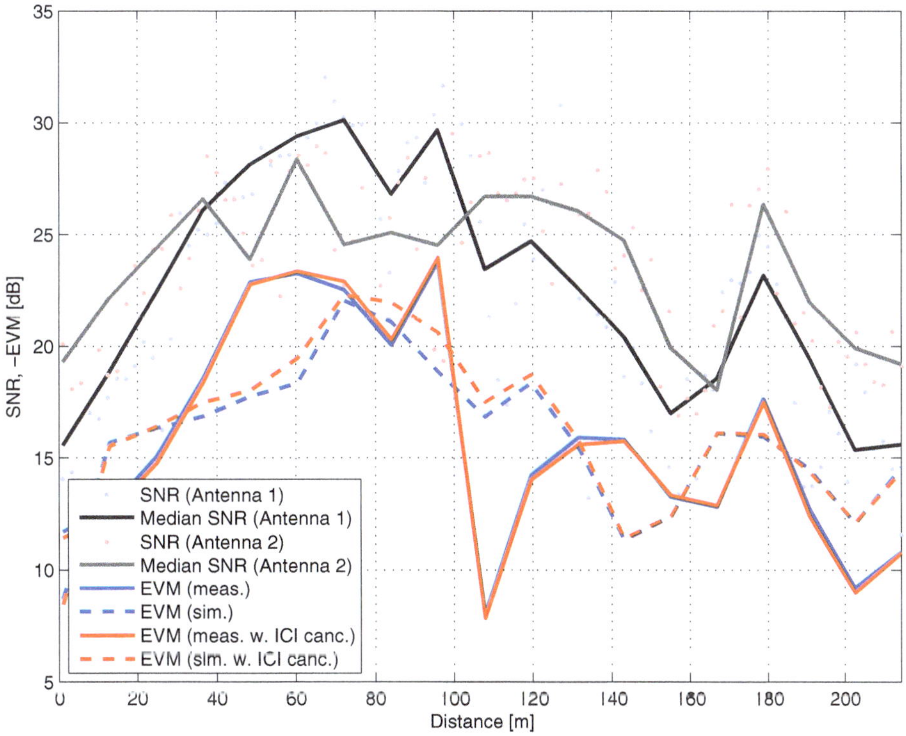

Figure 12 EVM and SNR for 2 × 2 and $v' = 120$ km/h. The receive antennas are placed outside the car.

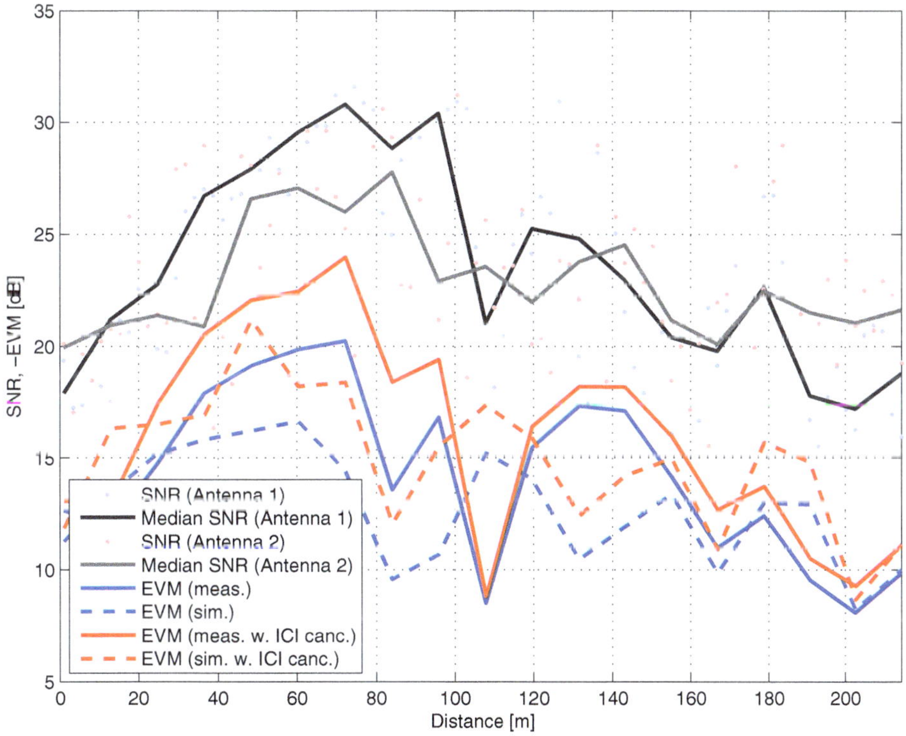

Figure 13 EVM and SNR for 2 × 2 and $v' = 360$ km/h. The receive antennas are placed outside the car.

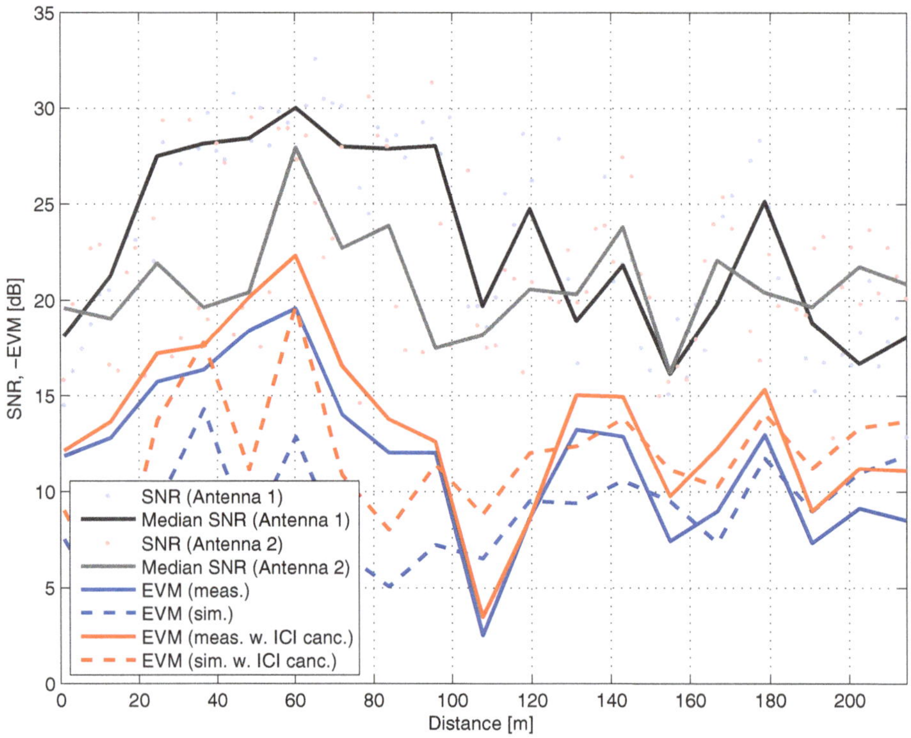

Figure 14 EVM and SNR for 2 × 2 and $v' = 600$ km/h. The receive antennas are placed outside the car.

assumption arises from the fact that for the highest interpolation factors, the bandwidth of the transmitted signals will be rather narrow, and therefore the frequency selectivity observed by the receiver would be negligible. As for the MIMO channel, the coefficients for each pair of antennas are drawn from independent Rice distributions, although not identically distributed, picking random complex mean values for each MIMO channel matrix entry.

Error vector magnitude and SNR

Error vector magnitude (EVM) is the first figure of merit considered in this work. EVM is measured as the mean squared error between the received constellation after equalization and the transmitted constellation, normalized by the transmitted constellation power, i.e.:

$$\mathrm{EVM}_F = \frac{\sum_{k=1}^{K}\sum_{n=1}^{N} \|\hat{\mathbf{s}}_k(n) - \mathbf{s}_k(n)\|^2}{\sum_{k=1}^{K}\sum_{n=1}^{N} \|\mathbf{s}_k(n)\|^2}, \quad (12)$$

being EVM_F the mean EVM for the Fth frame. Note that the original transmitted symbols are used to normalize the (EVM). Notice that if only symbol estimates were available, EVM measures would be less accurate.

Figures 6, 7, and 8 show the EVM and SNR results versus distance for SISO transmissions with outdoor antennas for interpolation factors $I = 4$, $I = 12$, and $I = 20$, respectively. The car moved at the actual speed of 30 km/h. The previous factors correspond to emulated speeds of 120, 360, and 600 km/h, respectively. To clarify the plots, the course of the car has been split in 20 equally long sectors, and the median EVM is determined for each sector. Also, as a reference, the median SNR for each sector is shown. Figures 6, 7, and 8 clearly show the performance gains obtained when ICI cancellation is performed at reception. The figures also show that comparable gains are obtained with simulation and real measurements, hence validating the performance of the ICI estimation and cancellation techniques utilized. Figure 6 shows a constant offset between simulations and measurements. Recall that

Table 2 EVM to burst-profile mapping for throughput estimations

Profile	-EVM [dB]
4-QAM 1/2	5
4-QAM 3/4	8
16-QAM 1/2	10.5
16-QAM 3/4	14
64-QAM 1/2	16
64-QAM 3/4	20

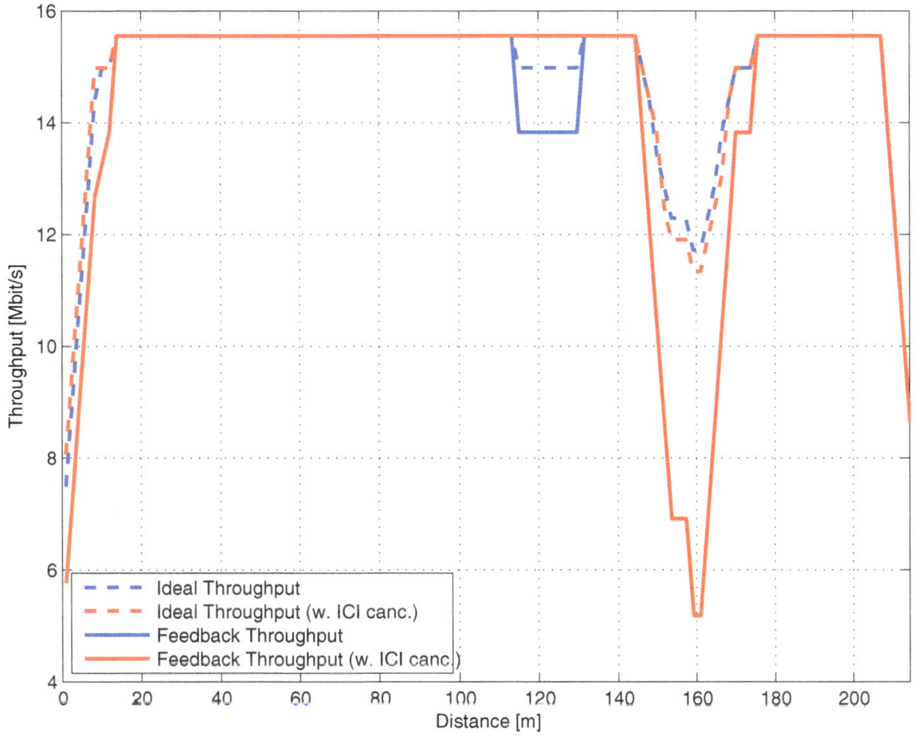

Figure 15 Throughput for 1 × 1 and v^{l} = 120 km/h. The receive antennas are placed outside the car.

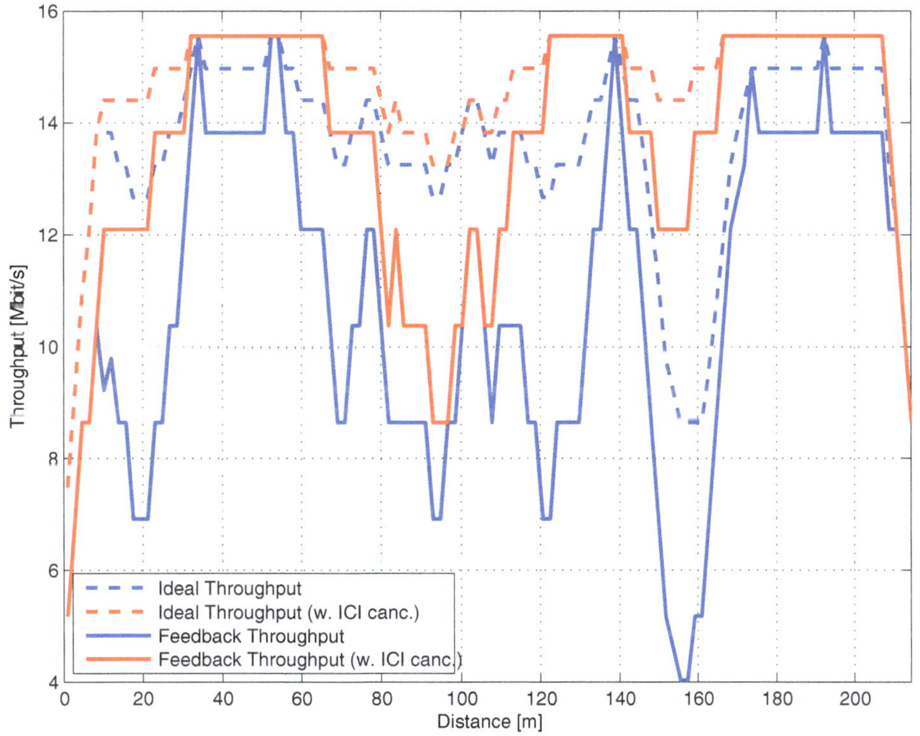

Figure 16 Throughput for 1 × 1 and v^{l} = 360 km/h. The receive antennas are placed outside the car.

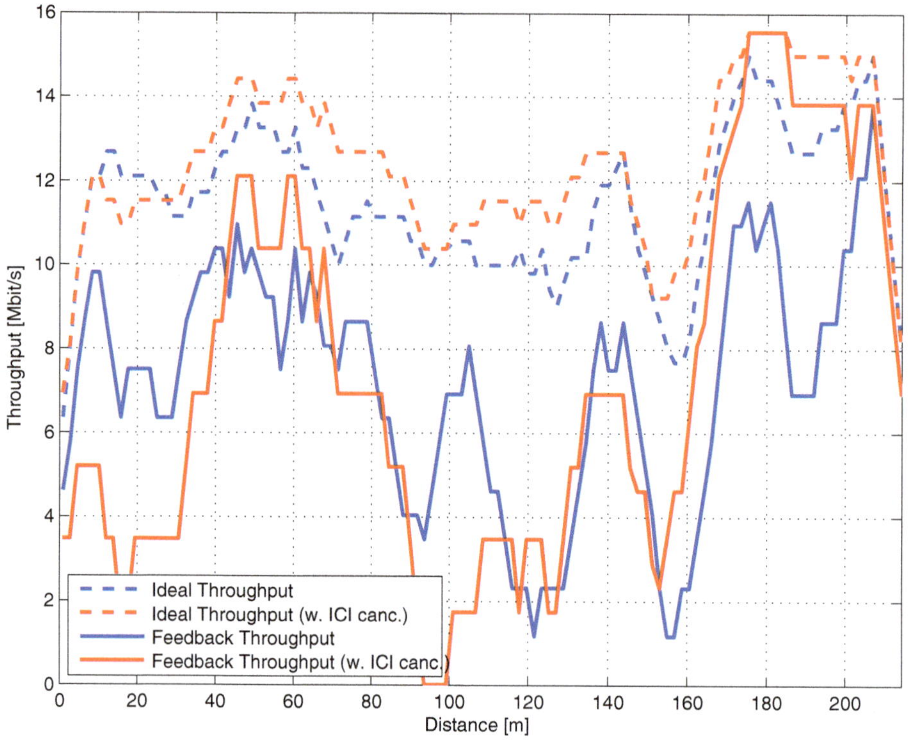

Figure 17 Throughput for 1 × 1 and $v^l = 600$ km/h. The receive antennas are placed outside the car.

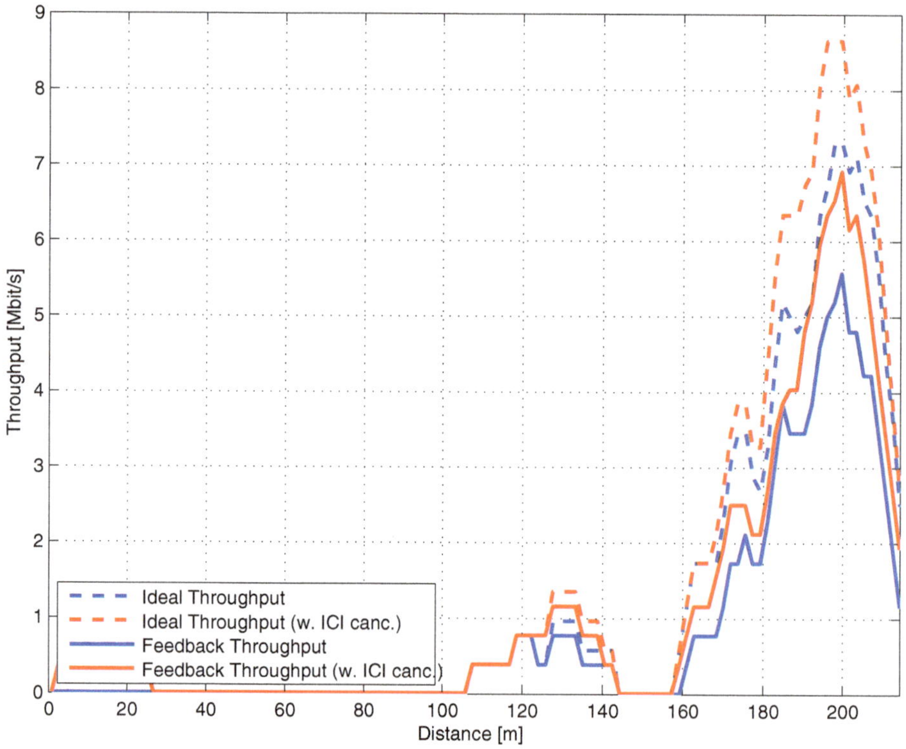

Figure 18 Throughput for 1 × 1 and $v^l = 120$ km/h. The receive antennas are placed inside the car.

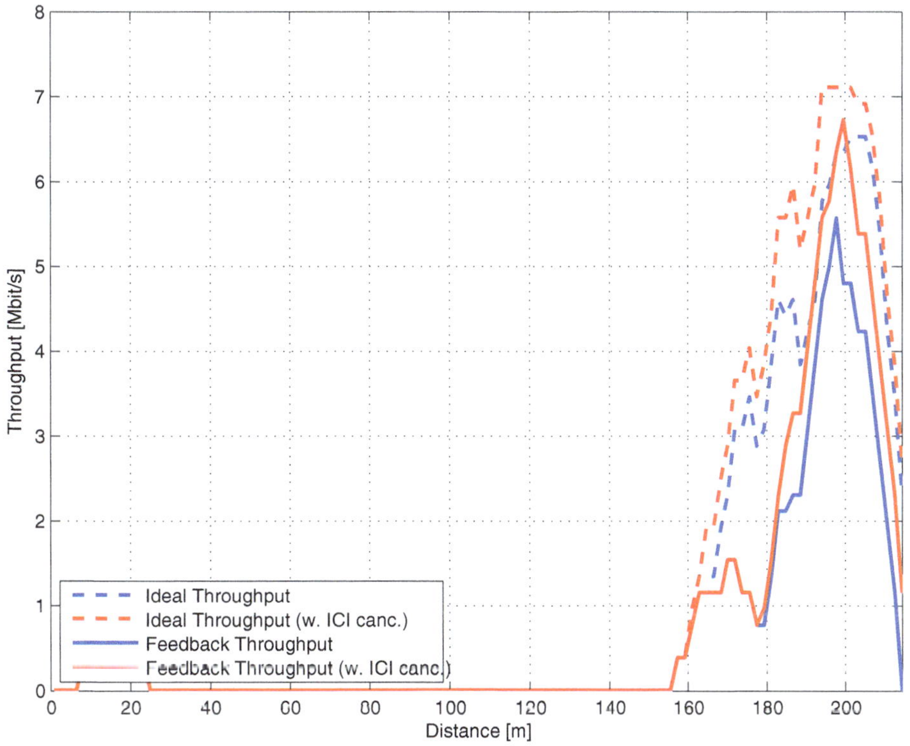

Figure 19 Throughput for 1 × 1 and $v' = 360$ km/h. The receive antennas are placed inside the car.

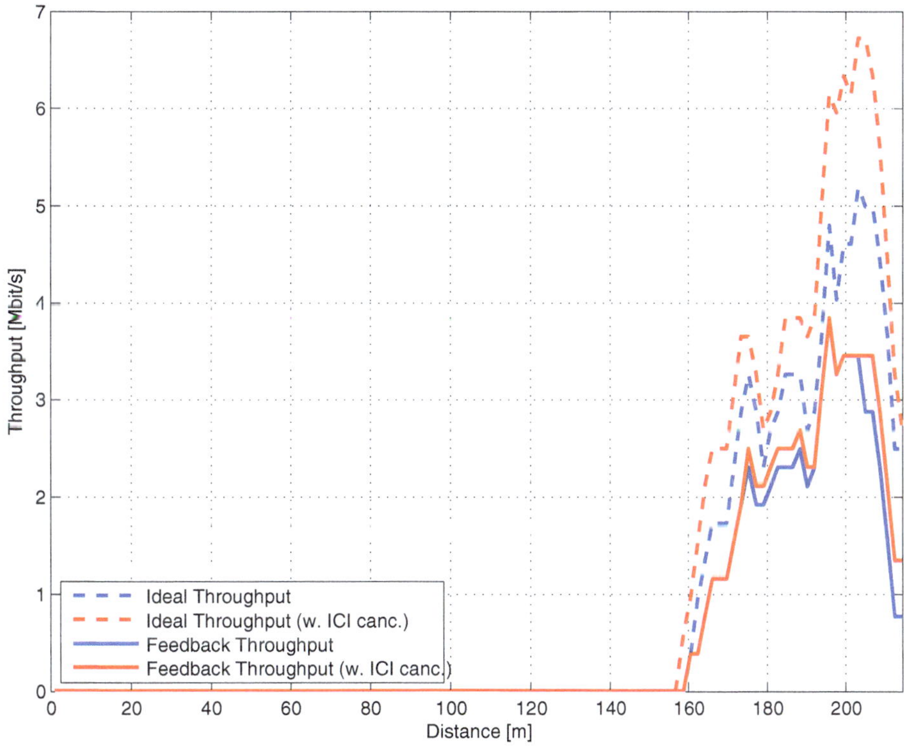

Figure 20 Throughput for 1 × 1 and $v' = 600$ km/h. The receive antennas are placed inside the car.

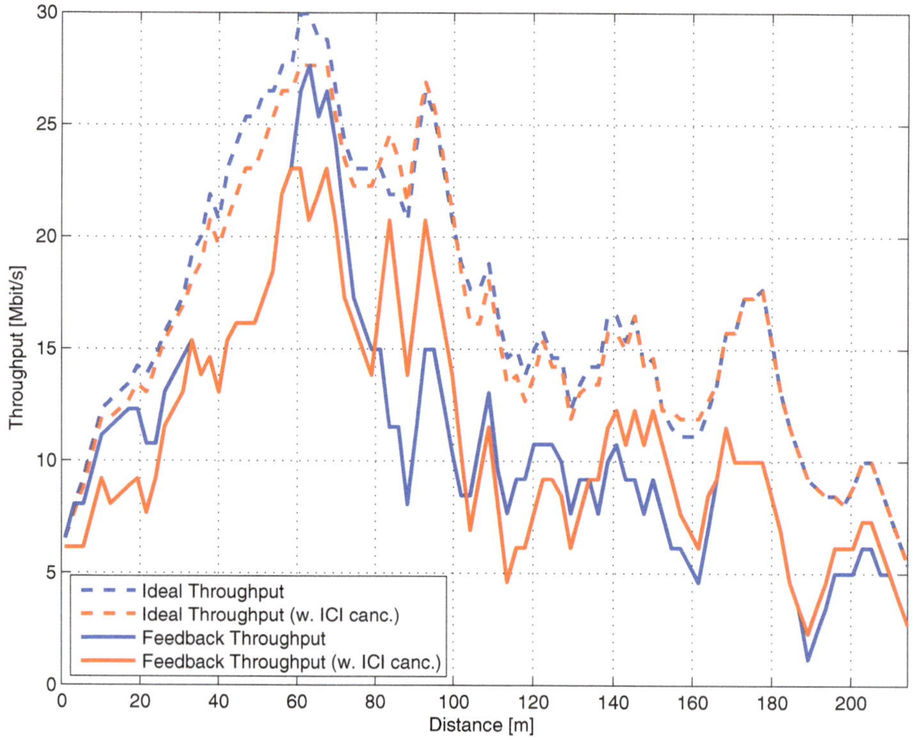

Figure 21 Throughput for 2 × 2 and $v' = 120$ km/h. The receive antennas are placed outside the car.

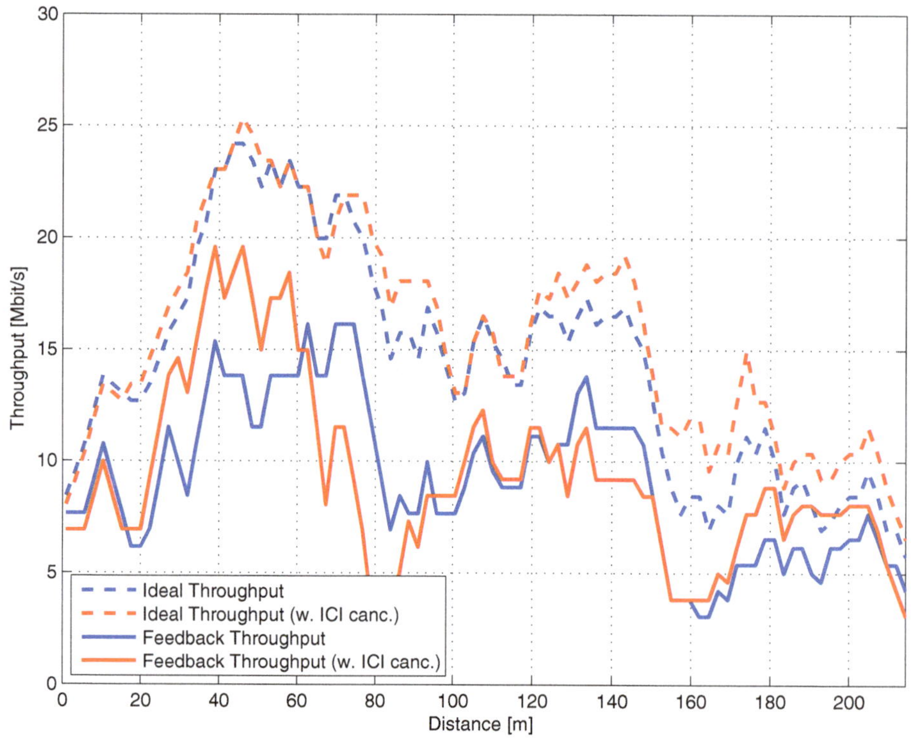

Figure 22 Throughput for 2 × 2 and $v' = 360$ km/h. The receive antennas are placed outside the car.

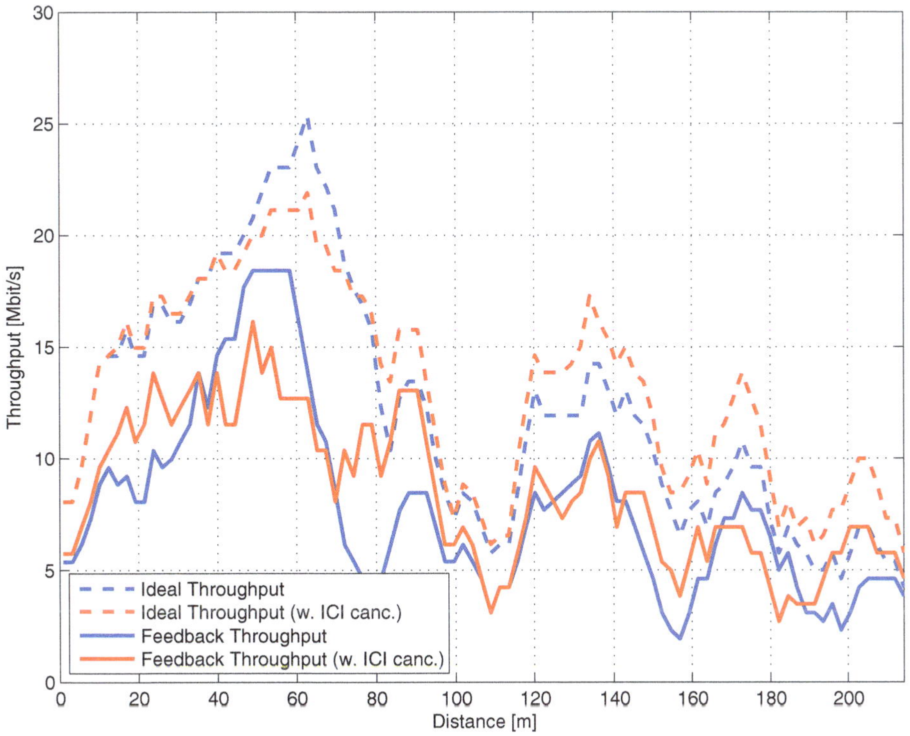

Figure 23 Throughput for 2 × 2 and $v^I = 600$ km/h. The receive antennas are placed outside the car.

simulation results were obtained assuming a frequency-flat channel. The soundness of this assumption weakens for low interpolation factors such as $I = 4$, so a significant difference between simulation and experimental results is to be expected. Conversely, better agreement between experiments and simulations is observed in Figures 7 and 8 which correspond to higher interpolation factors ($I = 12$ and $I = 20$, respectively). In such cases, the frequency-flat channel response assumption is tighter.

Figures 9, 10, and 11 show the results for the same measurements as before but with the mobile receiver inside the car. The most important difference compared to the previous figures is the lower SNR of this scenario, where median values range between 0 and 10 dBs, a large difference compared to the 30 to 35 dB achieved outside due to the strong attenuation introduced by the car structure. As a consequence, negligible performance gains are obtained with receivers that perform ICI cancellation because the noise at the receiver prevails over the interference of the neighbor subcarriers.

Figures 12, 13, and 14 show the results corresponding to 2 × 2 MIMO measurements with the receiving antennas placed outside the car for $I = 4$, $I = 12$, and $I = 20$, respectively. Contrary to the SISO case, ICI cancellation does not produce significant performance gains when $I = 4$. ICI cancellation produces significant performance gains for $I = 12$ and $I = 20$. Again, this is because the received SNR is significantly smaller and the Gaussian noise dominates over the ICI. Also, although spatially uncorrelated channels have been assumed in simulations, the real scenario will show some spatial correlation between antennas. Such correlation depends on the distance between antennas, and also on the distance between the transmitter and the receiver, an effect more evident when strong line-of-sight components are present [30]. This spatial correlation introduces interference between

Table 3 Mean throughput in Mbit/s for the 1 × 1 scenario when the receive antennas are placed outside the car

	Ideal throughput	Ideal throughput (with ICI cancellation)	Feedback throughput	Feedback throughput (with ICI cancellation)
$I = 4$	14.32	14.73	13.01	13.93
$I = 12$	12.22	13.78	8.81	11.39
$I = 20$	9.76	11.37	5.48	6.51

Table 4 Mean throughput in Mbit/s for the 1 × 1 scenario when the receive antenna is placed inside the car

	Ideal throughput	Ideal throughput (with ICI cancellation)	Feedback throughput	Feedback throughput (with ICI cancellation)
$I = 4$	3.15	3.48	2.11	2.45
$I = 12$	2.90	3.26	1.96	2.23
$I = 20$	2.67	3.03	1.82	1.97

antennas, which impairs the gain that could be obtained by the ICI cancellation technique.

Throughput

In order to evaluate the impact of the ICI cancellation methods on the final user quality of experience (QoE), throughput estimations have been carried out using the information conveyed in the bursts. Two types of throughput estimation have been considered. The first one assumes that the transmitter knows, for each frame, the optimal modulation and coding profile to be used. This case was modeled by estimating the bit error ratio (BER) of the six received bursts per frame and assuming that the whole frame was transmitted with the profile which carries more information bits without suffering any bit error. The second method assumes that the transmitter estimates the most suitable burst profile, and consequently only transmits the corresponding profile during all the burst. Errors are measured in this case assuming only the burst picked by the transmitter is available, regardless whether this burst has errors or not. This method is intended to determine the impact of the channel fast time-variations on this type of decisions.

To allow the transmitter to take these decisions, a frame by frame EVM metric is obtained in a similar way as the WiMAX standard proposes, i.e.:

$$\mathrm{EVM}_F = (1-\alpha)\mathrm{EVM}_{F-1} + \alpha\mathrm{EVM}_F, \tag{13}$$

where α is a forgetting factor. Along this work, we assume $\alpha = 1/4$. This EVM value is used to pick a modulation profile and burst according to Table 2.

Results for the 1 × 1 setup with the antennas placed outside the car are shown in Figures 15, 16, and 17 for interpolation factors $I = 4$, $I = 12$, and $I = 20$, respectively. When $I = 4$, the performance gain when using the receiver with ICI cancellation is not significant because the throughput already saturates at the maximum value 15 Mbit/s without cancelling the ICI (see Figure 15). On the contrary, significant performance gains are obtained for $I = 12$ and $I = 20$ when the ICI is cancelled. Nevertheless, note the large throughput degradation when the profile is determined from delayed EVM measurements rather than from the actual ones. In channels with low coherence times, estimations of the EVM according to Equation 13 are less reliable, thus increasing the probability of picking up an incorrect burst profile. This effect is more evident for the highest interpolation factor $I = 20$, where at some time instants even the ICI-cancelled signal provides lower throughput.

Figures 18, 19, and 20 show the results for the same measurement campaign as before but with the antennas placed inside the car. The most important effect is that the SNR is now much lower and the receiver is unable to detect any burst without errors during the first part of the measurement trajectory. The larger SNRs observed in the last sector of the measurement course allow for correctly decoding information bits but leading to throughput values far from those observed from the antennas placed outside the car. Also, the throughput gains obtained after ICI cancellation are quite low.

Corresponding results for the 2 × 2 MIMO setup are shown in Figures 21, 22, and 23. The first evident feature is the larger achievable throughput, since two streams are being transmitted in each frame. As with EVM, low gains when $I = 4$ are appreciated, but peak gains of 2 Mbit/s are observed when $I = 12$ and $I = 20$.

A more meaningful way to represent the mean throughputs obtained during all measurement campaigns is the one shown in Tables 3, 4, and 5. These throughputs show that in general, some gain is obtained by using receivers that cancel the ICI and how the throughput worsens with the interpolation factor. It is also important to note that in some scenarios at low-emulated speeds, negligible gains, or even less performance is obtained with these methods.

Table 5 Mean throughput in Mbit/s for 2 × 2 scenario when the receive antennas are placed outside the car

	Ideal throughput	Ideal throughput (with ICI cancellation)	Feedback throughput	Feedback throughput (with ICI cancellation)
$I = 4$	16.95	17.46	10.56	10.99
$I = 12$	15.74	17.48	9.20	11.01
$I = 20$	12.61	13.99	7.49	8.87

In general, a remarkable difference is observed between the ideal throughput and the realistic measure which considers the EVM metric to pick the burst profile. This is due to the inability of the EVM estimator to track the fast time-variations of the channel. Also, even though in the previous figures at some instants the ICI-cancelled throughput was lower, it can be seen that in average the ICI cancellation techniques provide a better performance. Finally, in the case of MIMO 2×2, although in terms of EVM almost no difference is appreciated when $I = 4$, in terms of throughput an average gain of approximately 0.50 Mbit/s has been obtained. And when $I = 12$ and $I = 20$, the average gains grow up to 1.8 and 1.4 Mbit/s approximately.

Conclusions

We have experimentally evaluated the performance of the WiMAX downlink physical layer in high-mobility scenarios. Both SISO and MIMO transmissions were considered, as well as placing the receive antennas outside and inside the car used for the experiments. We focused on the ICI caused by channel time variations in such scenarios. Cost-effective measurement campaigns were carried out where high-mobility scenarios are emulated with a vehicle moving at a low speed. The key idea is the enlargement of the symbol period prior to its transmission over the air. Such enlargement reduces the frequency spacing between the OFDM subcarriers in WiMAX transmissions and hence induces high ICI values on the received signals. Experiments illustrate the performance of WiMAX receivers with and without ICI cancellation in terms of EVM and throughput. ICI cancellation produces significant performance gains mainly when the received SNR is high. Otherwise, thermal noise dominates over the ICI, and gains are not so much appreciated. Furthermore, the spacing of subcarriers in this standard is sufficient to provide a robust behavior for moderate mobility environments, and the gain observed after ICI cancellation is not very significant for speeds up to 80 km/h. Note that the profile tested in this work is the WiMAX profile with the lower subcarrier spacing, so it is foreseeable that other profiles present better performance under these scenarios.

Competing interests

The authors declare that they have no competing interests.

Acknowledgements

The authors thank Ismael Rozas Ramallal for his support in developing and testing the testbed and conducting the measurement campaigns. This work was supported in part by Xunta de Galicia, MINECO of Spain, and by FEDER funds of the E.U. under Grant 2012/287, Grant IPT-2011-1034-370000, and Grant TEC2013-47141-C4-1-R.

References

1. UQ WiMAX gaining the wireless broadband market (2013). http://www.wimaxforum.org/LiteratureRetrieve.aspx?ID=177978
2. M Ehammer, E Pschernig, T Graupl, in *IEEE/AIAA 30th Digital Avionics Systems Conference (DASC)*. AeroMACS - an airport communications system (IEEE Seattle, USA, 2011), pp. 4–114116. doi:10.1109/DASC.2011.6095903
3. I Barhumi, G Leus, M Moonen, Optimal training design for MIMO OFDM systems in mobile wireless channels. IEEE Trans. Signal Process. **51**(6), 1615–1624 (2003). doi:10.1109/TSP.2003.811243
4. S Das, P Schniter, Max-SINR ISI/ICI-shaping multicarrier communication over the doubly dispersive channel. IEEE Trans. Signal Process. **55**(12), 5782–5795 (2007). doi:10.1109/TSP.2007.901660
5. Z Tang, RC Cannizzaro, G Leus, P Banelli, Pilot-assisted time-varying channel estimation for OFDM systems. IEEE Trans. Signal Process. **55**(5), 2226–2238 (2007). doi:10.1109/TSP.2007.893198
6. GB Giannakis, C Tepedelenlioglu, Basis expansion models and diversity techniques for blind identification and equalization of time-varying channels. Proc. IEEE. **86**(10), 1969–1986 (1998). doi:10.1109/5.720248
7. G Leus, in *Proc. of EUSIPCO*. On the estimation of rapidly time-varying channels (EURASIP Vienna, Austria, 2004), pp. 2227–2230
8. DK Borah, BT Hart, Frequency-selective fading channel estimation with a polynomial time-varying channel model. IEEE Trans. Commun. **47**(6), 862–873 (1999). doi:10.1109/26.771343
9. H Hijazi, L Ros, Polynomial estimation of time-varying multipath gains with intercarrier interference mitigation in OFDM systems. IEEE Trans. Vehicular Technol. **58**(1), 140–151 (2009). doi:10.1109/TVT.2008.923653
10. T Zemen, CF Mecklenbräuker, Time-variant channel estimation using discrete prolate spheroidal sequences. IEEE Trans. Signal Process. **53**(9), 3597–3607 (2005). doi:10.1109/TSP.2005.853104
11. KAD Teo, S Ohno, in *IEEE Global Telecommunications Conference, 2005. GLOBECOM '05*, vol. 6. Optimal MMSE finite parameter model for doubly-selective channels (IEEE St. Louis, USA, 2005), pp. 3503–3507. doi:10.1109/GLOCOM.2005.1578424
12. P Suárez-Casal, JA García-Naya, L Castedo, M Rupp, in *IEEE 14th Workshop on Signal Processing Advances in Wireless Communications. SPAWC '13*. KLT-based estimation of rapidly time-varying channels in MIMO-OFDM systems (IEEE Darmstadt, Germany, 2013)
13. A Stamoulis, SN Diggavi, N Al-Dhahir, Intercarrier interference in MIMO OFDM. IEEE Trans. Signal Process. **50**(10), 2451–2464 (2002). doi:10.1109/TSP.2002.803347
14. X Ma, GB Giannakis, S Ohno, Optimal training for block transmissions over doubly selective wireless fading channels. IEEE Trans. Signal Process. **51**(5), 1351–1366 (2003). doi:10.1109/TSP.2003.810304
15. F Hlawatsch, G Matz, *Wireless Communications Over Rapidly Time-Varying Channels*. (Academic Press, USA, 2011)
16. H Hijazi, L Ros, G Jourdain, in *Proceedings European Wireless Conference*. OFDM channel parameters estimation used for ICI reduction in time-varying multipath channels (IEEE Paris, France, 2007)
17. M Meidlinger, M Simko, Q Wang, M Rupp, in *IEEE International Conference on Communications 2013. ICC 2013*. Channel estimators for LTE-A downlink fast fading channels (IEEE Stuttgart, Germany, 2013)
18. J Rodríguez-Piñeiro, P Suárez-Casal, JA García-Naya, L Castedo, C Briso-Rodriguez, JI Alonso-Montes, in *IEEE 8th Sensor Array and Multichannel Signal Processing Workshop (SAM 2014)*. Experimental validation of ICI-aware OFDM receivers under time-varying conditions (IEEE A Coruña, Spain, 2014)
19. PS Casal, J Rodríguez-Piñeiro, JA García-Naya, L Castedo, in *The Eleventh International Symposium on Wireless Communication Systems (ISWCS)*. Experimental assessment of WiMAX transmissions under highly time-varying channels (IEEE Barcelona, Spain, 2014)
20. W Nam, YH Lee, Preamble-based cell identification for cellular OFDM systems. IEEE Trans. Wireless Commun. **7**(12), 5263–5267 (2008). doi:10.1109/T-WC.2008.071426
21. A Carro-Lagoa, P Suarez-Casal, J García-Naya, P Fraga-Lamas, L Castedo, A Morales-Mendez, Design and implementation of an OFDMA-TDD physical layer for WiMAX applications. EURASIP J. Wireless Commun. Netw. **2013**(1), 243 (2013). doi:10.1186/1687-1499-2013-243
22. Y Li, LJ Cimini, Bounds on the interchannel interference of OFDM in time-varying impairments. IEEE Trans. Commun. **49**(3), 401–404 (2001). doi:10.1109/26.911445
23. X Cai, GB Giannakis, Bounding performance and suppressing intercarrier interference in wireless mobile OFDM. IEEE Trans. Commun. **51**(12), 2047–2056 (2003). doi:10.1109/TCOMM.2003.820752

24. J Rodríguez-Piñeiro, JA Garcia-Naya, A Carro-Lagoa, L Castedo, in *Proc. of 16th Euromicro Conference on Digital System Design*. A testbed for evaluating LTE in high-speed trains (IEEE Santander, Spain, 2013)
25. Ettus USRP B210 (2013). https://www.ettus.com/product/details/UB210-KIT
26. Analog Devices AD9361 RFIC (2013). http://www.analog.com/en/rfif-components/rfif-transceivers/ad9361/products/product.html
27. Mini-Circuits high-power amplifier TVA-11-422 (2013). http://www.minicircuits.com/pdfs/TVA-11-422.pdf
28. Interline SECTOR IS-G14-F2425-A120-V vertically polarized antenna (2014). http://interline.pl/antennas/SECTOR-V120-14dBi-120deg-2.4-2.5GHz
29. Ubiquity AM-2G15-120 cross polarized antenna (2013). http://dl.ubnt.com/datasheets/airmaxsector/airMAX_Sector_Antennas_DS.pdf
30. M Matthaiou, DI Laurenson, C-X Wang, in *IEEE Wireless Communications and Networking Conference, 2008. WCNC 2008*. Capacity study of vehicle-to-roadside MIMO channels with a line-of-sight component (IEEE Las Vegas, USA, 2008), pp. 775–779. doi:10.1109/WCNC.2008.142

15

Energy optimisation of hybrid off-grid system for remote telecommunication base station deployment in Malaysia

Mohammed H Alsharif*, Rosdiadee Nordin and Mahamod Ismail

Abstract

Cellular network operators are always seeking to increase the area of coverage of their networks, open up new markets and provide services to potential customers in remote rural areas. However, increased energy consumption, operator energy cost and the potential environmental impact of increased greenhouse gas emissions and the exhaustion of non-renewable energy resources (fossil fuel) pose major challenges to cellular network operators. The specific power supply needs for rural base stations (BSs) such as cost-effectiveness, efficiency, sustainability and reliability can be satisfied by taking advantage of the technological advances in renewable energy. This study investigates the possibility of decreasing both operational expenditure (OPEX) and greenhouse gas emissions with guaranteed sustainability and reliability for rural BSs using a solar photovoltaic/diesel generator hybrid power system. Three key aspects have been investigated: (i) energy yield, (ii) economic factors and (iii) greenhouse gas emissions. The results showed major benefits for mobile operators in terms of both environmental conservation and OPEX reduction, with an average annual OPEX savings of 43% to 47% based on the characteristics of solar radiation exposure in Malaysia. Finally, the paper compares the feasibility of using the proposed approach in a four-season country and compares the results against results obtained in Malaysia, which is a country with a tropical climate.

Keywords: Mobile base station; Energy efficiency; Off-grid hybrid energy systems; Cost-effectiveness; Environmental impacts; HOMER

1 Introduction

The unexpected increase in subscribers and demand for high-speed data has led to tremendous growth in cellular networks in the last several years. In 2013, the number of mobile subscribers reached 6.8 billion [1], of whom 1 billion were in rural areas (off-grid sites) [2]. To access these new markets and provide a service to potential customers in remote rural areas, the number of BSs has been increased to fulfil the needs of mobile subscribers and increase the coverage area. According to [3], BSs are considered the primary source of energy consumption in cellular networks, accounting for 57% of the total energy used and putting the cellular operators under immense pressure to solve the problem of electricity supply in a reliable and cost-effective way. However, electrical grids are typically not available in these remote locations due to the geographical limitations (challenging terrain) that make access to these sites difficult. Therefore, supplying power to an off-grid BS is a significant challenge. Traditionally, a diesel generator (DG) is used to supply electrical power to a base station at an off-grid site [4]. Nevertheless, the concept of using DGs to power BSs has become much less viable for network companies looking to expand and deliver their services to potential new customers for the following reasons:

1. Economic aspect: most of the costs go to the fuel, which is expensive, and the fuel price will continue to increase in the future. Additionally, the transfer of fuel to these remote places characterised by challenging terrain may require special means of transport, such as helicopters, which increases the

* Correspondence: moh859@gmail.com
Department of Electrical Electronics and Systems Engineering, Faculty of Engineering and Built Environment, Universiti Kebangsaan Malaysia, 43600 Bangi, Selangor, Malaysia

cost of the operation [5]. Figure 1 presents the costs of operating DGs, which include the initial capital cost, installation cost, operating and maintenance (O&M) cost and fuel cost [6].

2. Environmental impact: air pollution by harmful components emitted from diesel fuels such as carbon dioxide, sulphur dioxide and nitrogen oxides causes global warming, the depletion of the ozone layer, cancer, genetic mutations and acid rain. According to [7], the ICT sector's CO_2 emissions will rise to 349 $MtCO_2$ by 2020, with 51% (179 $MtCO_2$) of the emissions coming from the mobile sector. Accordingly, increasing awareness of the environmental impact of greenhouse gas emissions and the exhaustion of non-renewable energy resources (fossil fuel) has highlighted the critical need to improve the energy efficiency of cellular networks.
3. Technical issues: the efficiency of the DG system is low, with only approximately 30% of the fuel energy being converted to electrical energy and the rest lost as heat [2]. In addition, BS sites supplied by DGs are not reliable; 65% of the telecom service losses are due to outages resulting from a variety of the failures from which these generators suffer [2]. When a failure occurs, the input of substantial time and money is generally required to bring a technician on site to perform the necessary repairs.

These drawbacks have motivated cellular network operators to search for solutions to both promote environmental conservation and reduce capital and operational costs. The specific power supply requirements for rural BSs, such as cost-effectiveness, efficiency, sustainability and reliability, can be met by utilising the technological advances in renewable energy. The network equipment deployed in these remote locations must be optimised so that they save energy while providing the best service

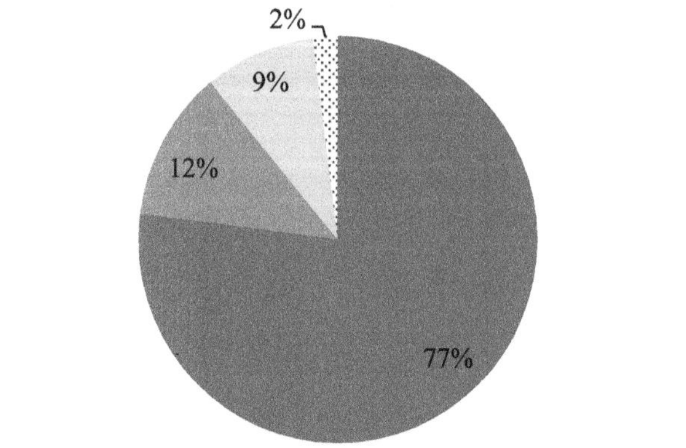

Figure 1 Net costs of operating a DG [6].

with the largest coverage feasible. If network companies can install renewable energy supplies for their remote sites, they would gain access to hundreds of millions of potential new customers. Furthermore, by adopting renewable energy sources, network operators can reduce their operating expenses and have a positive impact on the environment via the reduction of harmful gas emissions.

The rest of this paper is organised as follows. Section 2 presents studies that have investigated the possibility of reducing both the OPEX and greenhouse gas emissions through the use of renewable energy systems in different zones of the world. Section 3 discusses the potential for using renewable energy to supply the BSs in remote places in Malaysia, and Section 4 describes the use of solar energy in Malaysia, including the characteristics of the solar radiation of Malaysia and the barriers to using solar photovoltaic (SPV) panels in Malaysia, as well as some recommendations. In Section 5, the system architecture for the hybrid power model to supply long-term evolution (LTE)-BS is described, and Section 6 presents the mathematical model. Section 7 briefly introduces the hybrid optimisation model for electric renewables (HOMER) software, a hybrid power system modelling tool used in this study, and Section 8 includes the simulation configuration. Section 9 presents the results and discussion. Section 10 presents the comparison and estimation of the feasibility of using the solar energy approach in Germany by using Malaysia as the central point of reference, and Section 11 concludes this paper.

2 Related studies

Over the last 5 years, significant developments have been made in integrating electrical grids and renewable energy into a smart grid to manage the power supply of BS sites. In India, efforts have been made to optimise the size of wind turbine generators (WTGs), SPV arrays and other components for a hybrid power system; generator-based power supplies for global system for mobile communication (GSM, alternatively referred to as 2G) and code division multiple access (CDMA, 3G) standards have also been investigated [8,9]. Reference [10] studied the feasibility of implementing a SPV/diesel hybrid power generation system suitable for a GSM base station site in Nigeria. Waqas et al. [11] presented the design idea for a solar system with a diesel generator as a backup source for a GSM cellular network standard in Pakistan. Reference [12] studied the feasibility of implementing an SPV/diesel hybrid power generation system suitable for a GSM base station site in Bangladesh. Martinez-Diaz et al. [13] discussed a photovoltaic (PV)-wind-diesel-battery system for a station in Spain. In Nepal, reference [6] studied the optimisation of a hybrid PV-wind power system for a remote telecom station. Kanzumba et al. [2] investigated the possibility of using hybrid photovoltaic/wind renewable systems as primary sources of energy to supply mobile telephone base transceiver stations in the rural regions of the Republic of the Congo. Reference [14] discussed three types of renewable energy: (i) a SPV-battery system, (ii) an SPV-fuel cell (FC) system and (iii) an SPV-FC-battery system. The modelling and size optimisation of such hybrid systems feeding a stand-alone direct current (DC) load at a telecom base station have been carried out using the HOMER software. Vincent et al. [15] proposed a hybrid (solar and hydro) and DG system based on the power system models for powering stand-alone BS sites. Table 1 provides a summary of related works that have been investigated for green wireless network optimisation strategies within smart grid environments.

3 Potential for applying renewable energy as the energy supply for BSs in remote sites in Malaysia

Malaysia lies entirely within the equatorial region, between 1° and 7° N and 100° and 120° E, offering an abundant potential for the use of renewable energy resources, especially solar and wind power [16].

In the early 1980s, a study of Malaysia's wind energy was undertaken at University Kebangsaan Malaysia (UKM). The Solar Energy Research Group of UKM collected wind data from ten stations throughout the country over a period of 10 years from 1982 to 1991. The data studied include the hourly wind speed at the stations, which are located mostly at airports and near coasts, where the land and sea breezes may influence the wind regime [17]. The study showed that the mean wind speed is low, not exceeding 2 m/s. However, the wind does not blow uniformly, varying by month and region. The greatest wind power potential is that of Mersing and Kuala Terengganu, which are located on the East Coast of Peninsular Malaysia [18]. In 2014, reference [19] presented a study on predicting the wind speed in these states over the long-term using neural networks and a set of recent wind speed measurement samples from the two meteorological stations in these states. The results showed that the future mean wind speed will remain low, not exceeding 3 m/s. Therefore, wind turbines are rarely used in Malaysia and were excluded from this study.

Regarding solar energy, Malaysia has a stable climate throughout the year. Hence, the solar radiation in Malaysia is highly relative to global standards. Malaysia's solar power is estimated to be four times the power of the world's fossil fuel resources [20]. The global irradiation fluctuated in the range of 2 to 6 kWh/m^2/day. The second part of the year (October to February) is characterised by more cloud cover

Table 1 Summary of the approach that was discussed

Year of publication	Hybrid resources	Storage and support system	Cellular generation	Case study
2009 [14]	Photovoltaic	Battery	GSM	India
	Wind	Fuel cell		
		Diesel generator		
2010 [9]	Photovoltaic	Battery	GSM	India
	Wind	Diesel generator	UMTS	
2012 [8]	Photovoltaic	Battery	GSM	India
	Wind	Polar DC generator	CDMA	
2012 [2]	Photovoltaic	Battery	GSM	Congo
	Wind	Diesel generator		
2012 [12]	Photovoltaic	Battery	GSM	Bangladesh
		Diesel generator		
2013 [10]	Photovoltaic	Battery	GSM	Nigeria
		Diesel generator		
2013 [13]	Photovoltaic	Battery	GSM	Spain
	Wind	Diesel generator		
2013 [6]	Photovoltaic	Battery	GSM	Nepal
	Wind	Fuel cell	CDMA	
		Diesel generator		
2013 [15]	Photovoltaic	Battery	GSM	Nigeria
	Hydro turbine	Diesel generator		
2013 [11]	Photovoltaic	Battery	GSM	Pakistan
		Diesel generator		

and thus poorer solar potential than the first part of the year (March to October), and the average temperature ranges from 33°C during the day to 23°C at night [21]. Moreover, solar cells have low-maintenance needs and high reliability, with an expected lifespan of 20 to 30 years. One square metre of solar panelling in Malaysia is estimated to result in an annual reduction of 40 kg of CO_2 [18], making solar power a promising future energy source for telecommunication applications.

4 Solar energy in Malaysia

Solar radiation data in Malaysia have been the subject of earlier studies. Malaysia's climatic conditions are desirable for extending the utilisation of SPV systems due to the high amount of solar radiation received throughout the year. The northern region and a few places in eastern Malaysia receive the highest amount of solar radiation throughout the year. The lowest irradiance value is obtained for Kuching, whereas Kota Kinabalu has the highest measured solar radiation [16]. Figure 2 provides information on solar radiation in different states of Malaysia.

Figure 2 reveals that Sabah, Perlis and Kedah have sufficient solar resources to support solar energy applications. Figure 3 presents the daily solar radiation in these three states.

The average daily solar radiation in Kedah, Sabah and Perlis is equal to 5.48, 5.31 and 5.24 kWh/m^2, respectively. According to reference [22], the average solar radiation in these states indicates the high potential of SPV where the solar radiation is higher than 5 $kWh/m^2/day$. Moreover, Figure 4 shows Malaysia's average daily solar radiation, which is estimated to be 5.15 kWh/m^2 [18].

Accordingly, this study will investigate five average daily solar radiation values to cover all the states of Malaysia: 5.1, 5.2, 5.3, 5.4 and 5.5 kWh/m^2.

4.1 Barriers to using SPV panels for remote areas in Malaysia

The aforementioned discussion reveals that Malaysia has excellent solar energy potential. However, some barriers that affect the performance of SPV panels should be considered, as power shortages are not admissible in the cellular network sector.

One of the major drawbacks of using solar cells is shade, which may dramatically reduce the power produced by the solar cell. In some seasons, tropical countries experience heavy rain and cloudiness that may continue for several days, causing battery banks to run out of charge more quickly. To solve this problem, the present study suggests the hybridisation of the solar power system with existing backup DG in rural areas, which will provide BSs with a sustainable and reliable power supply, especially if the battery lifespan is short.

Dirt, dust, tree debris, moss, sap, water spots, mould, etc. on solar panels have a significant impact on the performance of solar power systems. Cleaning the panels is also a problem. The SPV panels are installed at relatively high sites to maximise their access to the sunlight, and moss and grass grow quickly on the panels, increasing the cost of cleaning the panels. Therefore, we recommend installing solar panels on tilted roofs (with tilts of as little as 5° to 10°) to allow wind, rain and gravity to remove most debris and dust naturally.

In addition, the rapid growth of the surrounding trees will shade the panels, decreasing the performance of PV panels. Solar PV panels mounted on high poles also attract lightning strikes, the destructive voltage of which destroys the electronic components. Furthermore, the bypass diodes, which are mounted in the termination box under each panel, may crack under high humidity, heat and short circuits. One of the solutions is to engage

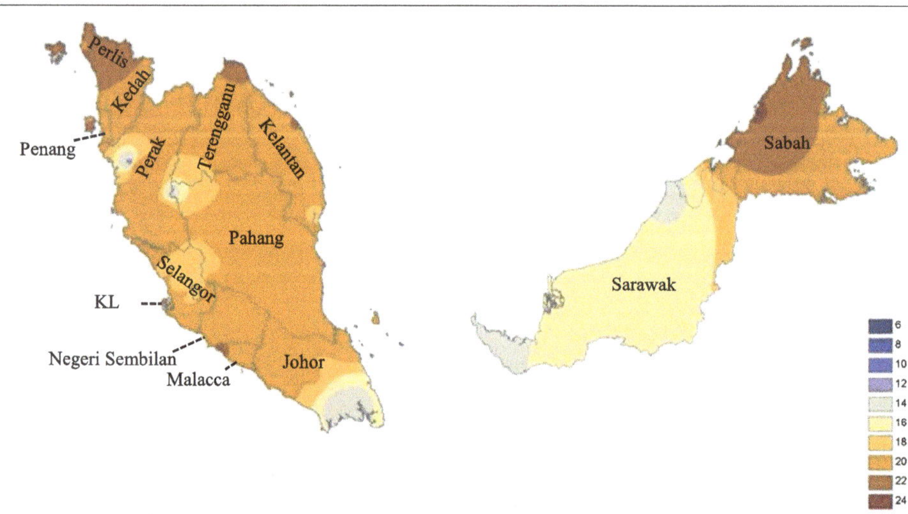

Figure 2 Annual average solar radiation ($MJ/m^2/day$) [16].

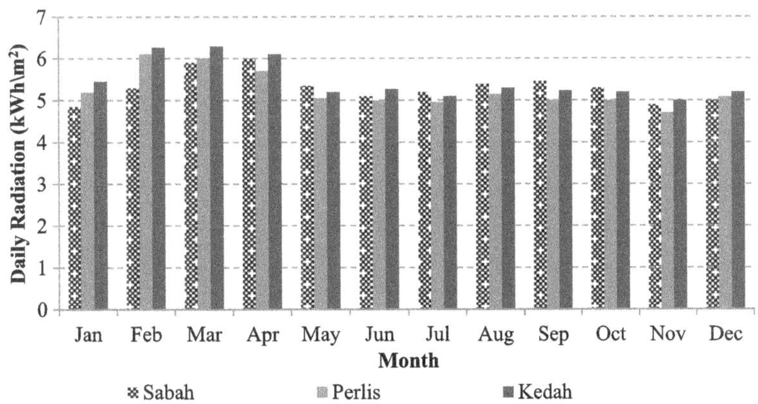

Figure 3 Average daily radiation (kWh/m²) [18].

trained educated villagers to operate, maintain and manage the solar panels.

5 System architecture

Figure 5 is a schematic showing two subsystems: the BS and the hybrid energy source.

5.1 Base station subsystem

The BS, a centrally located set of equipment used to communicate with mobile units and the backhaul network, consists of multiple transceivers (TRXs), which in turn consist of a power amplifier (PA) that amplifies the input power, a radio-frequency (RF) small-signal transceiver section, a baseband (BB) for system processing and coding, a DC-DC power supply, a cooling system and an alternating current (AC)-DC unit for connection to the electrical power grid. Table 2 summarises the power consumption of the different pieces of LTE-macro BS equipment for a 2 × 2 MIMO configuration, and three sectors where the total input power (*P*in) needed by the LTE-macro BS is 964.9 W. More details on the BS internal components can be found in [23].

5.2 Hybrid energy source subsystem

The main components of a hybrid energy source subsystem are listed below:

1. Solar panels: responsible for collecting sunlight and converting the sunlight into DC electricity.
2. Diesel generator: used as a secondary energy source during the peak demand or in the case of battery depletion. The optimum operation range for a diesel generator is between 70% and 89% of its rated power [24].
3. Battery bank: stores excess electricity for future consumption by the BS at night, during load-shedding hours, or if the available solar energy is not sufficient to feed the BS load completely. To protect the battery, inclusion of a charge controller is recommended. A charge controller or battery

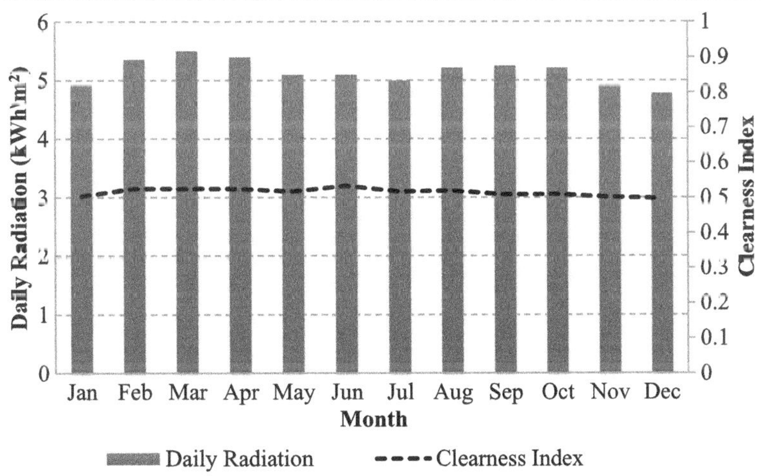

Figure 4 Average daily solar energy received in Malaysia [18].

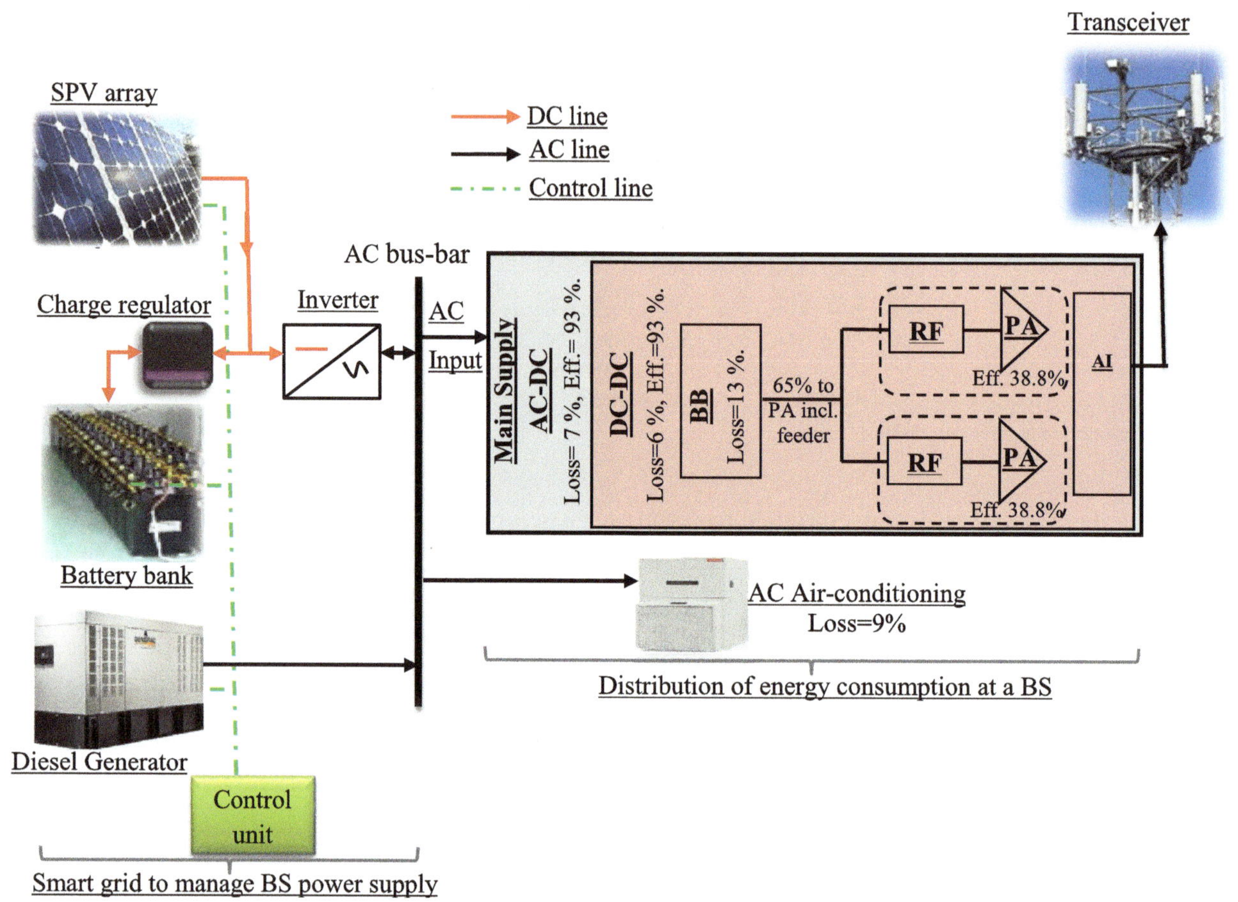

Figure 5 System model of an adaptive power management scheme for remote LTE-BS.

regulator limits the rate at which the electric current is added to or drawn from electric batteries, prevents overcharging and may protect against overvoltage, which can reduce battery performance or lifespan and may pose a safety risk. A charge controller may also prevent the complete draining ('deep discharging') of a battery or perform controlled discharges, depending on the battery technology, to protect battery life [24].

4. Inverter: converts the DC voltage from the load bus-bar and battery to AC voltage at a high efficiency to satisfy the (BS) requirement of the main load for uninterruptible AC power. The inverter is also able to log information such as system performance (e.g. electricity produced by the system on a daily, monthly or yearly basis) and safety measures to avoid electrical mishaps [24].
5. Control system: serves as the brain of a complex control, regulation and communication system. The most common communication units in the remote interface are wireless modems or network solutions. In addition to the control functions, data logger and alarm memory capabilities are of high importance. All power sources working in parallel are managed by a sophisticated control system and share the load with their capabilities to accommodate the fact that power shortages are not admissible in the cellular telephony sector.

6 Mathematical model

6.1 Photovoltaic system

The SPV generator contains modules that are composed of many solar cells interconnected in series/parallel to form a solar array. HOMER calculates the energy output of the SPV array (*E*SPV) by using the following equation [25]:

$$E_{\mathrm{SPV}} = Y_{\mathrm{SPV}} \times \mathrm{PSH} \times f_{\mathrm{SPV}} \tag{1}$$

where Y_{SPV} is the rated capacity of the SPV array (kW) and PSH is a peak solar hour which is used to express solar irradiation in a particular location when the sun is shining at its maximum value for a certain number of hours. Because the peak solar radiation is 1 kW/m^2, the number of peak sun hours is numerically equal to the daily solar radiation in kWh/m^2 [18] and f_{SPV} is the SPV derating factor (sometimes called the performance ratio), a scaling factor meant to account for effects of dust on

Table 2 Power consumption of the LTE-BS hardware elements [23]

Item	Notation	Unit	Macro
PA	Max transmit (rms) power, P_{max}	W	39.8
	Max transmit (rms) power	dBm	46.0
	PAPR	dB	8.0
	Peak output power	dBm	54.0
	PA efficiency, μ	%	38.8
	Total PA (P_{PA}) = $\frac{P_{max}}{\mu}$	W	102.6
TRX	P_{TX}	W	5.7
	P_{RX}	W	5.2
	Total RF (P_{RF})	W	10.9
BB	Radio (inner Rx/Tx)	W	5.4
	Turbo code (outer Rx/Tx)	W	4.4
	Processor	W	5.0
	Total BB (P_{BB})	W	14.8
DC-DC loss, σ_{DC}		%	6.0
Cooling loss, σ_{cool}		%	9.0
AC-DC (main supply) loss, σ_{MS}		%	7.0
Total per TRX = $\frac{P_{PA}+P_{RF}+P_{BB}}{(1-\sigma_{DC})(1-\sigma_{cool})(1-\sigma_{MS})}$		W	160.8
Number of sectors		#	3
Number of antennas		#	2
Number of carriers		#	1
Total number of transceivers (N_{TRX})		#	6
Total number of N_{TRX} chains, Pin = N_{TRX} × Total per TRX		W	964.9 s

the panel, wire losses, elevated temperature or anything else that would cause the output of the SPV array to deviate from the expected output under ideal conditions. In other words, the derating factor refers to the relationship between actual yield and target yield, which is called the efficiency of the SPV. Today, due to improved manufacturing techniques, the performance ratio of solar cells increased to 85% to 95%.

6.2 Diesel generator

The energy generated (*E*DG) by a DG with a given rated power output (*P*DG) is expressed in Equation 2 [25]:

$$E_{DG} = P_{DG} \times \mu \times t \tag{2}$$

where μ is the efficiency of the DG. Moreover, the fuel consumption (FC) is calculated as follows [25]:

$$F_C = E_{DG} \times F_{spe} \tag{3}$$

where E_{DG} is the energy production (kWh) and F_{spe} is the specific fuel consumption (L/kWh).

6.3 Battery model

The battery characteristics that play a significant role in designing a hybrid renewable energy system are battery capacity (Ah), battery voltage (V), battery state of charge (%), depth of discharge (%), days of autonomy (h), efficiency (%) and lifetime of battery (year). The 'Trojan L16P' battery model, which provides good characteristics as shown in Figure 6 combined with low cost, is used in this paper. More details can be found in [26].

The nominal capacity of the battery bank is the maximum state of charge SOC_{max} of the battery. The minimum state of charge of the battery, SOC_{min}, is the lower limit that does not discharge below the minimum state of charge, which is 30% in the 'Trojan L16P' battery model, as shown in Figure 6. The depth of discharge (DOD) is used to describe how deeply the battery is discharged and is expressed in Equation 4 [6]:

$$\mathrm{DOD} = 1\text{-}\mathrm{SOC}_{min} \tag{4}$$

Based on Equation 4, the DOD for the 'Trojan L16P' battery is 70%, which means that the battery has delivered 70% of its energy and has 30% of its energy reserved. DOD can always be treated as how much energy the battery delivered.

A battery bank is used as a backup system and is sized to meet the load demand when the renewable energy resources failed to satisfy the load. The number of days a fully charged battery can feed the load without any contribution of auxiliary power sources is represented by days of autonomy. The battery bank autonomy is the ratio of the battery bank size to the electric load (LTE-BS). HOMER calculates the battery bank autonomy (A_{batt}) by using the following equation [25]:

$$A_{batt} = \frac{N_{batt} \times V_{nom} \times Q_{nom}\left(1-\frac{SOC_{min}}{100}\right)(24\ \mathrm{h/day})}{L_{prim\text{-}avg}(1,000\mathrm{Wh/kWh})} \tag{5}$$

where N_{batt} is the number of batteries in the battery bank, V_{nom} is the nominal voltage of a single battery (V), Q_{nom} is the nominal capacity of a single battery (Ah) and $L_{prim,ave}$ is the average daily LTE-BS load (kWh).

Battery life is an important factor that has a direct impact on replacement costs. Two independent factors may limit the lifetime of the battery bank: the lifetime throughput and the battery float life. HOMER calculates the battery bank life (R_{batt}) based on these two factors as given in the following equation [25]:

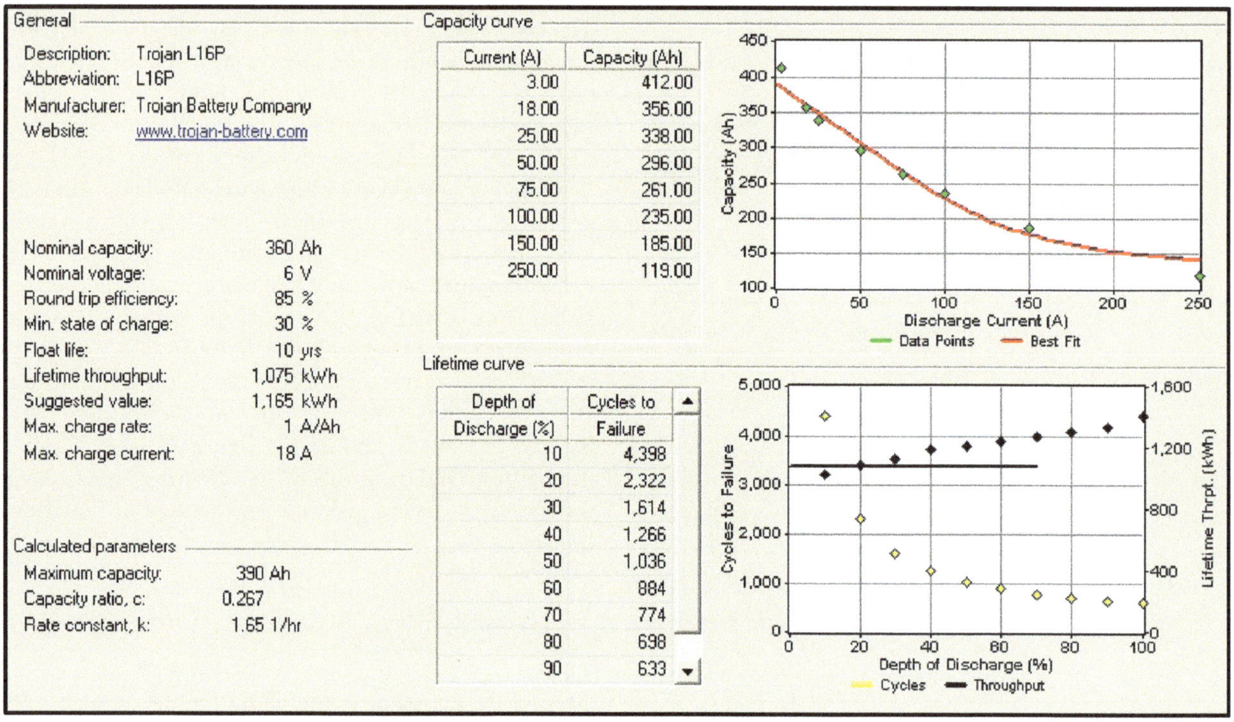

Figure 6 'Trojan L16P' battery model characteristics.

$$R_{\text{batt}} = \min\left(\frac{N_{\text{batt}} \times Q_{\text{lifetime}}}{Q_{\text{thrpt}}}, R_{\text{batt,f}}\right) \quad (6)$$

where Q_{lifetime} is the lifetime throughput of a single battery (kWh), Q_{thrpt} is the annual battery throughput (kWh/year) and $R_{\text{batt,f}}$ is the battery float life (year).

6.4 DC/AC inverter

Inverters convert electrical energy from the DC form into the AC form with the desired frequency of the load. The efficiency of the inverter is assumed to be roughly constant over the entire working range (e.g. 90%) [27].

The optimum criteria, including economic, technical and environmental feasibility parameters, were analysed using the HOMER software package developed by the National Renewable Energy Laboratory (NREL).

7 HOMER hybrid power system modelling software

HOMER [25] is an optimisation software package used to simulate various renewable energy source (RES) system configurations and scale them based on the net present cost (NPC). The NPC represents the life cycle cost of the system. The calculation assesses all costs that occur within the project lifetime, including initial setup costs (IC), component replacements within the project lifetime and maintenance. Figure 7 presents the architecture of the HOMER software.

HOMER calculates the NPC according to the following equation [25]:

$$\text{NPC} = \frac{\text{TAC}}{\text{CRF}} \quad (7)$$

where TAC is the total annualised cost ($). The capital recovery factor (CRF) is given by:

$$\text{CRF} = \frac{i(1+i)^n}{(1+i)^n - 1} \quad (8)$$

where n is the project lifetime and i is the annual real interest rate. HOMER assumes that all prices escalate at the same rate and applies an annual real interest rate rather than a nominal interest rate.

The discount factor (fd) is a ratio used to calculate the present value of a cash flow that occurs in any year of the project lifetime. HOMER calculates the discount factor by using the following equation [25]:

$$f_d = \frac{1}{(1+i)^n} \quad (9)$$

NPC estimation in HOMER also considers the salvage cost, which is the residual value of the power system

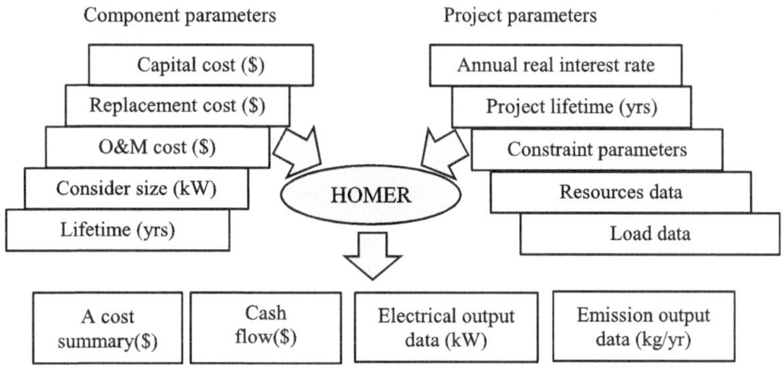

Figure 7 Architecture of HOMER software.

components at the end of the project lifetime. The equation used to calculate the salvage value (S) is as follows:

$$s = \text{rep}\left(\frac{\text{rem}}{\text{comp}}\right) \tag{10}$$

where rep is the replacement cost of the component, rem is the remaining lifetime of the component and comp is the lifetime of the component.

8 Simulation configuration

The lifetime of the project is 20 years, which represents the lifetime of the BS equipment and the solar cells. Between all the components in a base station, solar cells have been found to be the most expensive in terms of capital cost. Hence, solar cells have a lifetime of 20 years, which is the same as the project lifetime, so neither BS nor the solar cells require replacement during the 20-year period. In addition, the next-generation network, i.e. fifth generation (5G), is predicted to be implemented in the next 20 years, based on historical evolution from the previous legacy network cycle, e.g. 3G and 2G. The Malaysian annual real interest rate was 3.25% in 2014 [28]. HOMER makes a decision in each time step to meet the power needs at the lowest cost, subject to the constraints of the dispatch strategy chosen in the simulation and a set point of 80%. The system must supply electricity to both the load (base station system) and the backup power system each hour. In the present study, the backup power needs 10% of the hourly load requirement to retain enough spare capacity to serve the load even under a sudden 10% decrease in the renewable energy output within an hour. Moreover, several sets of sizes will be considered in the simulation, taking into account the SPV, inverter and number of batteries needed to achieve cost-effective, reliable and efficient performance in the optimisation process. The efficiency of the inverter is assumed to be roughly constant over the working range (e.g. 90%, based on [27]), and the battery efficiency is taken to be 85%. The DG configurations are based on [2], where the DG cost is \$660/kW. The replacement cost is assumed to be \$660/kW, and the operating and maintenance costs are \$0.05/h. The diesel price is \$0.7/L, and the carbon emission penalty is internationally set to \$2.25/t. The technical specifications, costs, economic parameters and system constraints that are used in the present study are given in more detail in Table 3 below.

9 Results and discussion

Different average daily solar radiation values of 5.1, 5.2, 5.3, 5.4 and 5.5 kWh/m^2 are used to simulate the application of solar energy across a wide range of Malaysian states (for a detailed discussion, see the first paragraph of Section 4). The total power consumption by the LTE-BS is 965 W (details given in Table 2). Additional configuration details are given in Table 3. The energy output, the economic analysis of the proposed hybrid systems and the related sensitivity analysis are provided in the following paragraphs.

9.1 Optimisation criteria

Table 4 includes a summary of the technical and economic criteria for the optimal design of the hybrid SPV/DG system at different daily radiation values.

The optimal size of the solar energy system is obviously the same for all solar radiation rates proposed (5.1 to 5.5 kWh/m^2/day) for the same capacity of the DG (1 kW). However, the energy contribution differs, with the contribution of energy from the solar power system increasing with increasing radiation rate. This increase will decrease the energy contribution of the DG, lowering its operating period and providing the benefits of reducing both the operating costs and pollution rate.

The system costs consist of the following: (i) the initial capital cost is paid at the beginning of the project and decreases with decreasing size of the elements of the project, with the largest proportion of these costs going

Table 3 Simulation setup of the SPV/DG hybrid system

System components	Parameters	Value
Control parameters	Annual real interest rate	3.25%
	Project lifetime	20 years
	Dispatch strategy	Cyclic charging
	Apply set point state of charge	80%
	Operating reserve: as percent of load, hourly load	10%
	Carbon emission penalty	$2.25/t
	Diesel price	$0.7/L
SPV	Sizes considered	1, 1.5, 2, 2.5 kW
	Operational lifetime	20 years
	Efficiency	90%
	System tracking	Two axis
	Capital cost	$4/W
	Replacement cost	$4/W
	O&M cost per year	$0.01/W
DG	Sizes considered	1, 2, 3, 4 kW
	Operational lifetime	25,000 h
	Intercept coeff.	0.08 L/h/kW
	Capital cost	$0.66/W
	Replacement cost	$0.66/W
	O&M cost	$0.05/h
Inverter	Sizes considered	0.5, 1, 1.5, 2 kW
	Efficiency	90%
	Operational lifetime	15 years
	Capital cost	$0.9/W
	Replacement cost	$0.9/W
	O&M cost per year	$0.01/W
Trojan L16P battery	Number of batteries	2, 4, 5, 6, 7, 8
	Round trip efficiency	85%
	Minimum state of charge	30%
	Nominal voltage	6 V
	Nominal current	360 Ah at 20 h
	Nominal capacity	6 V × 360 Ah = 2.16 kWh
	Lifetime throughput	1,075 kWh
	Max. charge rate	1 A/Ah
	Max. charge current	18 A
	Self-discharge rate	0.1% per hour
	Min. operational lifetime	5 years
	Capital cost	$300
	Replacement cost	$300
	O&M cost per year	$10

towards solar cells because of their high cost (approximately $4/W). Table 4 shows that the initial capital cost is fixed because the optimal system size is the same for all average daily solar radiation values studied. (ii) The operating cost is paid annually, and most of this cost goes towards operating and maintaining the DG. Table 4 indicates that the operating cost decreases with increasing solar radiation at the same optimal size of the system due to the increase in the energy contribution from the solar power system and the decrease in the energy contribution of the DG, which reduces the operating period of the DG. The NPC represents all costs that occur within the project lifetime, including initial setup costs, component replacements within the project lifetime and maintenance. More details will be provided in the next subsections.

9.2 Energy yield analysis

Figure 8 summarises the annual energy contribution of the solar electric system and the DG at different average daily solar radiation values. Higher solar radiation rates clearly correspond to higher annual energy contributions of the solar power system for the same SPV size stem series and less energy consumed by the generator. Thus, a higher solar radiation rate is desirable, as it decreases the maintenance and operational costs for the DG and the emission of polluting gases into the environment. The annual energy contribution of the solar electric system ranged from 43% to 47% for an average daily solar radiation value ranging from 5.1 to 5.5 kWh\m^2, respectively.

The following statistical analysis discusses the energy production based on an average daily solar radiation for Malaysia of 5.1 kWh/m^2 as a case study based on the equations that are given in Section 6. However, this analysis can be extended to other cases, yielding a slight difference in daily peak solar hours per case depending on the average daily solar radiation.

The average annual energy consumption by LTE-macro BS is 8,453 kWh, which is computed as an LTE-macro BS load of 965 W, as given in Table 2 × 24 h × 365 days/year. The annual energy contribution of the solar system to the total energy production is 4,290 kWh, which is computed based on Equation 1. SPV rated capacity is 2 kW × peak solar hours 5.1 h × SPV derating factor 0.9 × 365 days/year, which equals 3,351 kWh. However, the tracking system plays a role in increasing the total amount of energy produced by a solar system by approximately 20% to 25%. The present simulation adopted a dual axis tracker which increased the total amount of energy by approximately 21.90% to become the annual energy contribution of the solar system at 4,290 kWh. The DG covered the remaining portion of the energy by 5,573 kWh, which represents 57% of the total energy production.

Table 4 Summary of the technical and economic criteria for the optimal design of the hybrid SPV/DG system

Energy model					Economic factors			DG factors	
Daily solar (kWh/m²)	SPV (kW)	DG (kW)	Battery (unit)	Inverter (kW)	Initial Capital	Operating ($/year)	NPC ($)	Diesel (L)	DG (h)
5.1	2	1	4	1.5	$11,210	1,993	40,188	1,880	6,091
5.2	2	1	4	1.5	$11,210	1,980	40,000	1,866	6,039
5.3	2	1	4	1.5	$11,210	1,968	39,820	1,854	5,999
5.4	2	1	4	1.5	$11,210	1,956	39,653	1,840	5,954
5.5	2	1	4	1.5	$11,210	1,946	39,497	1,830	5,923

The total annual energy production of the hybrid system is 9,863 kWh (4,290 kWh of solar system + 5,573 kWh of DG), while the total annual energy needed by LTE-macro BS is 8,453 kWh. The difference between electrical production and consumption is equal to the excess electricity of 888 kWh/year, plus the battery losses of 112 kWh/year, plus inverter losses of 410 kWh/year.

The maximum energy contribution of the solar power system occurred in March and August. The average monthly energy contribution ranged from 510 to 564 kWh, and the average monthly energy contribution of the DG ranged from 675 to 636 kWh at 5.1 and 5.5 kWh/m^2/day, respectively. Meanwhile, the minimum energy contribution of the solar power system occurred in October and February. The average monthly energy contribution was 467 kWh at 5.1 kWh/m^2/day and 564 kWh at 5.5 kWh/m^2/day. These results are attributed to the differences in the average solar radiation rate, as the second part of the year (October to February) includes more cloud cover and thus lower solar potential than the first part of the year (March to October).

The number of batteries in a string size is four batteries in parallel arrangement where the battery annual energy-in is 770 kWh, while the annual energy-out is 655 kWh, where the round-trip efficiency was 85%. Batteries can supply LTE-BS load autonomy for 6.27 h, which is computed based on Equation 5, (number of the batteries is 4 × nominal voltage of a single battery 6 V × nominal capacity of a single battery 360 Ah × 0.7 × 24) divided by (daily average LTE-BS load 23.2 kWh). However, one battery can supply LTE-BS load autonomy 1.57 h. The battery expected life is 6 years, based on Equation 6, number of the batteries 4 × battery lifetime throughput 1,075 kWh divided by battery annual throughput (713 kWh). Moreover, the inverter annual energy-in is 3,545 kWh, while the annual energy-out is 3,191 kWh, with 90% efficiency and 6,047 h/year operation.

9.3 Economic analysis

The following analysis discusses the cash flow based on an average daily solar radiation for Malaysia of 5.1 kWh/m^2 as a case study, as shown in Figure 9. However, this analysis can be extended to other cases, yielding a slight difference in operating cost per case depending on the energy contribution of the solar power system, as explained in the above subsections.

The *initial capital cost*, paid once at the beginning of the project, is directly proportional to the size of the system. From Table 4, the initial capital cost is fixed at

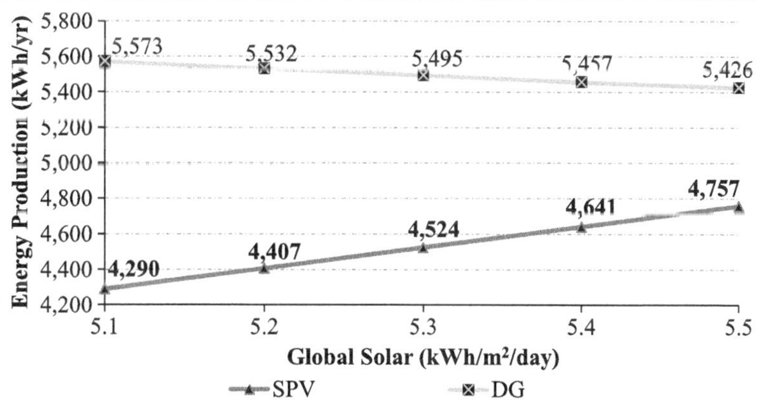

Figure 8 Annual energy contribution of various sources for different average solar radiation values.

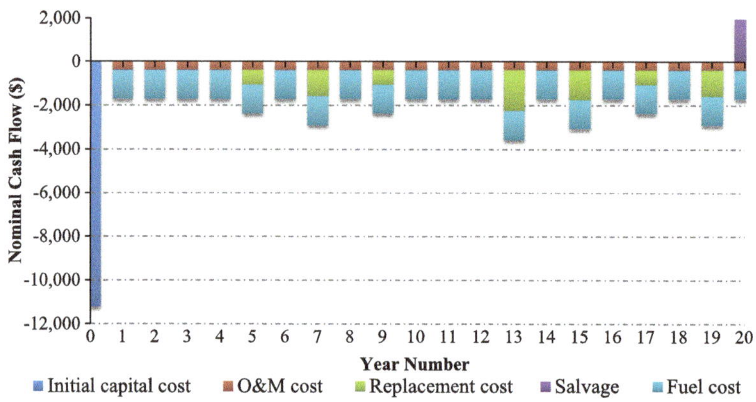

Figure 9 **Cash flow summary of the SPV/DG hybrid power system within the project lifetime for a solar radiation value of 5.1 kWh/m²/day.**

$11,210 because the optimal design found by the HOMER software for the hybrid power system is the same for all daily solar radiation values studied. The breakdown of this cost is as follows: (i) 71.4% for the SPV (size 2 kW × cost $4,000/1 kW = $8,000), (ii) 5.9% for the DG (size 1 kW× cost $660/1 kW = $660), (iii) 10.7% for the battery units (4 units × cost $300/unit = $1,200) and (iv) 12% for the inverter (size 1.5 kW × cost $900/1 kW = $1,350). In addition, the charger controller costs $2,000, and the control system cost depends on the system chosen. The SPV clearly represents the bulk of this cost.

The annual cost for the *maintenance and operation* of the system amounted to $1,707, as shown in Figure 9. Here, the DG represents the bulk of this cost ($1,621/year, including the fuel cost). A breakdown of this cost is $305 for DG maintenance per year based on a DG maintenance cost of $0.05/h (as given in Table 3) × annual DG operating hours 6,091 h (as given in Table 4) and a fuel cost of $1,316, equal to a total diesel consumption of 1,880 L per year, as shown in Table 4 (computed based on a specific fuel consumption of 0.388 L/kWh × annual electrical production of the DG of 5,573 kWh/year at 5.1 kWh/m^2/day, as given in Figure 8) multiplied by the diesel price of $0.7/L. Thus, lower DG energy consumption and increased reliance on the solar power system create more operational expenses and are achieved under a higher solar radiation rate. Figure 10 highlights on the annual diesel cost for each solar radiation rate.

Regarding the other components, the batteries, SPV and inverter cost $40/year, $20/year and $15/year, respectively.

Generally, the bulk of the *replacement cost* goes to the components with short operational lifetimes, which are the DG (4 years) and batteries (expected 6 years). The DG will be changed in years 5, 9, 13 and 17 at a cost of $660, while the batteries will be changed at a cost of $1,200 (4 units × $300/unit) in years 7 and 13. Moreover, the cost of replacing the inverter in year 15 is $1,350 (1.5 kW × $900/1 kW). The SPV array has a lifetime of 20 years, which is the same as the project lifetime, so neither requires replacement. Thus, the total replacement cost during the project lifetime is $7,590.

The economic analysis described above has been conducted on the basis of the nominal system. However, Figure 11, below, shows the discount factor for each year of the project lifetime based on Equation 9.

The NPC of a system is the present value of all the costs that it incurs over its lifetime, minus the present value of all the revenue that it earns over its lifetime. Costs include capital costs, replacement costs, O&M costs and fuel costs. Revenues include salvage value. The total NPC calculates by summing up the total discounted cash flows in each year of the project lifetime as follows: include capital costs $11,210 + replacement costs $5,198 + O&M costs $5,680 + fuel costs $19,138 – salvage $1,038, equal $40,188 for solar radiation rate 5.1 kWh/m^2. Figure 12 describes a comparison of the solar radiation rate and total NPC.

The NPC clearly decreases with increasing solar radiation at the same optimal size of the system due to the increase in the energy contribution from the solar power

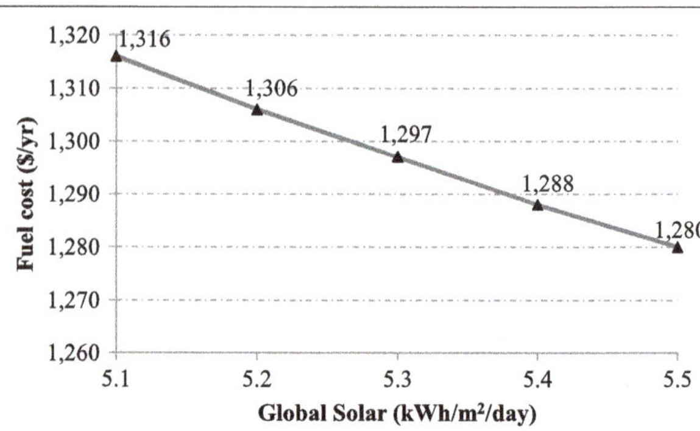

Figure 10 **Fuel cost vs. global solar radiation rate.**

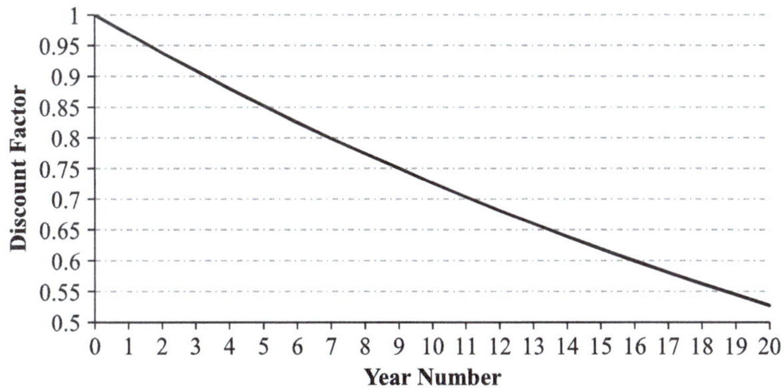

Figure 11 Discount factor for each year of the project lifetime.

system and the decrease in the energy contribution of the DG, which reduces the operating period of the DG, hence decreases the total operating and maintaining cost as well the fuel cost.

9.4 GHG emissions

Due to improved manufacturing techniques and higher volumes, the carbon footprint of solar panels is much lower than it once was. In general, as the solar radiation rate increases, the pollution rate decreases because the energy consumption of the DG and thus the operating period of the DG decreases. Figure 13 compares the solar radiation rate and annual pollutant emissions.

10 Comparison of the feasibility of using the solar energy approach between Germany and Malaysia

Germany is the world's top solar photovoltaic consumer. The annual number of daylight hours stands at approximately 1,300 to 1,900 h, and the annual average solar radiation is approximately 1,100 kWh/m^2, where the level of average solar radiation ranges from 2.6 to 3.7 kWh/m^2. Hence, the average daily solar radiation is estimated 3 kWh/m^2 [29]. In addition, Germany represents a four-season country climate, which is useful for comparison purposes to validate the feasibility of the proposed off-grid solar energy approach in two different climates, i.e. tropical (Malaysia) and four season (German). The comparison is based on three key aspects: technical criteria, energy yield analysis and economic analysis. By taking into account the main differences in the average daily solar radiation, annual real interest rate of 1% [30], SPV cost of $2/watt [31] and diesel price $1.65/L in Germany [32] and $0.70/L in Malaysia.

Table 5 includes a summary of the technical and economic criteria for the optimal design of the hybrid SPV/DG system for both Malaysia and Germany. The lower solar radiation rates lead to larger solar system size. For Germany, the average daily solar radiation is 3 kWh/m^2, which is equivalent to 58.8% of the average daily solar radiation in Malaysia. The designed system specification using the HOMER software consists of a 3.5 kW SPV array, a 1 kW DG, a 2.16 kAh/6 V storage battery

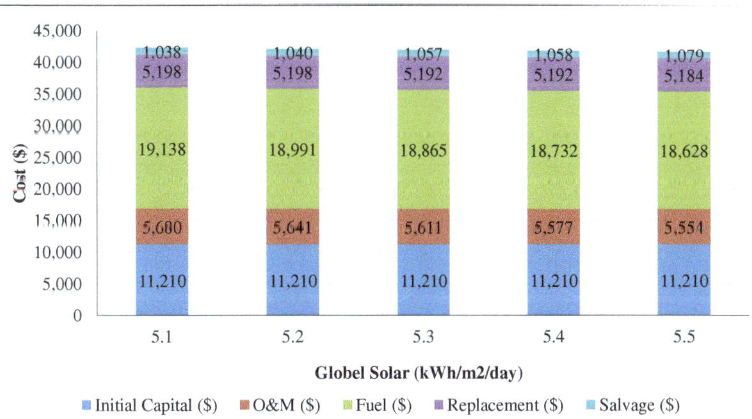

Figure 12 A comparison of the solar radiation rate and NPC.

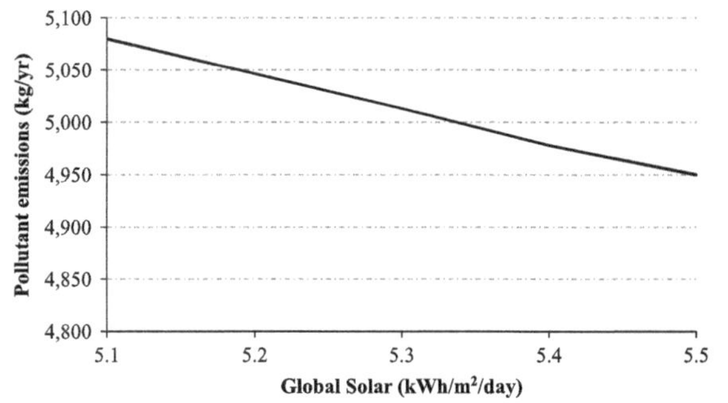

Figure 13 Pollutant emissions vs. global solar radiation rate.

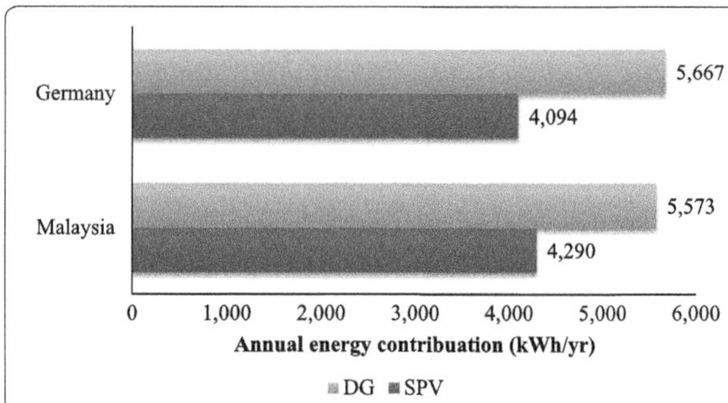

Figure 14 The annual energy contribution of different energy sources for Malaysia and Germany.

(360 Ah × 6 units) and a 1.5 kW inverter for the same LTE-BS load. Although the solar system size for Germany is larger than the solar system size for Malaysia (as shown in Table 5), the initial capital cost in Germany is less than the initial capital cost in Malaysia because the price of solar system components is cheaper in Germany. However, the NPCGermany is higher than NPCMalaysia, which is approximately doubled. The main reason behind the doubling is the price of fuel, as will be explained in the following paragraphs.

Figure 14 summarises the annual energy contribution of the solar electric system and the DG. The annual energy contribution reached 43% (4,290 kWh) of the solar system based on the characteristics of Malaysia's solar radiation exposure, and 57% (5,573) of the DG where the total annual energy production was 9,863 kWh. While, the annual energy contribution amounted to 42% (4,094 kWh) of the solar system based on the characteristics of Germany's solar radiation exposure, and 58% (5,667 kWh) of the DG where the total annual energy production was 9,761 kWh.

Figure 15 describes the comparison of all of the costs that occur within the project lifetime for both Malaysia and Germany. The initial capital cost is directly proportional to the size of the system. However, the initial capital cost of the project in Germany is lower than in Malaysia because the price of the solar system components in Germany is lower than in Malaysia. The total maintenance and operation costs for Germany are higher than for Malaysia because the size of the solar system larger than the Malaysian solar system leads to high operating and maintenance costs. Replacement costs depend significantly on the discount factor which is given in Equation 9. If the real interest rate increases, the discount factor will decrease, and the replacement cost will also decrease. The real interest rate for Malaysia is 3.25% and for Germany is 1% for quarter 3 of 2014, so the discount rate for Malaysia is higher than for Germany which is reflected in the total replacement costs. The fuel cost represents the bulk of the NPC, which amounted to 48% of the total project cost for Malaysia and 70% for Germany. Hence, NPCGermany is double the value of NPCMalaysia. NPC estimation in HOMER also considers the salvage cost (based on Equation 10), which is the residual value of the power system components at the end of the project lifetime.

11 Conclusions

This study examined the feasibility of the integration of a solar power system with a DG to supply power to remote BSs in off-grid sites of Malaysia to minimise both the OPEX and carbon emissions. Three key aspects have been investigated: (i) energy yield analysis, (ii) economic analysis and (iii) greenhouse gas emissions. When the solar radiation rates increase, the energy produced from the solar power system will increase, which has the positive effect of reducing the emission of polluting gases and the operating cost of the system. In summary, the bulk of the initial cost of the system is the cost of the SPV due to its inherent high cost. Meanwhile, the bulk of the annual operating and maintenance costs go to the DG, and the bulk of the replacement cost goes to the DG and batteries. However, the average annual OPEX savings of the hybrid system was 43% to 47% depending on the solar radiation rate. Comparison between Malaysia and

Table 5 Summary of the technical and economic criteria for the optimal design of the hybrid SPV/DG system

Energy model					Economic factors			DG factors	
Daily solar (kWh/m^2)	SPV (kW)	DG (kW)	Battery (unit)	Inverter (kW)	Initial Capital	Operating ($/year)	NPC ($)	Diesel (L)	DG (h)
5.1 (Malaysia)	2	1	4	1.5	$11,210	1,993	40,188	1,880	6,091
3 (Germany)	3.5	1	6	1.5	$10,750	4,436	80,044	1,886	5,862

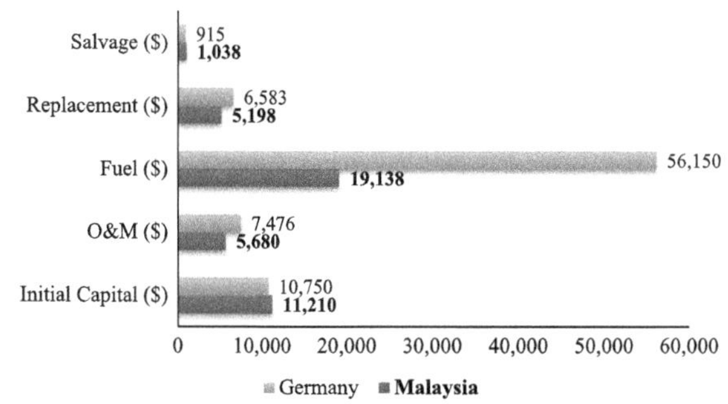

Figure 15 NPC comparison between Malaysia and Germany.

Germany shows that Malaysia's climatic conditions are desirable for wide utilisation of the proposed off-grid hybrid system due to the high amount of solar radiation received throughout the year, in addition to lower project cost in the long run.

Competing interests

The authors declare that they have no competing interests.

Acknowledgements

The authors would like to thank the Universiti Kebangsaan Malaysia for the financial support of this work, under the Grant Ref: ETP-2013-072.

References

1. International Telecommunication Union (ITU) statistics database; available: http://www.itu.int/en/ITU-D/Statistics/Pages/stat/default.aspx, accessed 26. Aug.2014.
2. K Kanzumba, JV Herman, Hybrid renewable power systems for mobile telephony base stations in developing countries. Renew Energ. **51**(2013), 419–425 (2013)
3. C Tao, Y Yang, H Zhang, H Kim, K Horneman, Network energy saving technologies for green wireless access networks. IEEE Commun. Mag. **18**(5), 30–38 (2011)
4. MH Alsharif, R Nordin, M Ismail, Classification, recent advances and research challenges in energy efficient cellular networks. Wireless Pers. Commun. **77**(2), 1249–1269 (2014)
5. H Ziaul, H Boostanimehr, VK Bhargava, Green cellular networks: a survey, some research issues and challenges. IEEE Commun. Surv. Tut. **13**(4), 524–540 (2011)
6. P Subodh, DS Madhu, A Muna, SN Jagan, *Technical and economic assessment of renewable energy sources for telecom application: a case study of Nepal Telecom, in Proc of 5th International Conference on Power and Energy Systems* (Nepal, Kathmandu, 2013), pp. 28–30
7. GeSI, Global e-sustainability initiative: SMART 2020: Enabling the low carbon economy in the information age. (2008); available: http://www.smart2020.org/_assets/files/02_smart2020Report.pdf, accessed 12. Nov. 2014
8. S Rath, SM Ali, MN Iqbal, Strategic approach of hybrid solar-wind power for remote telecommunication sites in India. Int. J. Sci. Eng. Res. **3**(6), 1–6 (2012)
9. P Nema, R Nema, S Rangnekar, Minimization of green house gases emission by using hybrid energy system for telephony base station site application. Renew Sustain Energ. Rev. **14**(6), 1635–1639 (2010)
10. AV Anayochukwu, EA Nnene, Simulation and optimization of hybrid diesel power generation system for GSM base station site in Nigeria. Elec. J. Energ. Environ. **1**(1), 37–56 (2013)
11. WA Imtiaz, K Hafeez, Stand alone PV system for remote cell site in Swat Valley, in *Proc of 1st Abasyn International Conference on Technology and Business Management, Pakistan*, 2013
12. S Moury, MN Khandoker, SM Haider, Feasibility study of solar PV arrays in grid connected cellular BTS sites, in *Proc of 2012 International Conference on Advances in Power Conversion and Energy Technologies (APCET), Mylavaram, Andhra Pradesh*, 2012, pp. 1–5
13. M Martínez-Díaz, R Villafáfila-Robles, D Montesinos-Miracle, A Sudrià-Andreu, *Study of optimization design criteria for stand-alone hybrid renewable power systems, in Proc of International Conference on Renewable Energies and Power Quality (ICREPQ'13)*, 2013
14. P Bajpai, N Prakshan, N Kishore, *Renewable hybrid stand-alone telecom power system modeling and analysis, in Proc of IEEE TENCON Conference*, 2009
15. VA Ani, AN Nzeako, Potentials of optimized hybrid system in powering off-grid macro base transmitter station site. Int. J. Renew Energ. Res. **3**(4), 861–871 (2013)
16. S Mekhilefa, A Safaria, WES Mustaffaa, R Saidurb, R Omara, MAA Younisc, Solar energy in Malaysia: current state and prospects. Renew Sustain Energ. Rev. **16**(1), 386–396 (2012)
17. K Sopian, MY Othman, A Wirsat, The wind energy potential of Malaysia. Renew Energ. **6**(8), 1005–1016 (1995)
18. H Borhanazada, S Mekhilefa, R Saidurb, G Boroumandjazib, Potential application of renewable energy for rural electrification in Malaysia. Renew Energ. **59**(2013), 210–219 (2013)
19. H Borhanazada, S Mekhilefa, R Saidurb, VG Ganapathy, Long-term wind speed forecasting and general pattern recognition using neural networks. IEEE T. Sustain. Energ. **5**(2), 546–553 (2014)
20. WA Azhari, K Sopian, A Zaharim, M Al Ghoul, A new approach for predicting solar radiation in tropical environment using satellite images e case study of Malaysia. WSEAS Trans. Environ. Dev. **4**, 4 (2008)
21. T Khatib, A Mohamed, K Sopian, M Mahmoud, Solar energy prediction for Malaysia using artificial neural networks. Int. J. Photoenergy, vol. 2012, Article ID 419504, doi:10.1155/2012/419504, 1-16 (2012)
22. Green Power for Mobile, GSMA, Community Power Using Mobile to Extend the Grid; Available: http://www.gsma.com/mobilefordevelopment/wp-content/uploads/2012/05/Community-Power-Using-Mobile-to-Extend-the-Grid-January-2010.pdf, accessed 18. Nov. 2014.
23. G Auer, O Blume, V Giannini, I Godor, MA Imran, Y Jading, E Katranaras, M Olsson, D Sabella, P Skillermark, W Wajda, *Energy efficiency analysis of the reference systems, areas of improvements and target breakdown. EARTH project report, Deliverable D2.3*, 2010
24. G Schmitt, The green base station, in *Proc. of 4th International Conference on Telecommunication - Energy Special Conference (TELESCON), 1–6*, 2009
25. T Lambert, P Gilman, P Lilienthal, Micropower system modeling with HOMER (2006); available: http://homerenergy.com/documents/MicropowerSystemModelingWithHOMER.pdf, accessed 18. Nov. 2014.
26. Trojan Battery Company; available: http://www.trojanbattery.com/, accessed 18. Nov. 2014.
27. S Rehman, LM Al-Hadhrami, Study of a solar PVedieselebattery hybrid power system for a remotely located population near Rafha, Saudi Arabia. Energ. **35**(12), 4986–4995 (2010)
28. Central Bank of Malaysia; available: http://www.bnm.gov.my/index.php?ch=statistic&pg=stats_convinterbkrates, accessed 28.Aug.2014.
29. International Energy Agency (IEA), Solar Energy Perspectives; available http://www.iea.org/publications/freepublications/publication/solar_energy_perspectives2011.pdf, accessed 18. Nov. 2014.
30. Organization for Economic Co-operation and Development (OECD); available: http://stats.oecd.org/index.aspx?queryid=86, accessed 18. Nov. 2014.
31. International Renewable Energy Agency (IRENA), Renewable Energy Technologies: Cost Analysis Series; available: http://www.irena.org/DocumentDownloads/Publications/RE_Technologies_Cost_Analysis-SOLAR_PV.pdf, accessed 18. Nov. 2014
32. Fuel prices europe information; available: http://www.fuel-prices-europe.info/, accessed 17. Nov. 2014.

16

Simultaneous power control and power management algorithm with sector-shaped topology for wireless sensor networks

Mohsen Nickray[1], Ali Afzali-Kusha[1*] and Riku Jäntti[2]

Abstract

In this paper, we propose a topology control technique to reduce the energy consumption of wireless sensor networks (WSNs). The technique makes use of both power control and power management methods. The algorithm uses the power management technique to put as many idle nodes as possible into the sleep mode while invoking the power control method to adjust the transmission range of the active nodes. On the contrary to earlier works in which both of these methods were used separately, in this algorithm, they are utilized simultaneously to decide about the sleep nodes and the ranges of active nodes. It is an approximation algorithm which is called simultaneous power control and power management algorithm (SPCPM). The performance bound of this centralized algorithm is determined analytically. Then, to make the proposed method practical for WSNs, a distributed algorithm based on SPCPM is introduced. To assess the efficiency of the proposed algorithm, we compare its average energy consumption with those of three existing topology control algorithms for a sector-based WSN. The simulation results which were obtained for different numbers of transmitting sensor nodes reveal less average energy consumptions for SPCPM compared to other algorithms.

Keywords: Power management; Power control; Wireless communication; Wireless sensor network; Network and architecture design; Low energy consumption

1 Introduction

Recent advances in wireless and electronic technologies have led to the emergence of wireless sensor networks (WSNs) with large-scale nodes. They are used in a wide spectrum of applications from industrial and military applications to health and environmental monitoring. In these kinds of networks, different small-size sensor types are deployed in the area to monitor, e.g., temperature, motion, noise, and seismic activities. A good survey of WSNs is presented in [1] and [2]. The sensors are wireless nodes that transmit the collected data (hop by hop) to a central station called base station or SINK node. Due to critical limitations of the node energy resources (e.g., battery), the nodes should minimize the energy consumption during their computation and communication. Therefore, low-power algorithms for WSNs are of prime interest. Among different WSN types, sector-shaped networks, where all the nodes send their data to a single base station, are the most common. They have various applications such as health monitoring, data gathering, and surveillance [3-5]. There have been several research efforts on reducing the overall transmission power of sector-shaped networks [5-10]. The efforts may be divided into two categories of power control and power management techniques. The key idea of the power control techniques is that instead of transmitting using the maximum power, the nodes in a WSN collaboratively reduce their transmission power by adjusting their transmission range while preserving some required properties (e.g., connectivity) [6-16]. The power control algorithms focus on reducing the power consumption when the nodes transmit or receive data in transmit and receive modes, respectively. Additionally, WSN nodes have a listen mode in which, while they are idle, they consume energy. The energy is comparable to the energy consumed in the other two modes. To reduce the energy consumption, the radio of the nodes in this mode may be turned off when they are not in use [17-20]. Turning

* Correspondence: afzali@ut.ac.ir
[1]Nanoelectronic Center of Excellence, School of Electrical and Computer Engineering, University of Tehran, P.O. Box 14395-515, Tehran, Iran
Full list of author information is available at the end of the article

off the radios when they are not in use is the basic idea of power management algorithms. Both of these techniques reside in the general category of the topology control algorithms.

It is expected that if these two types of techniques are invoked in coherence with each other in an algorithm, higher energy preservation will be achieved. This may be justified by noting that in this algorithm, the reduction of the energy consumptions of all the transmission, receive, and idle states are considered simultaneously. The energy conservation of wireless sensor networks may be considered as an optimization problem where the energy consumptions of different sources should be minimized. For this problem, each of the power control and power management algorithms may be used. Depending on the network load, one of these algorithms is used for the energy optimization. When the network load is low, the energy consumption of the network is dominated by the energy consumed by nodes in the idle state. For these loads, it is expected that fewer active (awake) nodes with long communication ranges lead to more energy conservation. In these cases, the power management algorithm should have a more determining role. In the cases where the network load is high, more active nodes with short communication ranges provide a better energy efficiency, and hence, the power control algorithm should play a major role in the optimization process. In this work, we propose an algorithm which integrates both of these techniques into one efficient algorithm.

1.1 Our contributions

In this paper, we make the following contributions. First, we present an analysis which shows that both the power control and power management techniques should be invoked in coherence to enhance the energy preservation of WSNs. Then, we introduce an approximation algorithm provable performance bound which is analytically shown. Finally, based on this centralized algorithm, we propose a practical distributed algorithm whose key advantages are given in the following:

- Its power management technique considers both time and space dimensions.
- It makes use of both power management and power control approaches simultaneously without giving any priority to any of them.

The rest of the paper is organized as follows. In Section 2, we briefly review the related works while Section 3 describes the network topology along with the energy model which is used in our study. In Section 4, we use an example to show the dependency of the minimum energy consumption to the network data traffic. In Section 5, we describe the proposed centralized Simultaneous Power Control and Power Management (SPCPM) algorithm in detail and compare its efficiency with similar techniques reported in the literature. Section 6 is devoted to the distributed SPCPM, and its performance is compared to those of existing techniques. Finally, the paper is concluded in Section 7.

2 Related works

In this section, we briefly review some of the previous works in the area of energy preservation of WSNs. These may be classified into two categories of power control and power management algorithms. The features of the algorithms are discussed while the main difference of our approach compared to these works is described at the end of the section.

2.1 Power control

Power control algorithms adjust the transmission power without adversely affecting desirable properties of the network (e.g., connectivity). Several empirical studies show that existing topology control solutions, which use static transmission power, transmission range, and link quality, might not be effective in the physical world (see, e.g., [21,22]). The authors in [22] propose a selective single-relay cooperative scheme, combining selective-relaycooperative communication with physical layer power control. Based on the medium access control (MAC) layer RTS-CTS signaling, a set of potential relays compute individually the required transmission power to participate in the cooperative communication. Based on the theory of stochastic fictitious play (SFP), a reinforcement learning algorithm to schedule each node's power level is proposed in [23]. Also, a transmission power self optimization (TPSO) technique is presented in [24]. It basically consists of an algorithm able to guarantee the connectivity as well as an equally high Quality of Service (QoS) concentrating on the efficiency of WSNs, while optimizing the necessary transmission for data communication in each node.

In the scheme proposed in [11], a node chooses its next node for relaying the information based on minimizing the total energy consumption for the whole transmission. It was shown that the network was strongly connected when every node had links only to those nodes that were within its *enclosure* defined by a relay region. In [12], the authors suggested two centralized algorithms to minimize the maximal power used per node while maintaining the connectivity of the network. Narayanaswamy et al. [14] and Kumar and Kawadia [15] propose algorithms to maintain network connectivity using a minimal transmission power. The authors in [16] proposed a minimum spanning tree (MST)-based topology control scheme which keeps the network connectivity and has bounded node degrees [16]. For detailed surveys on the power control techniques in WSNs, one may see, e.g., [25-27].

2.2 Power management

Power management algorithms focus on turning off the radios of the nodes which are not in use. There are two basic approaches, namely, scheduling-based [17], [18] and backbone-based sleep management [13,20]. In [28], a systematic and comprehensive survey of the power management schemes is presented. In the scheduling-based approaches, nodes turn on their radios only in scheduled slots. Backbone-based sleep management can improve network performance by maintaining a backbone composed of a small number of active nodes, while scheduling the other nodes to be active for small periods of time to conserve energy. From another aspect, the techniques may be divided based on the dimension that they use for the power management. The dimensions include *node* (e.g., [29]) or *time* (e.g., [13,20]). The approaches based on the node dimension, such as geographical adaptive fidelity (GAF) [29], mainly focus on detecting redundant nodes which can be turned off. In this algorithm, the network is divided into small square grids whose sizes are equal and fixed. All the nodes in a grid have the same routing role where at any time only one node is active and all the other nodes are in the sleep mode. The sleeping nodes periodically switch to the listen mode to guarantee that in each grid at least one node stays on. The algorithms based on the time dimension concentrate only on detecting the time when the nodes are idle and can be turned off [13,20]. In SPAN [13], a subset of nodes is adaptively selected to form a multi-hop forwarding backbone. The subset nodes, called coordinators, carry all the traffic of the network causing the energies of the other nodes which are in the sleep mode to be conserved. A technique, which is called STEM, defines two 'transfer' and 'monitoring' states for the nodes [20]. The approach is based on the fact that most of the time the sensor network is only monitoring its environment, waiting for an event to happen. In this state, the nodes can be asleep and the radio can be off. When an event is detected, the node becomes awake, changing its state to the transfer state.

2.3 Unified power control and power management algorithm

None of the power control or power management approaches optimizes the energy consumption of all the radio states. The power control techniques reduce the transmission energy of wireless nodes and do not consider the energy consumed during the idle state. The power management algorithms reduce the energy dissipation in the idle state while not considering the energy consumptions of the transmit and receive states. It has been shown that for an *ad hoc* network, when the network traffic is high, power control algorithms are effective while when the traffic is low, the use of power management algorithms provide us with more energy conservation [30,31]. In this work, we show that these findings hold true for a sector-shaped WSN too. Based on these facts, in this paper, we suggest a unified algorithm which makes a coherent use of both power control and power management algorithms to minimize the energy consumption both in the idle and transmit/receive states. In our algorithm, we use a cell-based approach similar to GAF. In our method, in addition to different forms of the cells, the transmission ranges of the nodes are adjusted dynamically. To the best of our knowledge, authors in [19] present the only similar approach. This approach is based on a joint duty-cycle optimization and transmission power control. By simultaneously adapting both parameters, the node can maximize the number of transmitted packets while respecting the limited and time-varying amount of available energy. Table 1 summarizes and makes a comparison between the related works and proposed algorithm in this work.

3 Network topology and energy model

3.1 Network topology

In a typical wireless sensor network, data flows toward the base station, and hence, the network area may be

Table 1 Comparison between the previous works and proposed algorithm in this work

Previous works	Power control algorithms		Power management algorithms			Simultaneous power control and management
	Static power control	Dynamic power control	Backbone-based		Scheduling-based	
			Node dimension	Time dimension		
[9,11-16]	√					
[23,24]		√				
[17,18]					√	
[21]			√			
[19]				√		
[20]			√	√		
Our work and [19]						√

assumed as a sector-shaped area [8,10]. Similar to [9], in this work, the whole area is divided into some cells with the same area. We model the network area denoted by $\psi(\alpha,R,N,K)$ using a sector with the angle of α and radius R which is divided into N cells and K annuluses. Figure 1 shows $\psi(\pi/6,R,16,4)$ which contains 4 annuluses and 16 cells with the angle of π/6. A sector area is divided into ring slices with the same ring width which are called network annulus. The areas of these network annuluses are not equal. Each network annulus may be divided into network cells with equal areas. Each annulus i may be divided into $2i$-1 cells with the same area [9]. Figure 1 shows the network $\psi(\pi/6,R,16,4)$ where each network annulus has been divided into network cells with equal areas.

3.2 Energy model

For the energy model, we adopt the first-order radio model described using [32,33]. Imagine u transmit data to node v. The energies consumed in the transmit state ($E^{tx}(d)$) and receive state ($E^{rx}(d)$) (transmit and receive energies, respectively) for a linear communication as a function of the distance (d) are, respectively, given by:

$$E^{tx}(d) = E_e^{tx} + E_r.d(u,v)^{\alpha} \tag{1}$$

$$E^{rx}(d) = E_e^{rx} \tag{2}$$

where $d(u,v)$ is the distance between transmitter u and receiver v, E_e^{tx} and E_e^{rx} are the is the energy consumed by the corresponding electronic circuits, Er is the energy consumed by the power amplifier, and α is the path loss factor. In this work, we assume that $E_e^{tx} = E_e^{rx} = E_e$.

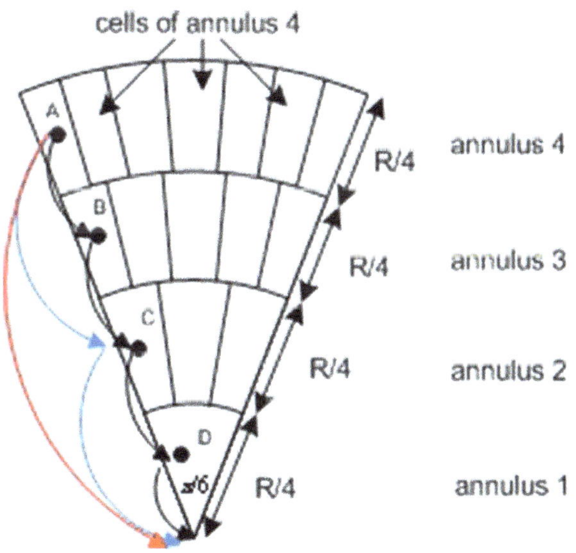

Figure 1 Network $\psi(\pi/6,R,16,4)$ which contains 4 annuluses and 16 cells with the angle of $\pi/6$.

4 Illustrating example

In this section, we illustrate the basic idea of our approach using a simple example. For the network shown in Figure 1, consider that the node A has some reports for the SINK node. The energy consumption used to transmit the data of the node includes the energies dissipated in each of the transmit, receive, idle, and sleep states. The average energy consumptions are denoted by $E_{tx}(d)$, E_{rx}, E_{id}, and E_s, respectively. As Figure 1 shows, the nodes A, B, C, and D are located in annuluses 4 to 1, respectively. Figure 1 shows three scenarios to transmit information to the SINK node. In one scenario (black arrows), first, the node A communicates with the node B, then the node B communicates with the node C, and after that, the node C transmit the data to the node D, and finally, the node D delivers the data to the SINK node all using a transmission range of $R/4$. In another scenario (blue arrows), the node A communicates with the node C and then the node C delivers the data to the SINK node both using a transmission range of $R/2$. The nodes B and D remain in the sleep state in this scheme. In the third scenario (red arrow), the node A directly delivers the reports to the SINK node using a transmission range of R. In this scenario the nodes B, C, and D remain in the sleep state. We denote the average energy consumptions of these three scenarios as E_1, E_2, and E_3, respectively. E_i denotes the sum of average energy consumption of all network nodes in scenario i. These energies may be obtained from:

$$E_1 = 4 \times \left(\frac{L}{L_{MAX}}\right) \times \left(E_e + E_r.\left(\frac{R}{4}\right)^2\right) + 4 \times \left(\frac{L}{L_{MAX}}\right) \times E_e + \left(5\text{-}\left(\frac{L}{L_{MAX}} + 3 \times 2 \times \left(\frac{L}{L_{MAX}}\right) + \frac{L}{L_{MAX}}\right)\right) \times E_{id} + 0 \times E_s \tag{3}$$

$$E_2 = 2 \times \left(\frac{L}{L_{MAX}}\right) \times \left(E_e + E_r.\left(\frac{R}{2}\right)^2\right) + 2 \times \left(\frac{L}{L_{MAX}}\right) \times E_e + \left(3\text{-}\left(\frac{L}{L_{MAX}} + 1 \times 2 \times \left(\frac{L}{L_{MAX}}\right) + \frac{L}{L_{MAX}}\right)\right) \times E_{id} + 2 \times E_s \tag{4}$$

$$E_3 = 1 \times \left(\frac{L}{L_{\mathrm{MAX}}}\right) \times \left(E_e + E_r \cdot \left(\frac{R}{2}\right)^2\right) + 1 \times \left(\frac{L}{L_{\mathrm{MAX}}}\right) \times E_e + \left(2 - \left(\frac{L}{L_{\mathrm{MAX}}} + 0 \times 2 \times \left(\frac{L}{L_{\mathrm{MAX}}}\right) + \frac{L}{L_{\mathrm{MAX}}}\right)\right) \times E_{id} + 3 \times E_s \quad (5)$$

Each term on the right hand side of the equations is the product of the energy consumption in a state and the time fraction that the nodes operate in that state. For example, the first term of $E2$ is the transmit energy of the nodes A and C with a transmission range of $R/2$ and L/L_{MAX} is the time fraction that these nodes operate in the transmission state. Note that we have assumed that transmit and receive times are linearly proportional to the packet length, L_{MAX} is the maximum packet length that a node can transmit during one period, and L is the length of the packet being transmitted. Since we have two nodes which take part in the packet transmission, a coefficient of two has been used. The second term corresponds to the receive energies of the node C and the SINK node. The third term is associated with the energy consumed when the nodes A and C as well as the SINK node are in the idle state. The coefficient (3-4 × L/L_{MAX}) is the fraction of time that these nodes are in the idle state (e.g., for the node C, the fraction is (1-2 × L/L_{MAX}). The forth term is the energy consumed by the nodes in the sleep state. For this term, since the nodes B and D are in the sleep states, the coefficient is two. Figure 2 part b depicts the terms of E_2 in details.

In Figure 2 part a, we have plotted E_1, E_2, and E_3 versus different packet sizes of the source node. The results have been obtained assuming L_{MAX}, R, E_r, E_{id}, E_e, E_s, and α to be 500 bytes, 400 m, 0.01 nj/bit/m^2, 50 nj/bit, 50 nj/bit, 12.5 pj/bit, and 2, respectively [34]. As the figure shows, when the network load is low, e.g., the packet size is 25 bytes, the minimum energy consumption will be achieved if the third scenario is used. In this scenario, the node A transmits data directly to the SINK node with the maximum transmission range while the other nodes are in the sleep state. When the network load is high, e.g., the packet size is more than 150 bytes, the first scenario will preserve energy more. For the medium range of traffic, e.g., the packet size of 80 bytes, the second scenario will lead to the minimum energy consumption.

As this example shows, when the network load is low, the energy consumption of the network is dominated by the idle energy. Therefore, scheduling more nodes to be in the sleep mode (power management task) with a long

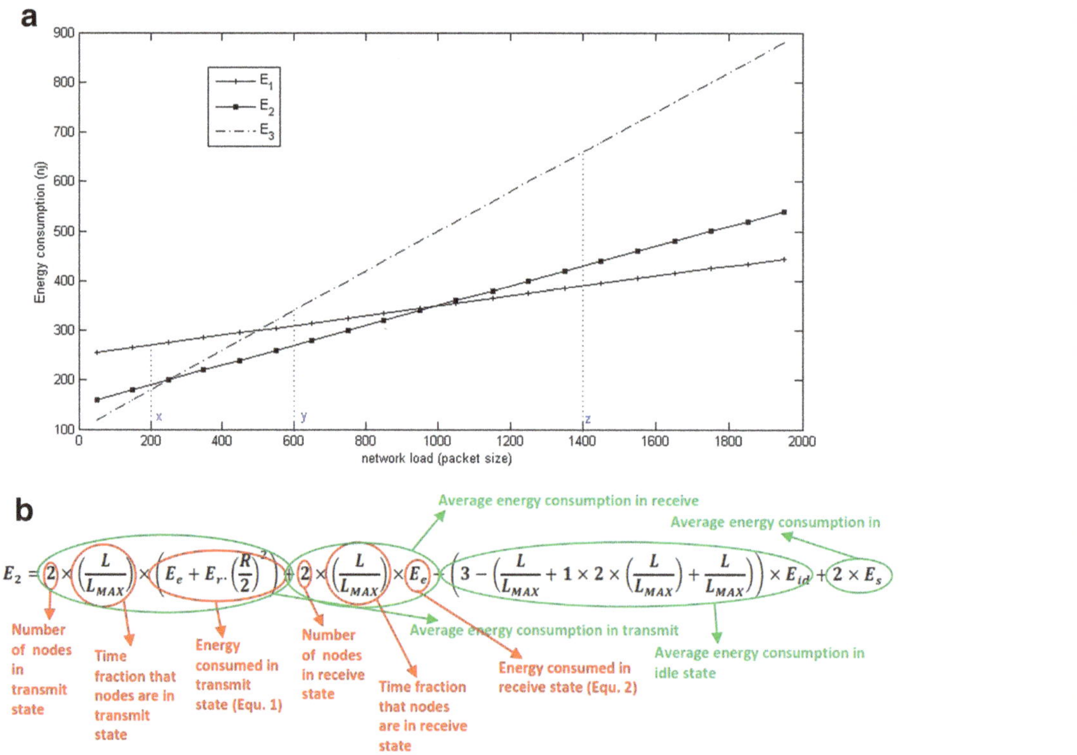

Figure 2 E_1, E_2, and E_3 versus packet size and E_2 in details. **(a)** $E1$, $E2$, and $E3$ versus packet size (network load) of the source node A. Packet size (bits) **(b)** depicts the terms of E_2 in details.

transmission range for the active nodes (power control task) will preserve more energy. When the load is high, the transmit energy dominates the energy consumption of the network. For this case, scheduling more nodes in the active mode with short transmission ranges will preserve more energy. Since the nodes B, C, and D first should receive the data and then relay it, they can have, at most, one half of the active period for receiving the packet and the other half of the period for transmitting the packet. Therefore, $L \leq L_{MAX}/2$ or $L/L_{MAX} \leq 1/2$. For this reason, in the case of the example discussed in Figure 2, we assumed L_{MAX} = 500 bytes × 8 bits = 4,000 bits and hence we plotted the energy consumptions versus L in the range of 50 to 2,000 bits.

We use the ratio L/L_{MAX} to denote the proportion of the network cycle time that is used by a node to transmit its data. This notation is only valid when the transmission rate is constant for all the nodes. Otherwise, L/L_{MAX} should be replaced by node transmission time divided by the network cycle period.

5 Proposed algorithm

5.1 Problem definition

In this section, we state the problem which is minimizing the energy consumption of the sector-shaped wireless sensor networks. We define our problem like [20]. Let us assume that we have a set of traffic demands denoted by $f(s_i)$ where s_i is the ith sensor source node which has some data to report to the SINK node. $t(\boldsymbol{s_i})$ shows the set of the edges used for transmitting data from the source s_i to the SINK node. Our problem is to find a subnet from our network nodes satisfying the traffic demands with the minimum energy consumption. Obviously, minimizing the total energy consumption may be achieved through minimizing the average energy consumption. The average energy consumption of node u may be computed as:

$$E(u) = \sum_{(u,v)\in t(s_i)} \frac{L_i}{L_{MAX}} \left(E_e + E_r \times d(u,v)^2\right) + \sum_{(v',u)\in t(s_i)} \frac{L_i}{L_{MAX}} . E_e + \left(1 - \sum_{(u,v)\in t(s_i)} \frac{L_i}{L_{MAX}} - \sum_{(v',u)\in t(s_i)} \frac{L_i}{L_{MAX}}\right) \times E_{id} \quad (6)$$

Here, the first summation in the first line is performed on all the nodes (v) that the node u transmit packet to them, and the second summation is performed on all the nodes (v') which they transmit to the node u. The average energy consumption of the node u is the weighted average of the transmit, receive, and idle energy consumptions. Also, since the energy dissipation of the node in the sleep state is very small, it has been ignored in the calculations [30].

Problem definition: Given a sector-shaped network which is represented by graph *G(V,E) (with vertices V and edges E)* and a set of traffic demands denoted by $t(\boldsymbol{s_i})$, find a subgraph $G'(V',E')$ such that the total average energy consumption $E(G')$ is minimal. This energy is given by:

$$E\left(G'\right) = \sum\nolimits_{u\in V'} E(u) = |V'|.W_c + \sum\nolimits_{u\in V'} \sum\nolimits_{(u,v)\in t(s_i)} f_i . W_v(u,v) \quad (7)$$

where:

$$W_c = E_{id}$$

$$W_v(u,v) = 2E_e + E_r \times d(u,v)^2 - 2E_{id}$$

$$f_i = \frac{L_i}{L_{MAX}}$$

From the above formulation, we can see that each node has a fixed cost of W_c. Also, for each edge (u,v), the cost is $W_v(u,v)$ which is independent from the packet size. The packet size (proportional to the traffic load) has been entered in the calculation using a data fraction factor denoted by f_i.

5.2 Centralized SPCPM

In this section, we introduce an approximation algorithm called SPCPM to maximize the energy preservation of a sector-shaped wireless sensor network. In this algorithm, based on a combined cost function, the algorithm decides about the state of nodes and the transmission range of the active nodes simultaneously.

5.2.1 A. Algorithm description

Let us describe the algorithm whose pseudo code is given in Algorithm 1. The input of the algorithm is the network $\psi(\alpha,R,N,k)$, and its output is the subnet graph $G'(V',E')$ where V' is the set of active nodes and E' is the set of the links between the active nodes. In step (1), we describe the network using the graph $G(V,E)$. To generate the graph G, for each node u, we should consider all possible routes that connect u to the SINK node. In SPCPM, we only consider transmission scenarios where the hops with about equal distances (in terms of the number of hops) are included. Figure 1 shows all the routes which may be used by SPCPM to transmit data from the node A to the SINK node. In scenario 1, 2, and 3, the node A transmits data by a 4-hop, 2-hop, and 1-hop transmission while communication scenarios like A → B → SINK and A → D → SINK are excluded. For some nodes (nodes with an odd annulus numbers), using equal hop distances may not be possible which in these cases, one of the hop distances will become one hop smaller than the others. For example, the node B in Figure 1, may take the route B → C → SINK in a two-hop scenario. As will be discussed in Section 6, the assumption of equal hop distances does not considerably affect the efficacy of the algorithm.

Algorithm 1 Centralized SPCPM

SPCPM

Input: $\psi(\alpha,R,N,k)$, source set S
Output: $G(V',E')$
(1) For each node u in annulus i
Assume a set of edges (u,v) that v is all the nodes in annulus j where $j = \text{i} - \left\lceil \frac{\text{i}}{\left\lceil \frac{\text{i}}{2^{\text{m}-1}} \right\rceil} \right\rceil \text{m} = 1, 2, 3, \dots \text{hop(i)} - 1,$ which $hop(i)$ is provided by (8)
(2) Label all nodes as sleep
(3) $W = S$
(4) While $W \neq \phi$
(a) Find $s_i \in W$ with minimum annulus number in $G(V,E)$ to SINK node using edge weights $g_i(u,v)$
(b) Add the shortest path from s_i to SINK in G'
(c) Label all nodes on the path as active
(d) $W = W - s_i$
(5) end

The number of different scenarios (inter-hop distances) for the node in the annulus i may be determined from:

$$\text{hop}(i) = log_2 i + 1 \tag{8}$$

Assume that the node u in the annulus i has j different scenarios which we denote them by sce(1) to sce(j). The number of hops for each sce(m) (symbolized by $|\text{sce}(m)|$) is calculated from:

$$[|\text{sce}(m)]| = \lceil i/2^{m-1} \rceil \text{where} \, m = 1, 2, 3, \dots, j \tag{9}$$

Also, the annulus number for the next hop of sce(j) is calculated from:

$$\text{Annulus\_Number\_of\_Next\_HOP}\,(i,j) = \text{Annulus\_Number\_of\_Current\_HOP} - i/|\text{sce}(j)|$$

As an example, consider the network of $\psi(\alpha,R,64,8)$. For this network, the number of transmission scenarios for a node in annulus 7 is 4 where the numbers of hops for these four scenarios will be 7, 4, 2, and 1. Considering sce(2), the sequence of hops identified by the annulus number are 7 (the source node), 5,3, 1, and 0 (the SINK node) ($|\text{sce}(2)| = 4$).

Having found G, in the next step, we assume that all the nodes are in the sleep state. Let us use a set denoted by W for the remaining source nodes which we should find a route for connecting them to the base station. This set is initialized to the set of all the source nodes (S). The routes are determined by finding the shortest path from each source to the SINK node such that the energy is minimized. From G, we have all the possible routes to the base station. In the iteration i, SPCPM selects the shortest path from G for connecting the source i to the SINK node using a weight function g_i. Each weight function $g_i(u,v)$ defines the cost of using the edge (u,v) for transferring the data of the source s_i to the base station. Obviously, the cost should depend on the data fraction (f_i) of the source and W_v based on Equation (7). On the other hand, based on Equation (7), in the expression for the energy, there is a constant (independent from f_i) energy cost for each node. Based on this observation, we should add this cost once for any node whose state is changed from sleep to active. Therefore, the weight function for each edge (u,v) is defined as:

$$g_i(u,v) = \begin{cases} f_i W_v(u,v) + W_c \, \text{when u is sleep} \\ f_i W_v(u,v) \, \text{when u is active} \end{cases} \tag{10}$$

After finding the shortest path for the iteration i, we set the state of all the nodes on this path to active. The output of SPCPM is the set of all the shortest paths found for all the iterations and the set of all the active nodes ($G'(V',E')$). It is apparent that the nodes which are not selected as active remain in the sleep state.

In Figure 3, we demonstrate how the centralized SPCPM works. For this example, we have a network with two annuluses and four cells (Figure 3a). Figure 3b shows $G(V,E)$ which is obtained in part 1 of the algorithm. There are two sources which are S(1) and S(2) with data fractions of 0.1 and 0.2, respectively. The numbers shown next to the edges in Figure 3b correspond to the values of W_v for the corresponding edges. Also, we have assumed $W_c = 1$. The results of the two iterations of the algorithm (corresponding to the two source nodes) are plotted in Figure 3c,d, where the numbers next to the edges are the weight functions of the source node. Note that, in each iteration, the weight of each edge is updated based on the data fraction as well as the change in the set of active nodes. Using Equation (7), we obtain the total average energy consumption for this example as 4.2.

5.2.2 B. SPCPM algorithm analysis

In this section, we study the complexity of SPCPM and also obtain the approximation ratio of the algorithm. The proposed problem in this work was defined in Equation (7). The problem is finding the minimum-weight Steiner tree in the graph $G(V, E)$ with the weight g_i given by Equation (12). Here, E and V are the sets of edges and nodes in G, respectively. Finding the minimum-weight Steiner tree is an NP-hard problem (see, e.g.,) [35]. We used the SPCPM algorithm to solve the problem. Assuming equal hop distances, the complexity of SPCPM becomes $O(|S|.log|E|.log|V|)$ where $|S|$, $|E|$, and $|V|$ are the numbers of sources, edges, and

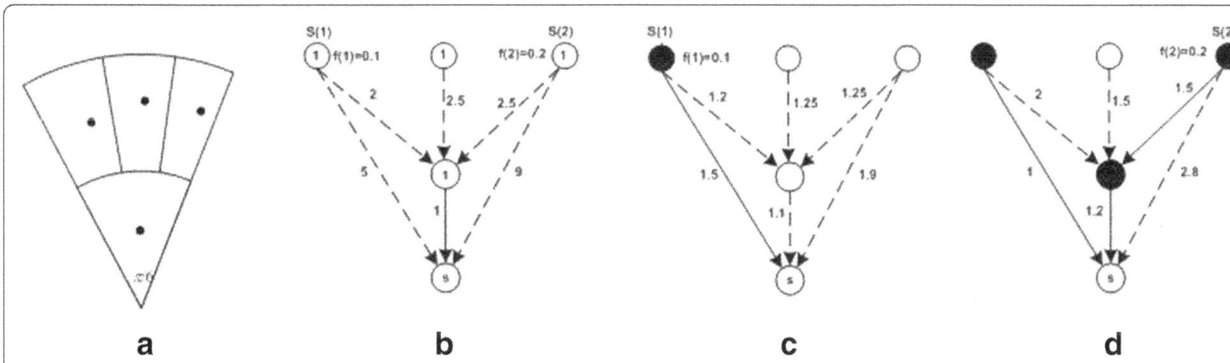

Figure 3 Demonstration of how the centralized SPCPM works. **(a)** Input network $\psi(\pi/6,R,4,2)$ **(b)** graph $G(V,E)$ along with$W_v(u,v)$ and two data source with $f(1) = 0.1$ and $f(1) = 0.2$. The outputs of the SPCPM algorithm for **(c)** shortest path for the S (1) (iteration 1) and **(d)** shortest path for the S (2) (iteration 2). For this graph, $W_c(u,v)$ has been assumed to be 1.

nodes in G, respectively. In the case of considering all possible routes, the complexity increases to $O(|S|.log|E|.log|V|)$ (see Equations (8)-(9)).

To calculate the approximation ratio of the algorithm, we follow the same approach as the one given in [30]. Let us define a set of weight function c_i and h_i as:

$$c_i = f_i \times W_v(u,v) \tag{11}$$

$$h_i = W_c + f_i \times W_v(u,v) \tag{12}$$

where $c_i(u,v)$ represents the cost of the edge (u,v) when the data of the source s_i travels from the node u to the node v. The proposed algorithm finds the shortest path from each source to the SINK node. The cost (average energy) for the path from the source s_i to the SINK node (SN) is denoted by $E_G^x(s_i, \mathrm{SN})$ under the weight function x. Based on these notations, Equation (7) maybe expressed as:

$$E\left(G'\right) = |V'|W_c + \sum_{s_i} E_{G'}^{c_i}(s_i, \mathrm{SN}) \tag{13}$$

Theorem: *The approximation ratio of the algorithm SPCPM is not greater than:* $\frac{|S|\times|K|}{|V'|}$.

Proof: Let $E(G')\left(=\sum_i E_{G'}^{g_i}(s_i, \mathrm{SN})\right)$ and $E(G'\text{min})\left(=\sum_i E_{G'_{\min}}^{g_i}(s_i, \mathrm{SN})\right)$ represent the total costs of G' found by SPCPM using the weight function $g_i(u,v)$ and the optimal solution given by Equation (13), respectively. To obtain the approximation ratio, let us first start with the SPCPM algorithm when the weight function $h_i(u,v)$ is used. Since $g_i \leq h_i$, the total cost of the shortest paths from all the sources to the SINK node in G' with the weight g_i is greater than the one with the weight h_i. Therefore, one may write:

$$\sum_i E_{G'}^{g_i}(s_i, \mathrm{SN}) \leq \sum_i E_{G'}^{h_i}(s_i, \mathrm{SN}) \tag{14}$$

When we use the weight h_i, W_c is always present in the weight function (regardless of the current state of the node being active or sleep), and hence, may be counted several times for each node in G'. Note that the maximum number of W_c which should be added to the total cost is $|S|.\ |K|$. The reason is that there are $|S|$ sources where, in the worst case, there should be one node from each annulus on the path from each source to the SINK node. The cell configuration in our algorithm is such that each route uses at most one node from each annulus. Thus, one may write:

$$\sum_i E_{G'}^{hi}(s_i, SN) \leq \sum_i E_{G'}^{ci}(s_i, SN) + |S|.|K|.W_c \tag{15}$$

Since $|V'| \leq |S|.\ |K|$, we have $\frac{|S|\times|K|}{|V'|} \geq 1$. Using this inequality and based on Equations (7) and (13), one may write:

$$\begin{aligned}\sum_i E_{G'}^{h_i}(s_i, SN) &\leq \frac{|S|\times|K|}{|V'|}\left(\sum_i E_{G'}^{c_i}(s_i, SN) + |V'| \times W_c\right)\\ &\leq \frac{|S|\times|K|}{|V'|} \times E\left(G'_{min}\right)\end{aligned} \tag{16}$$

Then, from Equations (14) and (17), we have:

$$E\left(G'\right) \leq \frac{|S|\times|K|}{|V'|} \times E\left(G'_{min}\right) \tag{17}$$

Therefore, the approximation ratio of SPCPM, in the worst case, is $\frac{|S|\times|K|}{|V'|}$.

5.3 Performance evaluation

To evaluate the effectiveness of the proposed algorithm, we have compared it with three other algorithms. They included the algorithms Transmission Power Minimum

Spanning Tree (TMST) [10], Transmission power Shortest Path Tree (TSPT) [36], and Minimum Power Configuration (MPC) [30]. All the algorithms were implemented in MATLAB. The first two algorithms are basic topology control methods while the last algorithm is similar to the SPCPM technique considering the minimization of both idle and transmit/receive energies. The TMST algorithm finds the minimum spanning tree of the network where each edge is weighed by the transmission power of the transmitting node. The TSPT technique finds the shortest path tree of the network when each edge is weighted by the transmission power. In the MPC algorithm which is more efficient compared to the two previous ones, each edge is weighted by a combination of the transmission and idle powers of the node. The parameters L_{MAX}, R, E_r, E_{id}, E_e, and E_s in our simulation were assumed to be 500 bytes, 400 m, 0.01 nj/bit/m^2, 50 nj/bit, 50 nj/bit, and 12.5 pj/bit, respectively [34]. We assumed that 100 nodes were randomly distributed in the sector-shaped area with $R = 800$ m and $\alpha = \pi/4$.

The total energy consumption of the network when the number of sources (|S|) varies from 1 to 50 is shown in Figure 4. The data fraction was assumed to be 0.1. The results shown in this figure reveal that MPC and SPCPM provide considerably higher energy preservations compared to TSPT and TMST. This may be justified by noting that these two techniques consider both the transmission and idle powers while TSPT and TMST ignore the idle power. The better performance of SPCPM compared to MPC could be attributed to two facts. First, the approximation for MPC ($|S|$) is higher than that for SPCPM ($|S|. |K|/|V'|$). As Figure 4 shows, when the number of sources increases, the performance of the SPCPM algorithm becomes better compared to that of the MPC technique. Second, SPCPM uses power management techniques in both the node and time dimensions while MPC considers only the time dimension. Note that even when the number of sources is very low, SPCPM outperforms MPC. For these cases, this outperformance was not initially expected due to the fact that the approximation ratios were about the same. This happens because the performance bound of SPCPM was derived for the worst-case scenarios. Our algorithm, however, worked better than the worst-case scenarios in these cases. This was achieved by taking advantage of employing the node dimension in the power management method.

6 Distributed SPCPM

In this section, the design and implementation of the distributed form of SPCPM are discussed. This distributed algorithm is based on destination-sequenced distance-vector (DSDV) method [37] which is a shortest path algorithm widely used in the literature. DSDV is based on a distributed implementation of the Bellman-Ford shortest path algorithm [38]. A node in DSDV advertises its current routing cost to the SINK (distance to the SINK node) by broadcasting update messages (control packets). Having received the messages, each node sets the neighbor with the minimum cost as its parent. The routing table for this node also is formed based on the broadcasted update messages where the table is periodically broadcasted to other nodes. In our work, we used an improved DSDV technique where the modifications (compared to the conventional one [37]) were described in [39-41]. Examples of these improvements include avoiding DSDV extra traffic using incremental updates and implementing multi-path routing.

In our algorithm, the routing cost of a node is calculated based on its annulus number (which determines the number of annuluses to the SINK node). When the network powers up, each node determines its position

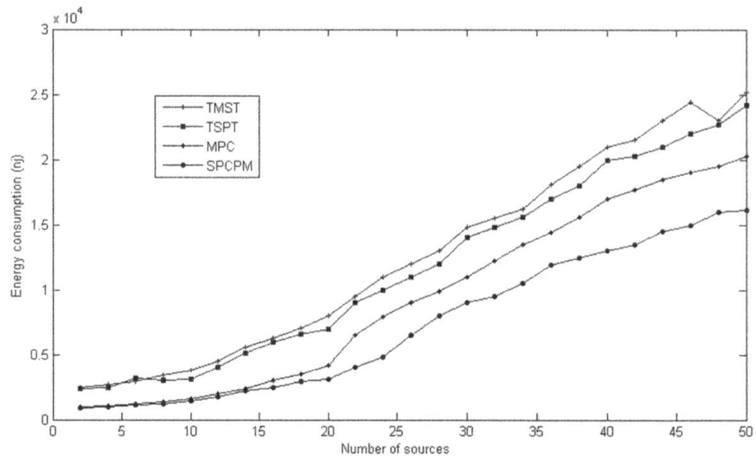

Figure 4 Energy consumption versus number of sources. The data fraction is 0.1.

and sends it to the SINK node. Then, the SINK node makes use of the reported positions of the nodes to determine the sectors and annuluses of the network. The information is used to specify the cell and annulus numbers of the nodes. The nodes are informed about their numbers using broadcast messages. Each node in SPCPM operates in either the active or sleep state. Initially, all the nodes operate in the sleep mode. For the sleep mode management, a scheduled-based algorithm like sensor medium access control (SMAC) may be used. A node in the sleep mode remains in this mode for a long time and wakes up periodically. When a source node starts transmitting data to the SINK node, a power-efficient route from the source to the SINK node is found. Similar to the centralized algorithm, all the nodes in the route which satisfies the distributed shortest path requirements become active while the other nodes remain in the sleep state. To find the distributed shortest path, periodic broadcasting of the routing table is required. Since in the algorithm the next hop is selected based on the data fraction, compared to the table of the DSDV, we need an extra entry for f_i in the routing table. Table 2 shows an example of a routing table in SPCPM for a given node, e.g., u. If the data fraction of u is less than 0.1, between 0.1 and 0.3, and more than 0.3, the packet should be sent to a node in the third cell of annulus 2, first cell of annulus 4, and first cell of annulus 7, respectively. The cost value indicates the cost of transmitting data from the node to the SINK, and the sequence number is for the sequence number originated by the SINK node. This number is used by the DSDV algorithm to prevent formation of routing loops [37].

It should be noted that when a new source node with a new data flow appears in the network, the data fraction (f_i) for some nodes may change. This change may cause the next hop for some nodes to alter too (see Table 2). In addition, the appearance of a new flow may activate a node previously running in the sleep mode. This reduces the cost of transmitting data for the neighboring nodes of the newly activated node. Therefore, the routing tables should be updated. The transmission costs are calculated from Equation (12). In addition to the appearance of a new data flow, the disappearance of an existing flow may also require a route update. In such a case, the active nodes on the routing path for this flow may switch to the sleep mode after some timeout. Another case where the routing tables should be updated is when a source node changes its data rate (significantly). This again changes the data fraction of some nodes and hence the routing tables should be updates. When the workload of the network is dynamic, multiple rounds of route updates may be initiated at the same time, resulting in high network contention. In the DSDV technique, to reduce the overhead of route updates in such a case, the SINK can include several default data rates in its initial route updates, based on the estimation of source rates.

Table 2 An example of a routing table for a node in distributed SPCPM

Data fraction	Cell number/annulus number	Cost	Sequence number
$f_i < 0.1$	3/2	5	100
$0.1 < f_i < 0.3$	1/4	4	20
$f_i > 0.3$	1/7	6	41

In our distributed algorithm, we have assumed that either a GPS or GPS-less mechanism may be used as a localization technique to obtain the location information of the nodes. In our algorithm, we supposed that each node knew its current location relative to the SINK node. The information is used to configure the network. Due to its importance, the localization (which is not the focus of this paper) is an active area of research (see, e.g., [42-46]). For example, the techniques presented in [44-46] determine the node location based on the information of the angle of arrival between neighbor nodes. These techniques which are accurate, decentralized, and GPS-less may be the better option for being used in conjunction with SPCPM.

6.1 Performance of distributed SPCPM

In order to evaluate the performance of the distributed SPCPM algorithm, we compare it with minimum transmission routing (MTR) [12], distributed minimum power configuration protocol (MPCP) [30], and the original DSDV [37]. A node in MTR chooses the next hop node based on minimizing the expected number of transmissions while considering the same transmission range for all the nodes. The MPCP algorithm is similar to the distributed SPCPM except for the fact that it does not have a cell-based configuration. To generate the results presented in this section, the simulator Prowler was used. Prowler is a Matlab-based network simulator that uses an event-driven structure. For the physical and MAC layer parameters, we used network parameters described in [47,48]. These parameters are used for the models concerning all levels of the communication channel and application. The nondeterministic nature of the radio propagation is characterized by a probabilistic radio channel model. The radio propagation model determines the strength of a transmitted signal at a particular point of the space. The deterministic part of the propagation function can be any user-supplied function where, in this work, we used the model explained in Subsection 3.2. The fading effect is also modeled by random disturbances. In our simulations, the required transmission power/energy for each node in an active period is estimated based on the distance to the next hop considering the radio environment (including fading). The simulator assumed a continuous adjustment of the transmission power. The same assumption has been considered for implementing all other algorithms in this paper.

For the MAC layer, we have employed a simple CSMA/CA scheme without RTS/CTS, which is based on SMAC. Like [30], as an automatic repeat request (ARQ) mechanism, we retransmit a packet if an acknowledgement is not received after timer preset. The maximum number of retransmissions before dropping a packet is 4. The results were obtained for two sector-shaped configurations. In the first one, a sector-shaped area with R = 400 m which was divided into 4 annuluses and 16 cells where 50 nodes were deployed. In the second configuration, the sector-shaped area had a radius of 800, 8 annuluses, 64 cells, and 200 nodes. The number of sources for these configurations changed from 5 to 25 and 10 to 100, respectively. The data packet size was 120 bytes, and the routing update packets were 40 bytes. For the energy parameters, we assumed the same values as those mentioned in section 4. The simulations were performed for one period of sending report by the WSN sources. It should be noted that in all simulations, to report the energy consumption like in Section 4, we got the sum of the average energy consumption of all network nodes. All the simulation results were obtained by averaging the results of ten simulation runs. We have also shown the corresponding 90% confidence interval for the results.

The energy consumptions of the protocols versus the number of sources are shown in Figure 5 which reveals that SPCPM consumes a considerably lower energy compared to those of the DSDV and MTR algorithms. This was expected because the idle power in DSDV and MTR was ignored. Note that both the SPCPM and MPCP algorithms attempt to optimize the transmission range and the idle power. Among the two, the SPCPM method performs better. This is may be attributed to the use of the cell-based topology which provides us with the ability to use the power management technique in the node dimension. Comparing the two network configurations indicate that as the numbers of nodes and the sources increase the efficiency of the SPCPM algorithm compared to that of MPCP improves. Also, note that the energy consumption of SPCPM slightly increases when the number of sources becomes larger than the number of cells (16 and 64, respectively). This is justified by considering that in our proposed algorithm, the number of active nodes is limited by the number of cells thanks to the use of node dimension power management technique (one active node in each cell). In the case of other algorithms, the energy consumptions increases more as the number of sources enlarges.

Figure 6 shows the data delivery rate of the distributed algorithms at the SINK node for different algorithms. As the results indicate, the delivery rate of all the algorithms decreases when the number of sources increases. The SPCPM algorithm outperforms other methods due to the use of a cell-based structure which decreases the contentions during the workload increase. A comparison between the two configurations demonstrates that for DSDV, MTR, and MPCP, the increase in the number of nodes and workload lowers the delivery rate considerably. In the case of SPCPM, however, the decrease is lower. The average end-to-end delay of data packets are plotted in Figure 7, which indicates that the SPCPM algorithm leads to a higher latency compared to those of the other approaches. The reason is that SPCPM reduces the idle power by taking some nodes to the sleep state, increasing the nodes which are shared among the routes between the sources and the SINK node.

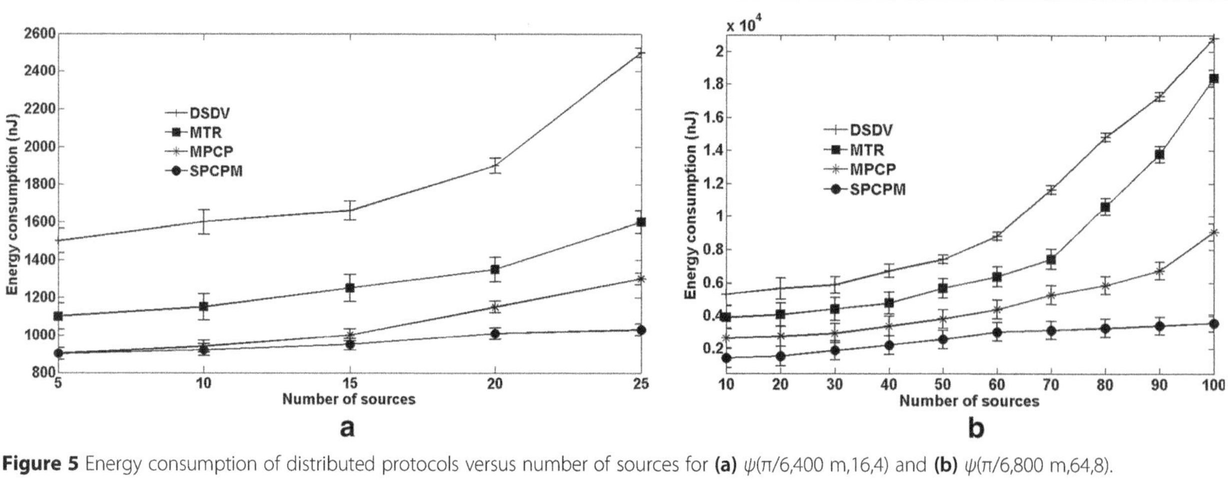

Figure 5 Energy consumption of distributed protocols versus number of sources for **(a)** $\psi(\pi/6,400\text{ m},16,4)$ and **(b)** $\psi(\pi/6,800\text{ m},64,8)$.

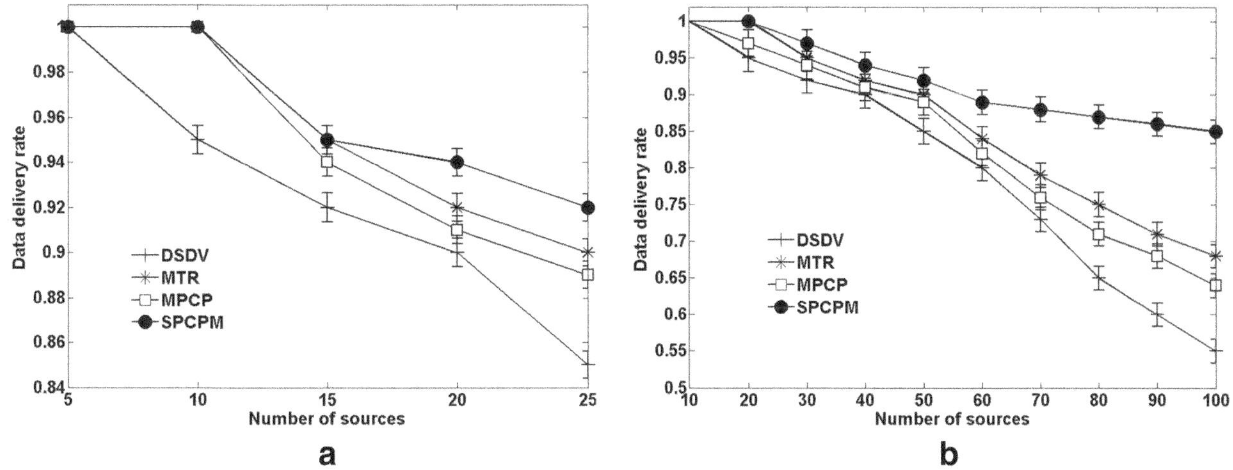

Figure 6 Delivery rate of distributed protocols versus number of sources for **(a)** $\psi(\pi/6,400\ \text{m},16,4)$ and **(b)** $\psi(\pi/6,800\ \text{m},64,8)$.

It should be mentioned that when the distributed algorithm is performed, the state of some nodes are changed. These changes are broadcasted in the network by the corresponding nodes via some small control messages. For this algorithm, we used the same message configuration as that of the MPCP algorithm. We have compared the message overheads of the algorithms in Figure 8. The results specify that the overhead of the SPCPM technique is higher. Note that the volume of the control messages is considerably smaller compared to the volume of the report data, and hence, the higher overhead for the proposed algorithm should not be a major concern. Also, we should mention that the overheads of SPCPM and MPCP increase as the number of sources enlarges. In the case of the other two approaches, the overhead almost remain unchanged with the number of sources. This may be explained by noting that for the case of SPCPM and MPCP, before a new source starts transmitting data, its state should be changed and its routing table along with those of the neighbors should be updated. These changes should be broadcasted for all the neighbors in the area via the control messages resulting in increase in the message overhead for the addition of each new source.

6.2 Impact of limiting transmission power on SPCPM

The SPCPM algorithm does not assume any transmission range limit (TRL). In practical cases, however, the nodes of WSN have limited transmission ranges. In this part, we study the impact of the limitation on the efficiency of the algorithm. It should be noted that decreasing the TRL of nodes lowers the efficiency of these types of algorithms. To study the effect of the TRL variation on the efficiency of the proposed algorithm, we have plotted the energy consumption of network versus the number of sources for the TRL values of 200, 400, and 800 m in the case of R = 800 m. The decrease in the efficiency may be attributed to the fact that, in our algorithm, to lower the energy consumption, we use longer transmission ranges for small data sizes. Reducing TRL

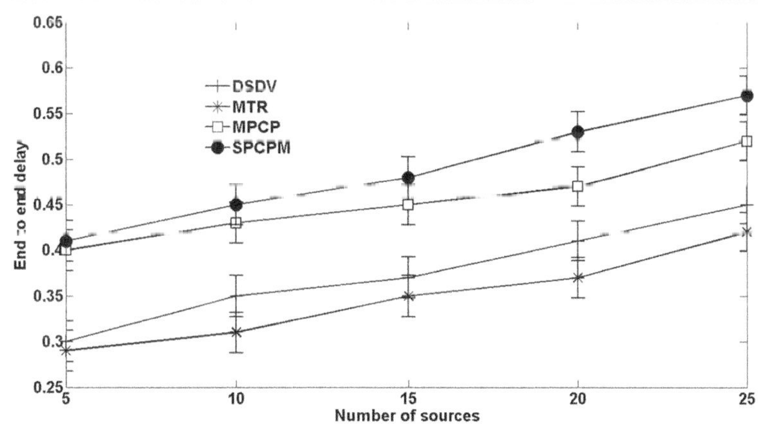

Figure 7 Average delay of a communication from source to SINK node versus number of sources.

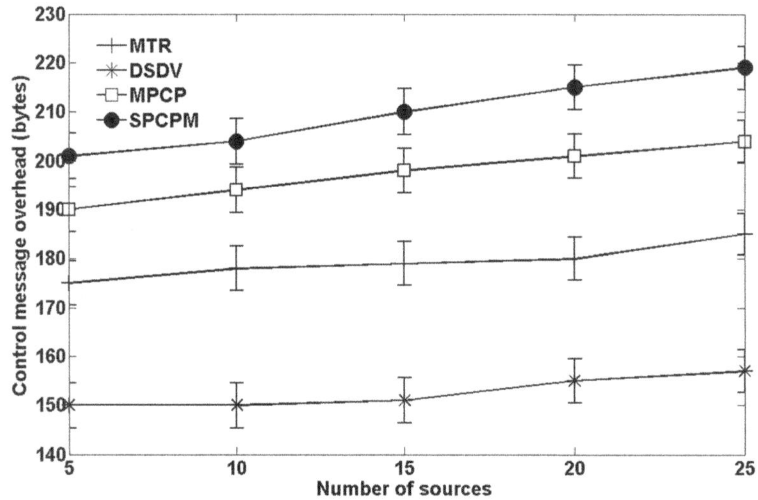

Figure 8 Message overhead of distributed protocols versus number of sources.

limits the transmission ranges, which may be used for these data sizes, lowering the algorithm efficiency. For example, in Figure 9a, when the TRL value is decreased from 800 to 400 m, the transmission ranges longer than 400 m are excluded. For example, in this case, the nodes in annulus 8 cannot transmit their reports directly to the SINK node. To compare the impact of the TRL changes on the efficiencies of the algorithms, we have plotted the energy consumption of network versus the number of sources for the TRL values of 200, 400, and 800 m in the case of, for example, $\psi(\pi/6,800\ \text{m},64,8)$ for both the SPCPM and MPCP algorithms. The results reveal that, on average, the amount of the energy increase with the number of sources is lower in the case of SPCPM.

6.3 Impact of assumption of equal distance hops on SPCPM

As mentioned in Subsection 5.2 A, to simplify the algorithm, we only considered the routes with equal distance hops to reduce the complexity (increase the speed) of the algorithm. Since some of the transmission scenarios are excluded, one may expect that the energy preservation of the algorithm is lowered. To study this, we compared the energy consumptions for $\psi(\pi/6,800\ \text{m},64,8)$ in the cases of including all the scenarios (complete SPCPM) and the scenarios with only (about) equal distance hops. The results, which include both the TRL values of 400 and 800 m, are shown in Figure 10. As is evident from the figure, the energy preservations in these cases are not considerably different. Specifically, in the maximum case, the energy consumptions for the complete SPCPM are 4, 5, and 9% lower than those of the SPCPM in the cases of TRL = 200 m, TRL = 400 m, and TRL = 800 m, respectively. The small differences between the efficiencies of the two algorithms may be justified by noting that the transmission energy constitutes a major part of the total energy consumption (the energy consumption given in Equation (7) includes both the idle as well as the transmission energies). As we have shown in [9], the maximum transmission energy preservation is achieved when we consider only the scenarios which have the hops with the same distance. Also, note that changing the TRL value does not change the difference much.

6.4 Impact of SPCPM on energy hole problem

The proposed algorithm, which is suggested for sector-based WSNs, suffers from the energy hole problem mentioned in [7-10]. We should mention that even though SPCPM did not directly deal with balancing the energy consumption, it is expected that the algorithm alleviate the energy hole problem. In conventional methods, e.g., [7-10], data are transmitted hop by hop (similar to the scenario 1 in Figure 1) toward the sink node. In the SPCPM algorithm, we replace some of these hop by hop transmissions with other transmissions which skip some of the hops including the hops close to the sink node (similar to scenarios 2 and 3 in Figure 1). The expectation may be validated by the results presented in Figure 11. First, in Figure 11a, we compare the lifetimes of the network normalized to that of the basic power control (BPC) algorithm [9] for the SPCPM, MPCP, and MTR techniques. The results were presented for $\psi(\pi/6,800\ \text{m},64,8)$ assuming a transmission range limit of 200 m. The BPC algorithm was used as a conventional hop by hop power control algorithm for a sector-shaped network. The results were extracted based on definitions and assumptions discussed in [9]. These results indicate that SPCPM leads to a longer lifetime for the network

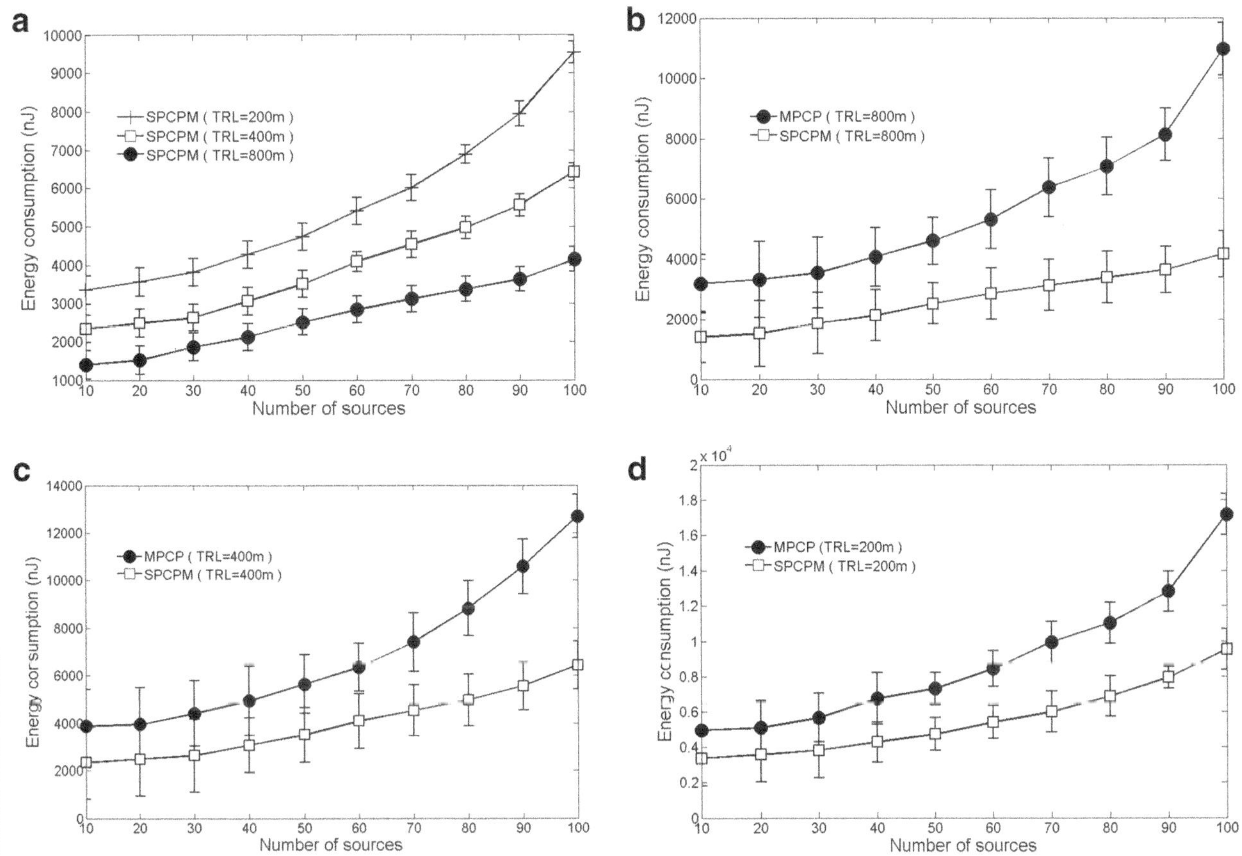

Figure 9 Plotted energy consumption of network versus the number of sources for the TRL values. **(a)** Energy consumption of network for the distributed SPCPM protocol versus the number of sources for $\psi(\pi/6,800\ \text{m},64,8)$ with TRL as the running parameter. Also, comparison of energy consumptions of network for SPCPM and MPCP algorithms versus the number of sources for $\psi(\pi/6,800\ \text{m},64,8)$ in the cases of **(b)** TRL = 200 m, **(c)** TRL = 400 m, and **(d)** TRL = 800 m.

compared to those of BPC, MPCP, and MTR. Next, in Figure 11b, we have compared the normalized lifetimes of the networks with the sizes of 400, 800, and 1100 m when the SPCPM algorithm is invoked. The results suggest that the efficiency of the SPCPM even becomes better when the network size increases. Finally, to study the impact of the transmission range limit on the efficiency of the algorithm in prolonging network lifetime, we have plotted the lifetimes for the TRL values of 200, 400, and 800 m in Figure 11c. Obviously, as TRL increases, there are more opportunities for skipping the nodes closer to the sink node, and hence, the longer lifetimes may be achieved. In conclusion, the SPCPM not only decreases the average energy consumption but also increases the

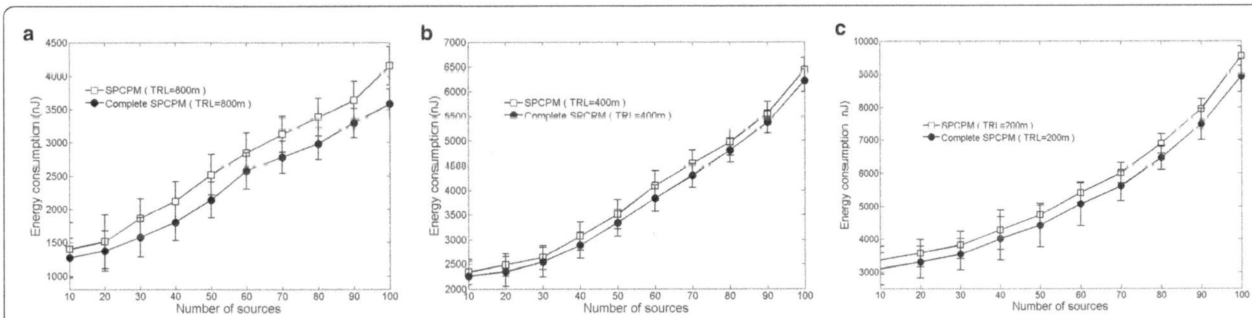

Figure 10 Energy consumptions of the SPCPM and *complete SPCPM* algorithms versus the number of sources for $\psi(\pi/6,800\ \text{m},64,8)$. **(a)** TRL = 200 m, **(b)** TRL = 400 m, and **(c)** TRL = 800 m.

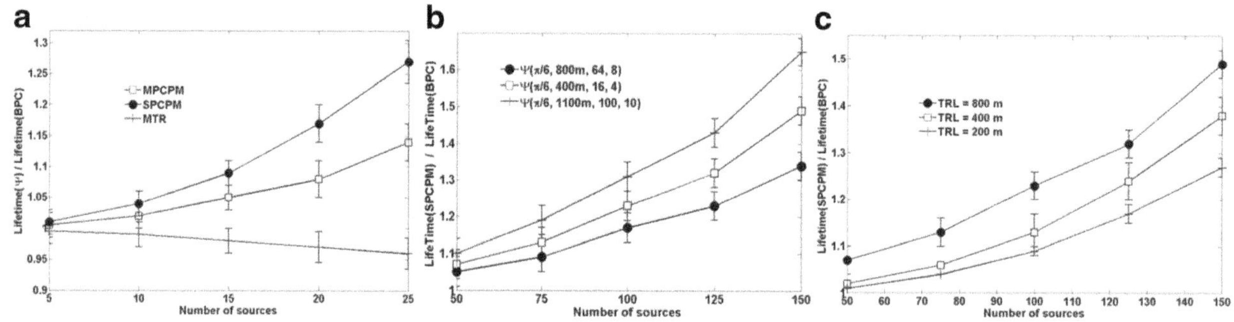

Figure 11 Lifetime comparison for **(a)** three algorithms of MTR, MPCP, and SPCPM for $\psi(\pi/6{,}800\ \text{m},64,8)$ assuming a TRL of 200 m, **(b)** three configurations of SPCPM with different radius sizes of $R = 400$ m, $R = 800$ m, and $R = 1100$ m assuming an TRL of 200 m, and **(c)** three different TRLs of 200, 400, and 800 m when SPCPM is used for $\psi(\pi/6{,}800\ \text{m},64,8)$.

network lifetime. Also, it should be mentioned that there are quite a number of works devoted to the energy hole problem (see, e.g.,) [9] which may be used with the SPCPM algorithm simultaneously.

7 Conclusion

In this paper, we proposed a topology control technique to reduce the energy consumption of wireless sensor networks (WSNs) based on simultaneous use of both power control and power management methods. The technique which was called SPCPM minimized the total energy consumption of WSNs. We discussed that none of the power control and power management approaches could optimize the energy consumptions of all the radio states. Also, it was analyzed that in a sector-shaped network, the power control algorithms were effective when the traffic of the network was high while the power management algorithms were effective when traffic was low. We presented an approximation algorithm with a provable performance bound and extended it to a distributed and practical algorithm. The algorithm dynamically updated the routes from the sources to the SINK node based on the data rates in each report period such that the energy preservation was maximized. The efficacy of the proposed algorithm was assessed by comparing its energy consumption, packet delivery rate, end-to-end delay, and control message overhead with those of the three other algorithms. The comparisons showed that the SPCPM technique conserved the energy considerably in both centralized and distributed forms. In addition, the packet delivery rate was higher while the end-to-end delay and control message overhead were slightly lower in the case of SPCPM.

Competing interests

The authors declare that they have no competing interests.

Acknowledgements

We would like to show our gratitude to Mehdi Kamal, Dr, University of Tehran, for sharing his pearls of wisdom with us during the course of this research.

Author details

[1]Nanoelectronic Center of Excellence, School of Electrical and Computer Engineering, University of Tehran, P.O. Box 14395-515, Tehran, Iran. [2]Department of Communications and Networking, Aalto University, P.O. Box 13000, FI-00076 Espoo Aalto, Finland.

References

1. IF Akyildiz, W Su, Y Sankarasubramaniam, E Cayirci, Wireless sensor networks: a survey. J. Complex Networks. **38**, 393–422 (2002)
2. G Anastasi, M Conti, M Di Francesco, A Passarella, Energy conservation in wireless sensor networks: a survey. J. Ad-hoc Networks. **7**, 537–568 (2009)
3. DJ Cook, JC Augusto, VR Jakkula, "Ambient intelligence: technologies, applications, and opportunities", in *Pervasive and Mobile Computing*, 2009, pp. 277–2985
4. DDL Mascarenas, EB Flynn, MD Todd, TG Overly, KM Farinholt, G Park, CR Farrar, Development of capacitance-based and impedance-based wireless sensors and sensor nodes for structural health monitoring applications. J. Sound Vib. **329**, 2410–2420 (2010)
5. E Prem, K Gilbert, B Kaliaperumal, EB Rajsingh, Research issues in wireless sensor network applications: a survey. Int. J. Inf. Electron Eng. **2**, 2012 (2012)
6. C Efthymiou, S Niksoletseas, J Rolim, Energy balanced data propagation in wireless sensor networks. Wireless Networks. J **12**(6), 691–707 (2006)
7. J Li, P Mohapatra, Analytical modeling and mitigation techniques for the energy hole problems in sensor networks. Pervasive Mob. Comput. **3**(8), 233–254 (2007)
8. M Nickray, A Afzali-kusha, R Jantti, MEA-An Energy Efficient Algorithm for dense sector based Wireless Sensor Networks. EURASIP J **1**, 1–13 (2012)
9. L Li, JY Halpern, P Bahl, Y Wattenhofer, "Analysis of a cone-based distributed topology control algorithm for wireless multi-hop networks", in *Proceedings of the 20th Annual ACM Symposium on Principles of Distributed Computing*, 2001
10. V Rodoplu, T Meng, Minimum energy mobile wireless network. IEEE J. Sel. Areas Commun. **17**, 1333–1344 (1999)
11. X Wu, G Chen, SK Das, Avoiding energy holes in wireless sensor networks with nonuniform node distribution. IEEE Trans. Parallel Distrib. Syst. **19**, 5 (2008)
12. A Woo, T Tong, D Culler, "Taming the underlying challenges of reliable multihop routing in sensor networks", in *Proceedings of the International Conference on Embedded Networked Sensor Systems (SenSys)*, 2003
13. B Chen, K Jamieson, H Balakrishnan, R Morris, *"Span: an energy-efficient coordination algorithm for topology maintenance in ad hoc wireless networks,"* (ProcMobiCom 2001, Rome, Italy, 2001), pp. 70–84
14. S Narayanaswamy, V Kawadia, R Sreenivas, P Kumar, *"Power control in ad hoc networks: Theory, architecture, algorithm and implementation of the COMPOW protocol,"* (European Wireless Conference Next Generation Wireless Networks: Technologies, Protocols, Services and Applications, Florence, Italy, 2002), pp. 156–162
15. PR Kumar, V Kawadia, *"Power control and clustering in ad hoc networks,"* (Proceedings of annual joint conference of the IEEE computer and communications societies (InfoCom), Sun Francisco, 2003), pp. 459–469

16. N Li, JC Hou, L Sha, *"Design and analysis of an MST-based topology control algorithm,"* (Proceedings of annual joint conference of the IEEE computer and communications societies (InfoCom), Miami, 2003), pp. 1195–1206
17. M Sha, G Xing, G Zhou, S Liu, X Wang, *C-MAC: Model-driven concurrent medium access control for wireless sensor networks,"* (Proc. of the 28th IEEE International Conference on Computer Communications, Rio de Janeiro, Brazil, 2009)
18. S-T Sheu, Y-Y Shih, W-T Lee, Csma/cf protocol for ieee 802.15.4 wpans. IEEE Trans. Veh. Technol. **58**(3), 1501–1516 (2009)
19. A Castagnetti, A Pegatoquet, T Nhan Le, M Auguin, "A joint duty-cycle and transmission power management for energy harvesting WSN". Ieee Transactions On Industrial Informatics **10**, 2 (2014)
20. G Xing, C Lu, Y Zhang, Q Huang, R Pless, "Minimum power configuration in wireless sensor networks, in *proceeding of the 6th ACM International Symposium on Mobile Ad Hoc Networking and Computing (MobiHoc)*, 2005
21. F Lavratti, AR Pinto, D Prestes, L Bolzani, F Vargas, C Montez, *"A Transmission Power Self-Optimization Technique for Wireless Sensor Networks," Journal ISRN Communications and Networking*, 2012
22. Z Zhou, S Zhou, JH Cui, S Cui, Energy-efficient cooperative communication based on power control and selective single-relay in wireless sensor networks. IEEE Trans. Wireless Commun. **7**(8), 3066–3079 (2008)
23. C Long, Q Zhang, B Li, HJ Yang, X Guan, Non-cooperative power control for wireless ad hoc networks with repeated games. IEEE J. Sel. Areas Commun. **25**(6), 1101–1112 (2007)
24. E Tsiropoulou, G Katsinis, S Papavassiliou, Distributed uplink power control in multi-service wireless networks via a game theoretic approach with convex pricing. IEEE Trans. Parallel Distrib. Syst. **23**(1), 61–68 (2012)
25. W Darqie, *"Dynamic Power Management in Wireless Sensor Networks: State-of-the-Art," IEEE Sensors Journal*, 2011, pp. 1518–1528
26. NA Pantazis, DD Vergados, A survey on power control issues in wireless sensor networks. Communications Surveys & Tutorials **9**(4), 86–107 (2007)
27. A Norouzi, A Sertbas, An integrated survey in efficient energy management for WSN using architecture approach. Int. J. Adv. Networking Appl. **03**(01), 968–977 (2011)
28. G Anastasi, M Conti, MDFA Passarella, Energy conservation in wireless sensor networks: a survey. J. Ad-Hoc. Netw. **7**(3), 537–568 (2009)
29. Y Xu, J Heidemann, D Estrin, "Geography-informed Energy Conservation for Ad Hoc Routing", in *Proc. of ACM Mobicom*, 2001
30. G Xing, C Lu, Y Zhang, Q Huang, R Pless, Minimum power configuration for wireless communication in sensor networks. ACM Trans. Sensor. Netw. **3**(2), 200–233 (2008)
31. Q Dong, S Banerjee, M Adler, A Misra, "Minimum energy reliable paths using unreliable wireless links", in *Proc. of 6th ACM International Symposium on Mobile Ad Hoc Networking and Computing*, 2005
32. M Boutabaand, L Sichitiu, Benefits of multiple battery levels for the lifetime of large wireless sensor networks. IFIP Int. Fed. Inf. Process, LNCS **3462**, 1440–1444 (2005)
33. V Raghunathan, C Schurgers, S Park, MB Srivastava, Energy-aware wireless micro sensor networks. IEEE Signal Process Mag. **19**(2), 40–50 (2002)
34. WB Heinzelman, AP Chandrakasan, H Balakrishnan, An application-specific protocol architecture for wireless microsensor networks. IEEE Trans. Wireless Commun. **1**(4), 660–670 (2002)
35. MR Garey, DS Johnson, *Computers and intractability: a guide to the theory of NP-completeness, 1990* (W. H. Freeman & Co, San Francisco, 1979)
36. S Singh, M Woo, SC Raghavendra, *"Power-Aware routing in mobile ad hoc networks", In Proceedings of the 4th Annual ACM IEEE International Conference on Mobile Computing and Networking*, 1998
37. CE Perkins, P Bhagwat, *Highly dynamic destination sequenced distance vector routing (DSDV) for mobile computers* (In Proceedings of the SIGCOMM Conference on Communications Architectures, Protocols and Applications, 1994)
38. D Bertsekas, R Gallager, *Data networks* (Prentice-Hall, Upper Saddle River, NJ, 1987)
39. NR Paul, L Tripathy, PK Mishra, Analysis and improvement of DSDV protocol. IJCSI Int. J. Comput. Sci. Issues **8**(5), 1 (2011)
40. L Feeney, M Nilsson, *"Investigating the energy consumption of a wireless network interface in an ad hoc networking environment", In Proceedings of INFOCOM 2001, volume 3, pages 1548–1557* (Anchorage, Alaska, 2001)
41. S PalChaudhuri, *Power mode scheduling for ad hoc network routing* (Master Thesis, Computer Science, Rice University, 2002)
42. N Bulushu, J Heidemann, D Estrin, GPS-less low cost outdoor localization for very small devices. IEE Pers. Commun. Mag. **7**(5), 28–34 (2000)
43. L Doherty, KSJ Pister, LE Ghaoui, *"Convex position estimation in wireless sensor networks,"* (Proceedings of the IEEE Infocom, pages: 1655-1663, Alaska, 2001)
44. R Peng, *"Angle of Arrival Localization for Wireless Sensor Networks", 3rd Annual IEEE Communications Society on Sensor and Ad Hoc Communications and Networks*, 2006
45. P Kułakowski, E Egea-López, J García-Haro, Angle-of-arrival localization based on antenna arrays for wireless sensor networks. Comput. Electrical Eng. Arch. **36**(6), 1181–1186 (2010)
46. D Niculescu, B Nath, *"Ad hoc positioning system (APS) using AOA"* (Proc. IEEE INFOCOM 2003, San Francisco, 2003), pp. 1734–1743
47. G. Simon, Probabilistic wireless network simulator. http://www.isis.vanderbilt.edu/projects/nest/prowler/.
48. G Simon, V Péter, M Miklós, L Ákos, "Simulation-based optimization of communication protocols for large-scale wireless sensor networks", in *IEEEAC, paper number*, 2002, p. 1136

A review on the recent energy-efficient approaches for the Internet protocol stack

Korhan Cengiz[1,2] and Tamer Dag[3*]

Abstract

The reduction of energy consumption has become a key research area for the information and communication technology (ICT) industry, due to economical, environmental, and marketing reasons. While the environmental direction aims at minimization of greenhouse gas emissions by enforcing the usage of renewable energy in the ICT industry, economical and marketing directions lead researchers to design low-power components or develop and enhance energy-saving protocols without an impact on the level of the performance. With the steady increase in the cost of energy, the expanding number of energy-hungry components and widespread usage of ICT industry, most of the protocols that have become an integral part of our lives but are yet developed without any energy constraints in mind in the past will need to be restructured or developed again. For this reason, researchers are studying on all layers of the Internet protocol stack to develop energy-efficient protocols and algorithms. This paper reviews recent approaches for energy efficiency studies for each layer in the Internet protocol stack from the physical layer to the application layer. It is expected that with the deployment of current research output, the studies performed at each layer will result in significant energy savings for the ICT industry which in turn will have a positive impact on our lives for their economical and environmental results.

Keywords: Green communications; Green networking; Energy-efficient protocols; Energy-efficient applications

Review

Introduction

Nowadays, the information and communication technology (ICT) industry represents an important source of energy consumption and greenhouse gas emissions. Although, it is extremely difficult to configure the exact amount of energy consumption for the ICT, various resources estimate these numbers from a conservative 3% of the worldwide electricity consumption up to two-digit figures [1,2]. Even if the above mentioned conservative figures are accepted, it is a well-known fact that the ICT industry will grow in an exploding manner and the trend for the rapid increase in the amount of energy consumption will continue. However, the excessive amount of energy consumption will not only increase greenhouse gas emissions but will also have a significant impact on global warming and climate changes.

*Correspondence: tamer.dag@khas.edu.tr
[3]Computer Engineering Department, Kadir Has University, Cibali, 34083 Istanbul, Turkey
Full list of author information is available at the end of the article

Green communication technologies are expected to reduce and/or solve the serious problems which we are very likely to embrace in the future with the emergence of new nature-friendly technologies. While the focus of the environmental view of green communications aims at minimizing greenhouse gas emissions and the focus of the engineering view of green communications tries to find ways of reducing the energy usage in the sector, both views expect similar outcomes.

Based on these facts, network and communication devices, protocols, and algorithms should be aware of the energy that they consume to make efficient and responsible decisions, as energy consumption and other green issues are fast becoming the ICT industry's biggest challenge. Many protocols that are currently in use were developed to achieve maximum performance, and the energy consumption issues were neglected. Given that the enormous growth of the ICT sector will continue, it has become extremely important to take precautions quickly and start implementing new protocols to reduce energy consumption. For this purpose, most of the existing protocols are being enhanced or new ones are being developed to achieve these goals in every layer of the Internet protocol stack.

The aim of this paper is to review and summarize the recent studies for energy-efficient approaches for all the layers of the Internet protocol stack starting from the physical layer up to the application layer. Table 1 summarizes some of the work done for each of the layers. A considerable amount of research is devoted for the physical layer for energy-efficient solutions. Developing energy-aware power amplifiers, modulation techniques are the focus of the physical layer studies. Most of the work completed on green communications for the data link layer is on the Media Access Control (MAC) sublayer since MAC protocols have a considerable impact on energy consumption. A MAC protocol reducing the number of collisions with effective usage of radio interfaces can result in significant savings. Another important area for conservation of energy is scheduling algorithms. Data centers are significant energy-hungry systems as their numbers, capacities, and sizes are increasing rapidly. It is estimated that data centers are responsible for approximately a quarter of the energy consumption of the overall network. Thus, many researchers have devoted their studies for minimizing energy consumption for the data centers by optimizing their locations, developing routing algorithms. Due to the bursty nature of the network traffic, a large proportion of network devices can be underutilized for some portion of their lifetimes. Thus, switching off those devices during off-peak hours or estimating the load on the devices is an important way of energy savings. Developing efficient algorithms for mobile *ad hoc* networks and wireless sensor networks also might help to reduce energy consumption. The transport layer energy efficiency studies are on optimizing the TCP. In the application layer, the development of mobile applications and usage of web services and real-time applications can contribute to a decrease in energy consumption. In addition to the studies in all layers, cross-layer energy efficiency research is also gaining strength.

The remainder of this review paper is organized as follows: A discussion on the impact of the energy consumption for the ICT industry is presented in the next section. By providing various examples from different sources, it is observed that ICT energy consumption has become an important portion of the total consumption, and thus, effective measures need to be taken. The latter sections are reserved for each layer of the Internet protocol stack: physical, data link, network, and transport and application layers. For each layer, recent studies for reducing energy consumption are described. Finally, the 'Conclusions' section summarizes and concludes the paper by specifying the importance of developing energy-efficient algorithms as they will have an impact not only on us but also on future generations.

Table 1 Summary of the recent energy-efficient approaches for all layers

Layers	Energy-efficient approaches
Application layer	Energy-aware usage of web services
	Energy-aware mobile application designs
	Periodic real-time applications
Transport layer	Energy-efficient TCP protocols
Network layer	Switching-off underutilized network devices
	Reducing energy consumption in data centers
	Energy-efficient approaches for mobile *ad hoc* networks
	Energy-efficient approaches for wireless sensor networks
Data link layer	Energy-efficient MAC protocols
	Energy-efficient scheduling
Physical layer	Energy-efficient power amplifiers
	Energy-efficient modulation schemes

The impact of ICT on energy consumption

Global warming is the increase in Earth's average surface temperature due to the effect of greenhouse gases, such as CO_2 emissions. Since the early twentieth century, Earth's mean surface temperature has increased by about 1.4°F, with about two thirds of the increase occurring in the last three decades [3]. A major portion of the global warming in the last half century has been caused by human factors as a result of emissions of greenhouse gases. Greenhouse gases can absorb and emit long wave (infrared) radiation in a planetary atmosphere. This phenomenon is known as the greenhouse effect. Energy-related CO_2 emissions, resulting from petroleum and natural gas, represent a significant amount of human-made greenhouse gas emissions [4]. Worldwide CO_2 emissions are expected to increase by approximately 2% annually in the next decade. Most of the increase in the emissions is expected to occur in the developing world in emerging economies [4].

According to scientific studies, if the rate of increase of greenhouse gas emission continues with the current trend, average global temperatures are projected to increase between 2°F and 11.5°F before the start of the next century [3]. The temperature increase can cause very serious and dangerous effects on the environment and human life such as accelerated sea level rise, reduction in the amount of fresh water, changes in the ecosystems, and increased amount and severity of floods and droughts, thus resulting in severe problems for all forms of life living on Earth [5,6].

The ICT sector which accounts for approximately 2% of global CO_2 emissions [7] has become an integral lifestyle commodity for human beings. ICT has played a central role in bringing the people of the globe together. People and the companies have become dependent on ICT for different kinds of reasons such as e-commerce, online banking, home shopping, and instant access to music, books, and videos. With the invention of new technologies and applications or widespread usage of emerging technologies such as Cloud Computing, Smart Cities, and Internet of Things, ICT's impact on human life will continue to increase rapidly. However, according to a report by Alcatel-Lucent [8], the network component of ICT is responsible for approximately 250 to 300 million tons of CO_2 emissions which is equivalent to 50 million automobiles or 20% of the cars in the US. Thus, the ICT sector has become a major contributor to global warming, and consequently, measures need to be taken for greening the ICT. Figure 1 shows how different ICT fields contribute to global CO_2 emissions.

Based on [9], ICT now uses approximately 1,500 TWh of power per year which is roughly equivalent to 10% of the world's total electricity generation or the combined power production of Germany and Japan. The power consumption of ICT is doubling every 5 years at an alarming rate. Another research by [10] illustrates the usage of phase electricity consumption for a number of ICT services as an overview of ICT energy consumption. According to this study, the collective electricity consumption of communication networks, data centers, and personal computers which are the major energy-hungry parts of the ICT is also growing rapidly. For example, in 2007, electricity consumption of these parts worldwide was less than 700 TWh. But in 2012, this amount has exceeded 900 TWh, about one-third increase only in 5 years. As the usage of ICT products and services is growing exponentially, electricity consumption will also increase in a similar manner. In 2012, phase electricity consumption of these parts exceeded 4.7% of the total electricity consumption.

In the UK, ICT equipment accounts for approximately 10% of UK's total electricity consumption according to [11]. Asami and Namiki [12] present the energy consumption for ICT equipment in Japan. The total power consumption of only the routers was 7.3 TWh in the year 2004, and in 2010, this value has reached 13 TWh. Another study [13] estimates that the electric power consumption of the 1.5 billion computers in the world is about 10% of the global consumption. Hinton et al. [14] also investigate the power consumption of ICT and illustrate the power consumption of the wired and wireless access network technologies per user as a function of the access rate and year. According to [14], wireless-based (WiMAX, 3G/UMTS) access networks are the most power-demanding technologies and the power consumption per user increases rapidly. For example, in 2010, while the power per user for WiMAX was about 8 W, this value is tripled to 24 W in 2014.

Nowadays, there are more than 4 million base stations serving mobile users, and each year, 120,000 new base stations are deployed servicing 400 million new mobile subscribers around the world [15]. Each base station consumes, on the average, 25 MWh per year. As a result, electricity has become a significant cost factor for both network operators and their users. Figure 2 shows the cellular network power consumption fields, and it can be clearly seen that the base stations and mobile switching are responsible for a significant portion of power consumption. Also, the most energy-consuming part of the

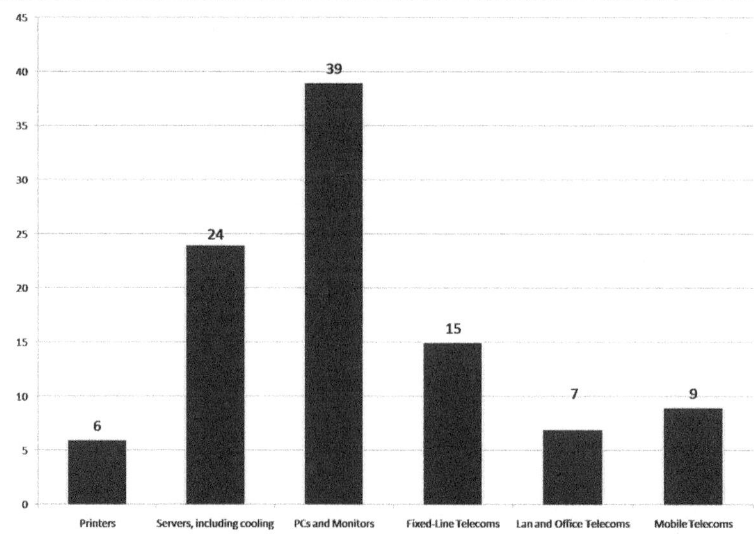

Figure 1 ICT fields' global CO_2 emissions.

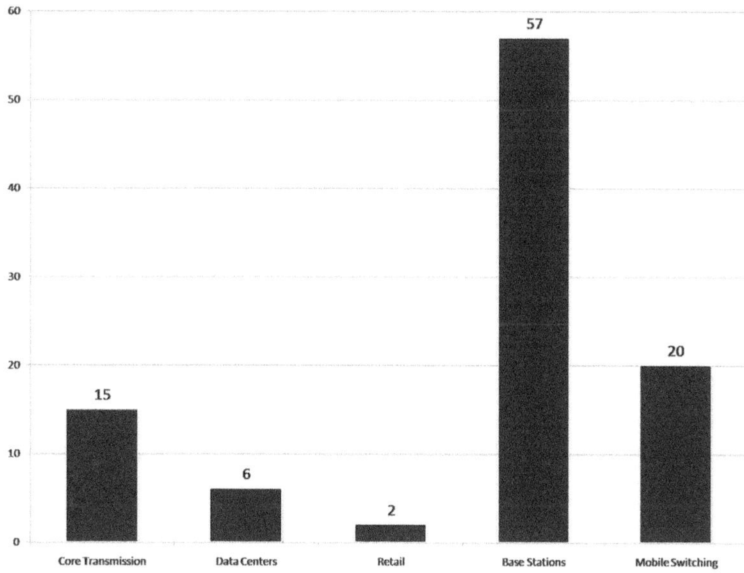

Figure 2 Cellular network power consumption.

base stations is the power amplifiers. According to the research of green wireless networks [16], power amplifiers cause more than half of the energy consumption of the base stations.

The Internet traffic is also rapidly increasing. Currently, the total annual traffic is at a level measured in exabytes (10^{18} bytes), and it is expected to reach zettabyte (10^{21} bytes) levels in the next 5 years. The energy consumption of the Internet has continued to increase as a result of demand growth of at least 50 to 100 times in the last decade. Further demand growth between 40% and 300% is estimated in the next 3 years [17]. According to a report by CNN [18], more power is pumped for the maintenance of Internet than the global automotive industry. Thus, Internet uses more electrical power than the entire automotive industry's combined production of cars and trucks. If we assumed that the Internet was a country, it would rank fifth in the world for the amount of energy consumption and carbon dioxide emissions. A report by Greenpeace states that the Internet consumes more power than Russia. Greenpeace estimates that when functioning at full capacity, Google's eight server farms could use up to 476 MW of electricity which would be enough energy to power San Diego [18] which had a population of about 1.32 million people in 2012.

The energy consumption of modern networks has reached remarkable levels in the recent years because of the increased pervasiveness of computer networks. European Telecoms nowadays consume approximately 21.4 TWh per year, and it is anticipated to increase to about 35.8 TWh by 2020 [19]. The IP networks currently are responsible for 20% of total energy consumption of the Internet, and the prediction shows that the energy consumption of IP networks in the year 2017 will be 12 times larger than that in 2009 [20].

With the entry of software-driven smart phones, the Internet is now accessible from mobile platforms which results in even greater demand for broadband. According to [9], last year, the average smart phone customer used 1.58 GB of data per month. Each GB requires approximately 19 kW. Thus, the average smart phone user consumes 361 kWh of electricity per year.

In addition to the global energy consumption perspective of the ICT, [21] points out that a desktop computer uses approximately 200 kWh of electricity annually, compared with 50 to 70 kWh for a notebook PC, 12 kWh for a tablet, or 2 kWh for a smart phone in a residential area.

As can be seen from the above examples, it is very difficult to find out the exact energy consumption of the ICT and the estimated figures are different from each other. However, on the other hand, it is very crucial to comprehend that these figures are extremely significant and they will continue to increase rapidly. Thus, urgent actions need to be taken to reduce the severity of this problem. For all layers of the Internet protocol stack, many researchers have started working on developing new technologies, enhancing current ones for energy efficient mechanisms. The rest of this paper will describe some of these research starting from the physical layer.

Physical layer

The physical layer consists of the basic networking hardware transmission technologies of a communication network. It is considered as the fundamental layer underlying the logical data structures of the higher level functions in a network and consists of radio frequency circuits, modulation, and coding. The physical layer provides an electrical, mechanical, and procedural interface to the transmission medium. A considerable amount of attention has been given to the physical layer for energy-efficient solutions.

Modulation is the process of varying one or more properties such as amplitude, phase, or frequency of a periodic signal which usually contains information. With modulation, a signal can be carried inside another signal that can be physically transmitted. Many different analog and digital modulation techniques constitute the basis for communication; thus, a considerable amount of effort is being spent for energy-efficient modulation techniques.

Power amplifiers (PA) consume large amounts of energy and dissipate heat, and since increasing demand of larger capacity, higher data rates will increase the burden on PAs; energy efficiency issues for PAs are another major research area. In this section, we discuss some of the recent energy-efficient approaches in the physical layer.

PAs have low energy efficiency because of the linearity requirement enforced by the amplitude phase modulation schemes. Introducing nonlinearity might be a way to overcome this problem; however, its consequence is distortion, and some studies for distortion cancellation [22] aim to enhance the energy efficiency of PAs by taking into consideration the polarization state of signals that is robust to nonlinear distortion and propose a polarization modulation design in wireless communications. In performance analyses, the energy efficiency of the proposed scheme is compared with phase shift keying (PSK) modulation and significant reduction in energy consumption is observed. Wei et al. [23] aim for an energy-efficient subcarrier-power allocation (EESA) scheme. EESA is implemented for the polarization-amplitude-phase modulation (PAPM) in the channel with polarization mode dispersion (PMD). In this scheme, the PA energy efficiency is optimized by distributing the PA input power on each subcarrier. According to the results, the energy efficiency of the PA is increased and significant energy gain is achieved.

Multiple-input multiple-output (MIMO) systems provide capacity and spectral efficiency improvements compared to single-antenna systems. Multiple RF chains cause an increase in energy consumption at the transceivers. Thus, decreasing the number of RF chains increases energy savings. The transmit antenna selection method is the fundamental method for single RF chains. In recent years, the spatial modulation procedure is used for utilizing multiple antennas with only one RF chain. In [24], a feedback-based spatial modulation technique is introduced by using only one RF chain. The proposed scheme results in energy savings compared to the single RF transmit antenna selection scheme.

In [25], adaptive modulation and dynamic signal-to-noise ratio (SNR) adjustment is combined for achieving an energy-efficient orthogonal frequency division multiplexing passive optical network (OFDM-PON). According to the results, a significant amount of energy efficiency is obtained for each bit in an FPGA-based receiver. In [26], linearly modulated signals are transmitted with peak power limitation. Several modulation formats are evaluated in terms of mutual information and compared from the energy efficiency perspective.

An ultra-compact, slight power modulation technique in a wireless transceiver by controlling transmit power for enhancing network power gains is designed in [27]. The on-off keying (OOK) modulator which provides low power dissipation underlies the proposed scheme, and significant energy savings are achieved.

An energy-efficient modulation consideration for wireless sensor communications is presented in [28]. In this study, an energy-efficient modulation scheme, AQ-DBPSK, is investigated. Its performance is evaluated in AWGN and Rayleigh fading conditions. It provides the best bit error rate (BER) performance; thus, the proposed scheme may be a good option for energy-efficient communication systems.

In [29], the cooperative diversity at physical layer is analyzed by using an energy-efficient model for the reach-back problem. In this problem, a large number of sensors are scattered on a field for measuring the state of some physical process that unfolds over the field and then cooperatively send this information back to a distant receiver for further processing. In this study, the BER of the destination is analyzed and the total energy consumption for each bit is obtained. The total energy absorption for achieving a given BER with ideal delay is obtained through numerical analysis under different simulation parameters. According to the simulation results, the total energy consumption of cooperative diversity with two nodes is more than that of the direct transmission for short transmission distances, because of the circuit energy consumption for short distance transmission. However, when the distance is long enough, cooperative diversity can save more energy than direct transmission.

Energy efficiency in wireless *ad hoc* networks with two realistic physical layer methods is discussed in [30]. These methods are the total energy consumption calculation and the reciprocal of the energy efficiency function. The authors investigate these methods in the Rayleigh fading and lognormal fading scenarios to derive the optimum transmission range in wireless *ad hoc* networks. In the two-node model, the hop-by-hop retransmissions (HHRs) model is used. In the HHRs model, a packet is retransmitted between two nodes until it is received and acknowledged correctly. The transmission ranges of the nodes are adjustable. Therefore, each node has different levels of transmission power, and the transmission power arrangement provides energy efficiency. An energy efficiency function which minimizes the energy consumption between the source and the destination nodes is also introduced. When this function is maximized, the optimal energy consumption between the source and the destination is obtained. The reciprocal of this function is linearly proportional to the total energy consumption value. According to simulation results, the two methods - the total energy calculation method for the end-to-end model and reciprocal of the energy efficiency function method - have the equivalent energy efficiencies.

In [31], an optimization scheme based on adaptive modulation and power control for a green routing protocol is used. In order to find the energy consumption of the data packet transmission, the energy consumption per bit for transmitting these packets is calculated. The derived equation demonstrates that the transmit power depends on the distance between the transmitter and the receiver. M-ary Quadrature Amplitude Modulation (M-QAM) provides better spectral efficiency, and the M-ary Frequency Shift Keying (M-FSK) warrants power efficiency. Using M-FSK is the most optimum solution for sensor networks because it requires the least circuit complexity, and for a given symbol error rate (SER), it needs less transmitted power. Based on this, the non-coherent M-FSK (NC M-FSK) is selected for modulation. Total energy consumption is calculated based on two energy spending reasons. The first reason is due to the RF signal generation and target distance. The second one is due to the hardware electronic components. In the network model, each node is stationary and knows its location by using GPS. The IEEE 802.15.4 standard is taken as a reference, and the ISM band is assumed for operating the modulation schemes. For evaluating the performance of the proposed algorithm over the multi-hop route, two scenarios are chosen: without optimization of power and modulation order and with optimization of power and modulation order. In the simulations, the impact of route length is analyzed. According to results, using maximum power while transmitting at a fixed rate causes the largest energy consumption. With optimization, the system achieves significant energy gains.

In [32], a method for a robust and energy efficient physical layer for wireless sensor networks is proposed. For assessing the radio link quality and calculating the energy per bit transmitted successfully, the BER is used. The results are presented for an indoor environment. For energy efficiency, transmission power evolution is tested for two transmission chains (with and without forward error correction), three modulation schemes (ASK, FSK, MSK), and two different frequency bands (433 and 868 MHz). The best configuration observed for optimal energy efficiency is a transmission system with MSK modulation, 433 MHz frequency band, and forward error correction (FEC) usage. It is also pointed out that the physical layer optimization alone is not suitable to achieve the desired lifetime for a wireless sensor network.

The authors of [33] had earlier introduced a novel recursive physical layer security scheme. In [33], a novel synchronization process is introduced where energy spending is minimized and the impact on memory and CPU is measured. The analysis which is confirmed by simulations sustains significant performance enhancements. The best physical layer authentication preamble (AP) consisting of a hash function using some secure keys and additional information is achieved. Through simulations, it is shown that the developed protocol is able to measure the optimum operational point where the energy dissipation of the system is minimum. The simulation results show that their system saves a significant amount of energy.

Some energy efficiency studies on the physical layer [34,35] also try to combine energy efficiency with security considerations. Zhang et al. [34] combine the physical layer secrecy and application layer secrecy requirements for characterizing the partial secrecy metric to provide an energy-efficient physical layer security in Gaussian fading channels. In [35], for compensating the relaying of the information in a secret and efficient scheme and taking into consideration the quality of service (QoS) limitations, the physical layer security technique with an energy-efficient power allocation is proposed.

Packet loss, delay, and bandwidth are the parameters which can affect the performance of VoIP services. Increasing the transmission power overcomes this problem, in the expense of more energy consumption. In [36], an energy-efficient transmission power algorithm (TPO) which improves the energy utilization by considering QoS limitations is proposed. The proposed algorithm adopts the lowest possible transmission power with the help of SNR approximation without affecting the VoIP service quality. The transmission power is managed in accordance with channel conditions. In simulations, significant energy gains are obtained especially in a noisy channel situation.

Data link layer

Most of the studies for green communications for the data link layer focus on the MAC sublayer. In this section, we discuss recent energy-efficient MAC protocols as well as Long-Term Evolution (LTE) solutions.

Media access control

MAC data communication protocol is considered as a sublayer of the data link layer. A MAC protocol defines the rules of how a frame is transmitted on a communication link by addressing and channel access control mechanisms. When multiple nodes share a single link, the MAC protocol coordinates the channel access. This channel may provide unicast, multicast, or broadcast communication services. In this section, recent energy-efficient approaches for the MAC sublayer are discussed.

For mobile devices, the most energy-hungry component is the radio interface. For example, the radio interface approximately consumes 70% of the total energy in a smart phone when the screen is off. The radio interface can be in transmit, receive, idle, or sleep mode, and the most power is used in the transmit mode. In the idle mode, a node requires to sense the medium and therefore the consumed amount of power is approximately equal to the receive mode. Collisions are another reason of high energy consumption in wireless devices. The reduction of collisions can enhance the throughput and decrease the energy consumption. As MAC protocols assign a channel to the nodes in the network, they directly control the radio interface operation and have a significant role in energy consumption for wireless devices.

In [37], a novel energy-efficient MAC scheme is studied. The proposed scheme schedules the active and sleep times of node radio interfaces in a distributed method. The collisions are reduced by the use of a temporary coordinator node. The network throughput is improved, and a lower packet delay than current power saving mechanisms is obtained. In performance analyses, the proposed scheme is compared with a power saving mechanism in IEEE 802.11 which is called as PSM where time is divided into beacon intervals that are used to synchronize the nodes and all the nodes must stay awake for a fixed time, and IEEE 802.11 without power saving mechanism (WoPSM). Aggregate throughput, power consumption, energy consumption, and packet delay are the performance measurement metrics. The power consumption analyses show that WoPSM is the most energy consumer because all the nodes are in the awake state all the time. The overall network power consumption increases as the load increases for all the schemes. It is clearly obtained that the proposed scheme outperforms both PSM and WoPSM in terms of energy efficiency and also in terms of throughput and delay.

A popular MAC standard described by IEEE 802.15.4 for wireless sensor networks and actuator networks defines two operating modes: non-beacon-enabled mode where the receivers are always awake to receive a frame and beacon-enabled mode which describes super frames where the nodes are only awake during a small part of a super frame and they sleep quite often. In the second mode, significant traffic results in frequent collisions and packet losses which causes an increase in energy consumption and low throughput. To operate IEEE 802.15.4 with low duty cycles needs MAC parameter adjustment to attain low energy consumption. The beacon-enabled mode behavior is analyzed in [38] to define the main performance bottlenecks. The adaptation method is referred to as ABE where a coordinator observes the number of idle slots before a transmission starts. However, a node consumes energy to observe idle slots because it needs to be awake. This problem is solved by delegating tracking of idle slots to the coordinator that always needs to stay awake during the active period. ABE is compared with the IEEE 802.15.4 standard configured with static backoff exponent which specifies the size of the contention window values. Results show that ABE provides optimal throughput without considering the number of nodes. The value of the contention window which optimizes throughput is found by the coordinator in ABE. The adaptation mechanism in ABE which adjusts the optimal contention window values yields high throughputs along with low duty cycles, leading to low energy dissipation.

Forcing the nodes to sleep adaptively while ensuring the continuation of communication is the purpose of [39]. The Base Station Controlled MAC (BSC-MAC) conserves energy by switching off the idle nodes. The source nodes generate the data and forward it to the base station through the root nodes. In BSC-MAC, the list of all nodes reaches the base station by flooding. The nodes with subsets are detected, and they are designated as the root nodes. The capture of a path causes the neighbor nodes to go to sleep mode. This mechanism provides power savings, reduces packet collisions, inhibits packet losses, and also prolongs the network lifetime. A NS-2 simulator is used for simulations. BSC-MAC is compared with Pattern Medium Access Control (P-MAC) which designates the sleep of sensor nodes adaptively, Timeout Medium Access Protocol (T-MAC) which is used for idle listening for wireless sensor networks (WSNs), and Sensor Medium Access Protocol (S-MAC) which puts the nodes into sleep mode periodically when the nodes unnecessarily listen to the environment to the point of total energy consumption for all scenarios. BSC-MAC performs better than S-MAC. It has worse results than P-MAC when the total number of generated packets increases. Also, BSC-MAC is compared with Adaptive Energy-Efficient MAC (AEE-MAC) which provides more energy efficiency than MAC by using additional improvements: adaptive sleep, reuse of the channel, and combined control packets. BSC-MAC can result in more energy savings compared with AEE-MAC.

Various MAC schemes have been proposed to enhance resource sharing and energy preservation for WSNs. However, these MAC mechanisms are complex and expensive. For this reason, simulators are required to understand the network behavior. The authors in [40] are interested in the performance of wireless MAC protocols for energy dissipation. A simulator (EnergySim) is developed to evaluate energy consumption of a network and nodes for any given wireless MAC protocol.

EnergySim provides position information of the nodes, nodes' residual energy, energy consumption, and the number of alive nodes. It is an event-driven simulator and specifies the shortest paths by the usage of the Dijkstra algorithm. It allows to make detailed energy analyses for any given wireless MAC protocol. Table 2 compares some CSMA/CA-based MAC protocols for their energy efficiency.

LTE (Long-Term Evolution)

Wireless access networks consume 80% of the total energy used by mobile communication network operators. Radio resource management and allocation functions for LTE have become an important research subject as energy consumption in base stations and downlink transmissions seems to be a major area where significant conservation can be achieved.

Instead of optimizing to improve spectral efficiency of access networks, in [41], energy efficiency of some well-known principal scheduling schemes is introduced. Three relevant energy-efficient schedulers are introduced: round robin (RR), best channel quality indicator (BCQI), and proportional fair (PF). RR is the one of the most popular principal schedulers. It assigns the same number of physical resource blocks (PRBs). BCQI and PF are channel-aware scheduling schemes. BCQI selects the user with the highest channel quality index (CQI), which measures the quality of a wireless communication channel; in other words, a high CQI is indicative of a high-quality signal at each transmission time interval (TTI) for each PRB. In PF, the users are selected based on a normalized achievable rate at each time slot. When the energy consumption model is examined, a linear power consumption model is considered. The bandwidth is traded off for energy whenever possible. According to derivations, the amount of energy consumed in each TTI is reduced so it provides overall reduction of the transmission energy. Finally, a maximum of one PRB is assigned to each user, and three energy efficient schedulers are obtained: energy-efficient round robin (EERR), energy-efficient proportional fair (EEPF), and energy-efficient best channel quality indicator (EEBCQI). In simulations, a network with a single base station with multiple users is used. The performance of the proposed schedulers (EERR, EEPF, EEBCQI) is compared with that of principal schedulers (RR, PF, BCQI) in terms of energy efficiency and QoS. According to results, when there is a low number of available resources (PRBs) in the system, the proposed schedulers save significantly more energy than the principal schedulers. By increasing the number of PRBs, this energy gain scales up because all of the PRBs are not occupied as ON/OFF traffic source is considered. So, if all the PRBs are assigned to the users, the energy gain will rise. In conclusion, this work shows that energy savings gain mostly depends on the available resources (PRBs) and number of users in each TTI.

3GPP Long-Term Evolution-Advanced (LTE-A) is one of the advanced technologies for mobile communications. In LTE-A, multiple QoS classes have been defined for different bit rate, packet delay, and packet loss rate characteristics. In order to decrease the power consumption of the user equipments (UEs), a discontinuous reception (DRX) mechanism has been designed in LTE-A. This mechanism switches off the radio interfaces of the UEs when evolved node B (eNB) does not have data to deliver to the UEs. Existing studies show that the usage of DRX can significantly reduce energy consumption in UEs. Although these studies have mentioned the DRX optimization problem, they did not consider QoS requirements and aimed to manage the DRX cycle length for UEs to enhance energy efficiency. Authors in [42] mention the DRX optimization problem. This study provides QoS satisfaction for UEs and saves energy by minimizing wake-up periods of UEs. A packet scheduling method and an efficient scheme are proposed. The proposed scheme is compared with the counter-driven DRX (CDD) which manages the length of the DRX cycle for each UE by using predefined thresholds, and multiple-threshold DRX (MTD) which designs multiple lengths and thresholds. The effects of packet loss rate, rate satisfaction ratio, and power consumption are examined. Reduced packet loss rates are obtained by using the proposed scheme since it extends wake-up periods and catches more packets. The proposed scheme possesses the highest satisfaction ratio when the network is saturated. Additionally, the power consumption performance of the proposed scheme has been observed to be in between the MTD and CDD. CDD outperforms the proposed scheme

Table 2 Comparison of MAC protocols

Protocols	Type	Latency	Throughput	Energy efficiency
P-MAC	Contention based	Medium	Medium	High
T-MAC	Contention based	High	Low	High
S-MAC	Contention based	Low	Low	Medium
AEE-MAC	Contention based	Low	Low	Medium/high
BSC-MAC	Contention based	Low	Medium	Medium/high

in terms of energy efficiency; however, it neglects the QoS satisfaction of UEs.

While some studies aimed to decrease the energy consumption of UEs in mobile networks, UE energy consumption is only a small fraction of the energy consumption in these networks. A larger fraction belongs to BSs. A study [43] is developed as part of the EARTH Project which aims to improve energy efficiency of LTE cellular networks while providing more wide coverage and to reduce the carbon footprint. Authors aim to enhance radio transmission techniques such as beamforming and MIMO antenna techniques to promote the objectives of the EARTH Project. Especially MIMO is used to enhance the spectral efficiency of the communication links. It is pointed out that in macrocells, significant reduction in power consumption can be obtained at high loads by using reconfigurable antennas. A suitable MIMO mode selection and proper precoding and scheduling can provide even more power savings. Also, significant power reductions are attained in picocells with convenient MIMO mode selection.

In [44], the design comparison between LTE and LTE-A is presented in terms of energy efficiency. LTE-A adds three new functionalities: carrier aggregation which increases the bit rate, heterogeneous networks which include combination of macrocell- and femtocell-based stations, and having multiple antennas in other words improved MIMO support to LTE. The energy efficiency in wireless access networks is formulated by specifying the most energy-efficient BS. The effects of the three new functionalities of the LTE-A on energy efficiency are investigated. The results show that especially LTE-A's carrier aggregation and MIMO enhancement provide significant increment on energy efficiency when compared with LTE. LTE-A also provides improved support for relaying, and this can also improve energy efficiency even more.

Network layer

The energy consumption of the ICT sector has grown rapidly in the recent years with the exponential increase in the number of people and devices that are connected to the network as well as the broad number protocols that have become key components of our daily lives. As a result, energy-efficient solutions have become a major concern and scientists have started to work on energy-aware protocols. With the next generation of networking, we will soon be using energy-efficient protocols which will replace the current protocols developed without any energy constraints.

Data centers are one of the largest and fastest growing energy consumers in the world as they have become the backbone of many institutions. The explosion of Internet traffic and its content pushes data centers' energy needs forward. Thus, a significant amount of effort is spent on designing energy-efficient data centers. Switching off existing devices when they are idle is also an important candidate for providing an immediate energy-efficient solution. However, the questions of when and how to make these decisions without an impact on the performance need to be answered. There are also energy efficiency studies on optical IP, mobile *ad hoc* networks, and wireless sensor networks. In this section, we try to describe some recent energy-efficient approaches for the network layer.

Switching off network devices

Switching off lightly loaded devices such as network links and/or nodes is one of the extensively studied energy efficiency methods as shown in Figure 3. Switching off network devices can result in immediate and significant energy savings provided that it does not have a negative impact on the network traffic. For this reason, traditionally used protocols need to be enhanced to compensate for switching off solutions.

With the Open Shortest Path First (OSPF) protocol, shortest path trees (SPTs) are computed and used by all routers in the network. Energy-Aware Routing (EAR) [45], on the other hand, uses only a subset of router SPTs for traffic routing. The EAR protocol classifies the routers into exporter, importer, and neutral routers. The exporters calculate their SPTs, and then the importers use these SPTs to construct their own modified path trees (MPTs) and to determine the links that can be switched off. Neutral routers act as in the traditional OSPF routing protocol. The performance of the protocol is measured by a using a performance index which shows the the percentage of links that the EAR protocol allows to switch off. The simulation results indicate that a significant number of links can be switched off and energy savings can be increased especially under light traffic conditions.

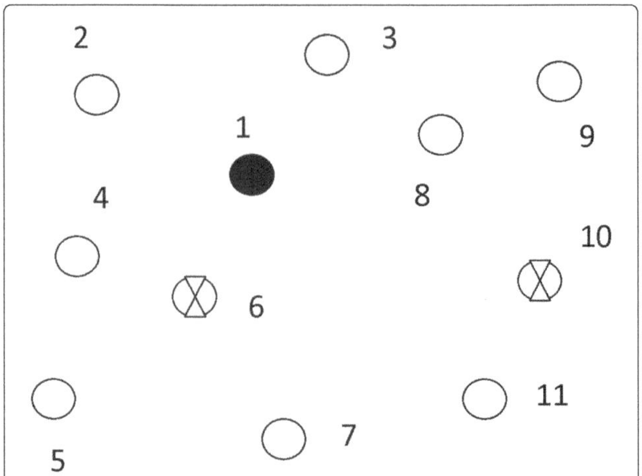

Figure 3 Switching off the nodes.

In [46], a distributed link management algorithm for energy-efficient IP networks is developed. The links in a network can be switched on/off based on the dynamically managed two threshold levels. The threshold levels are chosen based on the link flow amount. The distributed link management algorithm provides redirecting the flow on the fly without losses or loops during the long convergence of the routing protocols after switching on/off the links. In this work, the authors also implement an optimization tool which can specify the optimal power saving network topology.

A distributed approach [47] utilizes sleep mode abilities of links in an ISP network with the knowledge of current link load instead of current traffic matrix. The working states of the links are adjusted dynamically, and the network capacity is adapted to the current traffic load without overloading the resources. The optimization is to minimize the network power consumption under specific traffic conditions and QoS limitations. Distributed Least Flow (DLF) which selects the least loaded link and Distributed Most Power (DMP) which selects the most power-hungry link is used for algorithm design. These two algorithms are further divided into LastSleep which selects the last link entering a sleep state and Distance which selects the closer sleeping link to the congestion point. According to results, the DMP algorithms are more powerful than the DLF in terms of sleep attempts because DMP aims energy-hungry links which carry a lot of traffic. Furthermore, Distance policies save more power than LastSleep policies.

Minimizing energy consumption by putting idle routers and links into sleep mode and, furthermore, keeping the interference on network operation minimal is studied in [48]. On the contrary to the previously discussed works, this work can put both nodes and links into sleep mode. The optimization problem is formulated to maximize the number of sleeping nodes and links by taking into consideration the load balancing and the link utilization variation problems, and a heuristic Energy-Efficient Topology (EET) protocol is proposed. In the derivations, the energy efficiency of the nodes and the links is expressed separately, and the average energy efficiency of the overall network is obtained. The subset of routers and links that can be put into sleep mode is found. It is pointed out that more energy can be saved by switching off a network node than putting a single link into sleep mode. Thus, in EET protocol, the maximum number of nodes that can be switched off is determined first and then the links that can be switched off are identified. When the network load is low, a larger amount of routers can be selected to put into sleep mode.

A table lookup bypass (TLB) approach is proposed in [49]. Under suitable traffic conditions, the table lookup (TL) operation of a line card (LC) is bypassed and the received packets are deterministically forwarded through the next-hop router. Forwarding engine (FE) functionalities are frozen, and power dissipation is significantly reduced. By this way, the forwarding operation is independently achieved by each LC in core IP routers. The proposed approach also combines TLB and link switch off (LSO) approaches. There are two alternatives: to switch a whole link off or freeze out only the FE of LC. Especially in high traffic conditions, the proposed approach promises a lower energy consumption. It is pointed out that TLB is very suitable when LSO cannot be fulfilled; in other words, traffic load does not allow switching off of other links when the network is highly meshed.

In [50], a novel energy-efficient routing approach called Safe and Practical Energy-Efficient Detour routing (SPEED) is proposed. It provides power saving maximization with guaranteed connectivity by applying two key ideas: precomputing loop-free next-hops for each primary next-hop to forward traffic and selecting loop-free further next-hops then aggregating traffic to them without extra routing costs. Primarily, the proposed approach achieves energy savings by putting idle links into sleep mode after traffic aggregation. In simulations, real network topologies GEANT and Abilene are used. The performances of ORIG (routings computed by a traditional intradomain routing protocol), SPEED (proposed protocol), SPEED-C (proposed protocol with modified network configurations where the link weights are modified), OPT-MAX (the traffic over links with heavier loads will be pruned first), and OPT-MIN (traffic over links with lighter loads will be pruned first) are compared. According to the power saving ratio analyses of the simulations, it is proved that SPEED and SPEED-C are efficient for power savings in the traditional IP networks.

An energy efficiency and load balancing optimization algorithm called as Green-Load-balancing Algorithm (GLA) based on existing link switch off techniques is introduced in [51]. By optimizing the Interior Gateway Protocol (IGP) link weights of a network, energy saving gains are obtained by using the link sleeping and load balancing jointly. The algorithm aims to put more links to sleep without affecting the load balancing requirements. In simulations, the GEANT is used. The GLA is compared with Interior Gateway Protocol Weight Optimization (IGP-WO), and it is obtained that GLA shows slightly better performance than IGP-WO in terms of load balancing and results in energy savings at the same time. According to several experiment scenarios, the proposed algorithm improves the energy efficiency and enhances the performance of existing switch off algorithms in terms of energy efficiency.

A distributed energy-aware traffic engineering analysis DAISIES is proposed in [52]. DAISIES uses routing-based approach where the link switch off process is handled individually by each node. The switching off process is managed with respect to the traffic variations. A definite load-dependent link-weight function is proposed to perform DAISIES more efficiently. The proposed scheme is different from other studies in two important aspects: there is no off-line centralized component, and it searches for the best path for each traffic demand rather than search for the best set of links to switch off. It takes advantage of traffic engineering of the existing routers and modifies the path computation strategy, and it does not require significant changes in current routers. The event-driven simulation packet is used for simulations. For different traffic conditions, DAISIES performance is tested and it is observed that the proposed scheme provides significant energy gains.

In [53], switching off links and routers, managing the link weights for minimizing the energy consumption besides the network congestion with priority order is aimed. A Mixed Integer Linear Programming-based algorithm for Energy-aware Weights Optimization (MILP-EWO) which takes advantage of the IGP-WO algorithm is presented, and the link weight optimization is configured. Link capacities are considered, and the network congestion level is minimized for ensuring service quality. According to the simulation results for several real network topologies, significant energy savings can be gained by using the efficient configuration of link weights in IP networks with homogeneous and heterogeneous network equipments by switching off a considerable number of core nodes.

In [54], the green routing algorithms are analyzed based on realistic network scenarios. idleEnergy, fully proportional, and energy-agnostic models for a network device energy consumption are investigated with an objective to find the network configuration that minimizes the total energy consumption. The pan-European data network GEANT is used for comparing Green and IGP-WO routing algorithms for idleEnergy (realistic case), fully proportional (best case), and energy-agnostic (worst case) energy models. The results conclude that on all energy models, it is possible to achieve higher energy savings by switching off links rather than by switching off nodes. In addition, it is observed that the energy saving performance depends on the network topology, traffic requests, and the device technology rather than the maximum device utilization.

Optical IP

Telecommunication networks can be classified into access, metro, and core domains. Core networks consume a significant amount of energy as they support high capacities with very high data rates. Core networks depend on optical transmission. In particular, the IP over wavelength-division multiplexing (WDM) technology is used for compensating increased Internet bandwidth requirements. For reducing the power consumption, the optical switching solutions are chosen. The most significant advantage of the optical switches is that their power dissipation does not increase significantly as the data rate increases. Generally, three types of optical switching solutions - optical circuit switching (OCS), optical burst switching (OBS), and optical packet switching (OPS) - are considered. In OCS, low costs and power consumption reductions are proposed; however, flexibility and bandwidth utilization is restricted. OBS sets up data into bursts and decreases network costs and power, but results in high data losses at core nodes. The highest flexibility and resource utilization is obtained by OPS, and it can be considered as the best alternative for electronic switching. However, it needs complex optical components that lead to high power consumption.

For exploiting the advantages of OCS, OBS, and OPS on the same network, the hybrid optical switching (HOS) is implemented in [55] for achieving an energy-efficient Internet core. Three possible architectures - an all-optical HOS network, an optical/electronic HOS network, and all-electronic network - are proposed for future high-capacity networks. For comparison of the proposed network solutions, the Pan-European network is used. According to the results, the all-electronic network produces the highest carbon dioxide emissions. As the input load in HOS networks is lower, the HOS network emissions as well as the energy consumption are lower. But it can cause a decrease in the network throughput and larger data loss rates.

In [56], adaptive networks and mixed-line rate (MLR) IP networks are compared over DWDM networks in terms of cost and power efficiency. With a realistic IP traffic pattern, the network optimization scheme optimizes the IP layer and WDM layer by setting additional IP links to bypass IP routers and route them through suitable transmission links except for all links that minimize the power consumption of the whole network. The simulation results show that the MLR network achieves more power gain than the adaptive network which is more hardware efficient in terms of total number of transponders with hierarchical scenario (higher link load but less number of IP links) and flat scenario (additional shortcuts between ex-region nodes together with lower link capacities) in both optimized IP layer and optimized WDM layer.

In [57], the power savings achieved by optimization of the physical topology of IP over WDM networks is investigated in terms of non-bypass and bypass approaches under nodal degree constraints, traffic symmetry, and renewable energy availability. The results of simulations show that the optimized physical topologies provide significant energy gains when compared to the NSFNET topology, which covers the US and provides a realistic scenario, delay-minimized topology, and the virtual topology.

In [58], the energy consumption of IP over optical WDM networks is analyzed through an optical power-aware backbone network model. It is stated that the energy consumption of the IP over optical WDM networks can be reduced by bypassing the lightpath in the optical layer which causes a reduction in the number of IP router ports that consume maximum energy. In an energy-minimized design, the set of nodes corresponds to IP routers and optical switch nodes, and the link set symbolizes the physical fibers in the network. The proposed optimization model decreases energy dissipations in the network such as IP routers and WDM transponders to minimize overall network energy consumption. From the simulation results, it is observed that the proposed scheme consumes less power than the traffic grooming scheme when the lightpath bypass strategy is implemented. It is seen that the network size growth increases the power savings of the lightpath strategy because it provides longer lightpaths and allows bypassing of more intermediate switch nodes.

In [59], the energy efficiency of rate and modulation adaptive optical OFDM-based networks is investigated. OFDM promotes heterogeneous traffic demands. The distance-adaptive spectrum allocation which includes the adaptation of the modulation level of subcarriers by using optical paths of higher optical signal-to-noise ratios (OSNR) is taken into consideration with flexible bandwidth allocation. The proposed model differentiates the power-minimized and spectrum-minimized optimization schemes. In simulations, the optical OFDM-based networks are compared with conventional IP over WDM networks under the bypass approach. OFDM-based networks are also compared with MLR networks. The proposed model saves a significant amount of energy compared to IP over WDM networks and MLR networks. In addition, MLR networks also have higher energy efficiency than WDM networks.

Data center networks

Data centers are among the major contributors of energy consumption for the ICT sector due to the ever growing demand for video and data extreme applications in various fields such as informatics, genomics, finance, and medicine. To make things even worse, the cost of energy is steadily increasing. There are many studies that focus on minimizing the power consumption inside the data centers, as it is estimated that the data centers are responsible for approximately one quarter of the overall network power consumption. According to [60], the amount of energy consumed by data centers is growing by 12% per year. One data center can use enough electricity to power 180,000 homes; in other words, they can consume up to 100 times more energy than a standard office building [61]. In this section, we present recent studies that focus on decreasing power usage in data centers and reducing the power dissipation of data transportation between data centers and end users.

Air cooling represents an important share of the energy expenditure in data centers. If computational power, temperature, leakage, and cooling power needs can be reduced, major energy savings might be possible. In addition to air cooling, the location of data centers can have a significant impact on energy usage. For over 10 years, Google [62] has been building some of the most energy-efficient data centers in the world. Their data centers consume 50% less energy than the typical data centers, resulting in a savings of more than a billion dollars to date. By displacing the server farms to the banks of the Colombia River, the energy offered by the hydroelectric power plants nearby has been used with significant savings. The water flow provided by the river in addition is used within the cooling systems. An alternative cooling system designed by Microsoft in the Tent and Marlow projects [63] leaves servers in the open air so that heat can dissipate quickly.

Dong et al. [64] study the location of the data centers for minimizing the power consumption through mathematical optimization models. Two traffic scenarios: the traffic towards and from the data centers and the traffic between data centers with the non-bypass approach and the multi-hop bypass approach are analyzed. The optimal data center locations provide approximately one quarter more energy gain when compared to the worst locations in the non-bypass approach. The energy gain value is about one half of the non-bypass case with the multi-hop bypass approach. The problem of whether to locate data centers next to renewable energy or to transmit renewable energy to data centers is studied and concluded that a significant energy gain can be obtained in optimization of the data centers' locations with renewable energy.

As leakage power is exponentially dependent on temperature, higher temperatures can cause higher power consumption. Fan speeds affect the power dynamics of servers and can cause an increase in temperature and leakage power. In [65], a leakage and temperature-aware server control mechanism is presented for enhancing energy efficiency of enterprise servers in data centers. In order to determine the leakage-temperature effects on the energy efficiency of servers, two fan control schemes - the LUT-based Controller which periodically monitors the load utilization and provides fan speed management of the power supplies and the Bang-Bang Controller which keeps the temperature in between two desirable temperature values - are presented. According to the results, both fan control schemes provide energy savings in comparison to the original fan scheme. It is also observed that, the LUT-based scheme is better in energy efficiency because the runtime temperature values are lower and steadier and leakage is always kept low.

A study for temperature power management in [66] tries to optimize temperature sensor deployment in data centers to control cooling supply which is still a problem in the data center energy efficiency perspective. To solve this problem, authors deploy sensors in a single rack and aim to extend the algorithm to the whole data center. Then, an optimization problem is formulated to determine proper positions for temperature sensors to adjust cooling supply. In simulations, the relationship between the optimal temperatures and optimal sensor positions is investigated. It is observed that significant energy savings can be obtained when compared with random sensor deployment.

Authors in [67] enhance techniques to analyze and optimize energy management in multi-core servers by involving the dynamic power of the processor in addition to leakage power and switching overhead between speed levels. An objective function that is a linear combination of energy usage, queuing cost, and switching costs is established. The considered processor is multi-core where each core is running at a different speed, while previous studies considered only the case where all the cores were running at the same speed. According to the results, a significant part of the energy efficiency gain is obtained with one or two thresholds. Also, the hysteretic model, in which there are two threshold buffer levels and the queue is served with the lower service rate, outperforms the one-threshold model in terms of energy efficiency. Consequently, the most energy-efficient model is the two-threshold model for all scenarios.

A new throughput-guaranteed Power-Aware Routing algorithm (PRA) is presented in [68] which aims to reduce the total power consumption of network devices. PRA provides efficient power utilization in data center networks by exploiting sufficient network power to ensure the routing service while sustaining the target network throughput. In addition, the trade-off between power conservation and network fault tolerance is investigated. The proposed approach uses a different rate model where the transfer rate of each flow is bandwidth constrained and depends on network resource competition with other flows rather than the constant traffic demand. The proposed approach is pruning-based. The switches and the links are iteratively eliminated, and it is a significant alternative for existing routing algorithms. BCube and Fat-Tree topologies which are used for data center networks are considered in simulations. The results are obtained in the sense of various traffic patterns. It is obtained that the proposed algorithm decreases the power consumed by switches at low network loads, and as expected, the power saving percentage is inversely proportional to the network load. Besides, the proposed algorithm is compared with the ElasticTree scheme where every flow has a constant transfer rate. Results show that PRA obtains abundant power gains than ElasticTree.

Since networking, servers, and cooling are the major energy consumers in data center networks, most of the existing studies focus on energy-aware routing and topology designs rather than scheduling traffic flows in a more energy-efficient scheme. In [69], authors depart from this idea and propose a joint manner which combines preemptive flow scheduling and energy-aware routing. In the proposed algorithm, Flow Preemption and Energy-Aware Routing (FP and EAR), the higher priority owner flow is preferably scheduled and the lower ones are suspended until the flow ends. This scheme makes the links completely utilized and saves more energy. Then, the energy-aware routing assigns the paths sharing switches sufficiently to the traffic flows which are scheduled at the same time in the network. In the simulations, the performance of the FP and EAR is compared with ECMP routing which is traditionally used in data center network topologies and achieves load balance and high network performance. According to results, the network energy is saved significantly by the proposed algorithm under the impact of flow arrival interval. In addition, the effect of traffic flow distribution on the network energy consumption is also evaluated. It is seen that when the percentage of used servers is low, the impact of FP and EAR increases significantly.

Mobile ad hoc networks

Unreliable wireless links and multiple receivers associated with a single packet transmission can result in poor energy efficiencies for wireless networks. In this section, several studies aiming to overcome these difficulties and providing for energy-efficient solutions for mobile *ad hoc* networks (MANETs) are described.

The performance of several routing protocols such as AODV, DSR, DSDV, TORA, and EAODV for MANETs is compared in existence of frequent link failures due to the high mobility of the nodes in [70]. In energy efficiency analyses, it is pointed out that for measuring energy efficiency, the network lifetime can be used instead of power consumption. In simulations, four performance metrics - packet delivery ratio (PDR), normalized routing load (NRL), average end-to-end delay, and energy consumption - are used. Performance results of these metrics are obtained with the impact of three parameters: node velocity, packet sending rate, and pause time. The energy consumption of the on-demand protocols increases as the maximum node velocity increases. The results show that DSR has lower energy per user data than EAODV and DSDV. When velocity increases, the energy efficiency of the DSR is better than EAODV, TORA, and DSDV.

A scalable, opportunistic, and energy-efficient routing procotol (E2R) is proposed in [71] which employs a greedy forwarding algorithm by reducing the control message overhead. E2R protocol is highly distributed and scalable as it requires one-hop neighbor information. The protocol exploits spatial diversity in wireless networks rather than determine data delivery paths such as next hop. Thus, the discovery packets and data packets are delivered by broadcasting. The proposed protocol is compared with AODV and SBA-AODV (scalable broadcast algorithm AODV). In SBA, before rebroadcasting the first received packet, each node waits for a randomly assigned time interval. Packet delivery ratio, control overhead, and packet delivery delay are the metrics used for evaluating the protocols. The impact of node density and node failure is examined for these metrics, and it is observed that the control overhead is 90% reduced compared to AODV.

The energy-efficient routing protocols are generally classified into two categories in literature: Minimum Energy Routing Protocols and Maximum Network Lifetime Routing Protocols. Progressive Energy-Efficient Routing (PEER) presented in [72] is a minimum energy routing protocol. Contrary to other energy-efficient routing protocols that try to find the optimal path during route discovery process, PEER searches for the more energy-efficient path progressively and ensures the route continuously. A suboptimal path to the most energy-efficient path is formed between the source and the destination, and the transmission path changes when necessary with slight overhead to guarantee energy-efficient transmissions. PEER is a cost-based protocol. The cost of all the links are calculated, and the minimum cost path is selected. In simulations, the performances of PEER, Minimum Total Reliable Transmission Power (MTRTP), and AODV are compared in static and mobile scenarios. In the static scenario, PEER performs the best in terms of energy consumption per packet followed by MTRTP and AODV. PEER performs better than MTRTP because it uses more exact link costs, its routing overhead is lower, and it can adapt paths to the environment changes quickly. Also, PEER and MTRTP perform better than AODV as they search for the most energy-efficient path instead of the shortest path. In the mobile scenario, PEER has the best and MTRTP has the worst performance as MTRTP cannot adjust well to mobility. Consequently, PEER reduces energy consumption by approximately 25% compared to AODV and 40% compared to MTRTP in the mobile scenario.

In MANETs, nodes may act selfishly as the network does not rely on a pre-existing infrastructure. Topology control game is examined in [73] for attaining an energy-efficient topology in MANETs in the existence of selfish nodes. The proposed scheme uses the benefit of a node for joining a network as a function of its transmission power. However, the energy consumption of the nodes also depends on the volume of transferred traffic. A new objective function which takes into account both load and transmission power is described. For evaluating the energy efficiency of the communications, the energy usage of nodes is calculated. For global methods, each node is aware of the whole information about the connectivity of all other nodes. On the other hand, every node is only aware of topological information about nodes within a limited hop distance in local methods. Significant energy efficiency is obtained when each node merely removes one link. Local methods outperform the global methods in terms of energy efficiency.

A wireless mesh network (WMN) is a particular type of MANET which is reliable, offers redundancy, and consists of radio nodes organized in a mesh topology. It often includes mesh clients such as laptops, cell phones, mesh routers, and gateways. A routing algorithm based on Strength Pareto Evolutionary Algorithm (SPEA) is proposed in [74] for WMNs. SPEA aims to build the most efficient routes. These routes consider the shortest path, energy consumption, and QoS limitations. The main objective of the proposed protocol is to find an optimal route, which is energy efficient, by decreasing the possibility of packet loss and satisfying the maximum delay and minimum bandwidth requirements of the connection, for unicast or multicast schemes. In the work, a mathematical model is described with four objective functions: hop count, energy consumption, free space loss, and delay. For the energy consumption function, it is assumed that the network is formed by heterogeneous nodes and they can have different energy consumption for both transmission and reception. The proposed protocol selects the path with the lower overall energy consumption to increase the battery life of these devices. In simulations, different scenarios are configured to compare the solutions of the proposed algorithm and the traditional shortest path algorithm. When the power consumption is individually minimized by minimizing the energy consumption function, significant energy gain is achieved.

Delay-tolerant networks involve the technical issues of heterogeneous networks and are deprived of continuous network connectivity. They can operate in mobile or extreme terrestrial environments. Energy-efficient Routing and Rate Allocation (ERRA) protocol for a delay-tolerant MANET is described in [75]. The existing routing schemes for delay-tolerant networks (DTNs) mainly utilize metrics such as delay, hop count, and bandwidth. ERRA is designed for optimizing the energy efficiency with constraints of congestion, buffer, and delay bound for DTNs. In traditional networks, three events can occur: listening, sending, and receiving. In ERRA, if no events occur at a node, the node can switch to sleep state for saving energy. In energy consumption analyses, the energy cost function is calculated by considering the average amount of energy consumption per bit. In simulations, an *ad hoc* delay-tolerant network is used. A fixed static node is placed at the center and used as an access point. ERRA is compared with the Epidemic and modified Dijkstra algorithm. Epidemic routing uses a naive flooding method, but it wastes resources and reduces the network performance. The modified Dijkstra algorithm is based on a time-dependent graph forwarding scheme and maintains at most one copy of a packet in the network during the routing process. ERRA provides better energy efficiency because the Epidemic algorithm uses copy to deliver data packets so more energy is consumed. The modified Dijkstra algorithm uses forward method for minimizing the average delivery delay, but it does not consider energy wastings.

Wireless sensor networks

WSNs consist of distributed sensors which monitor physical or environmental properties, such as temperature, humidity, and pressure, and pass the collected data to a sink node. WSNs have a wide range of usability in many distinct areas, such as science, transportation, habitat, emergency, and military applications. The primary challenges in a WSN include energy efficiency, responsiveness, robustness, adaptation, and self-configuration. Similar to MAC protocols described before, routing protocols also have a significant impact on the energy consumption of WSNs. In this section, energy-efficient routing approaches for WSNs are discussed.

Limited energy availability in sensor nodes makes network lifetime an important issue in WSN applications. To extend the network lifetime, energy-efficient wireless sensor network protocols and algorithms have been devised in the literature. Node clustering, in-network data processing, data fusion, and network coding are some of the measures taken to reduce the amount of data that is processed, sensed, or transmitted. Minimization of energy spent in processing, sensing, and transmission of data allows sensor nodes to save energy. Such energy savings help to extend the lifetime of WSN applications.

The main goal of cluster-based routing protocol is to efficiently maintain the energy consumption of sensor nodes by involving them in multi-hop communication within a cluster and by performing data aggregation and fusion in order to decrease the number of transmitted messages to the sink and transmission distance of sensor nodes. Low-energy adaptive clustering hierarchy (LEACH) [76] forms clusters randomly in an adaptive and self-organized way. Figure 4 shows an example of a random cluster formation and cluster head (CH) selection with LEACH.

The LEACH protocol allows clusters and CHs to change dynamically based on the energy levels of the CHs, and Figure 5 shows another possibility for the network configuration with LEACH.

LEACH provides a significant decrease in the overall energy consumption of the network and improves network lifetime and decreases number of dead nodes. The randomly determined cluster formations of LEACH especially prolong the lifetime of the remote nodes to the base station (BS). Relay or route nodes are usually very energy harvesting, but changing the clusters and CHs in each round can extend the lifetime of relay and route nodes.

On the other hand, LEACH has some disadvantages. The random determination of clusters and CHs can

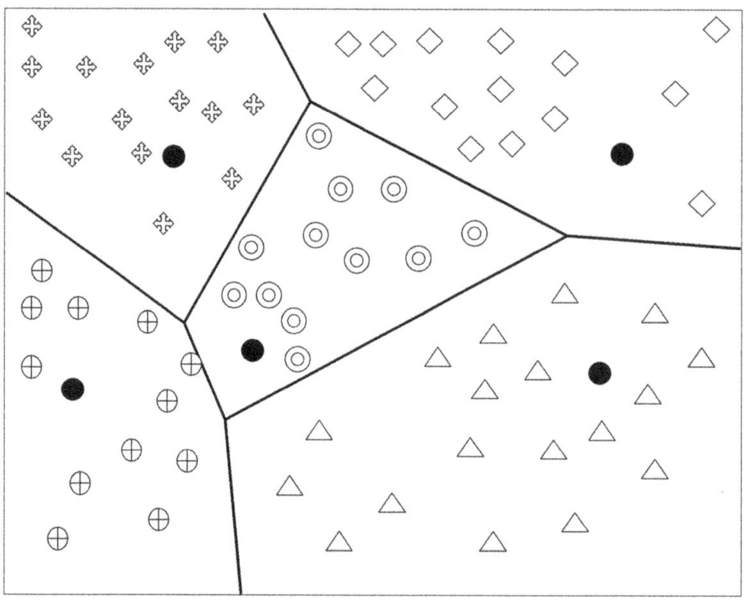

Figure 4 Cluster formations in LEACH, the cluster formation in the existing round.

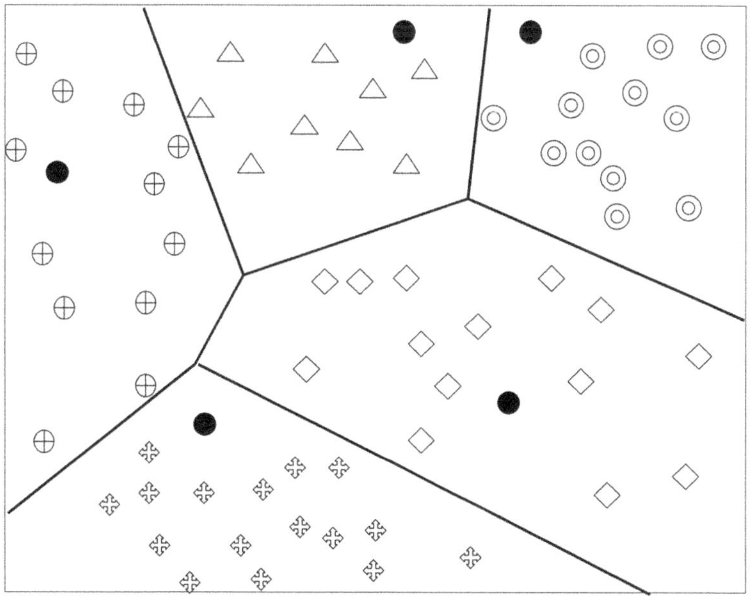

Figure 5 Cluster formations in LEACH, the cluster formation in the next round.

cause inefficient cluster formations and unbalanced cluster sizes. In some cases, the position of CHs cannot be optimal and they can be very far from the BS; thus, their batteries drain quickly and they can die prematurely, and hence, this can shorten the network lifetime. Also, the dynamic configuration of frequent cluster formations and CHs can induce more energy dissipation and can bring extra network costs while broadcasting the advertisement messages of the CHs. In recent years, numerous studies are proposed to solve the problems of LEACH and to improve its performance, such as MR-LEACH [77], I-LEACH [78], LEACH-IMP [79], V-LEACH [80], and C-LEACH [81]. Table 3 compares these protocols for their energy efficiency and lifetime.

In [82], an energy-efficient secure routing protocol for WSNs is presented. An integrated protocol which simultaneously takes into consideration the security and routing is proposed. The proposed protocol has three benefits. First, in order to offer a better delivery ratio, the protocol uses a sink-oriented grid structure originated from the grid design in Two-Tier Data Dissemination (TTDD) protocol. This protocol is a query-based routing protocol and designed to address the scalability and efficiency problem with multiple mobile sinks. Second, the benefit is that the proposed protocol searches the outermost highest energy emission node to find an efficient path from the source to the sink, resulting in a lower energy consumption. The last benefit is the security consideration. In the network model, a WSN with a stationary sink and stationary sensor nodes is considered. Without using GPS, each node knows its own position by using a localization discovery system. The nodes are homogeneous and conscious of their neighbors. From the source to the sink, multi-hop forwarding is used for data forwarding. In simulations, the comparison between the proposed protocol and TTDD is presented. According to results, the proposed protocol achieves significant improvement in delivery ratio because of the usage of the sink-oriented grid mechanism which is more

Table 3 Comparison of LEACH protocols

	Protocols					
	LEACH	MR-LEACH	I-LEACH	LEACH-IMP	U-LEACH	C-LEACH
Cluster formation	Each round	Each round	Each round	Fixed	Each round	Threshold
Self-organized	Yes	No	No	No	No	Yes
Location usage	No	Yes	Yes	Yes	Yes	No
CH selection	Random	By BS	Residual energy	Only once	Residual energy	Threshold
Energy efficiency	High	Very high	Very high	Very high	Very high	Very high
Network lifetime	High	Very high	Very high	High	Very high	Very high

successful in availability of the path from any node to the sink than in TTDD. The average energy consumption of TTDD is higher than that of the proposed algorithm, as in TTDD, the packet from the same source-destination pair travels through a longer path to the sink. The average hops in the proposed algorithm are relatively steady because of the estimation of the position of the sink through the farthest-highest energy dissemination node search, while in TTDD, they slightly increase with the node density. The longer the routed path, the more energy is consumed for transmission so the proposed algorithm is more successful than TTDD in energy conservation idea. Finally, in lifetime analysis, the proposed protocol provides significant improvement as compared to TTDD.

In [83], using the WSN, an energy-efficient routing approach in a VoIP network infrastructure is studied. Six steps - registering all mobile devices, arranging successful VoIP communication between mobile devices, clustering based algorithm, enhanced cluster head selection, registering mobile devices to their respective cluster heads, and selecting alternative cluster head - are defined. The outcome of the study shows energy efficiency in VoIP communication by using the clustering approach.

Pantazis et al. [84] provide a comprehensive survey for energy-efficient routing protocols in WSNs. The energy-efficient routing protocols are classified into four main types: network structure (flat, hierarchical), communication model (query, coherent, and negotiation based), topology based (location and mobile agent based), and reliable routing (multipath and QoS based). The performance comparisons between these types are presented in terms of energy efficiency. For small networks with stable nodes, the flat protocols are suitable. For large networks, hierarchical protocols solve the incommodiousness of the flat ones and provide efficiency by dividing the network into clusters. The location-based and mobile agent-based protocols can be a proper solution for highly dynamic networks. The negotiation-based ones provide optimum performance in both point-to-point and broadcast networks in terms of energy gains. Multipath-based protocols sustain multiple paths from nodes to sink; however, they suffer from the overhead of table and state maintanence. QoS protocols solve the delivery accuracy problem of the data. Query-based protocols can be used for dynamic network topologies, and they also promote multiple route replies. Finally, the coherent-based routing protocols include energy-efficient mechanisms and provide minimum processing cost to the sensor nodes.

Transport layer

The transport layer provides end-to-end reliable/unreliable communication service for the applications. The energy efficiency studies in the transport layer focus on increasing energy efficiency through new or enhanced transport layer protocols, but more commonly on TCP.

Authors in [85] investigate the effects of dynamic transmission control on the performance of the TCP. Measurement-based slow start (MSS) calculates the round trip time (RTT) to detect congestion and promptly enters into the congestion avoidance mode after finishing the slow start. The number of active links between switches is dynamically controlled by dynamic link aggregation protocol (dynamic LACP). MSS saves more power compared to compound TCP (CTCP) because the number of active links in MSS is less than that in CTCP and the available bandwidth is not used effectively in CTCP.

MPTCP is a coupled multipath-aware congestion control, and it proposes the resource pooling principle and provides a load balancing mechanism by moving more data traffic load into the less congested paths. MPTCP expands the standard TCP and allows data streams to be delivered across multiple connections and paths. This feature of MPTCP makes it quite charming for the recent consumer mobile devices that support multiple radio interfaces such as 3G, WiFi, Bluetooth, and LTE. However, an efficient multipath congestion control algorithm design that is both energy-aware and able to handle load balancing has not been addressed.

In [86], an energy-aware congestion control algorithm (ecMTCP) for multipath TCP is developed. In ecMTCP, the rate control depends on a traffic sharing principle between the paths and applied by its energy costs and traffic loads. The sender and receiver measure the energy costs of data and ACK packets, respectively. The energy costs of the receiver side is piggybacked to the sender periodically with ACKs, and finally, the sender calculates the overall end-to-end energy cost of the successful transmission of one data packet. Energy efficiency is provided by offloading traffic from higher energy cost paths to the lower paths and by using an energy-aware control mechanism where the congestion window increase is inversely proportional to the energy cost on each path. In simulations, energy efficiency is indicated by measuring the energy consumption difference of the proposed algorithm and MPTCP. According to results, the energy savings of ecMTCP is better than that of MPTCP.

Chen et al. [87] propose a novel energy-aware MPTCP-based content delivery scheme (eMTCP). eMTCP follows the channel status for each of the transmission paths and offloads traffic from the most energy-consuming interface to other interfaces for obtaining energy efficiency. In simulations, LTE and WiFi are considered. Data transmission performance is measured in terms of remaining energy, throughput, energy efficiency, estimated battery lifespan, and user-perceived quality. The performance of the eMTCP is compared with that of MPTCP and TCP. According to results, eMTCP achieves significant improvement on the battery lifespan when compared with the original MPTCP. Higher energy savings are achieved when compared with MPTCP and TCP.

Various versions of TCP such as AIMP, Cubic, Bic, Vegas, and CTCP are studied to analyze the congestion window and average throughput values in [88]. The ESTOP algorithm is used for decreasing the congestion window value. ESTOP switches off links in increasing order of an associated centrality level of the network by counting the number of shortest paths. In the algorithm, the switch off count is associated with a threshold value which preserves the network connectivity. After switch off operation, the link state update of OSPF is used for changing the topology. The ESTOP is simulated in the Abilene network. In simulations, a TCP flow is established between two distant nodes in the network. For the worst case scenario, the outermost two nodes are selected as TCP flow endpoints, and in these nodes, TCP flow is also affected in terms of available capacity after the links switch off. It is observed that, for the considered TCP versions, it is possible to gain energy in the network without a drastic compromise of the TCP performance. The switching off links obtained by using ESTOP produce higher energy savings.

The rapid growth of wireless Internet and broadband networking infrastructures, such as 3G, WLAN, WLAN-mesh, and WiMAX, makes multimedia information available anytime, anywhere, and on any device. An efficient transport protocol which reduces all types of losses by satisfying multimedia constraints is designed in [89]. Loss proportion increase and loss proportion decrease (PIPD) algorithm is used for defining burst period in order to increase efficiency. Efficient Transport Protocol (ETP) is implemented with NS2 to evaluate its performance over wireless Internet. Throughput, energy efficiency, fairness index, sending rate, data received with time, bandwidth, packets lost, and end-to-end delay are the performance evaluation parameters. ETP is compared with TCP Friendly Rate Control Protocol (TFRC), and it is observed that ETP performs better than TFRC for energy efficiency.

An energy-aware mechanism for the TCP-based transport protocol is proposed in [90]. The proposed mechanism aims to enhance energy savings at the end hosts by decreasing the multimedia traffic amount through handling loss tolerance of multimedia applications. This tolerance is utilized by reducing the offered QoS to the minimum required by the application. Three end host interfaces - IEEE 802.3 (for wired networks), IEEE802.11 (for wireless networks), and UMTS (for 3G mobile phones) - are considered. When the network is more reliable, the proposed mechanism is more efficient. The number of generated packets is reduced, energy savings are achieved, and the minimal QoS requirements are provided.

A Reliable and Energy-Efficient Transport Protocol (REETP) is developed in [91]. The proposed protocol implements an efficient node selection algorithm which determines a set of efficient nodes called as E-Nodes. These nodes form a near-optimal coverage set with largest area and highest discard energy level. REETP transfers encoded packets using Luby Transform (LT) Code from the source to the sink block by block, and each of these blocks is forwarded to an E-Node. After the reception of the encoded packets, the E-Node tries to reconstruct the original data packets and it encodes the original data packets again and relays them to the next E-Node until it reaches the sink. The results are compared with MAC-aware Energy Efficient Reliable Transport Protocol (MAEERTP) which provides the desired event reliability to the application by distributing the load at a sensor. The energy-related results are better than MAEERTP. The results show that the proposed protocol has higher packet delivery ratio with reduced packet loss and energy consumption than MAEERTP.

Based on the data reliability notion, the authors in [92] propose a Data-Reliable Energy-Efficient Transport Layer Protocol (DREET). As sensor data varies slightly in relatively short times unless certain events occur, a small subset of sensor data can be constructed as key data and provide reliable transport only for these key data instead of the whole sensor data. Their solution decreases retransmission and energy while keeping data reliability. Simulation results show that DREET provides an optimal event detection rate and relative standard deviation (RSD) between sensor data and sink data, by introducing a tiny increase in energy consumption. It can be obtained from this study that,DREET is a novel transport layer solution for reliable event detection in WSN with minimum energy expenditure.

In [93], a rate-based transport protocol for reducing energy consumption of nodes in MANETs is presented. The percentage of power consumption time is considered as the evaluation parameter, and NS2 simulator is used. The simulations are realized for different numbers of nodes and for various data rates. The number of flows through the intermediate node is changed, and its impact is analyzed. A higher number of flows increases the load on the node as well as the power consumption. According to the simulation results, the proposed approach consumes less power than the Ad hoc Transport Protocol (ATP) where the packets are carried upward to the transport layer only to update the delay stamping, and the flow control is dealt with only at the end nodes.

Application layer

The application layer is the highest layer of the Internet protocol stack. It contains protocols for process to process communication, and it is the only layer where an end user directly interacts with. The actual communication between processes on end users is initiated in this layer. Although a great number of energy saving mechanisms whose impact might be significant are conducted on lower layers, we believe that the studies for this layer are also valuable.

Browsing a single web page can consume a significant amount of power, and this amount increases rapidly with the usage of applications involving streaming video. For example, Google has determined the power consumed during an average Google search in 2009. Three hundred million searches per day can cause 90,000 kWh power consumption. The authors in [94] propose a model which supports energy-aware usage of web services. The proposed model estimates the power consumption during web service workloads. The factors which cause power consumption in web services such as nature of the I/O, server configuration, and web service design are investigated. In experiments, several measurable performance counters are determined, and several monitors are used. Multiple independent clients that invoke all of the web service operations are considered, and the samples of the actual power and performance counter measurements are collected. According to the results, a nonlinear model and CPU utilization-based power model segmentation increases accuracy. The proposed model significantly reduces the power consumption of the experimental web services and provides isolation of the power consumption of a particular web service within a regular server environment.

Mobile smart phones and other portable embedded systems such as PDAs and tablets are permeative devices. They are used for web browsing, multimedia, games, e-mail, and social networking. One of the most important parameters in mobile devices is battery lifetime. Some recent studies have discussed enhancing energy efficiency of applications in mobile environment, focusing on extending battery lifetime by using software optimizations. The authors in [95] present a novel application and user-interaction-aware energy management framework (AURA) for mobile devices. AURA exploits user idle time between interaction events of the spotlight application to optimize CPU and backlight energy consumption. It balances energy consumption and QoS demands. Based on user interaction, AURA dynamically classifies applications. After classification, the processor frequency and screen backlight levels are managed to reduce system energy consumption with Markov decision process-based power management algorithms. For user-device interaction field study, several Android applications from different categories are selected. In experiments, the performance of AURA middleware framework is compared with CHBL [96] which uses the changing blindness to increasingly ramp down CPU frequency and screen backlight levels over time. According to results, AURA presents higher energy savings because it dynamically adapts to the user interaction patterns and makes full use of user idle time. Contrary to this, CHBL is independent of user interaction patterns.

Another novel study for promoting energy-aware mobile application design is Anole [97]. It is a framework that aims to tell application designers the changing time of the energy consumption states of applications. The Anole framework involves three modules: energy profiler is a module used for estimating application energy consumption, energy-aware APIs that state applications by using the energy information and energy-aware policy which applies the energy adaptation of operating system by including service and hardware adaptations. The energy adaptation levels are predefined. This provides closing the services according to these levels. For example, when the level is six, the network service will be closed, and when the level is one, Bluetooth service will be turned off. The power consumption of the whole system is measured with voltage measurement of the resistor connected between the battery and mobile phone. To investigate the efficiency of application, energy adaptation, video player, and web browser applications are used and significant average energy savings are obtained.

In [98], user-aware streaming strategies for energy-efficient smart phone video playback applications are proposed. A Gaussian mixture model (GMM) is used for obtaining the distribution of the real playback length; thereby, the best buffering strategy for minimum energy losses is obtained. In experiments, an Android application is enhanced to monitor and collect the users' YouTube video playback activities. The results show that the proposed strategy ensures remarkable amount of reduction in energy usage when compared to the static strategies. The GMM model outperforms normal and exponential distribution models in terms of energy savings while downloading operations.

For periodic real-time applications, an energy-aware standby-sparing technique is presented in [99]. The use of Earliest Deadline First (EDF), which executes prepared tasks as soon as possible, and Earliest Deadline Late (EDL), which delays the tasks as much as possible, schedules on the primary and collateral CPUs, respectively. The joint use of EDF and EDL allows to minimize the overlap between the two copies at runtime and decrease the energy cost due to the backup implementations. This joint usage forms the Standby-Sparing for Periodic Tasks (SSPT) algorithm. The main idea in the algorithm is that if the primary processor is idle, the job is dispatched immediately. If not, a job is dispatched only if it has a superior priority than the currently executing job. The proposed algorithm is compared with reliability-aware power management (RAPM) techniques which assume that the system's original reliability is acceptable when all the tasks are executed at the maximum CPU frequency. The effect of the system load and workload variability is analyzed in simulations. According to results, the SSPT algorithm outperforms RAPM in terms of energy gain.

Conclusions

Energy conservation has become a very crucial research area to support the enormous growth of the ICT industry. Various analyses indicate that the ICT industry is now responsible for 2% to 10% of the global energy consumption and the weight of ICT is expected to increase as we continue to integrate ICT technologies in our daily lives. In this paper, we first review several indicators showing the impact of ICT on energy consumption from data centers to WSNs. Then, we try to describe some recent research studies for all layers of the Internet protocol stack which have addressed the issue of energy efficiency or in other words green communications.

Although the impact of energy consumption for some protocols may be larger than the others, it is very essential to work on all layers to develop less energy-consuming ways for the ICT industry as the increasing levels of energy consumption directly affects the greenhouse gas emissions, global warming, and thus our future.

Most of the energy efficiency studies are observed to be studied in the lower layers. The physical layer is historically the first layer concerned with the energy saving mechanisms and focuses on developing energy-efficient modulation schemes. The data link layer concentrates on MAC protocols for energy efficiency. Most of the research in the network layer focuses on developing energy-efficient routing protocols for both wired and wireless networks. Managing the network devices by turning them off (or on) depending on the traffic conditions and data center networks are other extensively studied ways for reducing energy consumption. While in the transport layer new versions of the TCP protocol are studied, the energy efficiency studies for the application layer concentrate on developing power-aware applications and software optimizations.

It is predicted that the growth rate of the ICT industry is going to continue with the invention and widespread usage of new technologies such as IoT (Internet of Things), Cloud Computing, Smart Cities, and Smart Homes. Thus, it is extremely important to develop energy-efficient protocols which can satisfy the growing demand with similar or better performance levels both for economical and environmental reasons. Although it might be very difficult, time consuming, and expensive to move from well-established protocols to new ones, it is highly likely that the traditional protocols that we are currently using will be forced to change in the future based on their energy impacts.

Competing interests

The authors declare that they have no competing interests.

Author details

[1]Electrical-Electronics Engineering Department, Trakya University, Prof. Dr. Ahmet Karadeniz Yerleskesi, 22030 Edirne, Turkey. [2]Electrical-Electronics Engineering Department, Kadir Has University, Cibali, 34083 Istanbul, Turkey. [3]Computer Engineering Department, Kadir Has University, Cibali, 34083 Istanbul, Turkey.

References

1. G Fettweis, E Zimmermann, in *The 11th International Symposium on Wireless Personal Multimedia Communications (WPMC 2008)*. ICT energy consumption - trends and challenges (Lapland, Finland, 2008)
2. M Pickavet, W Vereecken, S Demeyer, P Audenaert, B Vermeulen, C Develder, D Colle, B Dhoedt, P Demeester, in *The 2nd International Symposium on Advanced Networks and Telecommunication Systems (ANTS 2008)*. Worldwide energy needs for ICT: the rise of power-aware networking (Mumbai, India, 2008), pp. 1–3
3. Committee on America's Climate Choices, Board on Atmospheric Sciences and Climate, Division on Earth and Life Studies, National Research Council, *America's Climate Choices*. (National Academies Press, 2011)
4. Greenhouse gases, climate change and energy. http://www.eia.gov/oiaf/1605/ggccebro/chapter1.html. Accessed 6 April 2015
5. The current and future consequences of global change. http://climate.nasa.gov/effects. Accessed 6 April 2015
6. Effects of global warming. http://environment.nationalgeographic.com/environment/global-warming/gw-effects/. Accessed 6 April 2015
7. MA Yigitel, OD Incel, C Ersoy, Dynamic base station planning with power adaptation for green wireless cellular networks. EURASIP J. Wireless Commun. Networking. **2014**(1), 1–13 (2014)
8. TV Landegem, GreenTouch: Energy consumption in communication networks can be reduced by up to 90 % by the year 2020. http://www.alcatel-lucent.com/blog/corporate/2013/06/greentouch-energy-consumption-communication-networks-can-be-reduced-90-year-2020. Accessed 6 April 2015
9. Bryan Walsh, The surprisingly large energy footprint of the digital economy. http://science.time.com/2013/08/14/power-drain-the-digital-cloud-is-using-more-energy-than-you-think. Accessed 6 April 2015
10. C Marsden, T David-Barrett, E Pavan, in *Science.EU Project*. Overview of ICT energy consumption (The EINS Consortium, 2013)
11. T Restorick, An inefficient Truth, *Global Action Plan Report*, (2013)
12. T Asami, S Namiki, in *The 34th European Conference and Exhibition on Optical Communication (ECOC 2008)*. Energy consumption targets for network systems (Brussels, Belgium, 2008), pp. 1–4
13. Energy savings in ICT and ICT for energy savings (2009). http://ercim-news.ercim.eu/en79/special/introduction
14. K Hinton, J Baliga, MZ Feng, RWA Ayre, R Tucker, Power consumption and energy efficiency in the internet. IEEE Network. **25**(2), 6–12 (2011)
15. H Sistek, Green-tech base stations cut diesel usage by 80 percent. CNET News Green Tech, ed, (2008)
16. L Suarez, L Nuaymi, J-M Bonnin, An overview and classification of research approaches in green wireless networks. EURASIP J. Wireless Commun. Networking. **2012**(1), 1–18 (2012)
17. C Bianco, F Cucchietti, G Griffa, in *The 29th International Telecommunications Energy Conference (INTELEC 2007)*. Energy consumption trends in the next generation access network – a telco perspective (Brussels, Belgium, 2007), pp. 737–742
18. C Keenan, How much energy does the Internet consume? Accessed 6 April 2015. http://planetsave.com/2011/10/27/how-much-energy-does-the-internet-consume
19. R Bolla, F Davoli, R Bruschi, K Christensen, F Cucchietti, S Singh, The potential impact of green technologies in next-generation wireline

networks: Is there room for energy saving optimization? Commun. Mag. IEEE. **49**(8), 80–86 (2011)

20. C Lange, in *35th European Conference on Optical Communication (ECOC 2009)*. Energy-related aspects in backbone networks (Wien, Austria, 2009)
21. Stop worrying about IT power consumption. http://www.greenbiz.com/blog/2013/08/07/jonathan-koomey\-stop-worrying-about-it-power-consumption. Accessed 6 April 2015
22. D Wei, C Feng, C Guo, L Fangfang, A power amplifier energy efficient polarization modulation scheme based on the optimal pre-compensation. Commun. Lett. IEEE. **17**(3), 513–516 (2013). doi:10.1109/LCOMM.2013.012313.122639
23. D Wei, C Feng, C Guo, in *Wireless Communications and Networking Conference (WCNC 2013)*. An energy efficient subcarrier-power allocation scheme for polarization-amplitude-phase modulation in channel with polarization mode dispersion (Shanghai, China, 2013)
24. K Ntontin, M Di Renzo, A Perez-Neira, C Verikoukis, in *IEEE International Conference On Communications (ICC 2013)*. Antenna subset selection for spatial modulation: a novel and energy efficient single RF technique (Budapest, Hungary, June 2013)
25. H Kimura, K Asaka, H Nakamura, S Kimura, N Yoshimoto, in *39th European Conference and Exhibition On Optical Communication (ECOC 2013)*. First demonstration of energy efficient IM-DD OFDM-PON using dynamic SNR management and adaptive modulation (London, UK, 2013)
26. H Ochiai, in *10th International Symposium On Wireless Communication Systems (ISWCS 2013)*. On modulation selection for energy-efficient band-limited communication systems (Ilmenau, Germany, 2013)
27. D DiTomaso, S Laha, S Kaya, D Matolak, A Kodi, in *10th IEEE International Conference on New Circuits (NEWCAS 2012)*. Energy efficient modulation for a wireless network-on-chip architecture (Montreal, Canada, 2012)
28. MS Vasanthi, D Kumar, TR Rao, in *International Conference On Advances in Computing, Communications and Informatics (ICACCI 2013)*. Performance comparison study of AQ-DBPSK modulation scheme for energy efficient wireless sensor communications (Mysore, India, 2013)
29. Z Chen, C Yang, in *8th International Conference On Signal Processing (ICSP 2006)*. Energy efficiency of cooperative diversity at PHY layer in wireless sensor networks (Beijing, China, 2006)
30. W Feng, JMH Elmirghani, in *3th International Conference On Next Generation Mobile Applications, Services and Technologies*. Energy efficiency in ad-hoc wireless networks with two realistic physical layer models (Wales, UK, 2009)
31. A Chehri, HT Mouftah, in *25th IEEE Canadian Conference On Electrical Computer Engineering (CCECE 2012)*. Qos aware green routing protocol for wireless sensor networks (Montreal, Canada, 2012)
32. D Dessales, A Poussard, R Vauzelle, N Richard, F Gaudaire, C Martinsons, in *Conference On Design and Architectures for Signal and Image Processing (DASIP 2010)*. Physical layer study in a goal of robustness and energy efficiency for wireless sensor networks (Edinburg, Scotland, 2010)
33. A Bartoli, J Hernandez-Serrano, M Soriano, M Dohler, A Kountouris, D Barthel, in *IEEE Globecom Workshops (GC Wkshps 2012)*. Optimizing energy-efficiency of PHY-layer authentication in machine-to-machine networks (Anaheim, CA, 2012)
34. R Zhang, C Comaniciu, HV Poor, in *16th International Symposium On Wireless Personal Multimedia Communications (WPMC 2013)*. Outage capacity and partial secrecy for energy efficient physical layer security in gaussian fading channels (NJ, USA, 24)
35. X Chen, L Lei, Energy-efficient optimization for physical layer security in multi-antenna downlink networks with QoS guarantee. Commun. Lett. IEEE. **17**(4), 637–640 (2013)
36. I Mobin, RJ Mondragon, MM Khan, in *4th Conference on Computer Science and Electronic Engineering (CEEC 2012)*. Energy efficient transmission power estimation for WLAN VoIP (Colchester, UK, 2012)
37. KR Malekshan, W Zhuang, in *IEEE International Conference On Communications (ICC 2013)*. An energy efficient MAC protocol for fully-connected wireless networks (Budapest, Hungary, 2013)
38. N Abdeddaim, F Theoleyre, M Heusse, A Duda, in *IEEE International Conference On Distributed Computing in Sensor Systems (DCOSS 2013)*. Adaptive IEEE 802.15.4 MAC for throughput and energy optimization (Massachusetts, USA, 2013)
39. S Toklu, OA Erdem, BSC-MAC: energy efficiency in wireless sensor networks with base station control. Comput. Networks. **59**(0), 91–100 (2014)
40. A Nasir, B-H Soong, in *8th International Conference On Information, Communications and Signal Processing (ICICS 2011)*. Energysim; a novel, fast, extensible wireless sensor network mac protocol simulator for evaluating energy efficiency (Singapore, 2011)
41. MR Sabagh, M Dianati, MA Imran, R Tafazolli, in *IEEE International Conference On Communications (ICC 2013)*. A heuristic energy efficient scheduling scheme for voip in 3GPP LTE networks (Budapest, Hungary, 2013)
42. J-M Liang, J-J Chen, H-H Cheng, Y-C Tseng, in *9th International Conference on Wireless Communications and Mobile Computing (IWCMC 2013)*. Energy-efficient sleep scheduling with QoS considerations in 3GPP LTE-advanced networks (Cagliari, Italy, 2013)
43. FD Cardoso, S Petersson, M Boldi, S Mizuta, G Dietl, R Torrea-Duran, C Desset, J Leinonen, LM Correia, Energy efficient transmission techniques for LTE. Commun. Mag. IEEE. **51**(10), 182–190 (2013)
44. M Deruyck, W Joseph, B Lannoo, D Colle, L Martens, Designing energy-efficient wireless access networks: LTE and LTE-advanced. IEEE Internet Comput. **17**(5), 39–45 (2013)
45. A Cianfrani, V Eramo, M Listanti, M Marazza, E Vittorini, in *IEEE Conference on Computer Communications Workshops (INFOCOM 2010)*. An energy saving routing algorithm for a green OSPF protocol (San Diego, USA, 2010)
46. SSW Lee, P-K Tseng, A Chen, in *IEEE Global Telecommunications Conference (GLOBECOM 2011)*. A distributed link management algorithm for energy efficient IP networks (Texas, USA, 2011)
47. AP Bianzino, L Chiaraviglio, M Mellia, in *IEEE Conference On Computer Communications Workshops (INFOCOM 2012)*. Distributed algorithms for green IP networks (Florida, USA, 2012)
48. P Lai, Q Yang, C Wu, M Jiang, in *IEEE 3rd International Conference On Communication Software and Networks (ICCSN 2011)*. Configuring network topology towards energy-efficient IP networks (Xi'an, China, 2011)
49. A Coiro, M Polverini, A Cianfrani, M Listanti, in *International Conference On Computing, Networking and Communications (ICNC 2013)*. Energy saving improvements in IP networks through table lookup bypass in router line cards (San Diego, USA, 2013)
50. Q Li, M Xu, Y Yang, L Gao, Y Cui, J Wu, Safe and Practical Energy-Efficient Detour Routing in IP Networks. IEEE Trans. Networking. **22**(6), 1925–1937 (2013)
51. F Francois, N Wang, K Moessner, S Georgoulas, K Xu, in *IFIP Networking Conference (2013)*. Green IGP link weights for energy-efficiency and load-balancing in IP backbone networks (New York, USA, 2013)
52. A Coiro, M Listanti, A Valenti, F Matera, Energy-aware traffic engineering: a routing-based distributed solution for connection-oriented IP networks. Comput. Networks. **57**(9), 2004–2020 (2013)
53. E Amaldi, A Capone, LG Gianoli, Energy-aware IP traffic engineering with shortest path routing. Comput. Networks. **57**(6), 1503–1517 (2013)
54. AP Bianzino, C Chaudet, F Larroca, D Rossi, J Rougier, in *IEEE Global Telecommunications Conference(GLOBECOM 2011)*. Energy-aware routing: a reality check (Texas, USA, 2011)
55. M Fiorani, M Casoni, S Aleksic, Hybrid optical switching for an energy-efficient Internet core. IEEE Internet Comput. **17**(1), 14–22 (2013)
56. A Klekamp, U Gebhard, F Ilchmann, Energy and cost efficiency of adaptive and mixed-line-rate IP over DWDM networks. J Lightwave Technol. **30**(2), 215–221 (2012)
57. X Dong, TEH El-Gorashi, JMH Elmirghani, On the energy efficiency of physical topology design for IP over WDM networks. J Lightwave Technol. **30**(12), 1931–1942 (2012)
58. K Ramanujam, S Talbatulla, in *International Conference On Optical Engineering (ICOE 2012)*. Greening the IP over optical backbone network (Belgaum, India, 2012)
59. X Dong, TEH El-Gorashi, JMH Elmirghani, in *IEEE International Conference On Communications (ICC 2013)*. Energy efficiency of optical OFDM-based networks (Budapest, Hungary, 2013)
60. Data Center power consumption continuing to grow with Cloud Computing as an option. Accessed 6 April 2015. http://research.gigaom.com/2013/02/data-center-power-consumption-continuing-to-grow-with-cloud-computing-as-an-option
61. Federal Energy Management Program. Accessed 6 April 2015. http://www.energy.gov/node/782456/program
62. Google Green. Accessed 6 April 2015. http://www.google.com/green/bigpicture

63. R Barga, in *IEEE 9th International Conference On Peer-to-Peer Computing (P2P 2009)*. Cloud computing–a microsoft research perspective (Washington, USA, 2009)
64. X Dong, T El-Gorashi, JMH Elmirghani, in *13th International Conference On Transparent Optical Networks (ICTON 2011)*. Energy-efficient IP over WDM networks with data centres (Stockholm, Sweden, 2011)
65. M Zapater, JL Ayala, JM Moya, K Vaidyanathan, K Gross, AK Coskun, in *Design, Automation Test in Europe Conference Exhibition (DATE 2013)*. Leakage and temperature aware server control for improving energy efficiency in data centers (Grenoble, France, 2013)
66. J Qu, L Li, L Liu, Y Tian, J Chen, in *IEEE International Conference On Service Operations and Logistics, and Informatics (SOLI 2013)*. Smart temperature monitoring for data center energy efficiency (Dongguan, China, 2013)
67. NM Asghari, M Mandjes, A Walid, Energy-efficient scheduling in multi-core servers. Comput. Networks. **59**(0), 33–43 (2014)
68. M Xu, Y Shang, D Li, X Wang, Greening data center networks with throughput-guaranteed power-aware routing. Comput. Networks. **57**(15), 2880–2899 (2013)
69. Y Shang, D Li, M Xu, in *19th IEEE Workshop On Local Metropolitan Area Networks (LANMAN 2013)*. Greening data center networks with flow preemption and energy-aware routing (Brussels, Belgium, 2013)
70. BS Gouda, AK Dass, KL Narayana, in *International Multi-Conference On Automation, Computing, Communication, Control and Compressed Sensing (iMac4s 2013)*. A comprehensive performance analysis of energy efficient routing protocols in different traffic based mobile ad-hoc networks (Kerala, India, 2013)
71. T Zhu, D Towsley, in *IEEE Conference On Computer Communications Workshops (INFOCOM 2011)*. E2r: Energy efficient routing for multi-hop green wireless networks (Shanghai, China, 2011)
72. J Zhu, X Wang, Model and protocol for energy-efficient routing over mobile ad hoc networks. IEEE Trans. Mobile Comput. **10**(11), 1546–1557 (2011)
73. S Zarifzadeh, N Yazdani, A Nayyeri, Energy-efficient topology control in wireless ad hoc networks with selfish nodes. Comput. Networks. **56**(2), 902–914 (2012)
74. C Lozano-Garzon, M Camelo, P Vila, Y Donoso, in *International Symposium On Performance Evaluation of Computer and Telecommunication Systems (SPECTS 2013)*. Green routing algorithm for wireless mesh network: a multi-objective evolutionary approach (Toronto, Canada, 2013)
75. Y Zeng, J Wu, N Xiong, D Li, in *32nd International Conference On Distributed Computing Systems Workshops (ICDCSW 2012)*. Energy-efficient routing and rate allocation for delay tolerant networks (Macau, China, 2012)
76. WR Heinzelman, A Chandrakasan, H Balakrishnan, in *33rd Annual Hawaii International Conference On System Sciences*. Energy-efficient communication protocol for wireless microsensor networks (Maui, Hawaii, 2000)
77. MO Farooq, AB Dogar, GA Shah, in *4th International Conference On Sensor Technologies and Applications (SENSORCOMM 2010)*. MR-LEACH: multi-hop routing with low energy adaptive clustering hierarchy (Venice, Italy, 2010)
78. N Kumar, J Kaur, in *7th International Conference On Wireless Communications, Networking and Mobile Computing (WiCOM 2011)*. Improved leach protocol for wireless sensor networks (Wuhan, China, 2011)
79. X Hu, J Luo, Z Xia, M Hu, in *IEEE 3rd International Conference On Communication Software and Networks (ICCSN 2011)*. Adaptive algorithm of cluster head in wireless sensor network based on LEACH (Xi'an, China, 2011)
80. N Kumar, P Bhutani, P Mishra, et al., in *International Conference On Communication, Information & Computing Technology (ICCICT 2012)*. U-LEACH: a novel routing protocol for heterogeneous wireless sensor networks (Mumbai, India, 2012)
81. R Mehta, A Pandey, P Kapadia, in *International Conference On Computer Communication and Informatics (ICCCI 2012)*. Reforming clusters using C-LEACH in wireless sensor networks (Coimbatore, India, 2012)
82. H-W Ferng, D Rachmarini, in *IEEE Network Operations and Management Symposium (NOMS 2012)*. A secure routing protocol for wireless sensor networks with consideration of energy efficiency (Hawaii, USA, 2012)
83. M Tembhurkar, L Malik, N Chavhan, in *8th International Conference On Computer Science Education (ICCSE 2013)*. Energy efficient VoIP communication using WSN clustering approach (Colombo, Sri Lanka, 2013)
84. NA Pantazis, SA Nikolidakis, DD Vergados, Energy-efficient routing protocols in wireless sensor networks: a survey. IEEE Commun. Surveys Tutorials. **15**(2), 551–591 (2013)
85. Y Fukuda, S Kawamata, T Ikenaga, Y Oie, in *IEEE International Conference On Green Computing and Communications (GreenCom 2012)*. Performance evaluation of the transmission control protocols with the dynamic link control scheme for power saving (Besançon, France, 2012)
86. TA Le, C-S Hong, MA Razzaque, S Lee, H Jung, ecMTCP: an energy-aware congestion control algorithm for multipath TCP. IEEE Commun. Lett. **16**(2), 275–277 (2012)
87. S Chen, Z Yuan, G-M Muntean, in *IEEE Wireless Communications and Networking Conference (WCNC 2013)*. An energy-aware multipath-TCP-based content delivery scheme in heterogeneous wireless networks (Shanghai, China, 2013)
88. A Sassu, C Scarso, F Cuomo, in *The 2nd IFIP Conference on Sustainable Internet and ICT for Sustainability (SustainIT 2012)*. TCP behavior over a greened network (Pisa, Italy, 2012)
89. P Ramya, K Radha, Design of efficient transport protocol for multimedia through wireless networks. Int. J. Recent Technol. Eng. (IJRTE). **02**(03), 1–4 (2013)
90. M Oulmahdi, C Chassot, E Exposito, in *International Conference On Smart Communications in Network Technologies (SaCoNeT 2013)*. An energy-aware TCP for multimedia streaming (Paris, France, 2013)
91. KK Sharma, RB Patel, H Singh, A reliable and energy efficient transport protocol for wireless sensor networks. Int. J. Comput. Networks Commun. (IJCNC). **02**(05), 1–12 (2010)
92. G Han, J Lu, X Gan, H Li, W Dou, DREET: a data-reliable, energy-efficient transport layer protocol. J. Software. **07**(09), 1925–1931 (2012)
93. K Sasikala, R Wahidabanu, Rate based transport protocol using delay information for reducing energy consumption of nodes in manet. Int. J. Eng. Technol. **05**(05), 1–10 (2012)
94. P Bartalos, MB Blake, S Remy, in *IEEE International Conference On Service-Oriented Computing and Applications (SOCA 2011)*. Green web services: models for energy-aware web services and applications (Irvine, USA, 2011)
95. BK Donohoo, C Ohlsen, S Pasricha, in *IEEE 29th International Conference On Computer Design (ICCD 2011)*. AURA: an application and user interaction aware middleware framework for energy optimization in mobile devices (Amherst, USA, 2011)
96. A Shye, B Scholbrock, G Memik, in *42nd Annual IEEE/ACM International Symposium On Microarchitecture (MICRO-42 2009)*. Into the wild: studying real user activity patterns to guide power optimizations for mobile architectures (New York, USA, 2009)
97. H Chen, B Luo, W Shi, in *41st International Conference On Parallel Processing Workshops (ICPPW 2012)*. Anole: A case for energy-aware mobile application design (Pittsburgh, PA, 2012)
98. H Shen, Q Qiu, in *Design, Automation Test in Europe Conference Exhibition (DATE 2013)*. User-aware energy efficient streaming strategy for smartphone based video playback applications (Grenoble, France, 2013)
99. MA Haque, H Aydin, D Zhu, in *IEEE 29th International Conference On Computer Design (ICCD 2011)*. Energy-aware standby-sparing technique for periodic real-time applications (Amherst, USA, 2011)

Permissions

All chapters in this book were first published in EURASIP-JWCN, by Springer; hereby published with permission under the Creative Commons Attribution License or equivalent. Every chapter published in this book has been scrutinized by our experts. Their significance has been extensively debated. The topics covered herein carry significant findings which will fuel the growth of the discipline. They may even be implemented as practical applications or may be referred to as a beginning point for another development.

The contributors of this book come from diverse backgrounds, making this book a truly international effort. This book will bring forth new frontiers with its revolutionizing research information and detailed analysis of the nascent developments around the world.

We would like to thank all the contributing authors for lending their expertise to make the book truly unique. They have played a crucial role in the development of this book. Without their invaluable contributions this book wouldn't have been possible. They have made vital efforts to compile up to date information on the varied aspects of this subject to make this book a valuable addition to the collection of many professionals and students.

This book was conceptualized with the vision of imparting up-to-date information and advanced data in this field. To ensure the same, a matchless editorial board was set up. Every individual on the board went through rigorous rounds of assessment to prove their worth. After which they invested a large part of their time researching and compiling the most relevant data for our readers.

The editorial board has been involved in producing this book since its inception. They have spent rigorous hours researching and exploring the diverse topics which have resulted in the successful publishing of this book. They have passed on their knowledge of decades through this book. To expedite this challenging task, the publisher supported the team at every step. A small team of assistant editors was also appointed to further simplify the editing procedure and attain best results for the readers.

Apart from the editorial board, the designing team has also invested a significant amount of their time in understanding the subject and creating the most relevant covers. They scrutinized every image to scout for the most suitable representation of the subject and create an appropriate cover for the book.

The publishing team has been an ardent support to the editorial, designing and production team. Their endless efforts to recruit the best for this project, has resulted in the accomplishment of this book. They are a veteran in the field of academics and their pool of knowledge is as vast as their experience in printing. Their expertise and guidance has proved useful at every step. Their uncompromising quality standards have made this book an exceptional effort. Their encouragement from time to time has been an inspiration for everyone.

The publisher and the editorial board hope that this book will prove to be a valuable piece of knowledge for researchers, students, practitioners and scholars across the globe.

List of Contributors

Anabel Martínez-Vargas
College of Engineering, MyDCI, Autonomous University of Baja California (UABC), Blvd. Benito Juárez, Mexicali 21280, México

Ángel G Andrade
College of Engineering, MyDCI, Autonomous University of Baja California (UABC), Blvd. Benito Juárez, Mexicali 21280, México

Bingxin Yan
College of Electronic Information Engineering, Inner Mongolia University, Hohhot, China

Shubin Wang
College of Electronic Information Engineering, Inner Mongolia University, Hohhot, China

Yuanyuan Bao
College of Electronic Information Engineering, Inner Mongolia University, Hohhot, China

Gaofeng Nie
State Key Laboratory of Networking and Switching Technology, Beijing University of Posts and Telecommunications, Xitucheng Road, 100876 Beijing, China

Hui Tian
State Key Laboratory of Networking and Switching Technology, Beijing University of Posts and Telecommunications, Xitucheng Road, 100876 Beijing, China

Jiazhi Ren
State Key Laboratory of Networking and Switching Technology, Beijing University of Posts and Telecommunications, Xitucheng Road, 100876 Beijing, China

JingWang
DOCOMO Beijing Communication Laboratories Co., Ltd.,Kexueyuan South Road, 100190 Beijing, China

Liu Liu
DOCOMO Beijing Communication Laboratories Co.,Ltd.,Kexueyuan South Road, 100190 Beijing, China

Huiling Jiang
DOCOMO Beijing Communication Laboratories Co., Ltd.,Kexueyuan South Road, 100190 Beijing, China

Xiaolei Wang
Tsinghua National Laboratory for Information Science and Technology, Department of Electronic Engineering, Tsinghua University, 100084 Beijing

Jie Gong
Tsinghua National Laboratory for Information Science and Technology, Department of Electronic Engineering, Tsinghua University, 100084 Beijing

Congshi Hu
China Mobile Communication Corporation (CMCC), No. 29, Financial Street, Xicheng District, Beijing 100033

Sheng Zhou
Tsinghua National Laboratory for Information Science and Technology, Department of Electronic Engineering, Tsinghua University, 100084 Beijing

Zhisheng Ni
Tsinghua National Laboratory for Information Science and Technology, Department of Electronic Engineering, Tsinghua University, 100084 Beijing

Wan Hafiza Wan Hassan
College of Engineering and Science, Victoria University, Ballarat Road, 8001 Melbourne, Australia
School of Ocean Engineering, Universiti Malaysia Terengganu 21030 Kuala Terengganu, Malaysia

Horace King
College of Engineering and Science, Victoria University, Ballarat Road, 8001 Melbourne, Australia

Shabbir Ahmed
College of Engineering and Science, Victoria University, Ballarat Road, 8001 Melbourne, Australia

Mike Faulkner
College of Engineering and Science, Victoria University, Ballarat Road, 8001 Melbourne, Australia

Tsui-Tsai Lin
Department of Electronic Engineering, National United University, Miaoli 36003, Taiwan

Fuh-Hsin Hwang
Department of Optoelectronics and Communication Engineering, National Kaohsiung Normal University, Kaohsiung 82444, Taiwan

Rodrigo A Vaca Ramirez
Scientific Division, Federal Police, 01110, Constituyentes 947, Mexico City, Mexico

John S Thompson
The University of Edinburgh, Institute for Digital Communications, King's Buildings, Mayfield Road, EH93JL Edinburgh, UK

Eitan Altman
INRIA, BP93, 06902 Sophia Antipolis Cedex, France

Victor Ramos
Department of Electrical Engineering,Universidad Autónoma Metropolitana (UAM), 09340 Iztapalapa, Mexico City, Mexico

Jiying Huang
School of Electronic and Information Engineering, Beijing Jiaotong University, No. 3 Shangyuancun, Beijing 100044, China

Zhangdui Zhong
School of Computer and Information Technology, Beijing Jiaotong University, No. 3 Shangyuancun, Beijing 100044, China

Hongwei Huo
China Science and Technology Exchange Center, No. 54 Sanlihe Road, Beijing 100045, China

Jingxian Wu
Department of Electrical Engineering, University of Arkansas, Fayetteville, AR 72701, USA

Jing Yang
Department of Electrical Engineering, University of Arkansas, Fayetteville, AR 72701, USA

Ismail Butun
Department of Mechatronics Engineering, Bursa Technical University,Gaziakdemir M. Mudanya C. No:4/10, 16190 Osmangazi, Bursa, Turkey

In-ho Ra
Department of Information and Telecommunication Engineering, Kunsan National University, 573-701 Gunsan, South Korea

Ravi Sankar
Department of Electrical Engineering, University of South Florida, 4202 E. Fowler Avenue, Tampa, FL 33620, USA

Yuan Le
Department of Computer Science, The George Washington University, Washington, DC, USA

Xiuzhen Cheng
Department of Computer Science, The George Washington University, Washington, DC, USA

Dechang Chen
Uniformed Services University of the Health Sciences, Bethesda, MD, USA

Nan Zhang
Department of Computer Science, The George Washington University, Washington, DC, USA

Taieb Znati
Department of Computer Science, University of Pittsburg, Pittsburgh, PA, USA

Mznah A Al-Rodhaan
College of Computer and Information Sciences, King Saud University, Riyadh, Saudi Arabia

Abdullah Al-Dhelaan
College of Computer and Information Sciences, King Saud University, Riyadh, Saudi Arabia

Kyu-haeng Lee
Department of CSE, Seoul National University, 151-744 Seoul, Korea

Chong-kwon Kim
Department of CSE, Seoul National University, 151-744 Seoul, Korea

Fang Yuan
School of Electronics and Information Engineering, Beihang University, XueYuan Road No.37, 100191, Beijing, China

Chenyang Yang
School of Electronics and Information Engineering, Beihang University, XueYuan Road No.37, 100191, Beijing, China

Pedro Suárez-Casal
Department of Electronics and Systems, University of A Coruña, Facultade de Informática, Campus de Elviña, A Coruña 15071, Spain

José Rodríguez-Piñeiro
Department of Electronics and Systems, University of A Coruña, Facultade de Informática, Campus de Elviña, A Coruña 15071, Spain

José A García-Naya
Department of Electronics and Systems, University of A Coruña, Facultade de Informática, Campus de Elviña, A Coruña 15071, Spain

Luis Castedo
Department of Electronics and Systems, University of A Coruña, Facultade de Informática, Campus de Elviña, A Coruña 15071, Spain

Mohammed H Alsharif
Department of Electrical Electronics and Systems Engineering, Faculty of Engineering and Built Environment, Universiti Kebangsaan Malaysia, 43600 Bangi, Selangor, Malaysia

Rosdiadee Nordin
Department of Electrical Electronics and Systems Engineering, Faculty of Engineering and Built Environment, Universiti Kebangsaan Malaysia, 43600 Bangi, Selangor, Malaysia

Mahamod Ismail
Department of Electrical Electronics and Systems Engineering, Faculty of Engineering and Built Environment, Universiti Kebangsaan Malaysia, 43600 Bangi, Selangor, Malaysia

Mohsen Nickray
Nanoelectronic Center of Excellence, School of Electrical and Computer Engineering, University of Tehran, P.O. Box 14395-515, Tehran, Iran

Ali Afzali-Kusha
Nanoelectronic Center of Excellence, School of Electrical and Computer Engineering, University of Tehran, P.O. Box 14395-515, Tehran, Iran

Riku Jäntti
Department of Communications and Networking, Aalto University, P.O. Box 13000, FI-00076 Espoo Aalto, Finland

Korhan Cengiz
Electrical-Electronics Engineering Department, Trakya University, Prof. Dr. Ahmet Karadeniz Yerleskesi, 22030 Edirne, Turkey
Electrical-Electronics Engineering Department, Kadir Has University, Cibali, 34083 Istanbul, Turkey

Tamer Dag
Computer Engineering Department, Kadir Has University, Cibali, 34083 tanbul, Turkey